Plants of the
Baviaanskloof

Douglas Euston-Brown
& Magriet Kruger

With contributions by Ernst van Jaarsveld

BOTANICAL SOCIETY
OF SOUTH AFRICA

Mapula
Trust

Published by Struik Nature
(an imprint of Penguin Random House South Africa (Pty) Ltd)
Reg. No. 1953/000441/07
The Estuaries No. 4, Oxbow Crescent, Century Avenue, Century City, 7441
PO Box 1144, Cape Town, 8000, South Africa

Visit **www.struiknature.co.za** and join the Struik Nature Club
for updates, news, events and special offers.

First published in 2023

1 3 5 7 9 10 8 6 4 2

Publisher: Pippa Parker
Managing editor: Roelien Theron
Editor: Natalie Bell
Designer: Dominic Robson
Cartographers: Anisha Dayaram (maps of vegetation types),
Liezel Bohdanowicz (map inside front & back covers)
Proofreader: Emsie du Plessis

Reproduction by Studio Repro
Printed and bound by Shumani RSA, Parow, Cape Town

ISBN 978 1 77584 767 0 (Print)
ISBN 978 1 77584 768 7 (ePub)

Photographs

Front cover: *Leucospermum cuneiforme* in flower, with Kouga Mountains in
the background (RLR); **Title page:** *Gerbera piloselloides;*
This page (top to bottom): *Gazania lichtensteinii,
Anacampseros arachnoides, Helichrysum lancifolium* (MK), *Selago
myrtifolia* (MK); **Contents page:** Cockscomb Mountain;
Back cover: (top to bottom): *Pelargonium alchemilloides, Cyrtanthus
smithiae, Oedera speciosa* (MK); **Spine:** *Felicia ovata* (MK)

The geological diagrams on pages 8, 10 and 12 were reproduced from *The
Story of Earth & Life* (2005) by Terence McCarthy and
Bruce Rubidge, with kind permission of the authors.

Contents

Preface   4
Introduction   6
Geological history of the area   8
A variable climate   15
Vegetation types   16
About this book   38

Marchantiophytes   40

Pteridophytes   41

Gymnosperms   54

Palaeodicots   59

Monocotyledons   62

Dicotyledons   180

Illustrated glossary of terms   492
Glossary of terms   495
Photographic credits   498
Bibliography   499
Acknowledgements   501
Index to scientific names   502
Index to common names   511

Preface

To our knowledge, this is the first book about plants of the Baviaanskloof to be presented in a field guide format. It is the result of over two decades of dedicated study and record-keeping in the field, tracking our finds as we fell under the Baviaanskloof's spell.

We hope that this book does not spend long days on your bookshelf. Instead, we hope that it becomes worn-out from use, dog-eared and ingrained with good old-fashioned dirt as you explore the natural wonders of the Baviaanskloof. May this book be your preferred travelling companion, kept within arm's reach on your dashboard and in your backpack during hikes into the mountains.

It is our aim that this field guide will stimulate your appreciation of the plants that are found here and the ecosystems in which they live. (For studying plants in the field, we do recommend the use of a 10× magnification hand lens or loupe.)

As the world endures signs of global warming and the Baviaanskloof struggles through its umpteenth year of drought, this book feels even more significant and poignant as a record of flora found here. If our work has brought you just a little bit closer to enjoying, restoring or conserving the natural habitats found in this pristine and unique part of the natural world, then in our eyes it is a job well done.

Lastly, it is our dream that this book will be used by the farmers in Baviaanskloof as a management tool for conserving and restoring natural areas. Knowledge of what grows on the land can only enhance our collective stewardship of the area.

Doug, Magriet and Ernst

The Publishers and Authors extend grateful thanks to the Botanical Society of South Africa and to the Mapula Trust for their generous support of this publication.

Opposite: Nuwekloof Pass is the official gateway to the area from the west. The locals call the narrowest part of this ravine, just before the road reaches the valley, 'die sleutel' – the key to Baviaanskloof.

Introduction

A scenic three-hour drive north from Knysna or Gqeberha (Port Elizabeth) leads to one of South Africa's best-kept secrets — the Baviaanskloof. This narrow valley of just under 200 km supports an astonishing variety of plants and was declared part of the Cape Floral Kingdom World Heritage Site in 2004.

A rugged landscape of fold mountains, valleys and narrow kloofs, the Baviaanskloof is bound by two mountain ranges, the Baviaanskloof in the north and the Kouga to the south. Several biomes collide here, and unique plant and animal assemblages are common in these parts. Interestingly, the distribution range of a large proportion of plant species ends in the Baviaanskloof. The region represents the eastern limit of many Fynbos taxa and also the southwestern limit of several subtropical taxa from the northeast. This notion of the Baviaanskloof as the meeting point of many end points or boundaries extends beyond plants and is also true for birds and other life forms. The extreme edge of a taxon's distribution range is often the place where unusual varieties or even undescribed species are found, as life forms adapt and evolve in a relatively new and changing environment.

This guide describes over 1,100 plant species that occur in the Baviaanskloof. A total of ± 1,700 species have been found in the study area. Of these, ± 100 are exotics, ± 90 are endemic and ± 90 are near endemic. Over 120 of the species found in the Baviaanskloof were red-listed in the Threatened Species Programme (SANBI) in 2021, and several are Critically Endangered (CR).

A comprehensive plant species checklist for the Baviaanskloof is available online at: **www.baviaanskloof.com**.

View from Holgat Pass, facing west towards the confluence of the Kouga and Baviaanskloof rivers.

The study area

The study area covered in this book is larger than what is typically known as the Baviaanskloof. It encompasses an area of 10,000 km² and includes the mountains north of the Langkloof and south of the Steytlerville Karoo. The small towns of Uniondale and Willowmore are at the western extent, with Patensie (Khoekhoe for cattle resting place) and Loerie at the eastern extent. Another word with Khoekhoe origins is 'baviaan' – meaning baboon – an unsurprising name since many baboons are seen here. The study area is visually represented in a topographical map at the front and back of this book; it includes most place names referred to in the text.

Study area

The Baviaanskloof supports some of the largest and healthiest baboon troops in the country.

Looking into the Baviaanskloof valley from the Rus en Vrede 4x4 track on the lower slopes of the Kouga Mountains; the Baviaanskloof mountain range is visible in the distance.

Geological history of the area

The Baviaanskloof lies in a narrow valley, flanked by the Baviaanskloof and Kouga mountain ranges, which run west–east more or less parallel to each other. The landscape hosts a wide array of rock formations that are clad in a complex mixture of diverse vegetation. Insight into the area's geological history helps with understanding the vegetation types and landforms that are visible today.

The formation of Gondwana (460–450 mya)

The world's continents are constantly in motion, often colliding with each other and then rifting apart. Occasionally many continents come together to form supercontinents. The most recent of these is known as Gondwana. Africa, along with South America, Antarctica, India and Australia, constituted most of this large land mass. A broad, shallow, elongated basin, which

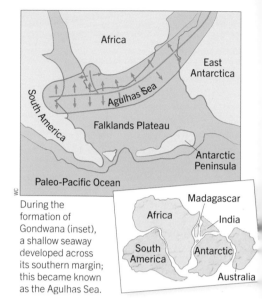

During the formation of Gondwana (inset), a shallow seaway developed across its southern margin; this became known as the Agulhas Sea.

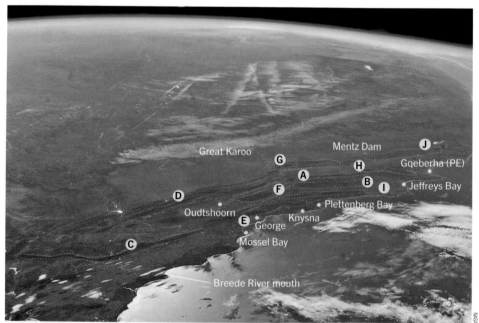

Mountain ranges surrounding the study area. This satellite photograph shows: **A** Baviaanskloof and **B** Kouga mountains, with their west–east orientation. The other ranges are: **C** Langeberg, **D** Swartberg, **E** Outeniqua, **F** Kammanassie, **G** Grootrivier, **H** Groot Winterhoek, **I** Tsitsikamma, **J** Zuurberg.

formed a seaway known as the Agulhas Sea, extended across the southern margin of what today is South Africa, and adjacent portions of South America and Antarctica.

The Cape Supergroup

Over a period of about 80 million years, the Agulhas Sea filled with layer upon layer of sediment that was eroded from the adjacent uplands. The sediment was converted to rock (lithified) and today forms the Cape Supergroup. The lower portions of the Supergroup consist of thick white to light grey sandstone layers, known as the Table Mountain Group, that flank the Baviaanskloof valley. These sandstones record the earliest sedimentation in the Agulhas Sea and were probably deposited by braided rivers that formed extensive plains around the shores.

The higher-lying parts of the Baviaanskloof, the Kouga and Groot Winterhoek mountains are part of the Table Mountain Group, and are mostly composed of Peninsula Formation Sandstone, one of the Group's subdivisions. This is the oldest of all sandstones and also the hardest, so it erodes away very slowly. These mountains, with their rocky, sandy and nutrient-poor soils, are the stronghold for Fynbos vegetation. The highest peaks are covered in Subalpine Fynbos, and the mountains receiving higher rainfall and/or the cooler south-facing slopes support Mesic Fynbos. Where rainfall is less, Arid Fynbos is more prevalent.

Other prominent sandstone subdivisions are the Goudini and Skurweberg formations. The Goudini Formation is relatively fine-textured and brownish in colour. It was prone to folding

Bosrug facing west along the top of the Baviaanskloof Mountains. Black dotted lines demarcate a shale band (Cedarberg Formation) that separates Peninsula Formation (above) from Goudini Formation (below), all of which are subdivisions of the Table Mountain Group. The Peninsula Formation is the oldest, buried under the other formations for millennia; today this rock forms the highest parts of these mountains, due to folding of the sedimentary layers. The high peak in the distance (trig beacon 59) is 1,451 m asl.

and warping, often resulting in small caves and 'holes', or 'bak kranse', in the cliffs and rocky slopes, where softer materials were eroded away. This sandstone type tends to support Grassy Fynbos.

Here, shale bands, comprising a thin layer of softer rock (20 to 100 metres thick), can be seen sandwiched between different types of sandstone. These bands run west—east, following the orientation of the Kouga and Baviaanskloof mountains, and are particularly prominent in the lower part of the Table Mountain Group. Composed of softer mudstone, the bands have eroded at a faster rate than the surrounding sandstone, giving rise to saddles and necks on the ridgeline. The finer-grained shale bands tend to support a grassier kind of Fynbos at higher altitudes, while at lower altitudes, Fynbos is usually replaced by Transitional Shrubland or Thicket vegetation.

Shales and fertile soils (400 mya)

Commencing about 400 million years ago, the Agulhas Sea underwent sudden deepening and lateral expansion. Mud and silt particles settled in the deeper water, where the flow disturbance was very weak or non-existent, and up to 2,000 metres of fine sediment accumulated in the deeper parts. This gave rise to the shales of the Bokkeveld Group. These shales decompose fairly quickly when exposed to the present-day atmosphere, producing fertile soils. Rocks of the Bokkeveld Group are best viewed at the extreme north of the Baviaanskloof Mountains, leading into Willowmore, where they overlie

the sandstones of the Table Mountain Group. The finer-textured and more fertile soils here support Succulent Karoo or Transitional Shrubland vegetation. However, at lower altitudes towards the east they support dense stands of Thicket.

Folding, deformation and breccia formation (330–250 mya)

Deposition of the Cape Supergroup ceased between 330 and 250 million years ago. At this time, a fragment of continental crust (probably a portion of present-day Patagonia) drifted northwards and collided with the portion of

Folds of sedimentary rock at Grootrivierpoort.

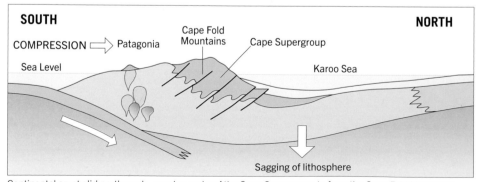

Continental crust slid northwards, causing rocks of the Cape Supergroup to form the Cape Fold Mountains.

The straight stretch of road between Kleinpoort and Zandvlakte lies more or less on the Baviaanskloof fault line.

Gondwana that accommodated the Cape Supergroup. The rocks of the Cape Supergroup were at the centre of this zone of convergence, and were folded and deformed to form the Cape Fold Mountains along the southern margin of Gondwana. In the Baviaanskloof area, folding and deformation was of such great intensity that it caused the strata to form tight folds with steeply dipping fold limbs.

Vensterklip is situated on the Baviaanskloof fault line and is made up of breccia (broken rock shards).

Along the northern flank of the Baviaanskloof valley, folding and deformation was so severe that thick slices of rock were thrust over each other from south to north. An important example of one of these thrust faults is known as the Baviaanskloof Thrust. Along this fault are angular fragments of broken rock, known as breccia, which formed due to crushing of the rocks in the fault zone. Soils tend to be better developed, and may support deeper-rooted, woody vegetation.

The Karoo Sea (300–180 mya)

The formation of the Cape Fold Belt resulted in thickening of the crust along the length of the belt. This led to an increase in the mass of the crust, which sank along this zone. The flanking depression was inundated by sea, forming the Karoo Sea. Again, sedimentary material eroded from the surrounding higher ground was transported to this sea, and it was filled with sediment over a period of about 120 million years.

The breakup of Gondwana (180–20 mya)

As is the nature of supercontinents, Gondwana eventually began to break apart, which commenced about 180 million years ago. At the southern edge of Africa, a large fault developed and a huge portion of the supercontinent detached from Africa by sliding westwards along the fault. Today, this portion of the South American continent, known as the Falklands Plateau, is largely submerged.

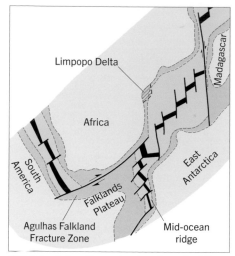

Rifts formed along the east and west coasts of the nascent southern Africa, creating the original Indian and Atlantic oceans (shown in green).

As the Falklands Plateau slid away to the west, large slabs of Cape Supergroup rocks were torn apart and collapsed, forming deep depressions. Rupture occurred along planes of weakness formed by earlier thrust faults, such as the Baviaanskloof fault. Material was rapidly eroded from the higher ground to fill the depressions, resulting in the coarse-grained

Enon conglomerate in the Baviaanskloof, directly after leaving Nuwekloof Pass.

Enon conglomerate at the western entrance to the reserve. The area has many remnants of this red rock, made up of round stones and red mudstone.

conglomerates and boulder beds of the Enon Formation. Remnants of this rock are scattered throughout the Baviaanskloof, and some have eroded into striking formations with characterful names like the Seven Dwarfs and Krokodil. Like breccia, this rock is more prone to erosion. However the parts left behind are extremely hard. Soils on the steeper slopes are poorly developed or completely absent, and in parts devoid of vegetation. The Enon Formations are generally associated with valleys where it s hotter and drier, and there is a prevalence of succulent plants such as Spekboomveld or Succulent Karoo. However, some parts are covered with more arid kinds of Fynbos or non-Fynbos shrublands.

Erosion after the breakup (120–20 mya)

The period following the breakup of Gondwana seems to have been geologically quiescent and was dominated by erosion in southern Africa. Erosion probably took place over the period between 120 and 20 million years ago. The region was bevelled to a vast plain, sloping gently towards the coast, probably much like present-day Australia. This plain is known as the African Erosion Surface. The climate seems to have been hot and humid, and deep, highly leached soils developed across the plain, mostly rich in aluminium, but occasionally iron-rich. In some areas, the upper soil layers became impregnated with silica, forming silcrete.

Uplift, ravines and 'oervlakte' (20 mya – the present day)

About 20 million years ago, southern Africa began to rise, with portions of the subcontinent rising by more than 1,200 m. Erosion by rivers draining the interior became more active, resulting in deep river gorges around the coast, such as along the Storms River and Oribi Gorge, and in the Baviaanskloof area, the numerous narrow side ravines cutting into the Kouga and Baviaanskloof mountains. They formed with hard and durable rock on either side. The magic of experiencing Baviaanskloof on foot lies in exploring these narrow cracks, the traces of time-worn river courses. At the base of the ravines are very deep, rocky soils and water is sometimes prevalent, allowing trees and forest to develop. Ravines with perennial streams or seepages support a diversity of ferns, reeds and unusual plants not seen elsewhere. The shaded cliffs support a succulent flora that is best ascribed to Thicket vegetation. However, on some of the steep slopes and ledges above these ravines, a Fynbos subtype mixed with trees, described as Fynbos Woodland can be seen.

In spite of the donga-like erosion by these rivers, patches of the former African Erosion

An old tree stump lodged in a ravine indicates the high point of a flood in times gone by.

Surface remain and form the tablelands, or 'oervlakte' (ancient plains), in the Baviaanskloof region, several of which retain the original soils of the African Erosion Surface. Soils here

Examples of 'oervlakte' or subplateaus seen facing east from Grasnek Pass. Landmarks are: **A** Mac Peak, the eastern extent of the Baviaanskloof Mountains; **B** Bergplaas, the R332 road passes over here; **C** southern extension of Elandsvlakte; **D** Elandsvlakte; and **E** Enkeldoring. The grey shrub in the foreground is renosterbos.

Facing up the Baviaanskloof valley from the slopes above Smitskraal. These alluvial gravels support healthy stands of sweet thorn (*Vachellia karroo*), favoured by a variety of browsing game.

are relatively deep, red, and clay-rich, richer in nutrients than the surrounding sandstone. These areas are almost entirely grassland, although, where a few thorn trees have established, there is a sense of savanna. These miniplateaus are easier to cultivate for grazing stock than hilly or rocky mountainous areas. The local names of the more significant oervlakte, such as Osberg, Elandsvlakte and Perdeberg, indicate that they were once favoured by grazing animals. Other common names pay tribute to the vegetation, such as Witruggens ('white ridge'), which from a distance, when the grassland dries in winter, appears much paler than the surrounding Fynbos. Transitional Shrubland is a collective term used to describe the non-Fynbos vegetation commonly found on these lower slopes and plateaus.

Topsoil erosion

One of the most intriguing features of this mountainous landscape, and one that has a profound influence on the vegetation, is the relative amount of topsoil erosion that has taken place over the ages. In general, the western parts of both the Kouga and Baviaanskloof ranges were heavily leached and eroded, leaving few remnants of soil above the sandstone. In the east, where the topography is slightly tamer and the terrain more gentle, there are visible remnants of old soils on some of the plateaus. This partly explains why the grassier kinds of Fynbos that prefer deeper and loam-rich soils are more prevalent in the eastern part of the study area.

Alluvial gravels

The alluvial gravels on the valley floor of Baviaanskloof are recent deposits. The soils here are deep and contain more nutrients than the surrounds, and groundwater surfaces in places. The Baviaanskloof and Kouga rivers also support a wide range of water-loving plant species, while the drier boulder beds host unusual mixtures of plant species from the Succulent Karoo and Thicket biomes.

A variable climate

The mountainous terrain makes for a vastly variable climate in the Baviaanskloof. Droughts are common in the study area, and yet rainfall events (anything from 50–100 mm in 24 hours) can happen at any time of the year. Mean Annual Precipitation (MAP) is as low as 200 mm in the kloof, whereas at higher altitudes in the mountains, the MAP can reach up to 800 mm.

There is no well-defined or predictable wet season here, although rainfall tends to be slightly higher in autumn or spring. The western part of the area can receive more winter rainfall than the eastern end, which in turn can receive more summer rainfall. Rain is also variable over space, and sometimes one part of the kloof will receive over 40 mm, while a mere 50 km away only a few millimetres will fall. The weather seems to follow ten-year cycles of dry periods followed by wetter spells that can last for a year to a few years.

The mountain ranges lie parallel to the coastline, and the variation in the height and breadth of each range has a profound influence on the distribution of rainfall. A fair proportion of the annual rain, especially in the upper mountains, is derived from moisture-laden winds coming off the Indian Ocean. The Tsitsikamma Mountains just north of Plettenberg Bay extend to about 20 km west of Humansdorp. Like the Kouga Mountains, the Tsitsikamma range is both wider and taller

in the west, and tapers away towards the east. This configuration results in the western end of the Baviaanskloof Mountains being situated in a rain shadow. The northwestern parts of the Kouga Mountains are also relatively dry. The eastern parts of both the Kouga and Tsitsikamma are relatively low in stature, so moisture derived from the coast can make it through to the eastern end of the Baviaanskloof range (east of Scholtzberg Peak), resulting in much higher rainfall there than in the west. The Groot Winterhoek Mountains (Cockscomb Peak, elevation 1,758 m) and Mac Mountain (elevation 1,562 m) are not situated in the rain shadow of any major mountain ranges and so, on average, they also receive more rainfall than the western parts of the Baviaanskloof.

In Baviaanskloof, the temperature regime is warm to temperate. The average midday temperature in the middle of winter is 16° C, and 28° C in summer. Minimum temperatures in winter drop to around 1° C. However, cold air sinks in the kloof itself, creating a temperature inversion on windless winter mornings when sub-zero temperatures and frost can be experienced. The higher-altitude, western part of the Kouga range is generally much cooler and can receive snowfalls, although this is not common. High peaks in both the Baviaanskloof and Groot Winterhoek mountains also receive occasional snow.

South-facing slopes of Scholtzberg Peak (1,626 m asl). The silhouette resembles a human face gazing skyward.

Vegetation types

The interaction between the geology and climate in Baviaanskloof is a key influencer of the distribution of vegetation types. In Baviaanskloof there are five major vegetation types. In this book, they roughly equate to the concept of biomes or ecosystems. Within these overarching groups are smaller vegetation subtypes comprising plants that share the same habitat or environment. The broader major types tolerate a wider range of environmental variables.

Vegetation types can be recognised in the landscape primarily by their structure and colour and also by the plant species that are common or prevalent in each type (indicator species). Some vegetation types are distributed in large expansive bands, while others occur as fragmented islands.

The Baviaanskloof vegetation types depicted were mapped between 1995 and 2006, in an attempt to capture vegetation patterns prior to recent human settlement, and to provide a snapshot of how a particular area looked at a specific time. Aerial and satellite images were utilised in the mapping process. Understanding if and how the boundaries shift, or how the plant species composition of the vegetation types changes, could help scientists in the future to understand and interpret the consequences of climate change on the vegetation.

Early morning haze in Baviaanskloof with *Aloe ferox* in the foreground.

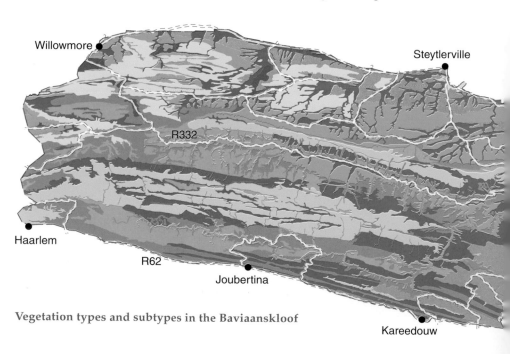

Vegetation types and subtypes in the Baviaanskloof

Vegetation types in the Baviaanskloof study area

Major vegetation type	Vegetation type	Approx. size of area covered	Percentage of study area	Number of patches	Average patch size in km²
Fynbos	Subalpine Fynbos	45 km²	0.4 %	19	2
	Mesic Fynbos	1,334 km²	13 %	41	31
	Grassy Fynbos	1,295 km²	13 %	124	10
	Arid Fynbos	686 km²	7 %	55	12
	Fynbos Woodland	537 km²	5 %	121	4
	Sour Grassland	861 km²	8 %	19	45
Forest	Forest	104 km²	1 %	150	1
Transitional Shrubland	Transitional Shrubland	1,530 km²	15 %	291	5
Thicket	Subtropical Thicket	508 km²	5 %	34	15
	Spekboomveld	1,353 km²	13 %	153	9
Succulent Karoo	Guarriveld	616 km²	6 %	28	22
	Apronveld	556 km²	5 %	80	7
	Doringveld	631 km²	6 %	31	20
	Gannaveld	106 km²	1 %	21	5

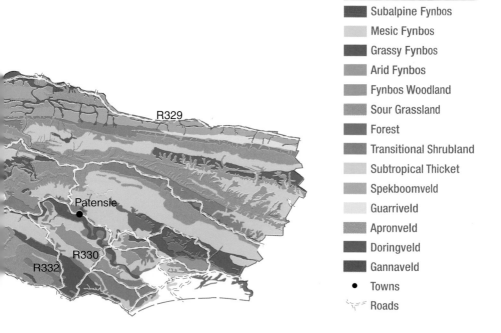

LEGEND

- Subalpine Fynbos
- Mesic Fynbos
- Grassy Fynbos
- Arid Fynbos
- Fynbos Woodland
- Sour Grassland
- Forest
- Transitional Shrubland
- Subtropical Thicket
- Spekboomveld
- Guarriveld
- Apronveld
- Doringveld
- Gannaveld
- ● Towns
- Roads

Fynbos

Fynbos is probably the best-known vegetation type of the Cape Floral Kingdom. Fynbos vegetation is characterised by the presence of any, or all three, members of the main Fynbos families, restio, erica and protea. It typically has a rich graminoid layer (grasses, sedges and restios), an abundance of fine-leaved shrubs (ericaceous or ericoid plants), and an overstorey of broad-leaved shrubs (typically proteas). Unlike ericas and proteas, restios occur in all Fynbos habitats and are thus the best indicator of this vegetation type.

Fynbos covers almost half (46 per cent) of the study area, but not much of it is visible from the road along the valley floor, giving the impression that it is quite rare. However, the mountains here are vast and huge expanses are covered with Fynbos. Many of these mountains are inaccessible – to even the most seasoned hiker – and some of the common plants remain uncollected.

Fynbos is confined to the sandy and/or loamy soils of the quartzitic sandstones of the Cape Fold Mountains, where rainfall is above 250 mm a year and falls mostly in winter. Fynbos also depends on being burnt in periodic fires to regenerate and persist.

Six different kinds of Fynbos are found in the area: Subalpine, Mesic, Grassy and Arid Fynbos, and Fynbos Woodland and Sour Grassland.

Grassy Fynbos; Rus en Vrede 4×4 track at Graskop. The high peak on the right is **A** Sipoonkop (1,485 m asl); **B** Tjandokloof is seen in the middle; the grey shrub in the foreground is *Aspalathus kougaensis*.

Subalpine Fynbos

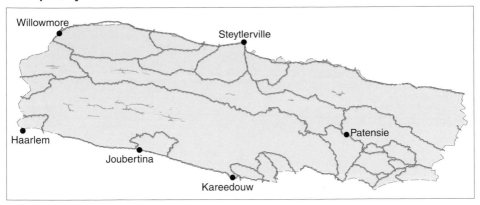

Subalpine Fynbos is restricted to the highest peaks, which reach over 1,400 m above sea level. These 'high-altitude islands' collectively cover less than half a per cent of the study area, making Subalpine Fynbos the smallest vegetation type in the area. The average size of the patches (2 km²) is also small.

From west to east, the most prominent peaks are Saptoukop, Smutsberg, Kougakop, Scholtzberg, Mac and Cockscomb. Winters see these mountaintops occasionally capped with snow. Certain plants are especially adapted to the unique climate around these high points and do not occur anywhere else. Where plants of the lower-lying Mesic Fynbos (see page 20) penetrate Subalpine Fynbos, they can appear quite different at higher altitude, and may be easily mistaken for undescribed Subalpine Fynbos species.

The soil here is typically shallow and rocky, and the plants are quite stunted. Even when mature, the vegetation seldom grows taller than 0.5 m.

There are likely to be several new and undescribed species waiting to be discovered on these seldom-visited and inaccessible high peaks.

Indicator species: *Mastersiella spathulata, Protea rupicola, Erica tragulifera*

Krantz sugarbush (*Protea rupicola*)

View from Scholtzberg Peak facing west along Baviaanskloof Mountains.

Mesic Fynbos

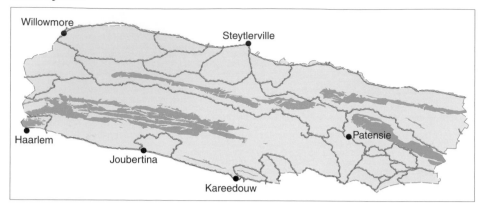

Mesic Fynbos occurs in the middle to upper reaches of mostly south-facing mountains, but sometimes extends quite far down the slopes, especially on mountains that receive slightly more rain. As its name suggests, this vegetation type is able to retain moisture, more so than other vegetation types in the area.

Mesic Fynbos is well represented in the mountains and covers about 1,334 km² or 13 per cent of the study area. The patches, averaging 31 km² in size, are relatively large.

The soil here is usually dark, rich in organic matter, and peaty in the seepage zones. Mesic species are exceptionally good at absorbing moisture trapped in the soil, be it from rainfall, mist or even dew. However, despite their efficiency at sucking up moisture, there is sometimes a surplus, and this gradually seeps into the porous bedrock before finding its way into rivers and dams, or deeper into aquifers. The gurgling mountain streams and perennial rivers further down the slopes of the Baviaanskloof Mountains are largely fed by this excess water.

The dominant species *Leucadendron eucalyptifolium* can form tall, crowded stands that can reach heights of up to 5 m when mature and may spread across large areas. Dense stands of *Protea neriifolia* and *P. punctata* are also typical. *Cannomois virgata* and *P. mundii* often dominate the seepage areas and shale bands on south-facing slopes. *Erica* species are commonly found in Mesic Fynbos, and most of the ericas in this book have been recorded in this vegetation type.

Indicator species: *Leucadendron eucalyptifolium, L. uliginosum, Elegia juncea*

Mesic Fynbos; Kougakop (1,719 m asl) in the distance, on the left.

Tsitsikamma conebush
Leucadendron uliginosum

Grassy Fynbos

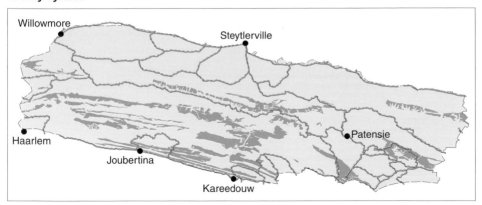

Grassy Fynbos is found on the warmer and drier mountain slopes below Mesic Fynbos. Almost 1,300 km² or 13 per cent of the study area is Grassy Fynbos, made up of 124 patches with an average size of 10 km².

The relative abundance of grasses, restios and sedges varies a great deal within this vegetation type. The most common grasses found include *Pentameris eriostoma*, *Capeochloa arundinacea*, *Themeda triandra* and *Urochloa serrata*. Dominant restio species include *Hypodiscus striatus*, *Rhodocoma fruticosa* and *Restio triticeus*.

At the boundary with Arid Fynbos (see page 22) the grass cover becomes very low and restios dominate the graminoid layer. This tends to happen in the western part of Baviaanskloof, where annual rainfall is lower and more prevalent in winter, and the soil sandier and rockier. The change in grass cover is most evident on the steeper slopes.

In the east, however, the opposite is found. Grassy Fynbos is replaced by Sour Grassland, with grasses dominating the graminoid layer and restios being almost entirely absent. The mountains in the east are not as high as those in the west and have a lower gradient. These factors, together with finer soil texture and more rain in summer, provide favourable conditions for grasses, and most restios struggle to compete with them here.

Several different kinds of Grassy Fynbos exist. The most common, Waboomveld (where *Protea*

nitida occurs), is found on lower slopes where the soil is deeper, loamier and slightly richer in nutrients; it is most prevalent in the Langkloof.

The composition and structure of this vegetation type has been detrimentally affected by periodic fires and grazing by stock (or historically by game). Grassy Fynbos seems to be especially vulnerable to vegetation changes brought about by fires that have burnt too frequently.

Indicator species: *Leucadendron salignum*, *Leucospermum cuneiforme*, *Erica demissa*, *Themeda triandra*

Grassy Fynbos on the Kouga Mountains. The yellow-orange flowers in the middle are cover species *Leucospermum cuneiforme*. The highest point in the distance is Scholtzberg Peak.

Arid Fynbos

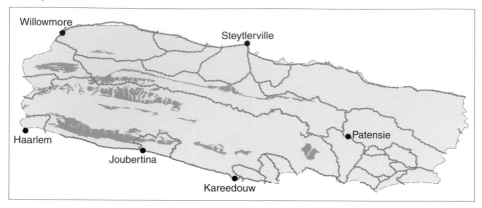

Arid Fynbos is most prevalent in the western part of the study area, on the hot and dry, steep, north-facing rocky mountain slopes, where the sandstone-derived soil is nutrient-poor and has a high gravel, stone and sand content. This stony, well-drained soil does not retain moisture in the way that clay soils and the finer-textured soil found in other Fynbos types do.

This vegetation type covers about 686 km², seven per cent of the study area, in 55 patches, with an average patch size of 12 km².

Arid Fynbos plants are perfectly adapted – both below and above ground – to thrive here. They tolerate hot and dry conditions, and respond swiftly when it rains, making use of the water before it drains away or evaporates. In

Strap-leaf sugarbush (*Protea lorifolia*)

Hypodiscus striatus

fact, attempts to grow these plants in soil with finer particles, or with more nutrients or water, are generally unsuccessful.

Restios, adapted to this low-nutrient, coarse, sandy substrate, dominate the graminoid layer, while grasses and sedges are very sparse. The most abundant restio species are *Hypodiscus striatus* and *Cannomois scirpoides,* but *Thamnochortus cinereus* is also often present on stony ridges. Proteaceous plants include *Protea lorifolia, Leucadendron nobile* and *Leucospermum wittebergense.* The erica family in Arid Fynbos is usually represented by *Erica pectinifolia, Erica umbelliflora* and *Erica rosacea.* Asteraceous shrubs include *Metalasia pallida, Oedera squarrosa, Oedera decussata* and *Athanasia tomentosa.* Other Fynbos shrubs that may be common include *Passerina obtusifolia,*

Euchaetes vallis-simiae, Muraltia juniperifolia, Aspalathus hystrix and *Aspalathus collina.*

In Arid Fynbos, the interaction of the vegetation and topography contributes to a much longer fire-return interval relative to other Fynbos types. Species such as *Protea lorifolia* and *Leucadendron nobile* succeed in this vegetation type as they take long to regrow and mature after a fire event. Shorter-lived species, found in other Fynbos vegetation types, require regular burns as their seeds depend on a shorter fire-return interval for germination. Such species often become locally extinct in infrequently burnt Fynbos, especially if their seeds are short-lived and prone to predation.

Indicator species: *Protea lorifolia, Hypodiscus striatus, Passerina obtusifolia*

Passerina obtusifolia (bright green shrub) and *Cannomois scirpoides* (restio to the left) are visible in the foreground. The distant mountain slopes may look gentle, but they can be steep, with some impassable ravines.

Fynbos Woodland

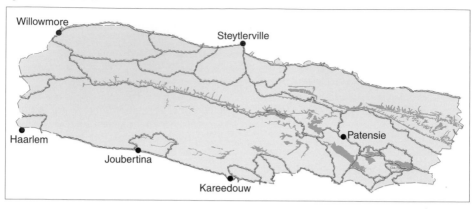

Willowmore

Steytlerville

Haarlem

Joubertina

Kareedouw

Patensie

Fynbos Woodland is a new term, introduced by the author, to describe vegetation that encompasses both Fynbos and woody Thicket, and is differentiated from other kinds of Fynbos by the presence of woody trees and bushes. It is often associated with steep, rocky slopes in the numerous kloofs that cut into the Kouga and Baviaanskloof mountain ranges.

The approximately 537 km² area that Fynbos Woodland covers is highly fragmented, with about 121 smallish patches of about 4 km² each. Together the patches cover approximately five per cent of the study area.

Fynbos Woodland is very variable in structure and floristics. Areas bordering on Fynbos are

An abundance of woody trees mixed with Fynbos vegetation is typical of Fynbos Woodland. This view from Combrink's Pass shows the silvery multi-stemmed tree known as teerhout or isibara (*Loxostylis alata*).

at risk of regular burning, but in lower-altitude regions, transition zones with Thicket are largely protected from fire. (Lightning tends to strike mainly higher-lying areas and then burn downwards, so the lower edges of this vegetation type are less exposed to fire than the upper slopes.) Fynbos Woodland on steep rocky slopes and cliffs is also protected from fire, by the topography, and can remain unburnt for hundreds of years. Such sheltered sites tend to lose some of their Fynbos plants, with woody trees and succulents becoming more abundant, but when they do burn, the Fynbos plants have a chance to win back some territory.

A characteristic feature of Fynbos Woodland is the unexpected assemblages of plants representing different biomes, with trees such as cedar found growing alongside spekboom, cycads or even proteas. Some fascinating plants found in Fynbos Woodland include the Willowmore cedar (*Widdringtonia schwarzii*), Thunberg's cycad (*Encephalartos longifolius*), and cliff-dwelling succulents such as *Haworthia cooperi* and *Tromotriche baylissii*. Restios, if present, are usually represented by *Restio gaudichaudianus*. In the eastern section, the attractive teerhout tree (*Loxostylis alata*) can be quite common. *Protea nitida* is sometimes present. Fynbos shrubs include *Passerina obtusifolia*, *Metalasia trivialis*, *Phylica axillaris* and *Agathosma ovata*.

Indicator species: *Widdringtonia schwarzii*, *Aloe perfoliata*, *Pterocelastrus tricuspidatus*

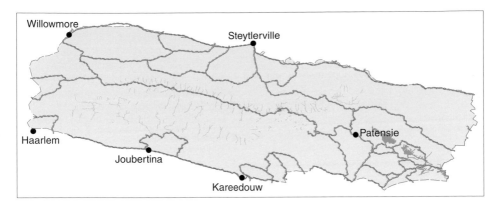

prevalent in the northwest, while Afromontane Forest is related to the Knysna forests in the southwestern section, and Subtropical Forest prevails in the east. Some kloofs in the heart of the area include all three forest types.

The number and variety of tree species vary from one kloof to the next. The kloofs in the northwest are the most species-poor, while those in the southern part of the Kouga range are the most diverse. This diversity could be a result of the southern Kouga being more moist, but it may also be ascribed to the proximity of the area to the Tsitsikamma and Knysna forests.

Another factor that influences the composition of the forests is the variable rainfall. The climate in the east is more subtropical and hotter than in the west. Species such as

Ficus sur and *Vepris lanceolata*, which are more typical of warmer climates, are more common in the eastern parts. Some rare endemics occur in these kloofs, including *Nemesia deflexa*, *Crassula cremnophylla* and *Cyrtanthus labiatus*.

The forested kloofs here are awe-inspiring and somewhat magical, and a trip to the area is not complete without venturing up one of the side ravines. Each one has something different or special to offer, be it a massive Cape fig (*Ficus sur*) or Cape holly (*Ilex mitis*), a crystal-clear rock pool with crabs, fish, frogs and limpets, or a shaded rock face with an amazing array of cliff-dwelling succulents.

Indicator species: *Podocarpus latifolius, Ilex mitis, Kiggelaria africana, Ficus sur*

Decaying and dead stumps of Cape fig (*Ficus sur*) are a favoured nesting site for the double-banded carpenter bee (*Xylocopa caffra*). The bee tunnel entrance holes are visible in this weathered trunk.

The Cape fig (*Ficus sur*) grows into a giant tree, with ripe figs that attract baboons, monkeys and fruit-eating birds, including the Knysna Turaco.

Transitional Shrubland

Transitional Shrubland is a transition zone between Fynbos and Thicket. It covers the largest surface area of all the vegetation types in the study area, 1,530 km² or 15 per cent, but it is also the most fragmented, comprising 291 relatively small patches, each with an average size of 5 km².

Shrubland is subject to periodic burns, after which grasses tend to dominate the ground cover for a few years until the shrub vegetation recovers.

There are several subtypes of Transitional Shrubland, each one identified by a dominant shrub, its occurrence governed by soil type. Asbosveld and Grassland are found in clay-rich soils, Renosterveld and Sandolienveld in loamy soils, and Passerinaveld in rocky and sandy soils.

Asbosveld is dominated by *Pteronia incana* and occurs on the 'oervlakte' — the flattish subplateaus where remnants of reddish, clay-rich soils occur. Thicket species are common

Transitional Shrubland subtype Asbosveld on the Kouga Mountains. Asbos (*Pteronia incana*) is the grey shrub in the foreground; the Baviaanskloof valley and mountain range shimmer in the hazy distance.

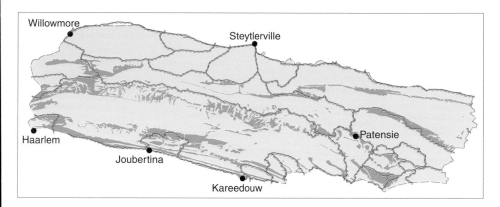

Willowmore
Steytlerville
Haarlem
Joubertina
Kareedouw
Patensie

here and often form a mosaic of patches. After fire, Asbosveld can become a grassland, especially in the eastern part of the region where the annual summer rainfall is higher. Common grasses include *Themeda triandra*, *Digitaria eriantha* and *Eustachys paspaloides*.

Renosterveld, dominated by renosterbos (*Dicerothamnus rhinocerotis*), occurs at the base of the mountains, often on shale-derived or loamy soils. It is mostly found to the west of Baviaanskloof and in the Langkloof.

Sandolienveld, dominated by the shrub sandolien (*Dodonaea viscosa* var. *angustifolia*), occurs in bands at the junction of shrubland with Fynbos. Here soils tend to be sandier and stonier.

Passerinaveld is dominated by *Passerina obtusifolia*. These plants favour hot, dry, steep and rocky mountain slopes with nutrient-poor soil – much like Arid Fynbos, although Passerinaveld is more arid and does not support restios, ericas or proteas. Turpentine grass (*Cymbopogon pospischilii*) is usually present and small succulents are also quite common.

Indicator species: *Pteronia incana*, *Dodonaea viscosa* var. *angustifolia*, *Dicerothamnus rhinocerotis* (= *Elytropappus rhinocerotis*), *Passerina obtusifolia*

Renosterbos (*Dicerothamnus rhinocerotis*)

Asbossie (*Pteronia incana*)

Thicket

Thicket is easily recognised as a dense entanglement of broad-leaved, mostly spinescent shrubs, climbers and succulents. It is remarkably resilient vegetation and well equipped to withstand extreme climatic events, from droughts, floods and heatwaves to extreme cold. Thicket plants are generally slow-growing and slow to react to external changes. Relative to plants in the other vegetation types, Thicket plants have almost stagnant internal dynamics.

If cut back to prevent coppicing, Thicket vegetation does not recover naturally and human intervention is required to initiate recovery.

Thicket can form small bush clumps in other vegetation types, and usually does so where it intersects with them and there is a mixture of plants. This creates a mosaic pattern, sometimes referred to as Bontveld. Such mosaic patches are found in Asbosveld (see page 28), but it is not unusual to find them in Grassy Fynbos too, where the grassy vegetation habitat and soils support termite populations. Termite mounds act as microsites for the establishment of Thicket pioneer species.

Browsing game then trample and graze the grass around the Thicket patches, creating a firebreak and space in which the Thicket can expand further to form mosaic patches.

Fire does not affect Thicket as much as it does other vegetation types. Historically, fires would seldom have penetrated Thicket, because much of it was sheltered in kloofs and other protected sites. Further to this, Thicket vegetation does not burn easily, so any fires reaching it are often extinguished naturally when they reach the Thicket edge.

When it has been necessary to clear Thicket to open up areas for grazing or to attract game, people have learned that, under suitable weather conditions, a fire started at the bottom of a slope and allowed to burn upwards does take hold and can successfully clear the vegetation on a grand scale.

Although many Thicket plants can resprout after fire, resprouting is more likely a response to heavy browsing by elephant, rhino and other animals.

In this guide, Thicket is divided into two types: Subtropical Thicket and Spekboomveld.

Thicket near Patensie. The naboom (*Euphorbia triangularis*) is seen protruding above the dense canopy.

Subtropical Thicket

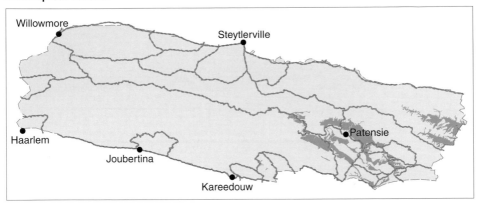

Willowmore
Steytlerville
Haarlem
Joubertina
Kareedouw
Patensie

Subtropical Thicket is the taller of the two types found in Baviaanskloof. It has a more easterly distribution and receives more summer rainfall. It could be viewed as an arid kind of Forest, but periodic droughts and high temperatures prevent true Forest species from establishing here.

Subtropical Thicket occupies 508 km^2, only five per cent of the study area, with an average patch size of 15 km^2. Here, cover and abundance of succulents is much less than in Spekboomveld.

In the past, Subtropical Thicket would have supported megaherbivores such as elephant and rhino. Their movement through the vegetation would have helped to open it up, creating sunny spots where sun-loving plant species could grow. Without these large browsers today, Subtropical Thicket has grown more dense, and any open spaces have resulted

Tree euphorbia (*Euphorbia triangularis*)

from human intervention. Sun-loving plants eke out an existence in spaces opened up by road building and the erection of fences.

Indicator species: *Euphorbia triangularis, Schotia latifolia, Scutia myrtina*

Bosboerboon (*Schotia latifolia*)

Cat-thorn (*Scutia myrtina*)

Spekboomveld

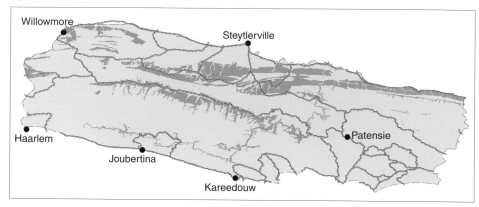

Spekboomveld is dominated by spekboom (*Portulacaria afra*), a lime-green succulent plant that clothes the lower, often steep and rocky mountain slopes of the Thicket. Spekboomveld covers 1,353 km², 13 per cent of the study area.

In harsh conditions, where most other plants would suffer and die, the hardy spekboom thrives in large expanses, and beneath its canopy creates a microclimate, allowing other Thicket plants to establish. Also found among this vegetation is the charismatic tree or valley-bush euphorbia (*Euphorbia grandidens*), which grows in sheltered sites and protrudes above the blanket of surrounding vegetation.

Indicator species: *Portulacaria afra, Putterlickia pyracantha, Euclea undulata, Pappea capensis*

Spekboomveld in the Baviaanskloof valley near Keerom. Spekboom (*Portulacaria afra*), is the bright green succulent shrub on the right and in the distance. To find out more about how this shrub comes to dominate these arid slopes, turn to page 332.

Succulent Karoo

Succulent Karoo vegetation is found in the most arid parts of the Baviaanskloof, where rainfall is less than 200 mm per year. The Succulent Karoo biome reaches its eastern limit in the Baviaanskloof. There is greater diversity and greater levels of endemism in this biome west of the Baviaanskloof region.

Succulent Karoo can be recognised by low (less than a metre) shrubs and, with the exception of a few microsites, the absence of trees. The only tree species able to grow here are the common guarri (*Euclea undulata*) and soetdoring or sweet thorn (*Vachellia karroo*). The expansive cover of succulent plants occurs in the lower valleys and on hills in shale-derived and nutrient-rich soils. Fire does not penetrate this vegetation, and it is almost impossible to get it to burn.

Four kinds of Succulent Karoo vegetation types have been recognised and mapped in the study area: Guarriveld, Apronveld, Doringveld and Gannaveld.

Guarri (*Euclea undulata*)

Gewone kapokbossie (*Eriocephalus ericoides*)

Karoo gold (*Rhigozum obovatum*)

Gombossie (*Pteronia glomerata*)

Guarriveld

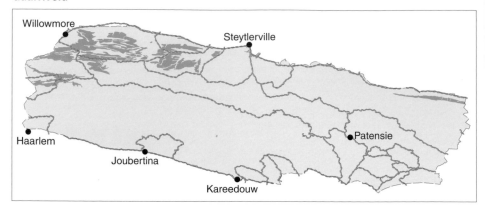

Guarriveld is named after the common guarri (*Euclea undulata*) that grows here on the shale hillsides. It is a remarkably tough and long-lived tree with an enormous, tuber-like root system.

Guarriveld occupies 616 km², only six per cent of the study area, with an average patch size of 22 km². Spekboom is absent from Guarriveld, probably due to frost and the subzero temperatures that occur here sometimes.

Guarriveld is dominated by small, tough shrubs, mostly from the daisy family.

Succulents are also present and include *Cotyledon orbiculata* var. *spuria* and *Crassula rupestris*. Many of the succulents grow under the shelter of small shrubs known as 'nurse plants'.

Grasses are generally quite sparse, but vingerhoedgras (*Fingerhuthia africana*) can become abundant after good summer rains.

Indicator species: *Euclea undulata, Fingerhuthia africana, Eriocephalus* spp.

Succulent Karoo; Guarriveld. A stand of guarri trees (*Euclea undulata*) is typical among the low shrubs.

Apronveld

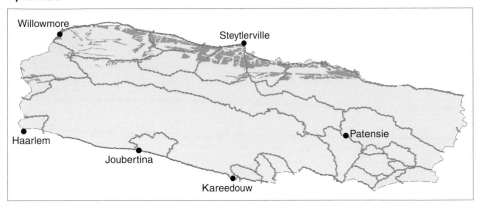

The name 'Apronveld' derives from the shape of the landscape where this veld type occurs: the lower, gentle slopes form 'skirts' or 'aprons' around the base of the koppies. Apronveld occupies 556 km², just five per cent of the study area, with patches averaging 7 km².

Apronveld is found on shale-derived soils, and features numerous exposed surface rocks, sometimes with quartz outcrops.

Small (under 50 cm high) shrubs dominate and typically include many *Eriocephalus* and *Pteronia* species. Larger shrubs, such as *Rhigozum obovatum*, are generally absent, but when they do occur they are a reliable indicator of proximity to Spekboomveld.

Browsing pressure by stock tends to keep the cover of grasses very low to almost absent in Apronveld. However, the shrubs in the veld offer good browsing for small game such as springbok, duiker and steenbok, and they are also favoured by domestic stock over other veld types.

Historically, Apronveld would have been exposed to herds of game passing through, and many of its plants are adapted to being browsed and trampled.

Indicator species: *Eriocephalus* spp., *Pteronia* spp., *Aizoon africanum* (= *Galenia africana*), *Rhigozum obovatum*

Succulent Karoo; Apronveld between Willowmore and Steytlerville. The low stature of the shrubs – which very seldom exceed knee height – is a typical characteristic of Apronveld. The north-facing slopes of the western end of the Baviaanskloof Mountains are seen in the distance.

Doringveld

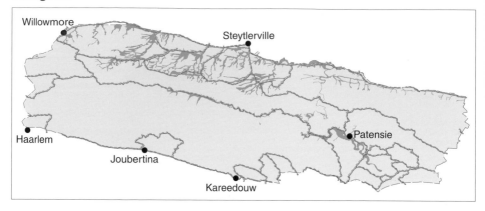

Doringveld covers the majority of the Baviaanskloof floodplain, including the riverbanks and bordering cobble beds. It takes its name from the stands of 'doringbome' (sweet thorn trees) that occur in some parts but proliferate along drainage lines as they access groundwater with their deep taproots.

Doringveld occupies 631 km², only six per cent of the study area. With an average patch size of 20 km², it supports a dense sward of grasses including *Panicum maximum*, *Eragrostis curvula* and *Digitaria eriantha*. The ground cover in some parts is dominated by tokkelossiebos (*Plumbago auriculata*). Slymbos (*Roepera foetida*) is found too, often clambering over thorn trees. Common shrubs include bitterbos (*Chrysocoma ciliata*), bakbesembossie (*Nidorella ivifolia*) and strooiblom (*Helichrysum* spp.). Succulents are also prevalent and include several species of crassula and some shrubby mesembs.

Parts of the cobble bed close to the river are periodically scoured in flooding events. These spaces support a low-growing shrubland that is unique to the area; the endemic *Amphiglossa callunoides* is found here in abundance.

Areas around the river support many plants that do not occur anywhere else in the region. Large shrubs and trees associated with riverbanks include pypsteelbos (*Cliffortia strobilifera*), river honey-bells (*Freylinia lanceolata*), taaibos (*Searsia tomentosa*), suurtaaibos (*Searsia rehmanniana*), karee (*Searsia lancea*) and Cape willow (*Salix mucronata* subsp. *mucronata*). Several alien invasives are associated with this riparian habitat; the more notable ones include the Belhambra tree or ombu (*Phytolacca dioica*), gum trees (*Eucalyptus* spp.) and oleander (*Nerium oleander*).

Game is abundant in Doringveld, the sweet thorn trees being favourite browse for kudu, rhino, bushbuck and bush pig.

Indicator species: *Vachellia karroo*, *Plumbago auriculata*, *Roepera foetida*

Doringboom (*Vachellia karroo*)

Tokkelossiebos (*Plumbago auriculata*)

Slymbos (*Roepera foetida*)

Gannaveld

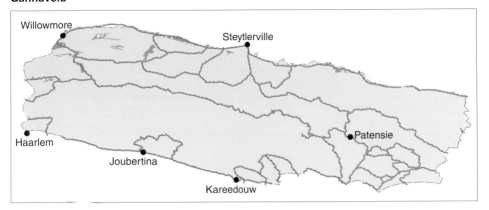

Gannaveld does not occur in the Baviaanskloof itself, but it can be seen along the drainage lines in the Steytlerville Karoo north of the Baviaanskloof Mountains. Gannaveld takes its name from the gannabos (*Salsola aphylla*), a prominent shrub that grows in the area.

Gannaveld occupies 106 km², only one per cent of the study area, with an average patch size of 5 km². It grows in deep, clay-rich soils on the valley floors. It is a relatively species-poor vegetation type, with no trees. Rather, it is dominated by succulent shrubs from the 'mesemb' family, with *Mesembryanthemum junceum* usually the most common.

Heavy grazing poses the greatest threat to this vegetation. Damaged veld is typically dominated by a few short-lived succulents such as bobbejaankos (*Augea capensis*), *Mesembryanthemum crystallinum* and *Malephora lutea*. Overgrazed areas have an extremely low cover of plants, which leads to desertification, exposing the ground to erosion. Rill and gully erosion are common in areas that have been exposed to prolonged browsing pressure by stock.

Indicator species: *Salsola aphylla*, *Mesembryanthemum junceum*

Succulent Karoo; Gannaveld. The dwarfed shrubs are often succulent and mostly upalatable to stock and some game. The thorn trees seen in the middle distance are in Doringveld; there are no trees in Gannaveld.

About this book

The species in this book represent all classes of plant found in the Baviaanskloof, from the most primitive liverworts and ferns to the more common grasses, reeds and flowering plants. This selection of over 1,100 species follows the standard taxonomic organisation into plant families and genera. Species accounts are arranged in six groups (Marchantiophytes, Pteridophytes, Gymnosperms, Palaeodicots, Monocotyledons and Dicotyledons) and are in alphabetical order by scientific name.

Family accounts include general information and give the genus and species counts worldwide, for southern Africa, and for the Baviaanskloof specifically.

Genus accounts cover genus names and derivations. Previous names or synonyms are in brackets, and common names and general information are provided where available.

Species accounts provide species names and derivations. Plant parts are described and include approximate measurements.

* Alien/exotic genera and species are indicated with an asterisk.

Common names are the ones most commonly in use in the Baviaanskloof area.

Other names in common use are listed with abbreviations for language (E = English; A = Afrikaans; X = Xhosa; S = Sotho; Z = Zulu).

Sandalwood

SANTALACEAE
A medium-sized family of hemiparasitic trees and shrubs with ± 43 genera and 1,000 species worldwide. The famous sandalwood tree (*Santalum album*) produces one of the most expensive kinds of timber in the world. Its oil is used in perfumes because the fragrance lasts for decades. The slow-growing trees in this family have been over-harvested for centuries and several species are now endangered. ± 6 genera and 226 species in southern Africa; 3 genera and 19 species in Baviaanskloof.

LACOMUCINAEA (= *Thesium*)
Named after Ladislav Mucina (1956–), a Slovakian-born botanist who co-authored a vegetation map of southern Africa; he is currently based in Australia.

1 species in southern Africa.

Lacomucinaea lineata (= *Thesium lineatum*) **Vaalstorm**
lineatus = marked with fine parallel lines; referring to the lines on the grooved branches

Dense, rigid shrub to 2 m tall, with spine-tipped, grey-green branches. **Leaves** Inconspicuous, linear, falling early. **Flowers** Small, on short racemes near the branch tips, whitish. Flowering in spring and summer. Fruit fleshy, 1 cm long, white. **Habitat** Infrequent in Transitional Shrubland, Succulent Karoo and Thicket. **Distribution** Widespread from Namibia to Baviaanskloof. **Other name** Witstorm (A).

parallel lines

Vlakkie, Zandvlakte 06.07.2011

OSYRIS (= *Colpoon*)
ozos = branch; referring to the branching habit of the plants; *kolpos* = breast; referring to the shape of the berries; alternatively, *colpus* = groove, which might refer to grooves in the ovary where the stamens pass through

Molecular studies have revealed that the genus *Colpoon* should be included in *Osyris*, however this has not been finalised. 8 species worldwide; 3 species in southern Africa; 1 species in Baviaanskloof.

Osyris compressa (= *Colpoon compressum*) **Cape sumach**
compressum = laterally flattened; referring to the arrangement of the leaves around the stems, with leaves sometimes appearing compressed

Resprouting, hemiparasitic shrub or small tree to 5 m. **Leaves** Opposite, ovate-elliptic, leathery, with obscure venation. **Flowers** In panicles at the branch tips, greenish. Flowering in summer and autumn. Fruit 1.5 cm long, red becoming black. **Habitat** Common in various habitats but mostly in Grassy Fynbos and Transitional Shrubland. **Distribution** Very widespread from Namaqualand to tropical Africa. **Other names** Pruimbas (A), intekaza (X), umbulunyathi (X, Z), ingondotha-mpethe (Z).

Waterkloof, Bokloof 26.04.2018

Labels on images point out features that are useful for identifying the species.

Footers indicate group and genera.

Flowering seasons are given:
spring (September – November)
summer (December–February)
autumn (March–May)
winter (June–August).

...hosma kougaense
...se = of the Kouga; referring to the origin
...lant in the Kouga Mountains. Kouga is
...ginal Khoisan name for a leopard.

...ding, slender shrublet to 25 cm. **Leaves** Linear-
...olate, keeled, 1 cm long, oil glands raised,
...ns finely-hairy. **Flowers** In pairs in the axils,
..., hairy. Flowering mainly in spring. Fruit 1- or
...ambered. **Habitat** Uncommon and localised
...e south-facing slopes of high peaks in Mesic
...Subalpine Fynbos. **Distribution** Endemic to the
...ga Mountains. **Status** Rare.

East of Kougakop 16.04.2012

...athosma mucronulata
...ron = pointed tip, ulata = possessing;
...rring to the leaves, which have a
...inct pointy, stiff mucron, making the
...ub prickly to the touch

...seeding, much-branched, rounded
...rub to 1 m, turpentine-scented.
...aves Ascending, closely packed, ovate, with a
...arp, sometimes recurved pointy tip (mucronulate),
...i 6 cm long. **Flowers** Clustered at the tips of the
...ranches, white with purple dots. Flowering in spring.
...ruit 3-chambered. **Habitat** Common on middle
...orth-facing slopes in Grassy Fynbos. **Distribution**
...ndemic to the Kouga and Baviaanskloof mountains.
...Notes Closely related to and possibly just an arid
...orm of A. martiana, which also grows in the area

Enyelandkop, Kouga 13.07.2011

Agathosma mundtii Jakkalspisbos
Named after JL Leopold Mund, 1791–1831, a Berlin
pharmacist who was sent to the Cape in 1815 to
collect plant specimens

Resprouting or reseeding wiry shrub to 1 m.
Leaves Finely-hairy and smelling extremely
unpleasant, to 1 cm long × 0.5 cm wide but
usually thinner, with margins strongly revolute.
Flowers Clustered into heads at the branch
tips, white. Flowering in winter and spring. Fruit
2-chambered. **Habitat** Frequent on upper slopes
in dry, rocky areas in Grassy Fynbos, but also in all
other Fynbos habitats. **Distribution** Widespread from
Witteberg to Baviaanskloof and Humansdorp.

Uitslag 4X4, Baviaans 09.07.2013

Agathosma | DICOTYLEDONS 451

Caption to species image includes the rough location of the plant and the date when it was photographed, where available. The map on the inside cover shows most places mentioned in the text.

Status refers to the conservation status of the species as listed in the IUCN's Red List of Threatened Species.

Kouga = Kouga Mountains

Notes identify uses of the plant, and subspecies or closely related species.

Photographic credits are listed in full on page 498.

Baviaans = Baviaanskloof Mountains

Habitat corresponds with the vegetation types described in the Introduction.

Distribution identifies the most western and northern town or mountain in the species range, followed by the most eastern or northeastern limit.

MARCHANTIOPHYTES Liverworts

Liverworts developed ± 500 million years ago, making them among the oldest living land plants. They evolved from algae and were the first plants to make the transition from aquatic to terrestrial habitats. This was a significant step in the evolution of plants.

Together with mosses and hornworts, liverworts are part of a larger group of non-vascular plants known as bryophytes. The absence of vascular cell tissue in the plant body (thallus) means that moisture or nutrients cannot be transported to the different parts of the plant. For this reason, they cannot grow stems or roots and thus are all small and flattish. Owing to the absence of roots, these plants grow mainly in damp areas and rely on moisture being available in their immediate environment. Structures called rhizoids hold the liverworts securely in place. Liverworts are the only extant group of plants without the minute 'breathing pores' (stomata) for exchange of gases. The gametophyte is the dominant stage in the life cycle, unlike ferns, where the sporophyte is dominant and the more visible part.

The liverworts' common name alludes to their resemblance to the human liver. In the past, according to the pseudoscientific doctrine of signatures — which proposed that a plant was able to heal the body part that it resembled — it was believed that liverworts could cure liver problems, but there is no scientific evidence for this.

There are ± 20,000 species of bryophytes worldwide, ± 9,000 of which are liverworts.

MARCHANTIACEAE Liverworts

1 genus only, worldwide, with ± 25 species.

MARCHANTIA

In 1713, Jean Marchant named the liverwort after his father, Nicolas Marchant, who was the first botanist to study mosses. Nicolas died in Paris in 1678 but his son continued his work describing plants in this genus.

5 species in southern Africa; 1 species recorded in Baviaanskloof.

Marchantia berteroana   Umbrella liverwort

Named after Luigi Carlo Giuseppe Bertero, 1789–1831, an Italian naturalist, botanist and doctor. He was lost at sea, presumably in a shipwreck, while sailing from Tahiti to Chile. The female reproductive organs of liverworts (called archegoniophores) resemble tiny umbrellas, thus the common name.

Dioecious, flat, with plant body (thallus) 2 cm long × 1.2 cm wide, glossy green above, with transparent scales below. **Parts** Female reproductive organs umbrella-like, 8 cm tall × 1 cm wide. Male reproductive organs flat and disc-like, 3 cm tall × 1 cm wide. Asexual reproductive organs ('gemma cups') resemble tiny suction cups, 0.3 cm tall × 0.4 cm wide, from which small new plants ('gemmules') are dispersed by raindrops. **Habitat** Uncommon on humus and debris near mountain streams in Forest. **Distribution** Widespread in southern Africa, Australia, New Zealand, South America and various smaller islands such as Java, Papua New Guinea and New Caledonia.

Waterkloof, Kouga Wildernis 14.05.2021

PTERIDOPHYTES Ferns

Before the evolution of gymnosperms and angiosperms, ferns were the dominant plants on Earth. These ancient plants do not produce flowers or seeds, rather they have a life cycle with two different stages. The sporophyte stage produces spores that are dispersed by wind and grow into small structures called gametophytes. The gametophytes produce eggs and sperm. Fertilisation occurs in water, after which a new sporophyte plant begins to grow, only for the cycle to begin again. The spore-producing stage is visible to observers and for this reason it is described in the species accounts below. Detailed descriptions of the gametophyte are beyond the ambit of this book.

Relative to other plant species, ferns tend to be more widespread across the globe. The same fern species can be found on islands around the world, a consequence of the lightweight nature of the spores, which makes them easily dispersed by wind across great distances.

Accurate fern identification is underpinned by knowledge of a plant's habitat, its frond structure, the arrangement of pinnae or pinnules and, most importantly, the arrangement of spores and sori on the undersurface of the pinnae.

A unique set of terms is used in the description of ferns and the basic ones are illustrated below. Readers can also consult the glossary at the end of this book for further clarity.

There are ± 50 families of ferns worldwide, with ± 12,000 species.

Fern fronds

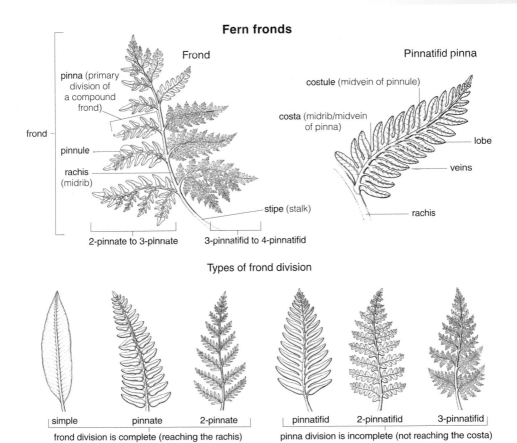

Frond

Pinnatifid pinna

pinna (primary division of a compound frond)

frond

pinnule

rachis (midrib)

costule (midvein of pinnule)

costa (midrib/midvein of pinna)

lobe

veins

stipe (stalk)

rachis

2-pinnate to 3-pinnate — 3-pinnatifid to 4-pinnatifid

Types of frond division

simple — pinnate — 2-pinnate

pinnatifid — 2-pinnatifid — 3-pinnatifid

frond division is complete (reaching the rachis)

pinna division is incomplete (not reaching the costa)

ANEMIACEAE

'Flowering ferns' and scented ferns

1 genus only, worldwide.

ANEMIA (= *Mohria*)

aneimon = naked, unclothed; referring to the uncovered sporangia, however, since *Mohria* has now been included in *Anemia*, this trait no longer holds true for the entire genus

± 139 species in the Americas, Africa, Madagascar and India; 8 species in southern Africa; 1 species in Baviaanskloof.

Anemia caffrorum (= *Mohria caffrorum*) **Scented fern**

caffrorum = from southeastern South Africa

A scented fern forming neat tufts. **Rhizomes** Creeping, covered with light brown lanceolate scales. **Fronds** Crowded, erect, to 16 cm; slightly dimorphic, with the fertile fronds longer than the sterile. Scaly and densely hairy when young, especially on the undersurface, but becoming glabrous with maturity. Pinnae 2-pinnatifid to 3-pinnate. Sterile fronds held flat with pinnules lobed with pointy tips or teeth; when fertile, teeth curl downwards and pinnules appear rounded. **Sporangia** Shiny, brown and round, marginal, set under the toothed lobes, without an indusium. **Habitat** Occurring at the base of rocks or in rocky places on shaded, south-facing mountain slopes in Grassy and Mesic Fynbos. **Distribution** Widespread from Namaqualand to Cape Peninsula and Eastern Cape. **Other name** Brandbossie (A). **Notes** This scented fern gives off a faint aroma when the fronds are crushed.

Zandvlakte, Baviaans 31.07.2011

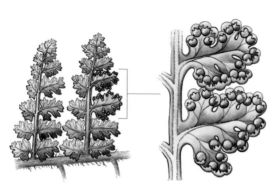

Undersides of fronds (left); enlarged view of marginal sporangia under the curled toothed lobes (right)

Rus en Vrede 4X4, Kouga 25.04.2018

ASPLENIACEAE Spleenworts

1 genus (formerly 10 genera) with 700 or more species worldwide.

ASPLENIUM

a = without, *splen* = spleen; pertaining to the old belief, based on the doctrine of signatures, that these could be used to reduce an enlarged spleen

40 species in southern Africa; 5 species known to occur in Baviaanskloof, possibly more.

Asplenium capense (= *Ceterach capense*) **Spleenwort**

capense = of the Cape; origin of the species

A small fern. **Fronds** Tufted, 10–15 cm. Stipe and undersurface covered with rust-coloured scales. Lamina pinnatisect to pinnate, sometimes shallowly 2-pinnatifid. Pinnae 1.7 cm long, joined to the rachis, with a decurrent base near the tip of the frond; pinnules with a slightly flared base and on a stalk at the lower part of the frond. **Sori** Linear, 0.3–0.6 cm long, 10 pairs per pinna, not obscured by the scales; indusium absent. **Habitat** Damp shaded areas, usually associated with riparian Forest. **Distribution** Widespread from the Cape to tropical Africa.

Zandvlaktekloof, Baviaans 18.02.2016

Asplenium rutifolium **Carrot fern**

Ruta = plant genus, *folium* = leaf; leaves look similar to those of some species of *Ruta* (Rutaceae), a genus of aromatic herbs from southern Europe.

Erect, neatly tufted fern, resembles carrot tops, grows to 35 cm. **Fronds** Lamina 2- or 3-pinnatifid, glabrous on both surfaces. **Sori** Oval, 1 per lobe, positioned close to the margin, about halfway up the lobe. **Habitat** Epiphytic or growing on rocks in Forest. **Distribution** Widespread from Swellendam, extending through tropical Africa, to Yemen, Madagascar and the Comores.

Poortjies, Cambria 26.10.2018

BLECHNACEAE
Hard ferns, deer ferns and chain ferns

9 genera and ± 250 species globally; 10 species in southern Africa.

BLECHNUM
blechnon = a fern; word used by Dioscorides to describe a fern

± 200 species globally; 9 species in southern Africa; 6 species recorded in Baviaanskloof, 2 of which are described here.

Blechnum attenuatum Large deer fern

attenuatum = narrowly tapering; an appropriate name for this species as the pinnae and fronds taper to a fine point

An attractive, shiny fern. **Rhizomes** Up to 50 cm long, with dark brown scales that are also present on the base of the stipe. **Fronds** Tufted and arching, over 1 m long. Pinnate fronds with basal pinnae reduced to a series of tapering lobes. Pinnae oblong-acute to linear-attenuate, with bases joined to the rachis, especially when fertile. Distinctive venation on undersurface of the infertile pinnae. **Sori** Covering the length of the pinna, with linear indusium. **Habitat** Cool, moist areas, often on seeping cliffs in narrow gorges in Forest, where it can form dense colonies. **Distribution** Widespread from the Cape to tropical Africa and Madagascar.

Uitspankloof, Baviaans 15.02.2016

Blechnum punctulatum

punctulatum = minutely-dotted; pertaining to the water-secreting gland along the pinna margin

Smaller than *B. attenuatum*. **Fronds** Arching, up to 70 cm long, sometimes reddish when young. **Fronds** pinnate with pinnae narrowly oblong, base usually flared, truncate or shallowly cordate with the lobes overlapping the rachis. Pinnae become narrower when fertile. **Sori** Parallel to the costa, continuing along the length of the pinna, with linear indusium. **Habitat** Forms disorganised colonies at the base of rocks in Mesic Fynbos, or in shaded kloofs in Forest. **Distribution** Widespread from the Cape to East Africa and Madagascar.

Zandvlaktekloof, Baviaans 19.01.2022

DENNSTAEDTIACEAE
Bracken and ground ferns

9 genera and ± 170 species globally; 5 genera and 7 species in southern Africa; 3 species in Baviaanskloof, 2 of which are described here.

HYPOLEPIS
hypo = under, *lepis* = scale; pertaining to the sori that are borne under a scale-like flap at the margin

2 species in southern Africa, both occurring in Baviaanskloof.

Hypolepis villoso-viscida
villoso-viscida = hairy-sticky; referring to the viscid, glandular hairs on most parts, making the plant sticky

An uncommon attractive fern. **Rhizomes** Widely creeping, with small pale brown hairs. **Fronds** 1–1.6 m long, widely spaced. Stipe 0.3–0.6 cm, purple-brown and viscid. Lamina triangular, 1 m long × 0.8 m wide, 2-pinnate to 4-pinnatifid. Pinnules with glandular hairs on both surfaces, sticky, margins crenate. **Sori** Round to oval, 0.05–0.1 cm wide, on the margins of the lobes. **Habitat** Prefers open boulder beds on stream banks in Forest. **Distribution** Known from only a few locations in the Western and Eastern Cape. In Baviaanskloof, only 1 small population known near the entrance to Zandvlaktekloof. **Notes** May be confused with *H. sparsisora*, but the latter is more widespread, is not sticky, and has softer fronds.

Zandvlaktekloof, Baviaans 18.02.2016

PTERIDIUM
pteridium = small fern; an inappropriate name for the genus as some of these ferns can reach 1.7 m

1 species in southern Africa, with ± 12 varieties.

Pteridium aquilinum Bracken fern
aquilina = of an eagle; the spreading pinnae resemble an eagle with extended wings

A colony-forming fern. **Rhizomes** Underground, widely creeping, with brown hair-like scales. **Fronds** Erect, stiff and hard, to 1.75 m. Stipe woody, to 50 cm long. Lamina 3-pinnatifid to 4-pinnate. Pinnae at right angles to the rachis. **Sori** Linear, just inside the in-rolled margins. **Habitat** Most common in sandy, well-drained areas that become damp or wet during the rainy season. **Distribution** Probably the most abundant, common and widespread fern in southern Africa. Widespread from the Cape to tropical Africa and Madagascar. It can cover vast areas and seems to indicate areas that have supported forest in the past. Not often seen in Baviaanskloof, most likely because of periodic droughts and relatively dry conditions. **Notes** Toxic to stock, yet eaten as a vegetable in Korea, Japan and parts of China. Contains the carcinogenic compound ptaquiloside. **Other names** Adelaarsvaring (A), ubende, ubulawu (X), bamagqira (Z).

Rooikloof, Kouga 28.04.2018

ELAPHOGLOSSACEAE

Deer tongue ferns

4 genera and ± 800 species globally; 2 genera and 9 species in southern Africa.

ELAPHOGLOSSUM

Tongue ferns

elaphos = stag, male deer, *glossis* = tongue; pertaining to the deer-tongue-like shape of the fronds

1 species only in Baviaanskloof.

Elaphoglossum acrostichoides

acrostichoides = resembling *Acrostichum*, a genus of fern that also has the entire undersurface covered with sori

A simple-leaved plant that does not resemble a fern, until the reverse of the **fronds** is revealed, covered with brown sori. **Rhizomes** Creeping, with pale to dark brown scales 0.3–0.5 cm. **Fronds** Less than 1 cm apart. Stipe 3–15 cm long; base with papery, ovate scales. Lamina linear-elliptic to very narrowly lanceolate, apex acuminate, base tapering, up to 35 cm long, but usually much shorter, fronds dark green, glabrous and shiny

Ridge west of Elandsvlakte, Baviaans 29.09.2011

above, paler below. **Sporangia** Covering the whole undersurface except the midrib. **Habitat** A wide variety of habitats. In Baviaanskloof, found in shady cracks in rocks on mountain tops in Mesic Fynbos. **Distribution** Widespread from the Cape to tropical Africa and Madagascar.

GLEICHENIACEAE

Coral ferns, forking ferns and fan ferns

6 genera and ± 125 species globally; 3 genera and 3 species in southern Africa.

GLEICHENIA

Named after Baron Wilhelm Friedrich von Gleichen-Ruswurm, 1717–1783, a German biologist

1 species only in Baviaanskloof.

Gleichenia polypodioides   Coral fern, creeping fern

polypodioides = resembling *Polypodium*, a genus of fern that also has the sori sunk into the frond surface

Forming big clumps to 1 m tall, with spreading rhizomes, and pale green fronds with dark wiry stipes. **Rhizomes** Widely creeping, covered in tiny brown scales. **Fronds** Stipe stiff, wiry, glabrous and shiny, up to 0.6 m long. Lamina a single axis with dichotomous pinnae also divided, thus the common name 'forking ferns'. **Sori** Sunk into 3-locular pits on each pinnule, creating a tiny bump on the upper surface. **Habitat** Forms dense stands in damp situations on shallow sand on bedrock, sometimes completely dominating and shading out

Uitspankloof, Baviaans 29.09.2011

all other vegetation. Not common in Baviaanskloof, where it is mostly confined to seepages on cliffs in narrow kloofs. **Distribution** Widespread from the Cape to tropical Africa and Madagascar. **Notes** On some mountain slopes in the southern Cape it forms huge, monospecific stands, seen from several kilometres away. Sometimes shows up light green on aerial photographs. Regarded as a pest in some forestry areas where it can smother saplings. **Other name** Kystervaring (A).

OSMUNDACEAE

Royal ferns and king ferns

4 genera and ± 20 species globally; 2 genera and 2 species in southern Africa, both in Baviaanskloof.

TODEA
Named after Heinrich Julius Tode, 1733–1797, a German cryptogamic botanist and clergyman

Todea barbara King fern

barbara = foreign, strange; Linnaeus regarded this fern as unusual when he described it in 1753.

A robust fern to 2 m tall × 3 m wide; it can be mistaken for a tree fern but the sporangia are very distinctive. **Rhizomes** Erect, up to 80 cm tall × 30 cm wide. **Fronds** In a neat tuft. Stipe 40 cm. Lamina lanceolate, 2 m long, 2-pinnate. Pinnules linear-oblong with margins crenate to bluntly serrate towards the apex. **Sporangia** Only present on the basal pinnules, covering most of the undersurface, leaving just the margin and the costule exposed. **Habitat** Water loving, usually found on quartzitic sandstone on the banks of perennial streams. In Baviaanskloof it is restricted to the narrow kloofs where perennial water is available. **Distribution** Widespread from the Kamiesberg to Chimanimani Mountains in Zimbabwe, also in Australia and New Zealand. **Notes** This fern has been propagated from spores and is available for sale from some nurseries.

Uitspankloof, Baviaans 15.02.2016

POLYPODIACEAE

Polygram ferns

Several genera were formerly placed in other families. ± 56 genera and 1,200 species globally; ± 9 genera and 16 species in southern Africa, with a further 2 genera and 3 species that have been introduced.

POLYPODIUM

poly = many, *pous* = foot; alluding to the creeping rhizome with its spaced leaf stalks

± 100 species worldwide; 2 species in southern Africa; 1 species in Baviaanskloof.

Polypodium ensiforme

ensiforme = shaped like a sword; referring to the long, narrow shape of the segments of the lamina

Perennial fern to 30 cm. **Rhizomes** Short, creeping. **Fronds** Stiff, erect, spaced up to 3 cm apart. Lamina deeply pinnatifid to pinnate, sometimes simple. Lobes linear, to 1 cm wide, flared at the base, angled forward at ± 45 degrees from the axis, glabrous, margins shallowly crenate. **Sori** Round, 0.2 cm wide, in a single row on either side of the costa. **Habitat** Found on cliffs, or epiphytic on trees, in Forest. **Distribution** Widespread from Betty's Bay to the Eastern Cape, mostly near the coast.

Boskloof, Kouga Wildernes
15.05.2021

PTERIDACEAE

Brakes and maidenhair ferns

17 genera and ± 400 species globally; 4 genera and 11 species in southern Africa; 4 genera and 9 species in Baviaanskloof, 6 of which are described here.

ADIANTUM

Maidenhair ferns

adiantos = dry or desiccated; the leaves repel water and remain dry, and can survive dry periods.

250 species worldwide; 6 species in South Africa, 1 naturalised; 1 species in Baviaanskloof.

Adiantum capillus-veneris

capillus-veneris = the hair of Venus; fronds resemble the hair of the Greek goddess.

One of the most well-known and attractive ferns. **Rhizomes** Creeping. **Fronds** Stipe 7–20 cm long, black, hairless and shiny. Lamina 2- or 3-pinnate, 20 cm × 18 cm. Pinnules like an inverted triangle, deeply incised at the end, bases wedge-shaped and attached at the end, veins ending in shallow marginal serrations. Petioles hair-like and black. **Sori** Oblong at the ends of the lobes, with a pale brown pseudo-indusial flap. **Habitat** Usually found hanging off cliffs in damp and shaded kloofs, which are not necessarily always wet. **Distribution** Widespread worldwide. **Notes** Common in nurseries and gardens. **Other names** Vrouehaar (A), umnambane (X).

Geelhoutboskloof, Kouga 20.04.2012

CHEILANTHES

Lip ferns

cheilos = lip, *anthos* = flower; referring to the sori, which are borne at the curled-over (lip) margin

± 200 species worldwide, 38 of which occur in southern Africa; 6 species are known to occur in Baviaanskloof, 4 of which are described here.

Cheilanthes contracta

contracta = compressed, narrowed or contracted; alluding to the habit of the pinnae to curl in on themselves during dry spells

Rus en Vrede 4X4, Kouga
25.04.2018

A small fern to 30 cm. **Rhizomes** Creeping with brown scales. **Fronds** Closely spaced and erect. Stipe to 1 cm long, dark brown with scales and hairs. Lamina linear-elliptic, to 30 cm tall, 2-pinnate to 3-pinnatifid. Pinnae lanceolate, set at an angle of 45–60 degrees from the axis. Pinnules with dense pale hairs that are cream to reddish brown, twisted and appressed. **Sori** Discrete, borne beneath the in-rolled margin, with minute or absent indusium. **Habitat** Rocky areas in the Cape Fold Mountains. Commonly encountered in Fynbos or close to Fynbos vegetation. **Distribution** Endemic to South Africa, mostly in the Fynbos biome. Dubious records in Mpumalanga.

Cheilanthes multifida subsp. *multifida*

multifidus = much divided; referring to the fronds

A tufted fern to 50 cm, with bright green, much-divided fronds and a dark stipe. **Rhizomes** Short and creeping, with brown scales that have a dark central stripe. **Fronds** Tufted and erect. Stipe to 42 cm long, dark brown or black. Lamina 3- to 5-pinnatifid, up to 80 cm long × 70 cm wide, but generally around half this size. Pinnules lanceolate-deltate, pinnatifid into ovate, obtuse lobes, margins crenate, veins obscure. Rachis and secondary rachides black and glabrous. **Sori** Marginal with a pale, semicircular indusium. **Habitat** Around the base of boulders and rock crevices in mountains. In Baviaanskloof it is associated with Arid Fynbos and Transitional Shrubland habitats and tolerates dry conditions. **Distribution** Widespread from Namibia to the Western and Eastern Cape.

Uitspan, Baviaans 19.02.2016

Cheilanthes parviloba

parviloba = small-lobed

A tufted fern to 50 cm with brittle fronds. **Rhizomes** Short and creeping, with linear, pale brown scales. **Fronds** Stipe to 15 cm with pale brown hairs. Lamina to 32 cm long × 7 cm wide, 2- to 4-pinnatifid, narrowly elliptic to oblanceolate. Rachis with gland-tipped hairs. **Sori** Marginal at the apices of the pinnule lobes, indusium lacking. **Habitat** Sandstone mountains in relatively dry or arid conditions. **Distribution** Widespread from Namibia to the Western and Eastern Cape and to Zimbabwe.

Zandvlaktekloof, Baviaans 01.03.2016

Cheilanthes viridis   Common lip fern, green cliff brake

viridis = green; pertaining to the colour of the lamina

A common fern to 40 cm. **Rhizomes** Creeping, with light brown scales. **Fronds** Tufted, arching, fresh green. Stipe same length as lamina, brown to blackish, with hair-like scales near the base. Lamina soft, deltate-pentagonal, 2- to 4-pinnate, with the basal pinnae the largest. Pinnules variable, with visible veins, margins entire to slightly crenate. Rachis with a wide groove, dark, always pubescent in axils. **Sori** On the edge, indusium continuous. **Habitat** Shaded forest floors, but it can grow in more exposed places among rocks. **Distribution** Widespread with some records in Namibia, but mostly from the Western and Eastern Cape, to tropical Africa, Madagascar, Yemen and India. **Other names** Iyeza ledliso, unomlindana (X).

Grootrivierpoort 25.10.2018

PELLAEA

pellos = dark-coloured; pertaining to the dark or black rachis of most species

± 35 species, mostly tropical; 7 species in southern Africa; 2 species in Baviaanskloof.

Pellaea calomelanos   Hard fern

calo = beautiful, *melanos* = black; an allusion to the black rachis that makes a striking contrast with the grey-green pinnules

A tufted fern, with thickly textured grey-green fronds with few rounded pinnules. **Rhizomes** Erect to suberect, with dark brown scales. **Fronds** Tufted, usually erect, with old, burnt stipe bases crowded at the bottom. Stipe up to 30 cm long, shiny black. Lamina ovate, to 25 cm tall × 15 cm wide, mostly 2- but sometimes 3-pinnate. Pinnules glabrous on both surfaces,

Bosrug, Baviaans 27.07.2011

grey-green on top, paler underneath. **Sori** Marginal with marginal indusium, both continuous. **Habitat** Common, frequently encountered in Baviaanskloof, in rocky areas in a wide variety of habitats in Fynbos and Transitional Shrubland. Often grows in sunny situations where other vegetation is sparse. **Distribution** Very widespread throughout central and East Africa. Also found in Sudan, Spain, India and Madagascar. **Notes** Resprouts after fire and dies back in droughts, with new fronds appearing after rain.

Pellaea leucomelas

leucomelas = white and black; alluding to the colour contrast between the pale frond and the dark stipe

Resprouting, tufted, erect fern to 25 cm. **Fronds** Stipe black and shiny. Lamina ovate, to 25 cm tall × 15 cm wide, mostly 2- but sometimes 3-pinnate. Pinnules glabrous on both surfaces, grey-green on top, paler underneath. **Sori** Continuous on leaf margins, but interrupted at the tip of the pinnule. **Habitat** Rocky and open areas in Grassy and Mesic Fynbos. **Distribution** Widespread from Montagu to Gqeberha (Port Elizabeth).

Rus en Vrede 4X4, Kouga 25.04.2018

PTERIS

pteryx = wing; alluding to the wing-like appearance of the frond

± 250 species in tropical and temperate regions of the world; 7 species in southern Africa; 1 native and 1 exotic species recorded in Baviaanskloof.

Pteris dentata Toothed brake

dentata = toothed, dentate; referring to the dentate margins of the sterile lobes near the base of the pinnules

A tufted fern to 1 m, with arching herbaceous fronds. **Rhizomes** Short, dark brown to black, with erect scales. **Fronds** Tufted. Stipe to 50 cm long. Lamina ovate, to 1 m long × 80 cm wide, 2-pinnatifid to almost 3-pinnate. Spines at the vein junctions along the upper surface of the costa. Pinnule entire except for infertile half, which has serrate-dentate margins. Rachis smooth and glabrous. **Sori** Linear, on the margins, with indusium. **Habitat** Found sporadically in narrow kloofs where perennial water is available, often in the shade of Forest, but also in more sunny openings. **Distribution** St Helena, Ascension, Cape Verde, Mascarene Islands and Madagascar, and on the Arabian Peninsula, down the Horn of Africa to the Cape Peninsula.

Uitspankloof, Baviaans 09.10.2016

---

*Pteris tremula**

tremula = trembling, shaking; referring to the quivering movement of the flimsy fronds in a light breeze, which assists with dispersal of spores

A tufted, light green fern to 75 cm, with softly herbaceous fronds. **Rhizomes** Short, erect, 0.4 cm long, with several crowns; scales dark brown. **Fronds** Tufted, soft, flimsy and delicate, light green fading to yellowish. Stipe to 60 cm long. Lamina triangular, up to 100 cm long and wide, 3- or 4-pinnate; basal pinnules of most pinnae pinnatifid, margins finely toothed. **Sori** Linear, on the margins, not continuous. **Habitat** Narrow kloofs within Forest. **Distribution** Native to Australia, New Zealand and Fiji. In South Africa it has become naturalised from the Cape Peninsula to Durban. Spreading in various parts of the world.

Waterkloof, Kouga Wildernis 14.05.2021

SCHIZAEACEAE

Comb ferns

2 genera and ± 30 species globally; 1 genus and 2 species in southern Africa.

SCHIZAEA

schizein = split; alluding to the alternating arrangement of spikes of the sporangia

Schizaea pectinata

pectinata = comb-like; referring to the comb-like portion of the frond

An attractive fern to 30 cm. **Rhizomes** Shortly creeping to suberect, with brown scales. **Fronds** Densely packed. Stipe to 7 cm long, dark brown. Sterile fronds simple, very narrowly linear (more like a winged rachis), to 30 cm. Fertile fronds rachis-like, but with a terminal comb-like portion of linear pinnae to 1 cm long. Rachis of fertile portion curving down, held perpendicular to the infertile portion. **Sori** In 2 rows within folded pinna. **Habitat** Quartzitic sandstones and impoverished soils, often at the base of rock slabs or cliffs. Also found in moist conditions that prevail at higher elevations in the mountains. **Distribution** Widespread from the Western and Eastern Cape to East Africa and to Madagascar. **Notes** Resprouts well after fire.

Cockscomb, Groot Winterhoek
17.02.2016

THELYPTERIDACEAE

Marsh ferns

5–30 genera, depending the classification followed, and ± 950 species globally. A somewhat spectacular family of ferns, many species of which are cultivated in nurseries around the world. 9 genera and 16 species in southern Africa; 2 genera and 3 species in Baviaanskloof.

CYCLOSORUS

kuklos = circle; referring to the round sori

3 species globally; 1 species in southern Africa.

Cyclosorus gueinzianum (= *Christella gueinziana*; *Thelypteris gueinziana*)

Named after Wilhelm Gueinzius, 1813–1874, a German naturalist who collected in KwaZulu-Natal

A water-loving fern to 1 m. **Rhizomes** Erect. **Fronds** Suberect to arching, tufted. Stipe to 50 cm long, pale brown, with minute white hairs and scales near the base. Lamina up to 1 m long × 40 cm wide, but mostly half this size, 2-pinnatifid, with the basal pinnae reducing. Pinnae linear-oblong, gradually narrowing, both surfaces hairy, more so along the midribs. **Sori** Closely spaced in 2 rows, indusium semicircular to kidney-shaped, with white hairs. **Habitat** Restricted to stream banks in Forest, but also grows in open grassland in areas where there is enough moisture. **Distribution** Widespread from Knysna region to tropical Africa.

Zandvlaktekloof, Baviaans 18.02.2016

Cyclosorus interruptus

interrupta = interrupted; describing the regularly incised margin of the pinna lobes

An aquatic fern that can form extensive colonies. **Rhizomes** Widely creeping, with scales dark brown to black. **Fronds** Spaced well apart. Stipe to 60 cm. Lamina oblong-lanceolate, pinnate to 2-pinnatifid, up to 1.5 m long in shade, 0.8 m in sun. Pinnae 10 cm long × 2 cm wide, with 7–15 pairs of veins. **Sori** Round, up to 18 per lobe, forming a zigzag pattern; indusium small and hairy, often falling early. **Habitat** Constantly wet soils along rivers and vleis. Tends to prefer warmer, sunny conditions – it becomes taller and softer in shady areas. **Distribution** Although rare in the Western Cape, it is a very widespread species, occurring sporadically in West and East Africa, Madagascar, Mauritius, and South and Central America. In Baviaanskloof it can be seen where the R332 crosses the Baviaanskloof River near the Smitskraal picnic area. **Notes** Used in herbal medicine for treating liver disease, malaria, gonorrhoea, coughs and sores.

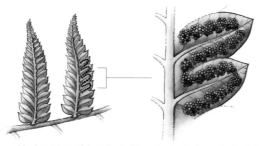

Undersides of fronds (left); enlarged view of zigzag pattern of sori on underside of pinnae (right)

Smitskraal, Baviaanskloof River 29.12.2017

THELYPTERIS

thelys = female, *pteris* = fern; Theophrastus (the 'father of botany') and Dioscorides (Greek botanist and physician) regarded this genus as a diminutive version of the similar-looking fern *Aspidium*.

± 3 species globally; 1 species in southern Africa.

Thelypteris confluens   Scaly lady fern, bog fern

confluens = converging, running together; referring to the rhizomes that creep and intertwine, resulting in healthy stands

A marsh fern that forms dense stands. **Rhizomes** Creeping underground. **Fronds** Erect, 5 cm apart, soft, pale green; 2-pinnatifid. Stipe non-hairy. Lamina lanceolate. **Sori** Circular, up to 18 per lobe, with glabrous indusium. **Habitat** Fairly common on damp stream banks and in marshes. **Distribution** Very widespread from the Cape Peninsula through Africa, and in Madagascar and Australasia. In Baviaanskloof, seen easily from the main road at Poortjies, near the eastern entrance to Baviaanskloof Nature Reserve.

Poortjies, Cambria 26.10.2018

GYMNOSPERMS — Cone-bearing plants

Gymnosperms evolved around 320 million years ago, making them some of the oldest living organisms on Earth, but not quite as old as the liverworts of some 500 million years ago. In contrast, angiosperms evolved from the gymnosperms more recently, about 140 million years ago.

Gymnosperms are woody, vascular, seed-producing plants that include cycads and conifers. The name gymnosperm means 'naked seed', and refers to the exposed or unenclosed seeds seen in this group of plants, contrasted with the seeds of angiosperms that are enclosed in an ovary or fruit. Conifers are the largest group of gymnosperms (600 species) followed by cycads (340 species). Three groups of the ancient gymnosperms are now extinct, but two well-known and charismatic survivors are the welwitschia (*Welwitschia mirabilis*) and the ginkgo (*Ginkgo biloba*).

There are ± 10 families of gymnosperms worldwide, with 83 genera and 1,000 species; 4 indigenous genera with 7 species, plus 1 exotic genus with 2 species in Baviaanskloof.

CUPRESSACEAE — Cedars

This well-known family of conifers includes the giant redwoods and junipers of the northern hemisphere. 30 genera and 140 species worldwide; 3 species in South Africa; 2 species in Baviaanskloof.

The Willowmore cedar — *Widdringtonia schwarzii*

The Willowmore cedar (*Widdringtonia schwarzii*), locally known as saprechout, is the largest and most useful of all the indigenous trees within the Baviaanskloof region. It is nearly endemic, with only a few populations in the Grootrivierberge northeast of Willowmore.

The Willowmore cedar is relatively common, despite extensive harvesting in the past. Younger trees are cone-shaped, but the adult trees grow in a spreading fashion. This species is one of three South African cedar trees belonging to the genus *Widdringtonia*. A second closely related species is the

Top: Female cones have four angled points. Above left: A Willowmore cedar stands sentry on rocky slopes above Zandvlaktekloof. Above right: Weathered fence pole of Willowmore cedar.

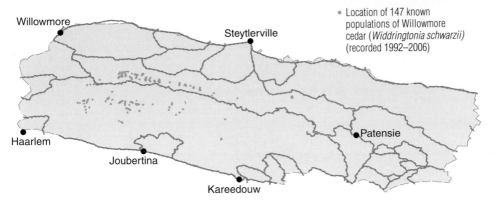

Willowmore cedar in Baviaanskloof study area

Clanwilliam cedar (*W. wallichii*) (= *W. cedarbergensis*). Both of these species, although growing in fire-driven Fynbos vegetation, are usually found only in sheltered rocky positions and will not resprout after a fire. The third species, the mountain cypress (*W. nodiflora*), is a smaller tree with a columnar shape. It grows in Fynbos and grassland and is adapted to sprout after a fire.

Of these three South African species, the Clanwilliam cedar is the most difficult to cultivate as its roots are adapted to grow among boulders and bedrock crevices. In a drive to grow this endemic tree, seedlings of both the Willowmore and Clanwilliam cedars were taken to a propagator in Betty's Bay, who grafted the Clanwilliam cedar onto the rootstock of the Willowmore cedar with some success.

The Willowmore cedar grows in deep, sheltered kloofs and on steep rocky slopes, sometimes in more exposed locations. When young, it has linear, needle-like leaves up to 2 cm long. As the seedling matures, it produces scale-like, greyish green leaves 0.2 cm long. Woody female cones are knobbly, 2 cm in diameter, with four rough, warty angled points. Male cones are shorter and are produced at branch ends. The cones are persistent on the tree and open spontaneously after a fire, releasing the flattened seed, which is dispersed by wind. This cedar is confined to rocky earth derived from the sedimentary rocks of the Cape Fold Belt; this soil is acidic and low in mineral content.

The timber of the Willowmore and Clanwilliam cedars is prized and valuable, because it is easily worked, durable, aromatic, resinous and insect-proof. This sought-after wood was used extensively by farmers in the past, mainly for construction purposes, leading to the demise of many large specimens that grew near these construction sites. Most of the accessible trees in the kloofs were removed, while those in inaccessible areas were left untouched. Transport of timber from unreachable spots was difficult, so trees were often felled and left in stream beds for flash floods to carry the logs to more convenient collection points. To this day, uncollected stump fragments can be found in some of the kloofs.

From 1978 to the early 1980s, a forester working in Baviaanskloof sent seed of the Willowmore cedar to Kirstenbosch National Botanical Garden. There the seed was sown in sandy soil and germination was rapid, with the seedlings growing quickly. They were planted out between the present-day Kirstenbosch Research Centre and the staff accommodation to the south. The trees have grown to 8 m in height, their circumference at chest height is 63–85 cm, and the tallest tree is 12 m. One of the specimens has an impressive stem circumference of 1.46 m at chest height. These dimensions are important indicators that the Willowmore cedar has potential for growth in forestry plantations – a fitting homage to this Baviaanskloof native.

WIDDRINGTONIA

Cedars

Named in honour of Samuel Edward Widdrington, 1787–1856, an English writer who lived in Spain for some time and took an interest in the coniferous forests there. He was the first to scientifically describe the conifers in Spain and published a book on European pines.

4 species in Africa; 2 species in Baviaanskloof.

Widdringtonia nodiflora Mountain cypress, mountain cedar

nodi = node, *flora* = flower; referring to the distribution of the female cones at nodes or points where branches arise off the stem

Multi-stemmed bush or small tree to 6 m, resprouting strongly. **Leaves** Adpressed in adults, keeled, scale-like, 0.2 cm long, dark green. **Cones** Male at the branch tips, 0.3 cm wide; female in bunches on the stems, 2 cm wide, roundish, wrinkled and knobbly, splitting into 4 segments to release winged seeds. **Habitat** Uncommon in Baviaanskloof, on south-facing slopes in Grassy and Mesic Fynbos. **Distribution** Very widespread from Cape Peninsula to tropical Africa. **Notes** On Mount Mulanje in southern Malawi, the Mulanje cedar (*W. whytei*) grows into a giant tree over 50 m tall that towers over the Afromontane forests; this tree was once regarded as a subspecies of *W. nodiflora*. **Other name** Bergsipres (A). **Status** *W. whytei* is Critically Endangered due to over-exploitation for its prized timber; harvesting was banned in 2007, but illegal logging continues.

Nooitgedagt, Kouga 12.05.2020

Widdringtonia schwarzii Willowmore cedar

Named after Friedrich (Fritz) Schwarz, 1900–1988, a German collector of cacti in Mexico

Tall, gnarled, slow-growing tree, usually 10–20 m tall, but can reach 40 m in sheltered kloofs. **Leaves** Adpressed, scale-like, to 0.2 cm, grey-green. **Cones** Male very small; female in clusters, warty, with winged seeds released in late summer. **Habitat** Occurring sporadically on rocky outcrops, and in rocky ravines that seldom burn, in Grassy and Mesic Fynbos and Fynbos Woodland. **Distribution** Nearly endemic to the Kouga and Baviaanskloof mountains, with small populations on the Grootrivierberge. **Other names** Baviaanskloofseder, sapreehout (A). **Status** Near Threatened.

Zandvlaktekloof, Baviaans 18.02.2016

PODOCARPACEAE

Evergreen coniferous trees and shrubs with ± 156 species in 19 genera, mostly in the southern hemisphere. 2 genera and 4 species in southern Africa.

AFROCARPUS

afro = of Africa, *carpus* = fruit

3 species in Africa; 1 species in Baviaanskloof.

Afrocarpus falcatus (= *Podocarpus falcatus*)   Outeniqua yellowwood

falcatus = sickle-shaped; referring to the shape of the leaves

Dioecious, massive tree, 20—40 m tall, with smoothish bark that peels off in flakes. **Leaves** Arranged spirally, linear-elliptic, twisted at the base, 4 cm long × 0.5 cm wide, bluish grey to dark green. **Seeds** 2 cm long, covered in a resinous skin, borne on short, leaf-like stalks; eaten and dispersed by bats. **Habitat** Forest in the sheltered ravines. **Distribution** Very widespread from Swellendam to tropical Africa. **Notes** Harvested for timber. Popular in gardens. Becoming invasive in Tokai and Constantia on the Cape Peninsula. Fruit attract fruit bats; bat droppings on walls are a familiar sight for those with a female tree in their garden. **Other names** Geelhoutboom, kalander (A), umsonti, umgeya, umkhoba (X), mogôbagôba (S). **Status** Protected in South Africa.

Nooitgedacht, Kouga 12.05.2021

PODOCARPUS

Yellowwoods

podo = foot, *carpus* = fruit; referring to the fleshy receptacle upon which the seed rests

± 100 species, mostly southern hemisphere; 3 species in southern Africa; 1 species in Baviaanskloof.

Podocarpus latifolius   Real yellowwood

lati = wide, *folius* = leaf; although the leaves are only 1 cm wide, they are wider than those of *Afrocarpus falcatus*.

Dioecious tree, to 20 m tall, with rough bark that peels off in strips. **Leaves** Linear-elliptic, 6 cm long × 1 cm wide, dark glossy green, young leaves paler or bronze. **Seeds** Round, grey-blue becoming purple, dispersed by bats and some birds. **Habitat** Uncommon in cool, damp Forest in narrow ravines in Baviaanskloof. **Distribution** Widespread from Cape Peninsula to tropical Africa. **Notes** Used for timber. **Other names** Opregte geelhout (A), umsonti, umgeya, umkhoba (X), mogôbagôba (S). **Status** Protected in South Africa.

28.04.2021

ZAMIACEAE   Cycads

Prehistoric plants with fern-like foliage, but leaves are hard, tough and sometimes spiny. Both males and females bear cones. Most cycads are toxic, especially the seeds. The stem core has been used as a food source. Poaching is a threat to wild populations as collectors prize bigger plants. Fortunately cycads are well established in gardens worldwide, although some species are so rare that only a few survive in the wild. 8 genera and ± 150 species worldwide. 2 genera and 38 species in South Africa.

ENCEPHALARTOS   Broodboom

encephala = in head, *artos* = bread; Khoisan used to make bread from the contents of the stem.

68 species in Africa; 37 species in South Africa; 3 species in Baviaanskloof.

Encephalartos horridus   Blue cycad

horridus = horrible; referring to the leaf tips that are armed with sharp, yellow spines

Stemless or low tree to 1 m. **Leaves** Lobed and sharply toothed, to 1 m long, grey-blue. **Cones** Female 40 cm long, bluish green. Coning in spring and summer. **Habitat** Uncommon and rare in arid Thicket (Noorsveld). **Distribution** Narrowly distributed from the Groot Winterhoek Mountains on the Karoo side of Steytlerville to Gqeberha (Port Elizabeth). Populations around Gqeberha and Kariega (Uitenhage) are now extinct due to over-harvesting. **Notes** Heavily poached in the past. **Status** Listed as Endangered in 2020.

Kariega (Uitenhage) 17.05.2006

Encephalartos lehmannii   Karoo cycad

Named after Johann Lehmann, 1792–1862, a German botanist. He established the genus *Encephalartos* and founded the Hamburg Botanic Garden.

Resprouting, with several stems growing from the base, to 2 m. **Leaves** 1.5 m long, glaucous becoming dark green. **Cones** Female 50 cm long, changing from green to brown. Coning in autumn. **Habitat** Uncommon on quartzitic outcrops in Succulent Karoo. **Distribution** Eastern Cape endemic from Steytlerville to Cookhouse. **Status** Listed as Near Threatened in 2021.

Jansenville 13.12.2014

Encephalartos longifolius   Thunberg's cycad

longi = long, *folius* = leaf; the leaves are longer than those of most other species.

Compact umbrella-shaped tree, resprouting, to 3 m tall, with several stems; bark grey or blackened by fire. **Leaves** In a rosette at the top of the stems, pinnate, hairy when young, 1–2 m long, blue-grey becoming dark green. Leaflets spine-tipped, oblong-lanceolate. **Cones** Female 60 cm long, olive-green. Coning in autumn. **Habitat** Sporadic on steep rocky slopes and in ravines in Fynbos Woodland and Thicket, and on rocky outcrops in Transitional Shrubland and Grassy Fynbos. **Distribution** Eastern Cape endemic from Joubertina to Somerset East. **Notes** Poaching has removed many, but still abundant in inaccessible terrain. Baboons remove leaf buds to eat the resinous gum that is exuded. **Status** Listed as Near Threatened in 2022.

Engelandkop, Kouga 09.07.2011

PALAEODICOTS   Basal angiosperms

The palaeodicots are a small group of ancient plants that diverged from the lineage that evolved to become the monocots and dicots. Palaeodicots are the ancestral angiosperms.

There are relatively few extant species and these survivors tend to be quite unusual plants. *Amborella trichopoda*, a monotypic genus confined to the main island of New Caledonia in the South Pacific, is regarded as the oldest living angiosperm, with a cell structure that is simpler than that of all other angiosperms. Some of the better-known palaeodicots include water lilies, magnolia, avocado, bay laurel, cinnamon, black pepper and nutmeg. Notable characteristics of this group are that the flowers are primitive and with the floral parts arranged in threes, the pollen has only one pore and the leaves have branching veins.

There are ± 9,000 species of palaeodicots in 30 families around the world.

HYDNORACEAE

A somewhat strange family of leafless root parasites with 2 genera and 11 species worldwide. *Prosopanche* has 4 species that are native to South and Central America. *Hydnora* has ± 7 species native to arid and semi-arid parts of Africa, Madagascar and the Arabian Peninsula.

HYDNORA

udnon = truffle; alluding to the fungus-like appearance of the flowers. Furthermore, *Hydnum* is a genus of fungus that has hairs resembling those on the flowers of *Hydnora*.

4 species in South Africa; 1 species in Baviaanskloof.

Hydnora africana   Bobbejaankos

africana = from Africa; the common name pertains to the edible fruit.

Leafless root parasite to 15 cm tall, parasitic on *Euphorbia*.
Flowers Single, on the ground, club-shaped, foetid-smelling, brown and scaly or warty outside, orange inside. Flowering in spring and summer.
Habitat Seldom seen in arid areas where euphorbias are present in Succulent Karoo and Thicket. **Distribution** Very widespread from Namibia to the Cape Peninsula and to KwaZulu-Natal. **Other names** Jackal food (E), idolo-lenkonyane, ubuklunga (X), umavumbuka (Z)

29.08.2015

LAURACEAE

Laurels

A family of 45 genera and ± 2,850 species growing in warm and tropical climates around the world. Laurels are mostly evergreen trees in the tropics, with *Cassytha* being the only genus in the family that is a parasitic vine. Avocados (genus *Persea*) belong in this family and stinkwood (*Ocotea bullata*) is also a member. Laurel forests in the tropics support several species of this family. The leaves of the bay tree (*Laurus nobilis*) are used for seasoning food, and cinnamon is derived from the wood of several species of tree in the genus *Cinnamomum*. The species *C. camphora*, used medicinally and as an ornamental, has become an invasive weed in subtropical parts of southern Africa. The fruit of this family is a berry that looks somewhat like an acorn in some species and is dispersed by birds and animals.

CASSYTHA

kesatha = a tangled wisp of hair; referring to the tangled masses of thin twining stems seen on this parasitic vine

16 species in tropical Africa, Asia and Australia; 3 species in South Africa; 1 species in Baviaanskloof.

Cassytha ciliolata   Dodder, devil's tresses

ciliolata = with fine hairs; referring to the whorls of anthers in the flower

Parasitic vine, yellowish, forming a tangled mass to 1 m wide. **Leaves** Scale-like. **Flowers** In small clusters, 0.3 cm wide, pale yellow. Fruit fleshy, yellow to red, with seed dispersed by birds. Flowering in spring and summer. **Habitat** Common in Fynbos and Transitional Shrubland. **Distribution** Widespread from Kamieskroon to Makhanda. **Notes** The flexible stems can be used as string. **Other names** Bobbejaantou, nooienshaar (A).

Kouga Wildernis 14.05.2021

NYMPHAEACEAE   Water lilies

A family of 5 genera and 70 species of rhizomatous aquatic herbs in tropical and temperate waters around the world. This family is regarded as basal in the evolution of angiosperms, being among the first to have evolved. The large leaves and flowers are dramatic and showy as they float on the water's surface. New leaves growing up under water accumulate gases that are released when they mature, bursting open in spring, causing bubbles to rise to the surface of the water. The sacred lotus (*Nelumbo nucifera*), the national flower of India and Vietnam, belongs in the Nelumbonaceae, not in the water lily family. 1 genus in South Africa.

NYMPHAEA
Named after Nymphaea, a nymph in Greek mythology

± 40 species around the world, mostly tropical; 2 species in South Africa.

Nymphaea nouchali var. *caerulea* Blue water lily

Noakhali = a region in Bangladesh where this species does not occur naturally, *caerulea* = blue, referring to the flower colour; there are 5 different varieties, but only var. *caerulea* occurs in Baviaanskloof.

Rhizomatous, aquatic herb with large floating leaves. **Leaves** Roundish and deeply notched at the base, with undulating or scalloped margins. **Flowers** 12 cm wide, blue to pink, closing at night. Flowering in late summer. **Habitat** Found in ponds, dams and lakes. **Distribution** Very widespread from Western Cape to tropical Africa and to India, Southeast Asia and Australia. **Notes** Cultivated in some parts of the world, making it difficult to differentiate between natural and introduced populations. The rhizomes are edible if boiled. Various parts of the plant are used medicinally. *N. lotus* has white flowers and only occurs further north. **Other names** Blouwaterlelie (A), ikhubalo lechanti, intekwane (X), amazibu, izeleba (Z).

Mistkraal Dam, Baviaans 06.04.2012

MONOCOTYLEDONS   Monocots

Of the two major groups of flowering plants, the monocotyledons — or monocots — are grass-like flowering plants with seeds that produce a single leaf (or cotyledon) when they germinate. The leaves of monocots always have parallel veins and are usually strap-shaped. Monocot flower parts are always in multiples of three and the pollen has a single pore. Vascular bundles or growth tissue inside the stems is scattered, unlike the arrangement of the dicots' growth tissue, which is in rings. Below the ground, monocot roots are mostly adventitious, whereas dicots have strong taproots with secondary roots.

Well-known members of the monocot group are irises, lilies, orchids, aloes, asparagus, grasses, sedges and restios. Relative to the dicots, fewer have developed into tall trees or shrubs, with notable exceptions being bamboo, palm trees, bananas, aloes and strelitzias.

There are ± 60,000 species of monocots in the world, making up one-quarter of all flowering plants.

AGAPANTHACEAE   Agapanthus

1 genus only, entirely southern African; 6 species in higher-rainfall areas.

AGAPANTHUS   Agapantha, bloulelie

agape = love, *anthos* = flower; referring to the striking flowers; the Afrikaans common name refers to the bluish flowers.

Very popular in gardens around the world, with many different cultivars and hybrids available. 1 species in Baviaanskloof.

Agapanthus praecox   Blue lily

praecox = early, premature; referring to the plant's tendency to flower early in summer compared with other *Agapanthus* species

Rhizomatous, evergreen geophyte to 1 m. **Leaves** Evergreen, strap-shaped and soft, in a rosette, 50 cm long × 2 cm wide, glossy green. **Flowers** Arranged in a neat umbel, up to 7 cm long, blue to white, broadly funnel-shaped. Flowering in summer. **Habitat** South-facing slopes in Grassy and Mesic Fynbos. **Distribution** Widespread from Knysna to KwaZulu-Natal. **Notes** Extremely popular in gardens. Used in traditional medicine to treat various ailments.

Kouenek, Kouga 20.01.2022

AGAVACEAE

Agaves, century plants

± 23 genera and 600 species in the world. The Joshua tree (*Yucca brevifolia*) and century plants (*Agave* species) are well-known members from dry parts of the Americas. Leaves are often in rosettes. There are ± 33 species of *Chlorophytum* in South Africa, several of which are naturalised.

AGAVE*

Sisal, garingboom

agauos = admirable; pertaining to the magnificence of the tree when in flower

± 250 species are native to hot and arid parts of the Americas. Some species are popular as ornamental plants. 2 species are naturalised in South Africa; 2 species in Baviaanskloof.

*Agave americana**

americana = from America

Robust succulent with a rosette of huge leaves reaching to 2 m; flowering pole may reach to 8 m. **Leaves** Strap-shaped, 150 cm long, greyish green with yellow, toothed margins, spine-tipped. **Flowers** Numerous in panicles on a long pole, flower 8 cm long, pale yellow. Flowering in summer and autumn. **Habitat** Valleys in disturbed areas, in Succulent Karoo and Thicket. **Distribution** Native to southwestern parts of the USA and Mexico. **Notes** Introduced to South Africa

Zandvlakte, Kouga 19.01.2022

for erosion control, now invasive throughout southern Africa. Used for fodder, protective hedging and medicine. Fibres from the leaves are used for twine and rope (sisal made with *A. sisalana*). Tequila is made by fermenting the liquid extracted from the roasted core or base of the plant.

CHLOROPHYTUM

Grass lily

chloros = yellow-green, *phyton* = plant; pertaining to the colour of the leaves and flowers

± 150 species, mostly in tropical Africa but also in Asia; ± 33 species in South Africa; 2 species in Baviaanskloof.

Chlorophytum comosum   Hen-and-chickens, spider plant

comosus = bearing numerous leaves; pertaining to the tufted habit. The common names refer to the habit of producing leafy tufts or propagules at the nodes, forming the 'baby chicks' or 'spiders'.

Resprouting rhizomatous geophyte up to 80 cm, with swollen roots. **Leaves** Abundant, tufted, with midvein and channelled, 30 cm long × 2 cm wide, usually pale green. **Flowers** Inflorescence lax, up to 1 m long; flowers star-shaped, growing in lateral clusters above the leaves, 2 cm wide, white. Flowering in summer. **Habitat** Damp, shaded rock ledges in Forest and Thicket. **Distribution** Very widespread from Swellendam in the Western Cape to Limpopo. **Notes** Possibly the most cultivated house plant in the world (several cultivars). The variegated form with white leaf margins is most popular. **Other names** Hen-en-kuikens (A), ujejane (X).

Zandvlaktekloof 19.01.2022

Chlorophytum crispum

crispus = crisped, kinked, curled;
pertaining to the margins of the leaves

Resprouting rhizomatous geophyte
to 50 cm. **Leaves** Grow flat in a basal
rosette, up to 7 cm long × 1.5 cm wide, fresh
green, with crisped margins fringed with
hairs. **Flowers** In a branched raceme, up to
1 cm long, white with green keels. Flowering
in spring, summer and autumn. **Habitat**
Scattered in Succulent Karoo, Transitional
Shrubland and Thicket. **Distribution** Fairly
widespread from Montagu to Kariega
(Uitenhage).

Uitspan, Baviaans 05.05.2006

ALLIACEAE   Onions

± 13 genera and 600 species in this commercially important family that is cultivated around the world.
Some of its better-known members are garlic (*Allium sativum*), onions (*A. sepa*), leeks, shallots and
chives (*A. schoenoprasum*); wildeui (*A. dregeanum*) is the only species native to South Africa, with a
few naturalised and cultivated species. The family is characterised by flowers in umbels, and bulbs with
narrow basal leaves that smell – unsurprisingly – of onions. In sub-Saharan Africa, there are 4 genera and
35 species, mostly in *Tulbaghia*.

TULBAGHIA

Named after Ryk Tulbagh, 1699–1771, governer of the Cape from 1751–1771. He sent plant material to
Linnaeus in 1763. The town of Tulbagh in the Western Cape bears his name.

± 20 species in southern Africa; 2 species in Baviaanskloof.

Tulbaghia violacea   Wild garlic

violacea = violet-coloured; pertaining to the flowers

Resprouting bulbous geophyte to
30 cm. **Leaves** Linear, smooth and
semi-erect, 30 cm long × 0.6 cm wide,
blue-grey. **Flowers** Several in an umbel,
each flower with 3 fleshy, paler lobes
at the throat, 2 cm long, mauve-purple.
Flowering in summer and autumn.
Habitat Transitional Shrubland
and Grassy Fynbos. **Distribution**
Widespread from the Little Karoo to
KwaZulu-Natal. **Notes** *T. capensis* has
6 orange lobes at the throat; tepals are
brown to purple, with contrasting green
calyx. **Other name** Wildeknoffel (A).

Combrink's Pass, Baviaans 02.02.2020

AMARYLLIDACEAE

± 60 genera and 850 species in the world. These bulbous or rhizomatous plants with spectacular flowers hail from the tropical and warm, temperate regions of the world, with centres of diversity in South Africa and South America. Members of this family have great economic value as ornamental plants. 210 species in southern Africa, with high levels of endemism in the Western Cape; 25 species in Baviaanskloof.

BOOPHONE
Oxbane, kopseerblom

bous = ox, *phonos* = murder, death; referring to the toxic properties of the bulb, with poison strong enough to kill an ox

2 species in Africa; 1 species in Baviaanskloof.

Boophone disticha Gifbol

disticha = two-ranked; referring to the symmetrical, fan-shaped arrangement of leaves

Resprouting, bulbous geophyte to 30 cm, with large, partly exposed bulb and erect leaves. **Leaves** 12–20, in a fan, 40 cm long × 3 cm wide, greyish, with undulating margins. **Flowers** Numerous in a spreading cluster, 2 cm wide, pink to red, on pedicels 10 cm long. Flowering in spring, summer and autumn. After pollination the pedicels elongate up to 30 cm and the dried inflorescence becomes a tumbleweed that disperses the seed as it rolls. **Habitat** Common on rocky, loam-rich soils in Transitional Shrubland and Grassy Fynbos. **Distribution** Very widespread from Bredasdorp to Africa's tropics. **Notes** Used in traditional medicine. The dry leaf bases that protect the bulb are used to dress wounds; also utilised by the Khoisan to cover and preserve the deceased before burial. **Notes** *B. haemanthoides* occurs along the West Coast from southern Namibia to Saldanha.

Patensie 05.02.2020

BRUNSVIGIA
Candelabra lily, kandelaar

Latin translation of Brunswick, a town in Germany, named after the Duke of Brunswick-Lüneberg, Karl Wilhelm Ferdinand, 1713–1780, for his promotion of the study of plants

The dried inflorescences are often called tumbleweeds because their light spherical form tumbles across the veld, dispersing seeds. Modern populations establish along fence lines if the habitat is suitable. ± 20 species in southern Africa; 2 species in Baviaanskloof.

Brunsvigia josephinae   Candelabra lily

Named after Empress Josephine, 1763–1814, born Marie Josèphe Rose Tascher de La Pagerie – Napoleon's first wife. She was passionate about roses and kept a magnificent rose garden.

Resprouting, bulbous geophyte to 60 cm, with bulb partly exposed. **Leaves** 8–20, oblong, recurved, 60 cm long × 20 cm wide, greyish, dry at flowering. **Flowers** Numerous, 8 cm long, in a wide head with tepals rolling back, red. Flowering in autumn. **Habitat** Infrequent and sporadic on shale soils in Transitional Shrubland and Grassy Fynbos. **Distribution** Widespread from Nieuwoudtville and Worcester to Baviaanskloof. **Notes** The bulb tunics are used in traditional medicine as a wound dressing. Xhosa men use them as plasters after ritual circumcision. **Other name** Lantanter (A). **Status** Vulnerable.

Joubertina 08.03.2009

Brunsvigia striata

striata = striped; referring to the slightly ribbed seed capsules and striated leaves

Resprouting, bulbous geophyte to 30 cm. **Leaves** 3–6, elliptical, 15 cm long × 6 cm wide, lying flat on the ground, fresh green with margins red and crisped. **Flowers** Relatively few, sometimes many, 4 cm long, in a round head to 30 cm wide, pink with darker mid-lines and brownish protruding anthers; seed capsules are 3-angled and ribbed. Flowering in autumn. **Habitat** Infrequent on warm, stony slopes on sandstone in Transitional Shrubland and Thicket. **Distribution** Widespread from Bokkeveld (Nieuwoudtville) to Baviaanskloof.

Rooihoek, Kouga 11.03.2012

CYRTANTHUS   Fire lily

kyrtos = curved, *anthos* = flower; alluding to the curved floral tube. The common name refers to the habit of some species to flower only after fire.

± 50 species in central and southern Africa; 14 species in Baviaanskloof.

Cyrtanthus collinus

collinus = pertaining to hills; referring to the preferred habitat of gently sloping plateaus

Bulbous geophyte to 30 cm tall, forming colonies. **Leaves** 3, narrow and strap-shaped, 20 cm long × 0.6 cm wide, grey, sometimes dry when flowering. **Flowers** 4–10, funnel-shaped and nodding, 4 cm long, bright orange-red. Flowering in late summer and autumn. **Habitat** Infrequent on grassy slopes in Grassy Fynbos and Transitional Shrubland. **Distribution** Widespread from Greyton to Zuurberg.

Bergplaas, Combrink's Pass 16.02.2007

Cyrtanthus flammosus Kouga flame lily

flammosus = like a flame; pertaining to the bright orange-red colour of the flowers

Bulbous geophyte to 25 cm; bulbs are partly exposed and covered with brown, papery scales. **Leaves** 2–4, lanceolate and thickly textured, spreading, 25 cm long × 2 cm wide, greyish green. **Flowers** 1 or 2, 10 cm wide, bright orange-red, trumpet-shaped. Flowering in autumn. **Habitat** Rare on shaded cliffs. **Distribution** Endemic to Baviaanskloof, known only from cliffs around the Kouga Dam and Grootrivierpoort. **Status** Critically Rare.

Kouga Dam 02.01.2004

Cyrtanthus labiatus

labiatus = lipped; referring to the tips of the tepals that have 2 lip-like lobes

Bulbous geophyte to 30 cm, producing bulbils. **Leaves** 2–4, linear, 30 cm long × 1.5 cm wide, dark green. **Flowers** About 8, tubular, 5 cm long, coral-red, with the upper 4 tepals overlapping to form a hood. Flowering in midsummer. **Habitat** Uncommon on shaded cliffs in narrow ravines in Forest. **Distribution** Nearly endemic to Baviaanskloof, from Klein Swartberg to Baviaanskloof.

Kouga Dam 10.02.2004

Cyrtanthus montanus

montanus = montane; referring to the habitat where the plants grow

Bulbous geophyte to 25 cm, producing bulbils. **Leaves** 3–5, 30 cm long × 2 cm wide, medium green, strap-shaped. **Flowers** 5–10, 5 cm long, red, funnel-shaped from a narrow tube, with stamens well exserted. Flowering in late summer and autumn. **Habitat** Uncommon in rocky crevices on upper slopes in Mesic Fynbos. **Distribution** Endemic to Baviaanskloof Mountains.

Kouga Dam 19.08.2009

Cyrtanthus obliquus Knysna lily, justafina

obliquus = oblique, slanting; perhaps referring to the tendency flowers have of hanging to one side only on the stalk

Robust, bulbous geophyte to 60 cm. **Leaves** About 6, 50 cm long × 5 cm wide, grey-green, erect, broadly strap-shaped, thick, leathery, twisted. **Flowers** 6–12, 7 cm long, orange and yellow, sometimes with green tips hanging down. Flowering in spring and summer. **Habitat** Occasional on open slopes in Grassy Fynbos and Transitional Shrubland. **Distribution** Widespread from Knysna to KwaZulu-Natal. **Notes** Used in traditional medicine. **Status** Declining.

Kareedouw 25.11.2008

Cyrtanthus sanguineus   Kei lily

sanguineus = blood-red; referring to the typical flower colour; the Baviaanskloof form has salmon-pink flowers.

Bulbous geophyte to 50 cm. **Leaves** 1–4, 30 cm long × 2 cm wide, bright green, linear-lanceolate, usually dry at flowering. **Flowers** 1–4, large, 6 cm long × 4 cm wide, salmon-pink, funnel-shaped. Flowering in late summer and autumn. **Habitat** Infrequently encountered on rocky, south-facing slopes in Grassy Fynbos and Fynbos Woodland. **Distribution** Very widespread from Baviaanskloof to East Africa and southern Sudan. **Notes** The form found in Baviaanskloof might be an undescribed species or a variety that requires further study.

Bruintjieskraal, Grootrivierpoort 05.04.2018

Cyrtanthus smithiae   Wilde-amarilla

Named after Lady Smith, 1798–1872, wife of Sir Harry Smith, governor of the Cape Colony from 1847–1852. She is said to have brought this species from the Eastern Cape to Cape Town, where it was successfully cultivated.

Bulbous geophyte to 30 cm, the bulb mostly submerged. **Leaves** 2–4, 20 cm long × 1 cm wide, dull green to glaucous, strap-shaped and spirally twisted. **Flowers** 1–few, 8 cm long, pale pink with darker red or brown stripes, trumpet-shaped and spreading to nodding. Flowering in late spring. **Habitat** Common in disturbed or open areas in Thicket and Transitional Shrubland. **Distribution** Widespread from Baviaanskloof to Graaff-Reinet and Peddie, Eastern Cape.

Zandvlakte bergpad, Kouga 07.07.2011

GETHYLLIS   Kukumakranka

gethyon = bulb, onion or leek, *ullus* = diminutive, a small onion; referring to the low stature of some species and the superficial resemblance to the growth form of some species of *Allium*. The common name is derived from the Khoisan name. The supposed Afrikaans derivation of kukumakranka from 'goed vir my krank maag' meaning 'good for my upset stomach' is apparently incorrect.

Early Dutch settlers used the aromatic fruit of *G. afra* to perfume cupboards and flavour brandy. ± 32 species, almost all in winter-rainfall southern Africa; 1 species in Baviaanskloof.

Gethyllis spiralis

spiralis = spiralled; pertaining to the spirally twisted leaves

Bulbous geophyte to 15 cm. **Leaves** 6–12, 15 cm long, greyish, linear, sometimes fringed with hairs, curled, dry at flowering. **Flowers** 5 cm long, white, pale pink below and pleasantly fragrant. Fruit is a pale brown elongated berry. Flowering in summer. **Habitat** Rocky sandstone ridges in Succulent Karoo and Transitional Shrubland. **Distribution** Widespread from Worcester to Addo.

Die Krokodil, Speekhout 24.12.2017

HAEMANTHUS

haima = blood, *anthos* = flower; pertaining to the red flowers of many species. The first species to be described was *H. coccineus*, with blood-red flowers.

These awe-inspiring bulbs have large 'elephant ear' leaves that vary in number from 1–6. Most species lose their leaves in summer and flower in autumn. Once the seeds are dispersed, new leaves begin to grow in winter. 22 species in southern Africa; 3 species in Baviaanskloof.

Haemanthus albiflos   Paintbrush lily

albus = white, *flos* = flower; pertaining to the white flowers

Robust, bulbous geophyte to 30 cm. **Leaves** 2–6, almost prostrate, tongue-shaped, smooth or hairy, up to 20 cm long × 8 cm wide, green when flowering, with no markings, margins hairy. **Flowers** Numerous in tight heads, 5 cm wide, white, with white, thinly textured bracts with green lines. Fruit is a berry that changes from white to vivid red. Flowering in winter. **Habitat** Frequent on rocky ledges in Thicket and Fynbos Woodland, and in Forest, especially in ravines. **Distribution** Widespread from the Little Karoo to KwaZulu-Natal.

Uitspankloof, Baviaans 02.05.2018

Haemanthus coccineus   April fool

coccineus = deep red; pertaining to the colour of the flowers. The common name alludes to the element of surprise when the bright flowers pop up out of seemingly bare earth in autumn; leaves die back in summer and the bulb persists, hidden underground until the next flowering.

Robust, bulbous geophyte to 40 cm, often forming clumps. **Leaves** 2, tongue-shaped and spreading, very variable in size and colour, up to 70 cm long × 20 cm wide, often speckled below, margins sometimes hairy and rolled back, dry at flowering. **Flowers** Tight flowerhead up to 8 cm wide with 25–100 flowers, each with yellow anthers; on bright red speckled stalk, which is topped with several fleshy scarlet bracts. Flowering in autumn. **Habitat** Rocky slopes in Transitional Shrubland and Thicket. **Distribution** Very widespread from Namibia to Makhanda. **Other name** Velskoenblaar (A).

Grasnek, Kouga 06.05.2006

Haemanthus sanguineus Velskoenblaar

sanguineus = blood-red; referring to the flower colour

Bulbous geophyte to 30 cm when flowering. **Leaves** 2, wider than long, sometimes with red margins, not speckled below, flat on the ground, dry at flowering. **Flowers** Numerous flowers in a tight head, red to orange with yellow anthers; bracts 5–11, leathery and red; stalk plain red and often compressed. Flowering in late summer and autumn. **Habitat** Stony soil on rocky ledges in Grassy Fynbos and Transitional Shrubland. **Distribution** Widespread from Clanwilliam to Port Alfred.

Langkloof 07.03.2009

NERINE

Named after Nerine, the Greek mythological sea nymph that protected sailors and their ships. Historical accounts tell of a ship transporting nerine bulbs that ran aground on Guernsey. It is said that nerine plants grew there, perhaps with thanks to Nerine who must have helped them ashore.

± 23 species in southern Africa; 1 species in Baviaanskloof.

Nerine humilis (= *N. peersii*) **Dwarf nerine**

humilis = low, low-growing; pertaining to the relatively low stature when in flower

Small, bulbous geophyte to 40 cm. **Leaves** 3–6, strap-shaped and spreading, up to 30 cm long × 1 cm wide, green at flowering. **Flowers** Several in a loose umbel, each flower to 3 cm long, white or pale pink, with a darker median line, or bright pink, with tepals undulating and apex of tepals recurved. Flowering in autumn. **Habitat** Common on rocky ledges in loamy soils on south-facing slopes in Grassy Fynbos and Fynbos Woodland. **Distribution** Widespread from Clanwilliam to Baviaanskloof.

Kleinpoort, Baviaans 20.04.2012

SCADOXUS

Blood lily, seeroogblom

skiadion = parasol, *doxa* = glory; referring to the magnificent umbel of flowers. The Afrikaans common name seeroogblom ('sore eye flower') alludes to some people experiencing sore eyes or a headache after exposure to the flowers in a confined space; it is also not advisable to touch one's eyes after contact with the flowers. It is likely to be the pollen that causes an allergic reaction.

Large flowers pollinated by birds; the ripe berries are eaten and dispersed by birds, baboons and monkeys. The poisonous bulb can cause death if consumed. The bulb and leaves are used in traditional medicine for a variety of ailments including stomach problems, and the plant is also used during childbirth. 9 species in southern and tropical Africa; 3 species in South Africa; 1 species in Baviaanskloof.

Scadoxus puniceus

puniceus = crimson, scarlet; referring to the colour of the flowers

Bulbous geophyte to 45 cm, with fleshy roots. **Leaves** Several, strap-shaped, erect with wavy margins and prominent midrib, 40 cm long, glossy green; leaf bases form a false stem that is often speckled with purple. **Flowers** Numerous small flowers in each 15 cm-wide brush-like head, bright coral-orange or reddish, surrounded by large bracts. Flowering in spring and summer. **Habitat** Shaded, damp slopes in Forest and Thicket. **Distribution** Widespread from De Hoop to the African tropics.

Ysrivier, Cambria 14.04.2021

STRUMARIA

Cape snowflake, tolbol

struma = cushion-shaped swelling, *aria* = possessing; pertaining to the swelling at the base of the flower

± 28 species in southern Africa; 1 species in Baviaanskloof.

Strumaria gemmata

gemmatus = with buds, budded; referring to the umbellate inflorescence that produces new flower buds at the centre

umbellate inflorescence

Bulbous geophyte to 40 cm. **Leaves** 2 or 3, 15 cm long × 1 cm wide, green, strap-shaped, spreading and usually hairy, dry when flowering. **Flowers** 10–20, each 1 cm wide in a neat, loose 10 cm-wide umbel, cream to yellow, star-shaped, tepals crisped, with a translucent swelling at base. Flowering in autumn. **Habitat** Locally abundant on alluvial flats in valleys in Succulent Karoo and Thicket. **Distribution** Widespread from Bredasdorp to the Great Karoo and Gqeberha (Port Elizabeth).

Zandvlakte, Baviaans 07.04.2012

APONOGETONACEAE

Cape pondweed

Aquatic plants with 1 genus, *Aponogeton*. Some species are grown as ornamentals in aquariums. Waterblommetjiebredie is a traditional South African stew prepared using the flowers of *A. distachyos* as a main ingredient.

APONOGETON

apon = water, *geiton* = near, neighbour; referring to the aquatic habitat

Over 40 species in tropical and temperate parts of Africa, Madagascar, Australia and Asia; 11 species in southern Africa; 2 species in Baviaanskloof.

Aponogeton desertorum Waterblommetjie

desertorum = desert; alluding to the habitat of this species in pools in desert areas

Rhizomatous aquatic with prostrate growth. **Leaves** 6–9, heart-shaped to oblong with a sharp pointed tip, floating, 15 cm long × 5 cm wide, fresh green. **Flowers** Numerous, crowded on forked spikes that are up to 10 cm long, each flower whitish with 2 tepals and 6 stamens. Flowering in summer and autumn. **Habitat** Occasional in perennial rock pools. **Distribution** Widespread from Namibia and Botswana to Kariega (Uitenhage).

Coleskeplaas, Baviaanskloof River 07.08.2010

Aponogeton distachyos Waterlelie

distichus = arranged in two opposite rows; pertaining to the inflorescence that has 2 opposite 'arms'; the tepals are also arranged distichously on the spike.

Rhizomatous, tuber-forming aquatic with prostrate growth. **Leaves** Many, narrowly oval, floating, 20 cm long × 6 cm wide on a leaf stalk up to 1 m long, fresh green, sometimes mottled. **Flowers** Several, symmetrically arranged on a forked inflorescence, white, with 1 tepal and up to 16 stamens. Flowering in winter, spring and early summer. **Habitat** Fairly common in pools and dams. **Distribution** Widespread from Nieuwoudtville to Baviaanskloof. **Notes** Tubers become dormant in summer and regenerate when rock pools fill with water after rains. This species has become naturalised in Australia, South America and Europe.

Poortjies, Baviaans 18.05.2021

ASPARAGACEAE

The best-known member of this family is a spring vegetable familiar to many tables and known as asparagus. *Asparagus officinalis* is native to Europe and Asia and is now cultivated around the world. ± 114 genera and 2,900 species worldwide; 1 genus and ± 86 indigenous species in southern Africa.

ASPARAGUS

asparag = sprout, shoot; referring to the plant's ability to resprout strongly after fire or other disturbance, also emphasising the plant's habit of sending out shoots of new growth, which are the edible parts in *A. officinalis*

Asparagus leaves are modified stems called cladodes. Many species are spiny and very strongly rooted and can persist in areas that have been overgrazed. ± 120 species mostly in Africa, but also Asia; 16 species recorded in Baviaanskloof.

Asparagus aethiopicus Wild asparagus

aethiopica = from sub-Saharan Africa; in classical times this region had not yet been explored by Europeans.

Resprouting, spiny climber to 3 m tall, with hooked thorns and pale, ribbed stems. **Leaves** In clusters, cylindrical, pointed, 4 cm long × 0.1 cm wide. **Flowers** Inflorescence a raceme of white, star-shaped flowers, each 0.5 cm long. Berry green becoming red, 0.5 cm wide. Flowering in summer, autumn and winter. **Habitat** Fairly common on rocky hillsides with bush or thicket in various habitats, but not Mesic Fynbos. **Distribution** Widespread from Namaqualand to Cape Peninsula and KwaZulu-Natal. An invasive weed in many parts of the world. **Notes** Often confused with *A. densiflorus*. **Other name** Haakdoring (A).

Nuwekloof Pass, Baviaans 05.05.2006

Asparagus africanus Bush asparagus

africanus = from Africa

Resprouting, spiny shrub to 1 m tall, with straight or spreading brown spines. **Leaves** ± 12 in clusters, 1.5 cm long × 0.05 cm wide. **Flowers** 1–6 in axils, on short thin pedicels 0.5 cm long; flowers small, 0.3 cm wide. Berry 0.4 cm wide. Flowering in spring and summer. **Habitat** Common on rocky slopes in most non-Fynbos vegetation. **Distribution** Widespread from Saldanha to KwaZulu-Natal. **Notes** The fresh shoots may be eaten as a vegetable.

Rus en Vrede 4X4, Kouga 28.07.2011

Asparagus asparagoides   Cape smilax

asparagoides = like asparagus; likely to be similar to the
asparagus with which Linnaeus was familiar

Resprouting, spineless scrambler to 3 m. **Leaves**
Ovate, dark green. **Flowers** Single in axils, hanging
down, tepals fused at the base. Berry black, 1 cm
wide. Flowering in winter and spring. **Habitat**
Frequent in shaded situations under thicket and bush,
particularly in Thicket. **Distribution** Very widespread
from Gifberg to tropical Africa.

Rooikloof, Baviaans 01.05.2018

Asparagus burchellii

Named after William John Burchell, 1781–
1863, an English explorer who collected plants
in southern Africa and Brazil. He amassed
over 50,000 plant specimens in southern
Africa and donated them to Kew Gardens,
each record with meticulous, detailed notes
on habit and habitat. He described his
journey in *Travels in the Interior of Southern Africa*, a
2-volume work published in 1822 and 1824.

Resprouting, usually tufted, with many stems
from the base to 1.5 m. Spines spreading and
slightly recurved, usually in groups of 3. **Leaves**
In clusters, cylindrical, 0.3 cm long. **Flowers**
1–3 on apical disc, often flowering profusely;
flowers small, to 0.3 cm wide, fragrant. Berry

Zandvlakte, Baviaans 07.04.2012

0.3 cm wide. Flowering in autumn. **Habitat** Common on alluvial soils in valleys in Succulent Karoo and
Thicket. **Distribution** Widespread from Stellenbosch to Komani (Queenstown).

Asparagus capensis   Katdoring

capensis = from the Cape; referring to the
geographical origin of the species

Resprouting, erect shrub to 1 m, with spines
in 3s. **Leaves** 5 in clusters, tiny, grey-green
and hairy. **Flowers** 1 or 2 on an apical disc,
white, fragrant. Berry red, 0.4 cm wide.
Habitat Common in all habitats except
Fynbos. **Distribution** Widespread from
Namibia to Graaff-Reinet and Makhanda.

Bokloof 4X4, Kouga 31.01.2020

Asparagus crassicladus

crassus = thick, *cladus* = branch, shoot; alluding to the relatively thick stems; the leaves are rather thick and somewhat succulent.

Resprouting, tough, sprawling shrub to 2 m, spines stout and hooked. **Leaves** In clusters, thick, curving upward, with pointed tips. **Flowers** Densely arranged on short racemes. Berry round, red becoming black, 0.5 cm wide. Flowering in summer. **Habitat** Occasional in Grassy Fynbos, Transitional Shrubland, Thicket and Succulent Karoo. **Distribution** Fairly narrow range from Oudtshoorn to Jansenville and Hlanganani.

Zandvlakte, Kouga 18.01.2022

Asparagus densiflorus   Asparagus fern

densiflorus = densely flowering; pertaining to some forms that have flowers closely packed around the stems, a form that is popular in gardens around the world

Upright or spreading shrub to 60 cm tall, with green striated stems and hooked spines. **Leaves** Single or paired, 1 cm long × 0.2 cm wide, fresh green, on short lateral shoots 10 cm long, sometimes densely leafy. **Flowers** On racemes 5 cm long, each flower 0.3 cm wide. Berry bright red, 0.7 cm wide. Flowering in spring, summer and autumn. **Habitat** Infrequent in Baviaanskloof on rocky slopes in Grassy Fynbos and Transitional Shrubland. **Distribution** Widespread from Kouga Mountains to Mozambique, mostly near the coast.

Zandvlakte 13.10.2010

Asparagus multiflorus

multi = many, *florus* = flowers; referring to the habit of flowering profusely

Resprouting, tangled climber to 2 m tall, spineless, branches velvety and with dark brown, blunt, triangular scales at the nodes. **Leaves** In clusters of 7, pointed, 1 cm long, lime-green. **Flowers** In the axils, 0.3 cm wide, few clustered on short pedicels. Flowering in summer. **Habitat** Loamy and clay soils in all habitats, but seldom in Fynbos. **Distribution** Widespread from Swellendam to Komani (Queenstown) and East London.

Nuwekloof Pass, Baviaans 16.01.2022

Asparagus setaceus Feathery asparagus

setaceus = bristle-like; alluding to the appearance of the soft leaves, arranged in clusters along the stem

Resprouting, tangled climber to 2 m tall, almost spineless, forming dense clumps. **Leaves** Very thin, 1.5 cm long, bright green, in dense clusters along green stems. **Flowers** Relatively few, single, white. Berry black. Flowering in spring, summer and autumn. **Habitat** Damp areas, usually near streams or dry riverbeds in kloofs under Forest and Thicket. **Distribution** Very widespread from Calitzdorp to tropical Africa.

Kouga Valley, Joubertina 22.06.2009

Asparagus striatus

striatus = striated; pertaining to the striated stems and leaves

Resprouting, tough, erect shrub to 75 cm tall, with spines only near the base. **Leaves** Single, linear, rigid, 4 cm long × 0.5 cm wide, striated. **Flowers** Several clustered on short pedicels from an apical disc; individual flowers 0.3 cm long. Flowering in spring and summer. **Habitat** Common on exposed and overgrazed slopes with clay in the soil, in Transitional Shrubland, Succulent Karoo and Thicket. **Distribution** Widespread from Namibia to Agulhas, Eastern Cape and Free State.

Zandvlakte, Kouga 18.01.202

Asparagus suaveolens

suaveolens = fragrant; referring to the sweet-smelling flowers

Resprouting, erect, spiny shrub to 1 m tall, branches sometimes spine-tipped with straight spines in 2s or 3s. **Leaves** 1–6, cylindrical, very thin, in clusters. **Flowers** 1–3 on an apical disc. Flowering in autumn, winter and spring. **Habitat** Common on sandy, stony or loamy soils in Transitional Shrubland, Succulent Karoo and Thicket. **Distribution** Very widespread in southern Africa to tropical Africa.

Zandvlakte, Baviaans 01.02.2020

Asparagus subulatus

subulatus = awl-shaped; pertaining to the finely pointed leaves

Resprouting, climbing shrub to 2 m tall, with shoots growing in an alternately branched and zigzag fashion, adpressed spines lower down. **Leaves** In clusters, 3–6, thread-like, 2 cm long × 0.1 cm wide. **Flowers** Several clustered on an apical disc. Flowering in spring and summer. **Habitat** Infrequent on loam-rich soils in Transitional Shrubland and Thicket. **Distribution** Fairly narrow distribution from Baviaanskloof to East London.

Elandsvlakte, Baviaans 01.11.2011

ASPHODELACEAE

Aloes

± 40 genera and 900 species. *Aloe* is the most diverse genus in the family, with over 550 species. Together with palms, bananas and bamboo, this is one of the few families of monocots that grow into trees. 10 genera and ± 350 species in southern Africa.

ALOE

Aloe, aalwyn

alloeh = bitter; referring to the bitter-tasting juice in the leaves

± 550 species in Africa, the Arabian Peninsula, Madagascar and Socotra. *A. vera*, originally from the Arabian Peninsula, has long been used for its healing properties and is now cultivated and naturalised in many parts of the world. The genus *Aloe* has recently been divided into several new genera, but of these, only the climbing aloe (*Alólampolos*) occurs in Baviaanskloof. 12 species in Baviaanskloof.

Aloe africana   Kariega aloe

africana = from Africa; imprecise, as this species is confined to a small coastal part of the Eastern Cape

Small tree, 2–4 m tall, usually single-stemmed, with leaves in a dense apical rosette. **Leaves** Curving downwards, 65 cm long, blue-grey with reddish marginal teeth; dried leaves persist. **Flowers** In dense racemes, each flower 5 cm long, yellow-orange, with tepals curving upwards. Flowering in winter and spring. **Habitat** Common on clay or loam-rich soils in Transitional Shrubland and stunted Thicket. **Distribution** Fairly narrow, near to the coast, from Humansdorp to Port Alfred.

Uitenhage 14.05.2006

Aloe ferox Bitter aloe, Karoo aloe

ferox = fierce, war-like; alluding to the leaves,
which are armed with sharp spines

A single-stemmed 2–5 m-tall tree, skirted with dried leaves.
Leaves Thick and erect, in a rosette, more than 1 m long,
grey-green, sometimes tinged reddish in drought. **Flowers** In
5–8 dense racemes up to 1 m long; florets 3 cm long, orange-
red, with stamens and stigma exserted. Flowering in autumn,
winter and spring. **Habitat** Common in Transitional Shrubland
and Grassy Fynbos. **Distribution** Widespread from Swellendam
to Port Edward. **Notes** Sap is extracted from harvested leaves
for medicinal and cosmetic use.

Geelhoutbos bergpad, Kouga 23.07.2011

Aloe lineata var. *lineata* Red-spined aloe

lineata = marked with fine parallel lines; leaf
markings run along the length of the leaf.

Small, stout tree, 1.5 m tall, usually single
stemmed, with persisting dry leaves sometimes
present, forming a firm trunk. **Leaves** 30 cm
long, bright green, distinctly marked with
lines, margins reddish brown with sharp red
teeth, spreading straight. **Flowers** In a dense,
unbranched raceme extending upwards to
1 m; individual florets 5 cm long, salmon-pink.
Flowering in late summer or early spring.
Habitat Infrequent, with small populations on
rocky slopes in Grassy Fynbos and Transitional
Shrubland. **Distribution** Widespread from
Riversdale and the Little Karoo to Makhanda.
Notes *A. lineata* var. *murii* is confined to the
Western Cape and has yellow-green leaves with red stripes and larger
teeth; flowering in winter and spring. **Other name** Streepaalwyn (A).

Sipoonkop, Kouga 30.04.2018

Aloe longistyla Ramenas

longi = long, *styla* = style; referring to the long styles that protrude
from the flower tube and extend beyond the anthers

Low, stemless, very spiny aloe to 25 cm, single to a few rosettes.
Leaves 15 cm long, waxy, crowded, grey-green, covered with
tough, white spines. **Flowers** In dense, unbranched racemes;
florets 5.5 cm long, salmon-pink to coral-red, mouth turned
upwards with anthers and style well exserted. Flowering in
early spring. **Habitat** Fairly frequent in Succulent Karoo in the
Steytlerville area, sometimes under small bushes. **Distribution**
Widespread from Calitzdorp to Makhanda and Middelburg.

Above Nuwekloof Pass 14.09.2010

Aloe microstigma subsp. *microstigma*   Karoo-aalwyn

micro = very small, *stigma* = stigma; referring to the tiny stigmas

Low, almost stemless aloe to 50 cm tall, sometimes taller, with single to a
few neat rosettes. **Leaves** 30 cm long, reddish green and white spotted, with
triangular teeth, reddish brown only on the margins.
Flowers In unbranched racemes, nodding, 3.5 cm
long, dull red fading to yellow. Flowering in autumn
and winter. **Habitat** Fairly common on dry slopes in
Succulent Karoo and exposed Thicket. **Distribution**
Widespread from Loeriesfontein near the Roggeveld
Mountains, Northern Cape, through the Karoo to
Cradock, Eastern Cape. **Notes** *A.microstigma* subsp.
framesii is confined to the West Coast from Port
Nolloth to Langebaan.

Zandvlakte 01.07.2011

Aloe perfoliata (= *A. comptonii*)   **Christmas aloe**

per = through, *folia* = leaf; referring to the manner in which the bases of the leaves
wrap around the stem, giving the appearance of the stem growing through the leaves

Low, short stemmed to 50 cm tall, seldom taller,
forming dense colonies. **Leaves** 30 cm long,
fairly wide, blue-grey tinged red, margins with
pale brown teeth. **Flowers** Bunched in rounded
heads on a branched inflorescence; individual
florets 4 cm long, dull red. Flowering in spring
and summer. **Habitat** Common on sandstone
rock outcrops and cliffs in Grassy Fynbos,
Fynbos Woodland and Thicket. **Distribution**
Widespread from Bokkeveld (Nieuwoudtville) to
Kariega (Uitenhage). **Other name** Kleinkaroo-
aalwyn (A).

Keerom, Baviaans 06.05.2006

Aloe pictifolia

picti = coloured, painted, *folia* = leaf;
referring to the varying leaf colours
giving a painted appearance

Low, short-stemmed aloe to 30 cm tall, with
persistent dried leaves. **Leaves** 17 cm long,
relatively narrow, blue-grey-green, white
spotted, margins with small red-brown teeth.
Flowers In dense, simple racemes, less than 2 cm long, red with
green tips. Flowering in winter and early spring. **Habitat** Uncommon
and seldom encountered on sandstone outcrops and cliffs in Thicket
and Grassy Fynbos. **Distribution** Endemic to Baviaanskloof in the
vicinity of Kouga Dam. **Status** Rare.

Kouga Dam 09.12.2006

Aloe pluridens French aloe

pluri = many, *dens* = teeth; referring to the
numerous teeth on the leaf margins

young plant

Tree, usually single stemmed, sometimes branched
above, to 5 m tall, with persistent dried leaves. **Leaves**
70 cm long, yellowish green and semi-transparent with
obscure lines, margins with incurved, triangular, white or
pale pink teeth. **Flowers** On dense, branched racemes;
florets 4 cm long, salmon-pink to dull red. Flowering in
autumn and winter. **Habitat** Well-developed Thicket or dry
Forest. **Distribution** Widespread from Plettenberg Bay to
Durban. **Notes** One of the few aloes with shade-adapted
seedlings and saplings, allowing it to regenerate in dense
thicket. **Other name** Garaa (A).

Hankey 25.10.2018

Aloe speciosa Spaansaalwyn

speciosa = splendid, showy; referring to the
tall stature of these plants, which stand
out in the surrounding thicket

Tree, usually single stemmed, sometimes branched,
to 5 m tall, with persistent dried leaves. **Leaves** 80 cm
long, blue-grey, margins with a few pale red teeth.
Flowers In simple, dense racemes, each flower 3.5 cm
long, red becoming greenish white. Flowering from
early winter to spring. **Habitat** Thicket, usually on
shale, where soil is deeper and richer. **Distribution**
Widespread from Montagu, Western Cape, to Peddie,
Eastern Cape.

Modderfontein, Zandvlakte 14.04.2008

Aloe striata Coral aloe

striata = striated; alluding to the
visible parallel lines on the leaves

Low, almost stemless, branching aloe, to
75 cm tall, with persistent dried leaves.
Leaves 50 cm long, blue-grey tinged
reddish, striated, margins spineless and
pale pink to red. **Flowers** In rounded heads
on a branched raceme, each flower 3 cm
long, red, base of tube globose. Flowering
in spring. **Habitat** Common in overgrazed
or exposed, rocky slopes in Thicket and
Succulent Karoo. **Distribution** Widespread
from Worcester to Komani (Queenstown).
Other name Makaalwyn (A).

Grootrivierpoort 26.10.2018

ALOIAMPELOS
Climbing aloe, scrambling aloe

Aloe = plant genus, *ampelos* = climbing plant; referring to the sprawling habit of this group of climbing aloes

7 species in South Africa; 1 species in Baviaanskloof.

Aloiampelos gracilis Scrambling aloe

gracilis = thin, slender; referring to the relatively thin stems that allow the plants to scramble and climb

Scrambling shrub to 2 m tall with slender branches, without persistent dried leaves. **Leaves** 25 cm long, dull green and unmarked, margins with white teeth; leaf sheath not eared and without hairy margins. **Flowers** On simple or branched racemes, each flower 4.5 cm long, bright red but yellowish at the mouth. Flowering in autumn, winter and early spring. **Habitat** Uncommon on rocky, south-facing slopes in Grassy Fynbos. **Distribution** Narrow distribution from Baviaanskloof and Kouga to Gqeberha (Port Elizabeth). **Other name** Rankaalwyn (A).

Zandvlakte, Baviaans 31.07.2011

BULBINE
Bulbine, kopieva

bulbus = an onion or bulb; superficially, these fleshy-leaved plants resemble onions, but they do not have bulbs, just fleshy roots

Species in this genus have distinctive feathery anthers. Fresh juice from the leaves is used to treat skin afflictions, especially burns. Bulbine is easy to grow and common in South African gardens. ± 75 species in Africa, mostly in South Africa; 10 species recorded in Baviaanskloof.

Bulbine abyssinica Bushy bulbine

abyssinica = Abyssinia (Ethiopia); referring to the region where the species was originally scientifically recorded

Succulent rhizomatous geophyte, to 60 cm tall, forming large tufts in colonies. **Leaves** Linear, soft and straight, 35 cm long, with papery yellow basal sheaths. **Flowers** Numerous, in a dense spike-like inflorescence 6 cm wide; individual flowers 2 cm wide, yellow. Flowering in spring and summer after rain. **Habitat** Occasional in all habitats except Fynbos. **Distribution** Very widespread from southern Namibia to Baviaanskloof and Limpopo, to Ethiopia. **Other name** Geelkatstert (A).

Uitspan, Baviaans 19.02.2016

Bulbine cremnophila

cremno = rock, *phila* = loving; referring to
the habitat of cliffs and rocky outcrops

Dwarf succulent to 30 cm, roots grey and fleshy.
Leaves Fleshy, 10 cm long, glaucous.
Flowers In a loose raceme, few open at
a time, each flower less than 1 cm wide,
yellow, on 2 cm-long pedicel. Flowering in
spring and summer. **Habitat** Uncommon
on shaded cliffs in narrow ravines in
Thicket. **Distribution** Endemic to Kouga
and Baviaanskloof. **Status** Rare.

Gert Smitskloof, Kleinpoort 19.08.2009

Bulbine frutescens

frutescens = bushy; referring to the tufted
and sometimes branching habit

Tufted succulent to 60 cm tall,
with wiry roots. **Leaves** Several to
many, subterete, 40 cm long, green,
sometimes glaucous. **Flowers** In a
dense raceme 5 cm wide; individual
flowers 1.5 cm wide, yellow, seldom
with orange or white. Capsules
spreading and upcurved. Flowering
from spring to autumn. **Habitat**
Common in Transitional Shrubland,
Succulent Karoo and Thicket.
Distribution Very widespread across
South Africa.

Enkeldoring, Baviaans 21.09.2011

Bulbine latifolia Broad-leaved bulbine

lati = wide, *folia* = leaf; referring to the relatively wide leaves

Stout, tufted succulent with a thick rhizome and wiry roots, to 70 cm
when flowering. **Leaves** Broadly lanceolate, in a rosette, 15 cm
long × 5 cm wide, bright green. **Flowers** In a dense raceme, 40 cm
long × 5 cm wide; florets 1.5 cm wide, yellow. Flowering in spring.
Habitat Rocky slopes and outcrops in Transitional
Shrubland and Grassy Fynbos. **Distribution** Widespread
from western Kouga to Mpumalanga. **Notes** Roots used
in traditional medicine to ease vomiting and diarrhoea,
and to treat diabetes and rheumatism. In Baviaanskloof
used as a remedy for prostate cancer. The leaf sap can
be applied to wounds and skin irritations. **Other names**
Rooiwortel (A), incelwane (X), ibhucu (Z).

Rus en Vrede 4X4, Kouga 09.03.2010

Bulbine narcissifolia   Strap-leaved bulbine

narcissus = daffodil, *folia* = leaf; the leaves resemble those of some species of daffodil.

Robust perennial succulent to 50 cm with rhizomatous base. **Leaves** Strap-shaped, slightly twisted, 35 cm long, grey-green with yellowish sap. **Flowers** Many in a dense spike, yellow with white membranous bracts. Flowering in spring and summer. **Habitat** Localised in grassy Transitional Shrubland, often in disturbed places. **Distribution** Very widespread from Langkloof to Limpopo and to Ethiopia. **Notes** Can be used medicinally to treat a variety of ailments including skin problems. **Other names** Geelslangkop, wildekopieva (A), umalala (X), khomo-ea-balisa, serelelile (S).

Kouga Valley, Joubertina 15.01.2011

GASTERIA

gaster = abdomen, belly; referring to the swollen base of the tubular flowers, a feature that differentiates members of the genus from aloes

Compact succulents that are easy to grow from cuttings. Popular in pots and gardens. ± 30 species in southern Africa; 8 species in Baviaanskloof, 6 of which are endemic.

Gasteria brachyphylla

brachy = short, *phyllon* = leaf; pertaining to the short leaves that appear to end abruptly

Stemless succulent, branching at the base, to 1 m tall when in flower. **Leaves** Distichous, 20 cm long × 7 cm wide, green to purple-brown with transverse lines and white spots, margins crenulate and white, tip pointed. **Flowers** On a long, sometimes branched, nodding raceme, salmon-pink with green tips. Flowering in winter, spring and summer. **Habitat** Well camouflaged under karroid shrubs in Succulent Karoo and Thicket. **Distribution** Not very widespread from Barrydale to western Baviaanskloof and Steytlerville.

Witberg, Uniondale 28.10.2018

Gasteria camillae

Named after Camilla Christie, who first recorded the plant on cliffs in a shaded kloof near Cambria on 1 December 2019

Short-stemmed succulent to 20 cm. **Leaves** Smooth, keeled, arranged in dense rosettes, each leaf 15 cm long × 2 cm wide. **Flowers** In a raceme 30 cm long, each flower 3 cm long, orange-pink with green striations. Flowering in early summer. **Habitat** Seldom seen on cool, shaded cliffs in Forest and Thicket. **Distribution** Endemic to the eastern Baviaanskloof Mountains, in the kloofs draining Mac Mountain.

Cambria 15.12.2019

Gasteria glomerata

glomeratus = collected closely together into a head; referring to the clumped habit with numerous rows of leaves packed together

Stemless dwarf succulent to 20 cm. **Leaves** Distichous, short and stiff, with several rows clumped together, surface rough to the touch. **Flowers** In a suberect raceme, nodding, red and green. Flowering in spring. **Habitat** Uncommon on vertical cliffs of Enon conglomerate in Thicket. **Distribution** Endemic to the Gamtoos Valley near Patensie. **Status** Critically Rare.

Gamtoos Valley, Andrieskraal 10.12.2006

Gasteria pulchra

pulcher = beautiful; referring to the flowers, but the long tapering leaves are also strikingly attractive

Stemless, elongated succulent to 1.5 m. **Leaves** In a spiral rosette, linear-triangular, with toothed margins and keel. **Flowers** Reddish and nodding on a sometimes branched raceme. Flowering in spring and summer. **Habitat** Found on rocky slopes under shrubs in Thicket and Transitional Shrubland. **Distribution** Endemic to the Gamtoos Valley around Hankey and Patensie.

Hankey 25.10.2018

Gasteria rawlinsonii

Named after SI Rawlinson, a succulent collector from Bloemfontein

Stemmed succulent with hanging branches to 1 m long. **Leaves** Distichous, neatly rowed on the stem, rough, margins slightly toothed. **Flowers** On a suberect raceme, nodding, well inflated below, pink or white and green. Flowering in spring. **Habitat** Uncommon, though locally abundant on cliffs in a few of the narrow ravines, notably in Gert Smitskloof at Kleinpoort. **Distribution** Endemic to Baviaanskloof. **Status** Rare.

Gert Smitskloof, Baviaans 17.01.2022

HAWORTHIA

Named after English naturalist Adrian Hardy Haworth, 1768–1833, who collected over 40,000 insects and published books on the butterflies of England and the succulents of America

Miniature aloes with tiny white flowers and succulent leaves, often in a neat rosette, with notable markings and soft teeth. Popular in pots. ± 70 species in dry parts of southern Africa; 5 species in Baviaanskloof.

Haworthia cooperi

Named after Thomas Cooper, 1815–1913, an English botanist and explorer. He came to South Africa in 1859 and collected specimens mainly in the Drakensberg. About 20 plants have been named after him.

Succulent with soft leaves, often half buried under the soil. **Leaves** Distinctly windowed and veined; some varieties are toothed. **Flowers** White. Flowering in early summer. **Habitat** Sheltered south-facing slopes and rocky outcrops and cliffs in Grassy Fynbos, Fynbos Woodland and Forest. **Distribution** Eastern distribution from Joubertina to Makhanda.

Witruggens, Kouga 06.02.2012

Haworthia decipiens

decipiens = deceiving; referring to the close resemblance to another species, making accurate identification tricky

Soft-leaved succulent, sometimes half sunken. **Leaves** Thin, almost papery, relatively broad, with soft spines on margins and keel. **Flowers** White. Flowering in early summer. **Habitat** Grassy Fynbos, Transitional Shrubland and Succulent Karoo. **Distribution** Fairly limited distribution from the Little Karoo to Gqeberha (Port Elizabeth).

North side Groot Winterhoek 17.05.2006

Haworthia monticola

monticola = mountain dwelling;
referring to the mountainous habitat

Soft-leaved succulent, forming small clumps.
Leaves Relatively small and narrow, tips
erect or curved in, dark green to reddish,
margins with white teeth. **Flowers** White.
Flowering in spring. **Habitat** Uncommon on rocky outcrops
and cliffs in Grassy Fynbos and Arid Fynbos. **Distribution**
Limited distribution from Outeniquas and Oudtshoorn to
Baviaanskloof. Easily seen at Raaskrans in Nuwekloof Pass.

Bosrug, Baviaans 12.03.2012

HAWORTHIOPSIS

Haworthiopsis = like *Haworthia*; referring to the similarities between these 2 related genera

These plants differ from *Haworthia* in having a thicker and tougher leaf surface. ± 18 species in
southern Africa, mostly in the Eastern Cape; 6 species recorded in Baviaanskloof.

Haworthiopsis fasciata

fasciatus = transversely banded; referring to the
transverse rows of white dots on the leaves

Hard-leaved succulent, sometimes stemmed,
forming dense rosettes, often clumped.
Leaves Green, incurved, with bands of white
tubercles. **Flowers** White. Flowering in early
summer. **Habitat** Uncommon on rocky slopes
in Grassy Fynbos. **Distribution** Fairly narrow
distribution from Baviaanskloof to Gqeberha
(Port Elizabeth). **Status** Near Threatened.

Drinkwaterskloof, Kouga 09.02.2012

Haworthiopsis longiana

Named after Frank Reginald Long, 1884–1961,
an English horticulturist from Kew who lived
and collected succulents in South Africa.
Coincidentally, the leaves are relatively long.

Hard-leaved succulent, stout, short
stemmed, forming clumps. **Leaves** Often
twisted sideways, rigid and almost smooth,
light green to yellowy-brown. **Flowers** White.
Flowering in summer. **Habitat** Grassy areas
in Grassy Fynbos, Transitional Shrubland and
Thicket. **Distribution** Narrow distribution
from Humansdorp to Kariega (Uitenhage).
Status Endangered.

Patensie 15.05.2006

Haworthiopsis scabra

scabra = rough; referring to the rough leaf
surface covered with tubercles

Hard-leaved succulent, stemless and small. **Leaves** Few in an
untidy rosette, dark brownish green, surface usually scabrid and
with tubercles. **Flowers** White. Flowering in summer. **Habitat**
Uncommon and seldom detected on shallow loam-rich soils
on rocky slopes in Grassy Fynbos and Transitional Shrubland.
Distribution Fairly narrow from the Little Karoo to Baviaanskloof.

Kammanassie, 04.2015

Haworthiopsis viscosa Koedoekos

viscosa = glutinous, sticky; referring to the leaves, which are
sometimes covered with a sticky resin. The plants are browsed
by kudu, hence the Afrikaans common name ('kudu fodder').

Hard-leaved succulent to 50 cm, rigid, with several closely packed
stems. **Leaves** Symmetrically arranged and forming a triangular
column, brownish, tinged green-
brown, tips pointed. **Flowers** White.
Flowering in late spring. **Habitat**
Common and frequently encountered
on arid, rocky, sandstone outcrops
in Thicket and Succulent Karoo.
Distribution Widespread from the
Little Karoo to Baviaanskloof and
Graaff-Reinet.

Geelhoutbos bergpad, Kouga 08.07.2011

KNIPHOFIA   Red-hot poker, vuurpyl

Named after Johann Hieronymus Kniphof, 1704–1763, a German professor of medicine. He produced
botanical images by soaking dried specimens in ink and pressing them on paper.

± 65 species in Arabia and sub-Saharan Africa; 1 species in Baviaanskloof.

Kniphofia uvaria

uvarius = like a bunch of grapes; referring to the fruits that vaguely
resemble a bunch of grapes

Resprouting, rhizomatous, tufted, forming clumps to 1.2 m. **Leaves**
Strap-shaped, folded, 2 cm wide × 100 cm long, bright green.
Flowers Numerous in dense, oblong to globose racemes, orange to
yellow. Flowering in summer, especially after fire. **Habitat** Localised
in damp seepages usually on or near shale bands in Fynbos
and Transitional Shrubland. **Distribution** Very widespread from
Namaqualand to Hogsback. **Notes** An infusion of the roots has been
used to treat chest disorders.

Coleskeplaas, Baviaans 06.03.2012

COLCHICACEAE

± 16 genera and 290 species worldwide. Herbaceous plants with a corm or rhizomes. Floral parts in sixes and stigma usually divided into three. In South Africa there are ± 8 genera and 150 species. *Colchicum* is the most diverse genus in the family, followed by *Wurmbea*. Some species contain colchicine, a toxic substance extracted to treat gout. 2 genera in Baviaanskloof.

COLCHICUM (= *Androcymbium*)   Men-in-a-boat, cup-and-saucer, patrysblom
Named after Colchis, an ancient region on the Black Sea

Low, cormous geophytes with few leaves spreading on the ground, and with colourful upright bracts surrounding the flowers. ± 150 species in Africa, Europe and Asia. Almost 50 species in South Africa, mostly in Namaqualand and the Little Karoo; 2 species in Baviaanskloof.

Colchicum longipes   Men-in-a-boat

longus = long, *pes* = foot; describes the 2 long leaves at the base of the plant

Resprouting cormous geophyte to 10 cm. **Leaves** 2, spreading, channelled, 20 cm long × 2 cm wide. **Flowers** Yellow-green, with leaf-like bracts. Flowering most of the year except late summer. **Habitat** Seasonally damp areas in Grassy Fynbos, Transitional Shrubland and Succulent Karoo. **Distribution** Widespread from the Little Karoo to Gqeberha (Port Elizabeth). **Other name** Inokam (X).

Rooikloof, Kouga 28.04.2018

ORNITHOGLOSSUM

ornithos = bird, *glossa* = tongue; pertaining to the tapering tepals that resemble a bird's tongue

8 species in tropical and southern Africa; 1 species in Baviaanskloof.

Ornithoglossum undulatum   Snake lily

undulatum = undulating; pertaining to the undulating, wavy leaves

Cormous geophyte to 20 cm. **Leaves** Few, 10 cm long × 2 cm wide, blue-grey, usually crisped or undulating. **Flowers** Showy, nodding, 4 cm wide, white to pink with purple or maroon tips. Flowering in winter and spring. **Habitat** Rocky areas with clay in Succulent Karoo. **Distribution** Very widespread from Namibia to Eastern Cape. **Other name** Slangkop (A).

03.07.2015

COMMELINACEAE

Dayflowers, spiderworts

41 genera and 731 species. Better-known genera in this family include *Commelina* (dayflowers) and *Tradescantia* (spiderworts). Many are cultivated as ornamentals. There is considerable variation in flower morphology and the flowers do not produce nectar. Flowers are mostly open for a day only or just part of the day. ± 7 genera and 48 species native to southern Africa.

COMMELINA
Dayflower

Named after 2 Dutch botanical authors and spice merchants: Johannes Commelin, 1629–1692, and his nephew Caspar, 1667–1731. The Amsterdam City Council commissioned them to create the garden that became 'Hortus Botanicus', one of the oldest botanical gardens in the world.

In some species the flower is open for 1 day only, thus the common name. ± 230 species worldwide; 14 species in South Africa; 2 species in Baviaanskloof.

Commelina africana   Common yellow commelina, geeleendagsblom

africana = from Africa; referring to the geographical origin of the species; not entirely accurate because the species also occurs in Arabia and Madagascar

Soft, sprawling, perennial herb to 50 cm, resprouting from a woody rootstock. **Leaves** Oblong to linear, 2 cm long × 1 cm wide. **Flowers** To 1 cm wide, yellow. Flowering in summer and autumn. **Habitat** Occurring in the shade of other shrubs in most habitats except the upper mountains. **Distribution** Very widespread from the Cape Peninsula through Africa to Arabia and Madagascar. **Other names** Lekzotswana (X), idangabane (Z).

Damsedrif, Baviaans 17.01.2022

Commelina benghalensis   Blouselblommetjie

benghalensis = from the Indian state of Bengal; an introduced species native to Bengal

Spreading annual weed to 30 cm. **Leaves** Ovate, hairy, 6 cm long × 2.5 cm wide, pale fresh green, margins sometimes wavy; petiole sheathing the stem, with 0.3 cm-long bristles at the mouth of the sheath. **Flowers** Single or paired at the branch tips, 1 cm wide, blue. Flowering in summer and autumn. **Habitat** Shady and damp sites in various habitats but seldom in Fynbos. **Distribution** Naturalised from the Cape Peninsula to tropical Africa and other tropical and subtropical parts of the world. **Other names** Uhlotshane (X), idangabane (Z).

Modderfontein, Zandvlakte 05.02.2022

CYANOTIS

Doll's powderpuff

kyanos = dark blue, *otis* = ear; pertaining to the pair of blue tepals, which spread sideways like ears

± 50 species in the palaeotropics; 5 species in South Africa; 1 species in Baviaanskloof.

Cyanotis speciosa Bloupoeierkwassie

speciosa = splendid, showy; referring to the striking blue flowers and feathery filaments

Resprouting from an abruptly bent rhizome, tufted perennial to 50 cm. **Leaves** Lanceolate, hairy below, variable in length, up to 12 cm long. **Flowers** In dense clusters, each flower 1 cm wide, blue to mauve, filaments blue and feathery with yellow anthers. Flowering in summer and autumn. **Habitat** Grassy Fynbos and grassy Transitional Shrubland. **Distribution** Very widespread from Riversdale to southern Tanzania and Madagascar. **Other names** Umagoswana, umadlowana, iyeza lesisu (X).

Zandvlakte bergpad, Kouga 11.12.2011

CYPERACEAE

Sedges

Sedges – or reeds – are a large family of monocotyledons comprising ± 108 genera and > 5,000 species worldwide. The plants are obscure and easily overlooked in the veld, so they are under-collected and relatively poorly studied; they are also notoriously difficult to identify. But these plants are good indicators of habitat type. The most reliable feature to examine is the nutlet (seed), but unfortunately this can only be achieved under a microscope. However, there are other features that are visible to the naked eye. Generally, sedges are tufted plants. The leaves, crowded at the base, often resemble the stems and often have fibres or membranes at the base of the culms, which can be useful for identification. Sedges do not have a leaf sheath, rather the leaves are attached directly to the culm, a feature that distinguishes them from grasses. Many sedges have culms that are angular in cross section, whereas grasses only have cylindrical culms. Sedges usually have bracts or spathes (green leaf-like structures) protruding above the inflorescence. Unlike grasses and restios, sedges do not have nodes. ± 40 genera and 400 species in southern Africa; 15 genera and 56 species in Baviaanskloof.

BULBOSTYLIS

bulbo = bulb, *stylis* = style; referring to the base of the style which is thickened or swollen

hairy tufts on leaf sheath

Plants usually have 2 hairy tufts at the opening of the leaf sheath. ± 200 species worldwide; 14 species in South Africa; 1 species in Baviaanskloof.

Bulbostylis humilis

humilis = low, low-growing; pertaining to the small size of this sedge

Densely tufted annual or short-lived perennial to 10 cm. **Leaves** Base pale brown, leaf sheath with 2 tufts of long silky hairs. **Inflorescence** Spikelets greenish to pale brown, 0.6 cm long. Flowering in early spring. **Habitat** Rocky ledges and stony ground in Grassy Fynbos and Transitional Shrubland. **Distribution** Very widespread from the Cape Peninsula to tropical Africa.

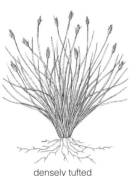

densely tufted

Zuuranysberg, Kareedouw 11.05.2021

CAREX

carex = Latin name for a sedge, originally derived from *kupeiros* – which is Greek for sedge

Over 2,000 species worldwide; 18 species in South Africa; 6 species in Baviaanskloof.

Carex aethiopica

aethiopica = from sub-Saharan Africa; in classical times this region had not yet been explored by Europeans.

Tufted perennial to 1.2 m. **Leaves** Flat or keeled, 100 cm long × 1.2 cm wide, dark green to glaucous. Culms triangular. **Inflorescence** A raceme of up to 6 spikes, each spike 7 cm long × less than 1 cm wide, greenish. Flowering in spring and early summer. **Habitat** Occurring sporadically on shaded, marshy stream banks under Forest. **Distribution** Widespread from the Cape Peninsula to Gqeberha (Port Elizabeth).

Kaseykloof, Baviaans 26.12.2017

Carex capensis (= *Schoenoxiphium ecklonii*)

capensis = from the Cape; referring to the geographical origin of the species

Resprouting tufted perennial to 30 cm. Base slightly swollen, with pale to dark brown dry leaf sheaths. **Leaves** Grass-like, folded, softish, 20 cm long × 0.3 cm wide, lime-green. **Inflorescence** In panicles, spikelets lime-green to gold, 0.3 cm long × 0.2 cm wide, several clustered on short spikes, 1 cm long. Flowering in spring. **Habitat** Occasional on open slopes in Grassy and Mesic Fynbos. **Distribution** Widespread from the Cape Peninsula to Kariega (Uitenhage).

Rus en Vrede 4X4, Kouga 25.04.2018

Carex glomerabilis   Foxtail sedge

glomerabilis = rounded; alluding to the
shape of the inflorescence

Tufted, mat-forming perennial to 70 cm, with
long rhizomes. **Leaves** Flat or channelled,
to 30 cm long × 0.5 cm wide, yellow-green.
Culm triangular. **Inflorescence** A glomerate
spike, to 6 cm long × 1.5 cm wide, yellowy-
green with olive-green streaks, with a few
whisker-like bracts to 3 cm long. Flowering from spring to
autumn. **Habitat** Marshes with perennial water. **Distribution**
Widespread from Calvinia, Northern Cape, to the Western
and Eastern Cape and Gauteng. **Notes** Easily seen at the
entrance to the boardwalk at the Sewefontein fig forest.

Sewefontein, Studtis 26.12.2017

Carex uhligii   (= *Schoenoxiphium lehmannii*)

Named after Viktor Karl Uhlig, 1857–1911, a German geologist
from the University of Vienna who collected in Tanzania

Tufted perennial to 40 cm. **Leaves** Soft, slightly
channelled, 20 cm long × 0.4 cm wide, pale green.
Inflorescence Small bunches of spikelets, well spaced
on a long arching culm, green to brownish. Flowering in
summer. **Habitat** Uncommon in damp seepages under
Forest. **Distribution** Very widespread from the Cape
Peninsula to the Eastern Cape, and to Ethiopia.

27.11.2021

CARPHA

carphos = dry straw, chaff; pertaining to the dry, straw-coloured spikelets

16 species worldwide; 5 species in South Africa; 1 species in Baviaanskloof.

Carpha glomerata   Vleibiesie, vleiriet

glomerata = compactly clustered; referring to the
aggregation of spikelets into heads

Resprouting, robust, tufted perennial to 2 m.
Leaves Folded, soft to touch, 70 cm long ×
1 cm wide. **Inflorescence** Numerous spikelets,
straw-brown, clustered into heads clumped
along the end of the triangular culm. Flowering
in spring and summer. **Habitat** Found in
perennial streams, usually stable cobble beds,
in Fynbos. **Distribution** Widespread from the
Cederberg and Cape Peninsula to Mpumalanga.

Poortjies, Cambria 26.10.2018

CHRYSITRIX

chrysos = gold, *thrix* = hair; referring to the spikelet made up of scales, with prominently protruding anthers at the top resembling a short haircut

4 species in the Cape and Australasia; 1 species in Baviaanskloof.

Chrysitrix capensis

capensis = from the Cape; referring to the geographical origin of the species

Tufted, thick-rooted perennial to 40 cm. **Leaves** Leaves and stems, flattened and rigid, 0.5 cm wide. **Inflorescence** A single terminal spikelet, 1.5 cm long, black and rusty-brown, with a prominent tuft of anthers, scales and filaments (the golden hair). Flowering from autumn to spring. **Habitat** Damp seepages in Mesic and Subalpine Fynbos. **Distribution** Widespread from the Cape Peninsula to Humansdorp.

07.12.2021

CYPERUS   Matjiesgoed

kupeiros = the ancient Greek name for sedge; the abbreviated common name is 'cyp'.

The genus *Pycreus* has been included in *Cyperus*. ± 815 cosmopolitan species in the genus; ± 28 species in South Africa; 10 species recorded in Baviaanskloof.

Cyperus congestus   Hedgehog sedge

congestus = brought together; referring to the spikelets clustered together

Tufted perennial to 60 cm, with a slightly swollen base. **Leaves** Flat, stems triangular. **Inflorescence** A simple or compound umbel or head, with individual heads 3 cm wide. Spikelets flattened, 1.5 cm long × 0.1 cm wide, green and reddish brown. Flowering in summer and autumn. **Habitat** Sandy stream banks. **Distribution** Widespread in southern Africa and other parts of the world, weedy.

Rooikloof, Kouga 28.04.2018

Cyperus difformis   Small-flowered nutsedge

difformis = differing from the usual; referring to the flowers of this species, which are unusually small in the genus

Slender, tufted annual to 50 cm. **Leaves** Flat, 20 cm long × 0.5 cm wide. Stem triangular, slightly winged. **Inflorescence** A simple or compound umbel of dense, globose heads, individual heads 1.5 cm wide, pale green. Numerous spikelets, 0.4 cm long × 0.1 cm wide. **Habitat** Damp stream banks. **Distribution** Widespread throughout the world. A problem weed in some areas, especially in rice paddocks.

Rooikloof, Kouga 28.04.2018

Cyperus mundii

Named after Johannes Ludwig Leopold Mund, 1791–1831, a Berlin-based pharmacist, sent to the Cape in 1815 to collect plant specimens

Creeping, mat-forming perennial to 60 cm. **Leaves** Leaf blades flat, 15 cm long × 0.5 cm wide, with sheaths around the stems. **Inflorescence** A compound umbel, with spikelets clustered into loose heads, pale brown. Spikelets 0.6 cm long × 0.3 cm wide. Flowering from spring to autumn. **Habitat** Very wet marshy stream banks. **Distribution** Very widespread from the Cape to tropical Africa, the Mediterranean and Mascarene Islands.

Baviaanskloof River, Studtis 27.04.2018

Cyperus polystachyos   Bunchy sedge

poly = many, *stachyos* = spikes; referring to the numerous spikelets

Tufted perennial with a short rhizome to 50 cm. **Leaves** Linear, tufted at the base, 10 cm long × 0.4 cm wide. **Inflorescence** An irregular cluster with numerous spikelets in a dense yellow-green head, stem triangular in cross section. Spikelets 1 cm long × 0.2 cm wide. Flowering in summer and autumn. **Habitat** Wet areas, usually on riverbanks. **Distribution** Very widespread in southern Africa and with a worldwide distribution.

Kouga Kliphuis, Kouga River 13.05.2021

Cyperus semitrifidus

semi = half, *tri* = three, *fidus* = divided; referring to the almost triangular stem or the 3 leafy bracts at the base of the inflorescence

Tufted, low perennial to 25 cm tall, with a slightly swollen base. **Leaves** Short, 10 cm long × 0.3 cm wide, pale green. Culms triangular. **Inflorescence** Usually a pair of stellate heads, 2 cm wide. Spikelets flattened, 1 cm long × 0.3 cm wide, reddish brown. Flowering in summer and autumn. **Habitat** In shallow soil on damp slopes in Mesic Fynbos. **Distribution** Widespread from Baviaanskloof to Gauteng and Limpopo.

Elandsvlakte, Baviaans 16.02.2016

Cyperus sphaerospermus

sphaera = ball, *spermus* = seed; referring to the round seeds of this species

Tufted perennial to 60 cm. **Leaves** 20 cm long × 0.4 cm wide. Stems triangular. **Inflorescence** A compound umbel of several to many heads. Each head with a few spikelets. Spikelets flattened, 1 cm long × 0.2 cm wide, golden to reddish brown. Flowering from spring to autumn. **Habitat** Marshy stream banks. **Distribution** Widespread throughout southern Africa to tropical Africa.

Rooikloof, Kouga 28.04.2018

Cyperus textilis   Umbrella sedge, mat sedge

textilis = woven; this species is used to make woven articles such as mats and baskets.

Robust, rhizomatous, mat-forming reed to 1.2 m. **Leaves** Reduced to sheaths clasping the stem; stems relatively thick and fleshy, ending with a striking whorl of leaf-like bracts at the base of the inflorescence. **Inflorescence** A compound umbel of numerous heads with several small, flattened, narrow, rust-red spikelets. Flowering from spring to autumn. **Habitat** Marshy stream banks where the strong root system can sometimes withstand scouring effects during floods. **Distribution** Widespread from the Cape to KwaZulu-Natal. **Other name** Matjiesgoed (A).

Baviaanskloof River, Doringkraal 26.10.2018

Cyperus uitenhagensis

The plant was first recorded near Kariega (Uitenhage), but it has since been found to be widespread in southern Africa.

Small, tufted perennial to 25 cm, with a swollen base. **Leaves** Short, curling and twisted. **Inflorescence** 1 to a few short spikes, each 1 cm long, greenish, made up of several small 0.3 cm-long spikelets. Flowering from spring to autumn. **Habitat** Uncommon on open, dry, stony ground in Spekboomveld and Transitional Shrubland. **Distribution** From the Little Karoo and Baviaanskloof northwards. Widespread in southern Africa.

Drinkwaterskloof, Kouga 09.11.2011

ELEOCHARIS

eleo = oil, *helos* = marsh, *charis* = grace, beauty; referring to the plant's habitat in stagnant water, or marshes that sometimes have an oily surface

± 255 species, cosmopolitan; 1 species in Baviaanskloof.

Eleocharis limosa

limus = mud, *osa* = abundance; referring to the muddy substrate of the marshes where this plant grows

Aquatic perennial to 60 cm tall, with creeping rhizomes. **Leaves** Reduced to sheaths at the base, culm round. **Inflorescence** A simple spike, 2 cm long × 0.4 cm wide, brown, but may appear white due to anthers or stigmas. Flowering in spring and summer. **Habitat** Aquatic, in pools or muddy dams and vleis. **Distribution** Widespread throughout southern Africa and other parts of world, except the tropics and Australia.

Ysrivier, Cambria 22.01.2022

FICINIA
Named after Heinrich David August Ficinus, 1782–1857, a German doctor and naturalist from Dresden

Plants are frequent in Fynbos and more common after fire. The seeds are dispersed by ants. ± 75 species, southern and tropical Africa; ± 60 species in South Africa; 14 species in Baviaanskloof.

Ficinia acuminata
acumen = sharp point, *ata* = possessing; referring to the sharply pointed spikelets

A hanging, tufted, tussock-forming perennial to 60 cm, often found on rocky ledges. **Leaves** Filiform, very thin and relatively flimsy. **Inflorescence** A cluster of flattened, reddish-brown, acuminate spikelets. Flowering in spring and summer. **Habitat** Shaded rock crevices and ledges, also below boulders in Fynbos. **Distribution** Widespread from Cederberg to Baviaanskloof.

Dam se Kloof, Baviaans 27.04.2018

Ficinia albicans
albico = to make white; referring to the greenish-white spikelets

Small, tufted perennial to 15 cm. **Leaves** Base slightly swollen, with light brown to greyish fibres. **Inflorescence** Spikelets ribbed, 1 cm long × 0.3 cm wide, creamy white-green, turning brown with age, with obtuse glumes. Flowering in spring and summer. **Habitat** Grassy Fynbos. **Distribution** Widespread from the Western and Eastern Cape to Kariega (Uitenhage).

Guerna, Kouga 16.02.2016

Ficinia brevifolia
brevis = short, *folium* = leaf; pertaining to the leaves that are short relative to the culms

Robust, thick-stemmed, tufted perennial to 80 cm. **Leaves** Base covered with soft, pale membranes. **Inflorescence** Head 1.2 cm long, comprising several, densely packed spikelets that are pale green when young, becoming dark brown. Flowering in spring and summer. **Habitat** Damp, rocky places next to streams in narrow ravines. **Distribution** Widespread from Namaqualand to the Cape Peninsula and Baviaanskloof.

Uitspankloof, Baviaans 09.10.2016

Ficinia deusta

deustus = burned; pertaining to the plant's ability to resprout after fire; also due to the presence of charred leaves visible at the base

Tough, resprouting, stout perennial to 40 cm. **Leaves** Hard, 0.4 cm wide, with pale, transparent membranes at the base. **Inflorescence** Spikelets single, dark, 3 cm long × 0.7 cm wide, with 2 erect leaf-like bracts, glumes with membranous margins. Flowering in autumn and winter. **Habitat** Rocky slopes in Grassy and Mesic Fynbos. **Distribution** Widespread from Namaqualand to the Cape Peninsula and to Baviaanskloof.

Rus en Vrede 4X4, Graskop, Kouga 25.04.2018

Ficinia fascicularis

fascicularis = fascicled, clustered; referring to the bunched appearance of the spikelets

Tufted perennial to 60 cm, with thin stems that tend to droop. **Leaves** Base with reddish-brown membranes, splitting; leaves long and narrow, often arching, 0.2 cm wide, fresh green, round in cross section. **Inflorescence** Several spikelets clustered into an elongated head, each spikelet 0.7 cm long, pale brown. Flowering in spring. **Habitat** Uncommon in Baviaanskloof, in damp areas on stream banks at the margins of Forest. **Distribution** Widespread from Caledon to East London.

Kaskloof, Kouga Wildernes 16.05.2021

Ficinia gracilis

gracilis = slender; referring to the filiform, thread-like leaves or the slender tussocks

Tufted perennial to 20 cm, but can reach 50 cm. **Leaves** Base with brown membranes, splitting; leaves shiny, slightly sticky, thin, 0.1 cm wide. **Inflorescence** A stellate head, 1 cm wide, with numerous tightly packed spikelets. Flowering in spring and summer. **Habitat** Occurring mostly in Grassy Fynbos. **Distribution** Widespread from the Cape Peninsula throughout higher-rainfall South Africa to the highlands of East Africa.

Combrink's Pass, Baviaans 26.10.2018

Ficinia nigrescens

nigresco = becoming black; referring
to the spikelets that sometimes
turn black after maturing

Tufted perennial to 40 cm, tussock forming.
Leaves Base with leathery, brownish
sheaths; leaves thin, relatively long and
drooping. **Inflorescence** With 2 or 3 erect,
leafy bracts. Heads a single dark, brown
or pale spikelet, with glume tips paler and
recurved, especially when mature. Flowering from
autumn to spring. **Habitat** Occurring mostly in Grassy
Fynbos and other Fynbos habitats, with a preference
for rocky ridges on well-drained stony sands.
Distribution Common and widespread from southern
Namibia to the Cape Peninsula and Makhanda.

Rus en Vrede 4X4, Graskop, Kouga 25.04.2018

Ficinia ramosissima

ramus = branch, *issima* = most; referring
to the much-branched habit

Erect, much-branched, mat-forming perennial
to 20 cm. **Leaves** Numerous, in bunches along
erect stems, flattened, 3 cm long × 0.2 cm wide.
Inflorescence Spikelets single, light brown, 1 to a
few on short peduncles. Flowering in spring. **Habitat**
Rocky ledges in Fynbos. **Distribution** Widespread
from the Cape Peninsula to Makhanda.

Quaggasvlakte, Kouga 12.02.2016

Ficinia stolonifera

stolo = shoot, *fero* = carry or bear;
pertaining to the well-developed
underground stolons or roots

Loosely tufted, short-rhizomatous,
resprouting perennial to 15 cm.
Leaves Base slightly swollen, with
brown fibres and transparent
membranes; leaves 8 cm tall × 0.5 cm wide,
in neat tufts. **Inflorescence** A few spikelets
clustered in 1-sided heads, each spikelet
0.4 cm long, rusty-brown. Flowering in spring
and early summer. **Habitat** Shallow sandy
soil in open, rocky areas in Grassy and Arid
Fynbos. **Distribution** Very widespread from
the Cape Peninsula to tropical Africa.

Rus en Vrede 4X4, Graskop, Kouga 25.04.2018

Ficinia trispicata

tri = three, *spica* = point, *ata* = possessing; alluding to
the 3 pointy spikelets on some specimens

Attractive, densely tufted, tussock-forming, softly
drooping perennial to 60 cm. **Leaves** Strong base; leaves and
stems break off individually when pulled. Each stem with a
rust-red membrane at the base. **Inflorescence** 1–3 spikelets,
each 0.5 cm long × 0.2 cm wide, white to light brown, leafy
bracts absent. Flowering in spring and summer. **Habitat**
Kaseykloof, Baviaans 26.12.2017

Uncommon and sporadic in well-shaded seepages in narrow
ravines under Forest. **Distribution** Widespread from Swellendam to Mthatha. **Notes** Similar in growth
form to *F. acuminata*, but the damp habitat and pale cream spikelets of this species are distinctive.

FUIRENA

Named after Georg Fuiren, 1581–1628, a Danish botanist and doctor

± 60 species, almost cosmopolitan; 1 species in Baviaanskloof.

Fuirena hirsuta

hirsuta = hairy; referring to the hairs all over the plant

Resprouting, clump-forming, densely tufted perennial to
50 cm. **Leaves** Hairy, 0.5 cm wide, basal and along the stem.
Inflorescence A spherical mass of spikelets, 1.5 cm wide, blue-
green. Flowering in spring and summer. **Habitat** Marshy stream
banks along perennial rivers. **Distribution** Very widespread from
Bokkeveld (Nieuwoudtville) to Mpumalanga and Mozambique.

Kleinrivier, Kouga Wildernis 18.12.2011

ISOLEPIS

iso = equal, *lepis* = a scale; referring to the spikelets made up of similar-sized bracts

These are often tiny to small annual sedges that grow in damp or wet areas, especially after rain.
± 75 species, cosmopolitan; ± 30 species in South Africa; 4 species recorded in Baviaanskloof.

Isolepis hystrix

hystrix = hedgehog; referring to the numerous
awns on the spikelet, reminiscent of a hedgehog

Small annual to 10 cm, densely tufted. **Leaves** Lime-
green, soft. **Inflorescence** Spikelets singular heads, each
spikelet 0.4 cm wide, pale green, glume apices minutely
awned with recurved tips. Flowering in spring. **Habitat**
Damp, sandy seepages along streams. **Distribution** Very widespread from
Namaqualand to Cape Peninsula and Baviaanskloof. Also found in Australia.

Rooikloof, Kouga 28.04.2018

Isolepis prolifera

proles = offspring, *fera* = carry or bear; referring to the runners that root at the nodes, forming new plants

A tufted perennial to 60 cm, grows in a spreading habit in marshes. **Leaves** Reduced to reddish-brown basal leaf sheaths. **Inflorescence** Spikelets 1 cm long × 0.2 cm wide, light brown, in small clusters at the top of the culms; spikelets replaced by new plants that root when the culms droop, making contact at the ground surface.

Zandvlaktekloof, Baviaans 18.02.2016

Flowering from spring to autumn. **Habitat** Seasonal pools in streams and rivers. **Distribution** Widespread from Namaqualand to KwaZulu-Natal. Now a weed in southwestern Australia and New Zealand.

Isolepis sepulcralis

sepulcralis = of a tomb; referring to the flowers being buried under the bracts

Small, tufted annual, to 20 cm. **Leaves** Elongated peduncles act as leaves, soft and flimsy, very narrow, to 0.1 cm wide, fresh green. **Inflorescence** Spikelets small, 1 to a few, 0.2 cm long, green with reddish brown. Flowering in spring and summer. **Habitat** Sandy, marshy stream banks. **Distribution** Very widespread from Namaqualand to Limpopo, East Africa and Australia.

Rooikloof, Kouga 28.04.2018

PSEUDOSCHOENUS

pseudo = false, *Schoenus* = plant genus; the plant may look superficially like a *Schoenus*, but close inspection reveals that it is different.

1 species in southern Africa.

Pseudoschoenus inanis (= *Scirpus inanis*)

inanis = empty; referring to the deciduous nature of the glumes and fruit; mature inflorescences are often 'empty'.

Robust, tough, reed-like, rhizomatous perennial to 1 m. **Leaves** Reduced to dark brown or blackish basal sheaths. Reed-like stems or peduncles act as the leaves, 0.6 cm wide, round in cross section. **Inflorescence** A rigid panicle, to 10 cm long, pale brown, with numerous spikelets each 0.7 cm long. Flowering in summer. **Habitat** Dry stream banks at high altitude. **Distribution** Widespread from Namibia to Cederberg, Baviaanskloof and Lesotho. **Notes** An unusual sedge, easily seen on dry stream banks next to the road at the start of Nuwekloof Pass – gateway to the Baviaanskloof.

Nuwekloof Pass, Baviaans 13.10.2016

Schoenus graciliculmis (= *Tetraria cuspidata*)

gracilis = slender, thin, *culmis* = culm; referring to the slender culms; a recently described taxon (2020)

Resprouting tufted sedge with thin leaves and culms, basal sheaths reddish black, to 60 cm. **Leaves** Over half the length of the culm. **Inflorescence** Spikelets with bristled awns. Flowering in winter, spring and summer. **Habitat** Rocky, south-facing slopes in Grassy and Mesic Fynbos. **Distribution** Widespread from the Outeniqua and Swartberg mountains to the Amatolas. **Notes** Easily confused with *S. submarginalis*, but that species has streaked spikelets. This species differs from *S. cuspidatus* by the awned spikelets. It could also be confused with *S. galpinii*, but that species occurs on the escarpment and has not been recorded in Baviaanskloof.

Cockscomb, Groot Winterhoek 17.02.2016

Schoenus megacarpus (= *Tetraria compar*)

mega = large, *carpus* = fruit; referring to the large nut

Robust, resprouting, to 90 cm, neatly tufted. **Leaves** Base with pale yellow to brown sheaths, base of culms often tinged reddish. Relatively short and thick. Culms relatively thick, 0.2 cm wide. **Inflorescence** Few spikelets in a dense cluster, 1 cm long, pale brown. Flowering in autumn. **Habitat** Fairly uncommon on rocky ridges in Grassy and Mesic Fynbos. **Distribution** Widespread from Worcester to Makhanda.

Engelandkop, Kouga 19.04.2012

Schoenus quadrangularis (= *Epischoenus quadrangularis*)

quadrus = square, *angulus* = angle, *aris* = pertaining to; alluding to the 4-angled culms

Resprouting, densely tufted, blue-grey to 75 cm. **Leaves** Culms quadrangular, sometimes slightly flattened. **Inflorescence** Spikelet singular, brown, with an erect, flattened, leaf-like bract. Flowering in summer and autumn. **Habitat** Occasional on rocky slopes, usually in or next to streams or marshes in Grassy and Mesic Fynbos. **Distribution** Widespread from the Cape Peninsula to Gqeberha (Port Elizabeth).

Cockscomb, Groot Winterhoek 17.02.2016

Schoenus schonlandii

Named after German botanist Selmar
Schonland, 1860–1940, curator and director
of the Albany Museum, and founder of the
Department of Botany at Rhodes University

Resprouting, loosely tufted, to 100 cm
tall, with basal leaves rarely present and
rudimentary if so. **Leaves** Leaf sheath
and ligule without membranous margins.
Culms cylindrical, ridged, but not 4-angled.
Inflorescence With few spikelets scattered

Ridge west of Elandsvlakte, Baviaans 16.02.2016

throughout the panicle, without pendulous spikelets; spikelet lanceolate, glumes sometimes with
narrow membranous margins. Flowering in summer and autumn. **Habitat** Uncommon in Mesic Fynbos.
Distribution Eastern distribution from George to Makhanda.

Schoenus selinae

Named after Selina Muthaphuli,
1954–2015, Venda-born mother of
Ngalirendwe Muthaphuli, a plant
taxonomist at the University of Cape Town

Resprouting, tufted, to 40 cm tall, with red
basal sheaths. **Leaves** Basal leaf sheaths
with membranous margins, ribbed, and
with a short apical mucron that elongates
with age. Culms slightly curved to straight,
round, striated, almost ribbed, blue-grey.
Inflorescence Spikelet flattened, bracts
red-brown with pale membranous margins,
fading to light brown, bracts fraying.

Quaggasvlakte,
Kouga 13.02.2016

Flowering in winter and spring. **Habitat** Occurring at the base of rock slabs on mountain peaks in Mesic
and Subalpine Fynbos. **Distribution** Widespread from Gifberg and Cape Peninsula to Baviaanskloof.

Schoenus submarginalis

sub = in part, somewhat, *marginalis* = marginal; referring
to firm leaf sheaths that do not have membranous margins

Resprouting, neatly tufted, thin culmed, with red
basal sheaths, to 40 cm. **Leaves** Usually straight;
leaf sheaths firm, without membranous margins or
streaks, red to brown. **Inflorescence** Spikelets 0.4 cm
long, with reddish-purple streaks on the margins
of the glume. Flowering from late summer to early
winter. **Habitat** Stony soils on gentle southern slopes
in Grassy Fynbos. **Distribution** Narrow distribution
from Montagu to Baviaanskloof.

Bokloof 4X4, Kouga 31.01.2017

TETRARIA
tetra = four, *aria* = concerning; the first species scientifically described had floral parts arranged in 4s.

± 38 species, mainly in South Africa, to Africa and Australasia; ± 12 species recorded in Baviaanskloof.

Tetraria bromoides Bergpalmiet
bromos = oats, edible grain, *oides* = resembling; referring to the tall and sometimes hanging inflorescence that vaguely resembles some edible grasses

Robust, resprouting, tufted perennial to over 1.5 m. **Leaves** Base swollen, covered in old, dark sheaths and brown netted fibres. Leaves numerous, flat, drooping, 80 cm long × 0.7 cm wide. **Inflorescence** An elongated panicle with many, relatively large, long-awned, pinkish to brown spikelets. Flowering in spring and summer. **Habitat** Rocky upper slopes, mainly in Mesic Fynbos. **Distribution** Widespread from Porterville to Cape Peninsula to Kariega (Uitenhage).

Quaggasvlakte, Kouga 12.02.2016

Tetraria burmannii
Named after Johannes Burman, 1706–1779, a Dutch botanist and physician, and erstwhile employer of the young Carl Linnaeus

Resprouting, tufted perennial to 30 cm. **Leaves** Base with pale brown sheaths and netted fibres. Leaves both basal and off the culm, thin. Several sheaths, tightly cylindrical, spaced out on the culm, 1 cm long, dark brown. **Inflorescence** Paniculate. Spikelets dark brown, 0.5 cm long × 0.2 cm wide, awnless, without leafy bracts. Flowering from spring to autumn. **Habitat** Occasional on upper rocky slopes in Grassy and Mesic Fynbos. **Distribution** Widespread from Kogelberg to Bathurst.

Bosrug, Baviaans 27.10.2018

Tetraria capillacea

capillis = a hair, *acea* = resembling; the tufted
leaves are fine, flimsy and hair-like.

Resprouting, densely tufted perennial to
80 cm, with long thin leaves. **Leaves** Base
dark brown to black with dark brown netted
fibres; leaves thin and rolled, 25 cm long,
culm sheaths black. **Inflorescence** An
elongated panicle, 15 cm long. Spikelets
0.6 cm long, dark brown. Flowering mainly
in spring. **Habitat** Localised in Mesic Fynbos
on damp, upper, south-facing slopes,
usually associated with the peaty soil of
seepages. **Distribution** Widespread from
Cape Peninsula to Eastern Cape.

Cockscomb, Groot Winterhoek 16.02.2016

Tetraria fimbriolata

fimbriae = fringe, *iolata* = possessing; referring to the narrow
fringe of white-woolly hairs on the edge of the glumes

Resprouting, neatly tufted perennial to 40 cm. **Leaves**
Base slightly swollen and covered in pale brown netted
fibres. Leaves thin and short, tending to curl downwards.
Inflorescence A panicle. Spikelets short and squat,
0.2 cm long, light brown, with a distinctive dense fringe
of tiny, woolly white hairs along the edge of the glumes.
Flowering in late summer. **Habitat** Uncommon on
loamier soils in Mesic Fynbos. **Distribution** Widespread
from the Cape Peninsula to Kariega (Uitenhage).

13.06.2014

Tetraria maculata

macula = spot, *ata* = possessing; referring to the
spikelets that are dotted with small white spots

Resprouting, tufted, sparsely leafy perennial
to 40 cm. **Leaves** Base slightly swollen,
dark brown, with netted fibres. Leaves thin,
15 cm long. **Inflorescence** Relatively simple
with a few clusters of spikelets. Spikelets
dark brown, 0.7 cm long, margins pale, with
tiny white specs. Flowering in late summer.
Habitat Uncommon in Mesic and Subalpine
Fynbos, on high peaks at the base of rocky
slabs in shallow sandy soil. **Distribution**
Widespread from Cederberg to Kouga
Mountains.

Quaggasvlakte, Kouga 13.02.2016

Tetraria triangularis

tria = three, *angulus* = angle, *aris* = pertaining to; referring to the base of the culms that are triangular in cross section

Resprouting, leafy, tufted perennial to 60 cm. **Leaves** Base swollen, covered in a soft, gummy, pale brown substance. Leaves V-shaped in cross section, 0.8 cm wide. Culms triangular. **Inflorescence** Spikelets relatively large, rust-brown turning black. Flowering in late summer. **Habitat** Occurring locally in Mesic and Subalpine Fynbos on the damp slopes of high peaks. **Distribution** Widespread from Cederberg to Cape Peninsula to KwaZulu-Natal.

East of Kougakop 16.04.2012

Tetraria ustulata

ustulo = crisped hair, *ata* = possessing; referring to the hair-like awns on the spathes and glumes of the inflorescence

Resprouting, densely tufted, leafy, forming tussocks to 90 cm. **Leaves** Base swollen, with brown sheaths and brown netted fibres. Leaves to 60 cm long × 0.6 cm wide. **Inflorescence** Spikelets brown and covered with several brown, awned bracts. Flowering in late summer and autumn. **Habitat** Mesic Fynbos on open, rocky slopes in slightly deeper soils. **Distribution** Widespread from Namaqualand to Cape Peninsula to Baviaanskloof.

Bosrug, Baviaans 14.03.2012

DIOSCOREACEAE
Yams

One well-known species in this family, *Dioscorea alata*, produces the yam — a starchy tuber that is consumed as a root vegetable staple by many of the world's inhabitants. It is not to be confused with the comon sweet potato (*Ipomoea batatas*), and the New Zealand yam (*Oxalis tuberosa*). 9 genera and 700 species in the world.

DIOSCOREA
Yam

Named after Pedanius Dioscorides, c. 40–90, a Greek doctor and author of *De Materia Medica*, an important historical record of medicines used in the Middle Ages

± 400 species in pantropical and warm, temperate parts of the world; 15 species in South Africa; 1 species in Baviaanskloof.

Dioscorea elephantipes   Elephant's foot

elephantipes = elephant's foot; referring to the bark covering the short and thick tuberous stem (caudex), which is reminiscent of the skin on an elephant's foot

Deciduous and dioecious climber, resprouting from a woody caudex (tuber), covered with thick, tortoise-shell-like plates of bark. About ¾ of the tuber is exposed above ground. **Leaves** To 3 cm long, alternating on the twining stems, heart-shaped, fresh green becoming bluish with age, folding in droughts. **Flowers** Several on an inflorescence to 8 cm long, small, to 1 cm wide, yellowish green to mustard-coloured. Seed pods triangular, 2 cm long, winged and persisting. Flowering in summer. **Habitat** Uncommon on dry rocky slopes in Transitional Shrubland, Thicket and Succulent Karoo. **Distribution** Widespread from Richtersveld to Graaff-Reinet.

Zandvlakte, Baviaans 31.07.2011

HEMEROCALLIDACEAE
Day lilies

This is a relatively small plant family, with between 5 and 19 genera and ± 70 species. Exactly which plants belong in this family is largely undecided, so the exact number of genera and species worldwide is unresolved. *Hemerocallis* is the type genus, and also the second-largest genus, with ± 15 species, but it is absent in Africa. *H. fulva* is a popular ornamental. *Dianella* is the largest genus in the family with 20 species, *Caesia* is the third-largest genus with 12 species.

CAESIA
Blue grass-lily

Named after Federico Cesi, 1585–1630, Italian botanist and scientist. He was the first person to notice, with the help of a microscope donated to him by Galileo, that ferns have spores.

12 species in southern Africa, Madagascar and Australia; 1 species in Baviaanskloof.

Caesia contorta Sokkiesblom

contortus = powerful, strong; derivation unresolved, but possibly refers to the rhizome and resprouting, tufted evergreen habit. The Afrikaans common name draws attention to the banded filaments that look like sock cuffs.

Resprouting geophyte, evergreen, forming loose tussocks to 40 cm. **Leaves** Linear to strap-shaped, folded at base, 30 cm long × 1 cm wide. **Flowers** Nodding on lax, sprawling racemes, 2 cm wide, blue, filaments banded with dark blue. Flowering in summer and autumn. **Habitat** Fairly frequent on stony, sandstone slopes in Fynbos and Transitional Shrubland. **Distribution** Very widespread from Namaqualand to KwaZulu-Natal.

16.11.2021

HYACINTHACEAE   Hyacinths

A family of bulbs with ± 70 genera and 900 species worldwide. The mostly deciduous leaves are lance-shaped and soft, with a slimy sap. The flowers can be prolific and are arranged on a spike or raceme. Individual flowers are radially symmetrical with the floral parts in two whorls of three (six tepals). Well-known genera in South Africa include *Albuca*, *Drimia*, *Lachenalia* and *Ornithogalum*. Some genera are toxic. Several species are cultivated as ornamentals. Not an easy group to identify and with considerable taxonomic confusion.

ALBUCA   Slime-lily, slymlelie, tamarak

albicans = becoming white; referring to the flower colour of some species. The common name refers to the slimy leaf sap. Tamarak is derived from the Khoisan name.

Bulbs are usually covered with layers of dry onion-like leaves called tunics. A distinguishing feature of most albucas is that the 3 inner tepals remain closed, or almost so. ± 60 species from South Africa to Arabia; 50 species in South Africa; 10 species in Baviaanskloof.

Albuca bracteata (= *Ornithogalum longibracteatum*) **Pregnant onion**

bracteate = with bracts; referring to the numerous prominent bracts below each flower on the spike-like inflorescence. The common name refers to the bulb's capacity for making bulblets.

Bulbous geophyte to 1.5 m tall in flower, bulb mostly above ground. **Leaves** Lanceolate, fleshy, 60 cm long × 3 cm wide. **Flowers** Numerous on a long raceme, each flower 1.5 cm wide on short stalks 1.5 cm long; 6 tepals white with a green midrib. At the base of each flower is a long, narrow, greenish, lanceolate bract, up to 4 cm long. Flowering from spring, through summer, to autumn. **Habitat** On rocks and cliffs in the narrow, shaded ravines in Forest or Thicket. **Distribution** Widespread from Mossel Bay to tropical East Africa. **Notes** The bulb is eaten by baboons, but it is toxic to humans. Popular in pots and grows well indoors. **Other names** Ibucu, umredeni omhlope (X).

Gert Smitskloof, Baviaans 17.01.2022

Albuca canadensis Slymstok, wittamarak

Derivation unclear as the plant is not from Canada

Bulbous geophyte to 1 m tall, with
slightly fibrous tunics. **Leaves** Numerous,
lanceolate, channelled , fleshy, 60 cm
long × 3 cm wide. **Flowers** Numerous on
a raceme 3 cm wide, hanging down, each
flower white with green keels. Flowering in
spring. **Habitat** Open and disturbed areas in
Thicket and Succulent Karoo. **Distribution**
Widespread from Namaqualand to
Makhanda.

Zandvlakte 16.09.2010

Albuca cremnophila

cremno = rock, phila = loving; referring to the
habitat on rocks or cliffs

Bulbous geophyte to 50 cm tall, with dark,
flaking, tunics covering the exposed bulb.
Leaves Several, long, fleshy, channelled
, clasping at the base. **Flowers**
White with green keels. Flowering
in winter and spring. **Habitat**
Uncommon on cliffs in the
shaded narrow ravines in Thicket.
Distribution Narrow strip along
the coast from Baviaanskloof to
East London.

Kouga Valley, Joubertina 23.10.2011

Albuca longipes

long = long, pes = foot; pertaining
to the long flower stalks

Bulbous geophyte to 30 cm
tall, with dry, wrinkled bulb
tunics. **Leaves** Several, linear,
not clasping, 30 cm long ×
1 cm wide, dry at flowering.
Flowers Held erect on long
peduncles, each flower
3 cm wide, white with green
keels. Flowering in spring.
Habitat Loam-rich soils in
Grassy Fynbos. **Distribution**
Widespread from Namibia
to Baviaanskloof.

Guerna, Kouga 09.02.2012

Albuca schoenlandii

Named after German botanist Selmar Schonland,
1860–1940, curator and director of the Albany Museum, and
founder of the Department of Botany at Rhodes University

Bulbous geophyte to 30 cm tall, with dry, firm bulb
tunics. **Leaves** Flat, oblong, 3 cm wide, dark green,
with hyaline margins; leaves dry at flowering. **Flowers**
Held erect on long peduncles, 3 cm wide, white or pale
yellow with green keels. Flowering in spring. **Habitat**
Loamy soils in Transitional Shrubland. **Distribution**
Widespread from Oudtshoorn to the Eastern Cape.

19.09.2021

Albuca setosa Diktamarak

setosus = bristly; referring to the dry,
fibrous, often scorched tunics that form a
neck around the stalk. The common name
alludes to the relatively robust or thick
('dik' in Afrikaans) bulb of this species.

Large-bulbed geophyte to 60 cm tall,
tunics fibrous, and forming a thick neck.
Leaves Several, fleshy, not clasping,
sometimes channelled , margins hyaline and sometimes
minutely hairy. **Flowers** Held upright on long pedicels, 3 cm
wide, white or yellow with green keels. Flowering in spring.
Habitat Finer-textured red soils on quartz outcrops and rocky
sandstone slopes in Succulent Karoo, Thicket, Transitional
Shrubland and Grassy Fynbos. **Distribution** Widespread from
Namaqualand to Eswatini.

Enkeldoring, Baviaans 22.09.2011

Albuca virens Bosui

virens = green; referring to the
green keels on the tepals

Bulbous geophyte to 30 cm.
Leaves Few, thin, 20 cm long,
linear, fleshy. **Flowers** In a spike-
like raceme, each flower 1 cm
wide, usually white with green
keels, bracts long and narrow.
Flowering in summer and
autumn. **Habitat** Loamy soils in
Grassy Fynbos and Transitional
Shrubland. **Distribution**
Widespread from Baviaanskloof
to tropical Africa.

Bokloof 4X4, Kouga 31.01.2020

BOWIEA

Rankbol

Named after James Bowie, 1789–1869, an English botanist and plant collector who was sent to South Africa by the Royal Botanic Gardens, Kew. He wrote the earliest guide to the Cape flora, printed in South Africa (1829), and he provided Europe with many new succulent species.

1 species in southern and tropical Africa.

Bowiea volubilis subsp. *volubilis* Climbing onion

volubilis = twining; referring to the climbing habit

Bulbous geophyte, with green bulb partly above ground. **Leaves** Filiform and thread-like, dry at flowering. **Flowers** In climbing, twining, laxly branched racemes, soft and fleshy, green, with tepals bending backwards. Flowering in late summer. **Habitat** Infrequently found among rocks in dense bush in the narrow ravines in Thicket. **Distribution** Very widespread from Namibia to tropical Africa. **Notes** 2 subspecies: subsp. *gariepensis* occurs in Northern Cape; has been overharvested for medicinal use. Popular to grow in pots. **Status** Vulnerable. **Other names** Umagaqana (X), iguleni (Z).

Gert Smitskloof, Kleinpoort 26.04.2018

DIPCADI

Slangui

dipcadi = Turkish name for the grape hyacinth *Muscari*, from the northern hemisphere; *Dipcadi* bears a vague resemblance to *Muscari*.

± 30 species in Africa, Eurasia and India; 13 species in South Africa; 1 species recorded in Baviaanskloof.

Dipcadi viride Skaamblommetjie

viridis = green; referring to the colour of the flowers. The Afrikaans common name means 'shy little flower', alluding to its habit of hanging downwards.

Bulbous geophyte to 20 cm. **Leaves** Few, linear to lanceolate, 30 cm long × 1 cm wide. **Flowers** 1.5 cm long, green to brown; tepals with filiform appendages to 2 cm long. Flowering in spring and summer. **Habitat** Occasional on exposed stony ground in Succulent Karoo. **Distribution** Very widespread in southern Africa and to Ethiopia. **Other names** Dainty green bells (E), gifbolletjie (A), lephotoana (S), ikhakhakha eliluhlaza (Z).

Uitspan, Baviaans 15.02.2016

DRIMIA

Poison squill

drimys = acrid, pungent; pertaining to the sap, which is toxic in most species

Onion-like bulbs with leaves often dry at flowering. Individual flowers in the inflorescence only last for a day. ± 100 species in Africa, the Mediterranean and Asia; 60 species in South Africa; 8 species in Baviaanskloof.

Drimia anomala Ungcana (X)

anomalus = unusual, abnormal; the single, firm-fleshy, cylindrical leaf is unusual in the genus.

Bulbous geophyte to 50 cm tall, with banded, papery sheaths around the exposed bulb. **Leaves** Single, firm, erect, cylindrical, to 30 cm long, but often bitten off and short-stubby. **Flowers** Numerous on a long, narrow raceme, 0.6 cm wide, whitish with green, tepals reflexed, slightly scented. **Habitat** Occurring in cracks on rocky ridges and cliffs in exposed situations in Fynbos, Transitional Shrubland and Thicket. **Distribution** Widespread from the Little Karoo to KwaZulu-Natal.

raceme

Scholtzberg, Baviaans 23.12.2011

Drimia capensis Maerman

capensis = from the Cape; referring to the geographical origin of the species; the Afrikaans common name 'thin man' pertains to the typically tall and thin inflorescence.

Large, bulbous geophyte to 2 m. **Leaves** Several, spreading, oblong, 35 cm long × 6 cm wide, dark green, dry at flowering. **Flowers** In a tall, whorled raceme, 1 cm wide, whitish with green anthers, tepals curved back. Flowering in late summer. **Habitat** Deeper, heavier soils in Thicket and Succulent Karoo. **Distribution** Widespread from Namaqualand to Gqeberha (Port Elizabeth).

Keerom, Baviaans 20.03.2010

Drimia ciliata

ciliata = fringed with fine hair;
referring to the margins of the leaves

Bulbous geophyte to 15 cm tall, with a shallow
bulb 3 cm wide, bulb slightly wider than long.
Leaves In a rosette, flat on the ground, stiff and
leathery, 2.5 cm long × 2 cm wide; thickened
margins are hairy and papillate, dry at flowering.
Flowers Star-shaped, in loose racemes on
wiry peduncles, white, opening late afternoon,
sweetly scented. Flowering in midsummer.
Habitat Open areas in shallow, loamy soils on
sandstone-rock ledges and outcrops in Grassy
Fynbos and Fynbos Woodland. **Distribution**
Widespread from Bredasdorp to Gqeberha
(Port Elizabeth).

Nooitgedacht, Kouga 12.05.2021

Drimia haworthioides

Haworthia = plant genus, *ioides* = resembling;
referring to the flowers that slightly resemble
those of *Haworthia*, or alluding to the exposed
bulb that could be mistaken for it

Bulbous geophyte to 40 cm, sometimes with
bulbs exposed, with loose scales. **Leaves**
Lanceolate, spreading, soft, 5 cm long × 1 cm
wide, margins sometimes hairy. **Flowers** On lax
racemes, at the end of 30 cm-long flower stalks,
greenish white, tepals reflexed. Flowering in
summer and autumn. **Habitat** Loamy or clay-
rich soils in Transitional Shrubland and Thicket.
Distribution Widespread from Worcester
to Makhanda.

Kouga Kliphuis, Kouga 13.05.2021

Drimia intricata

intricatus = entangled; pertaining to the unusual,
divaricately branched inflorescence, sometimes with
finely entangled, delicate side branches

Delicate, bulbous geophyte to 20 cm. **Leaves** 4–10, terete,
8 cm long, dark green, dry at flowering. **Flowers** Few,
tiny, usually whitish or yellow, at the tips of a divaricately
branched inflorescence. Flowering in late summer.
Habitat Seldom seen on rock ledges and cliffs in Fynbos,
Transitional Shrubland and Thicket. **Distribution** Very
widespread from southern Africa to tropical Africa.

Kouga Kliphuis, Kouga Valley 12.05.2021

Drimia uniflora Fairy snowdrop

uni = one, *flora* = flower; referring to the single, tiny flower

Miniature, bulbous geophyte to 8 cm. **Leaves** Filiform, dry at flowering. **Flowers** Single, seldom 2, nodding, white to pale pink. Flowering in late summer. **Habitat** Found on moss banks on cliffs and rocky outcrops in shaded situations in Grassy Fynbos, Transitional Shrubland and Thicket. **Distribution** Very widespread from Namaqualand to Zimbabwe. **Other name** Khoho-ea-lefika (S).

Nuwekloof Pass, Baviaans 23.04.2018

LACHENALIA

Named after Werner de Lachenal, 1736–1800, a Swiss professor of botany at the University of Basel and a friend of Linnaeus. He was well acquainted with European plants and took great care to maintain the botanical garden at the University.

± 110 species mainly in winter-rainfall parts of South Africa; 3 species in Baviaanskloof.

Lachenalia ensifolia

ensifolia = sword-leaved; referring to the lanceolate leaves that taper to a sharp point

Low, bulbous geophyte, to 5 cm. **Leaves** 2, very variable, spreading to prostrate, lanceolate to ovate, 2 cm wide. **Flowers** Several in a basal corymb between the leaves, usually white or pale blue, with anthers exserted. Flowering in late autumn. **Habitat** Various habitats, often in damp situations; absent from the upper mountains. **Distribution** Very widespread from Kamiesberg to Kenton-on-Sea. **Notes** Forms from damper habitats tend to have longer, softish, striated leaves, while those in karroid areas have tougher, more ovate leaves without striation.

Above Nuwekloof Pass, Baviaans 19.05.2006

Lachenalia latimerae

Named after Marjorie Courtenay-Latimer, 1907–2004, a South African naturalist and museum curator who worked in the Eastern Cape. She is especially remembered for bringing the coelacanth to the world's attention after it was found in a catch of fish off East London. The coelacanth genus *Latimeria* was named after her.

Bulbous geophyte to 30 cm. **Leaves** 1 or 2, erect, narrow, to 1 cm wide, with purple spots. **Flowers** In a raceme, flower stalks to 0.5 cm long, each flower 1 cm long, pale pink with greenish brown markings, anthers exserted. Flowering in late winter. **Habitat** Sandy loam soil in Transitional Shrubland. **Distribution** Fairly narrow distribution from the Swartberg to the Baviaanskloof.

Vlakkie, Zandvlakte 27.08.201

LEDEBOURIA

Named after Carl Friedrich von Ledebur, 1785–1851, a German–Estonian botanist. He was the first to describe the species *Pyrus sieversii*, the ancestor of today's cultivated apple.

Bulbs with attractive, spotty leaves. An identifying feature is the property of the papery sheaths around the bulb to produce fine threads when torn. ± 50 species in Africa and India; 40 species in South Africa; 3 species in Baviaanskloof.

Ledebouria revoluta   Ubuhlungu (X)

revolutus = edges rolled backwards; pertaining to the leaves that have edges rolled back

Bulbous geophyte to 15 cm tall, with scales producing fine threads when torn. **Leaves** Several, lanceolate, spotted purple-red. **Flowers** Numerous in a broad raceme, with pedicels longer than the flowers, each flower purple and greenish. Flowering mainly in early summer. **Habitat** Sporadic and infrequent on richer soils in Grassy Fynbos and Transitional Shrubland. **Distribution** Very widespread from Langeberg through eastern Africa to India. **Notes** Used medicinally. **Other names** Inqwebebane, ikreketsane (X).

Rooihoek, Baviaans 02.02.2020

ORNITHOGALUM   Chincherinchee, tjienk

ornithos = bird, *gala* = milk; origin unresolved, however 'bird's milk' was an expression by the ancient Greeks for something wonderful, which is appropriate for the striking, porcelain-like flowers of several species. The common name alludes to the 'squeaky' sound produced when the stems are rubbed together.

Many species are poisonous and toxic to stock. ± 160 species in Africa, Eurasia and India; 80 species in South Africa; 4 species in Baviaanskloof.

Ornithogalum dubium   Geeltjienk

dubius = doubtful; alluding to the difficulty of differentiating between species in this genus and of making an accurate identification

Bulbous geophyte to 50 cm tall, with dark, leathery bulb tunics. **Leaves** In a rosette, each leaf 20 cm long × 1 cm wide, usually drying at flowering. **Flowers** Several on a rounded raceme, each flower 2.5 cm wide, white, yellow or orange with a dark centre. Flowering in spring and summer, especially after rain and/or fire. **Habitat** Common on slightly richer soils on gentle slopes and plateaus in Grassy Fynbos and Transitional Shrubland. **Distribution** Widespread from Bokkeveld (Nieuwoudtville) to Gqeberha (Port Elizabeth). **Other name** Itsweletswele lasethafeni (X).

Zandvlakte bergpad, Kouga 02.11.2011

Ornithogalum juncifolium   Grass-leaved chincherinchee

juncifolius = rush-leaved; referring to the slender,
tufted leaves that are somewhat reed- or grass-like

Bulbous geophyte to 40 cm tall, with
dried leaves forming a fibrous sheath
at the base. **Leaves** Basal, in a tuft,
filiform, slender and ribbed. **Flowers**
In a slender, spike-like raceme, only a
few flowers open at a time, each flower
1.5 cm wide, white with green keels.
Flowering in summer and autumn.
Habitat Fairly common on exposed
rocky slopes at lower altitudes in
Transitional Shrubland, Succulent Karoo
and Thicket. **Distribution** Widespread
from Caledon and the Little Karoo to
KwaZulu-Natal. **Other names** Lijo-tsa-
noko (S), indlolothi encane (Z).

Joubertina, Langkloof 02.01.2013

VELTHEIMIA

Named after August Ferdinand von Veltheim, 1741–1801, a German mineralogist and geologist. He was
the first to write that granite rock was derived from volcanic processes. He maintained a public garden at
Urabke Castle, where he retired.

2 species in southern Africa.

Veltheimia bracteata   Forest sand-lily

bracteata = with bracts; pertaining to the narrowly lanceolate floral bracts on this species

Bulbous geophyte to 40 cm
tall, with fleshy bulb tunics.
Leaves 40 cm long × 10 cm
wide, glossy green, some
green at flowering, with
undulating margins; less
deciduous than the other
species. **Flowers** In a dense,
egg-shaped raceme, tubular,
nodding, pink or pale yellow,
sometimes speckled with
red. Flowering in spring.
Habitat Localised in damp,
shaded situations in Thicket.
Distribution Widespread from
Baviaanskloof to Gqeberha
(Port Elizabeth) and Morgans
Bay. **Other name** Sandui (A).

Erasmusboskloof, Zandvlakte 21.08.2015

Veltheimia capensis   Sandlelie

capensis = from the Cape

Bulbous geophyte to 40 cm tall, with papery bulb tunics. **Leaves** 30 cm long × 8 cm wide, greyish green, with undulating or crisped margins, dry at flowering. **Flowers** In a dense raceme, tubular, nodding, 3 cm long, pinkish. Flowering in autumn and winter. **Habitat** Localised in damp, shaded situations in Thicket and Forest, usually on cliffs in the narrow kloofs. **Distribution** Widespread from Namaqualand to Baviaanskloof. **Other name** Qarobe (X).

Kouga Valley, Suuranysberg 11.05.2006

HYDROCHARITACEAE   Tape grasses

A family of freshwater and marine plants with 16 genera and ± 135 known species worldwide. Some species are used as ornamental plants and have become problematic aquatic weeds around the world, especially in the genera *Egeria*, *Elodea* and *Hydrilla*. 6 genera and 13 species native in southern Africa, with 3 genera having 3 exotic species that have become naturalised; 1 invasive weed recorded in Baviaanskloof.

EGERIA*

Egeria = a mythological water nymph

2 species native to South America, 1 species in Baviaanskloof.

Egeria densa*   Brazilian waterweed

densa = dense; referring to the habit of forming a dense tangled mass

Submerged aquatic with stems to 3 m long. **Leaves** In whorls of 4 or 5, green, 3 cm long, with finely serrated margins. **Flowers** With 3 tepals, on long stalks exserted 2 cm above the water's surface, each flower 1.5 cm wide, white or yellow. **Habitat** Forming dense colonies in slow-flowing rivers and pools. **Distribution** Native to South America, and an invasive weed in South Africa and many parts of the world. It has invaded gently flowing parts of the Kouga and Gamtoos rivers.

Kouga Kliphuis, Kouga River 21.01.2022

HYPOXIDACEAE

Star grasses

Herbaceous geophytes, often with hairy leaves, star-shaped flowers, and roots with scaly plates. 9 genera and 130 species in southern Africa, South America, Australia and tropical Asia; 6 genera and ± 88 species in South Africa, with 5 endemic genera; 3 genera and 5 species in Baviaanskloof.

EMPODIUM
Autumn star, ploegtydblommetjie

em = within, *pod* = foot; referring to the ovary that is found underground, below the solid floral tube

± 9 species in southern Africa, mostly in winter-rainfall areas; 1 species in Baviaanskloof.

Empodium plicatum

plicatum = folded into pleats or furrows; referring to the leaves that have a pleated appearance

Cormous geophyte to 30 cm tall, with membranous sheaths at the base. **Leaves** 1–4, narrowly strap-shaped, pleated, 20 cm long × 0.4 cm wide, emerging at flowering, hairy on ribs. **Flowers** 3 cm wide, yellow above, pale green below, tepals spreading, on a solid neck to 10 cm. Flowering in autumn. **Habitat** Common on clay or loamy flats in Transitional Shrubland and Succulent Karoo. **Distribution** Widespread from Namaqualand to Baviaanskloof.

Vlakkie, Zandvlakte 03.04.2012

HYPOXIS
Stargrass

hypo = below, *oxy* = pointed; referring to the pointed base of the ovary or fruit

Cormous perennials with over 90 species found in the world. An important source of traditional and modern medicine in South Africa, *Hypoxis* species are used in immune-boosting drugs for patients with HIV and AIDS; marketed as the Africa potato. ± 40 species in South Africa, mostly in the grassland biome; 3 species in Baviaanskloof.

Hypoxis villosa   Golden winter star

villosa = villous; referring to the long shaggy hairs on the plant

Cormous geophyte to 15 cm tall, with a fibrous neck. **Leaves** 4–7, lanceolate, sickle-shaped, silky-white-hairy below. **Flowers** Few, but sometimes up to 10 in a corymbose inflorescence, yellow, hairy. Flowering in summer and autumn. **Habitat** Common on loamy slopes in grassy Transitional Shrubland and Grassy Fynbos. **Distribution** Widespread from Baviaanskloof to East London. **Other name** Inongwe (X).

corymb

Guerna, Kouga 09.02.2012

Empodium • Hypoxis | MONOCOTYLEDONS   117

PAURIDIA

Klipsterretjie

pauros = small, *idia* = diminutive; referring to the small size of some species. Revised in 2014 to include genus *Spiloxene*. ± 30 species in southern Africa and a few in Australasia; 3 species in Baviaanskloof.

Pauridia trifurcillata

trifurcillata = three small forks; referring to the 3 stigmas that have slender, downturned extensions

Small and delicate cormous geophyte to 15 cm tall, with soft brown tunics. **Leaves** 5–8, thin-textured, narrow and keeled, 7 cm long × 0.2 cm wide. **Flowers** Small, 1 cm wide, yellow, tepals pale green below. Flowering in late summer and autumn. **Habitat** Common on seasonally damp south-facing slopes and rocky ledges in Grassy Fynbos and Fynbos Woodland. **Distribution** Eastern distribution from Baviaanskloof to Somerset East.

Uitspan, Baviaans 05.05.2006

IRIDACEAE

Irises

The largest and best-known plant family in Europe with ± 65 genera and 1,800 species worldwide; over half of the diversity occurs in southern Africa. The name is derived from the Greek goddess Iris, who brought messages to Earth from Olympus via a rainbow. Linnaeus, who coined the name in 1753, saw all colours of the rainbow in the spectacular flowers.

Members of this family are mostly perennials with corms or rhizomes. They have symmetrical, colourful flowers with 6 tepals. The flowers are usually arranged in a spike. Leaves are often flattened, with a raised midrib and arranged in a fan or tuft.

The most diverse genera in southern Africa are *Gladiolus*, with ± 255 species, and *Moraea*, with ± 200 species; 27 genera and just over 700 species in the Cape flora; 16 genera and over 50 species have been recorded in Baviaanskloof.

ARISTEA

Blousuurkanol

arista = a point or beard; pertaining to the pointed leaves

Rhizomatous perennials with leaves – tough and without a midrib – arranged in a fan. Flowers are a striking blue or mauve. ± 55 species in Africa and Madagascar; 45 species in South Africa; 6 species in Baviaanskloof.

Aristea anceps

anceps = two-edged; pertaining to the flattened and 2-winged stems

Rhizomatous perennial, usually to 15 cm tall, with winged stems. **Leaves** Straight, pointed, 15 cm long × 0.4 cm wide. **Flowers** 3 cm wide, blue, with prominent rusty spathes. Capsule oblong, 3-lobed and without wings. Flowering in spring and summer. **Habitat** Grassy Transitional Shrubland on clay slopes and flats. **Distribution** Widespread from Baviaanskloof to Mthatha.

Engelandkop, Kouga 19.04.2012

Aristea cuspidata

cuspidata = cuspidate; referring to the leaves that taper gradually to a rigid point

Rhizomatous perennial to 60 cm tall, with slightly compressed stems. **Leaves** Linear and relatively narrow. **Flowers** 3 cm wide, blue, the rusty spathes have transparent, membranous margins. Capsule 3-winged. Flowering in early summer. **Habitat** Upper sandstone mountains in Mesic and Grassy Fynbos. **Distribution** Widespread from the Cape Peninsula to Baviaanskloof.

Bergplaas, Baviaans 23.09.2011

Aristea ensifolia

ensifolia = sword-leaved; referring to the narrowly tapering leaves ending in a sharp point

Rhizomatous perennial to 50 cm tall, with flattened, 2-winged stems. **Leaves** Relatively soft and narrow. **Flowers** Blue, with membranous spathes. Capsules elongated, 3-angled, not splitting open with age. **Habitat** Localised in damp, shaded Forest. **Distribution** Widespread from Riversdale to Kariega (Uitenhage); easily seen at Poortjies.

10.11.2017

Aristea pusilla

pusilla = very small; alluding to the small stature of this species

Rhizomatous perennial to 20 cm tall, with flattened, 2-winged stems. **Leaves** Soft and sword-shaped, 12 cm long × 0.5 cm wide. **Flowers** Blue, spathes with membranous margins. Capsules without wings, trigonous, elongated, 2.5 cm long. Flowering in spring. **Habitat** Occasional in Transitional Shrubland and Grassy Fynbos. **Distribution** Widespread from the Little Karoo to the Eastern Cape.

Kouenek, Kouga 26.09.2011

BABIANA

baviaantje = Dutch for baboon; baboons like to eat the corms.

Cormous perennials with hairy, pleated leaves and usually 2-lipped flowers. The bracts are hairy with dry tips. Corms eaten by porcupines and francolins, and also enjoyed roasted by people. 86 species in southern Africa; 1 species and subspecies in Baviaanskloof.

Babiana sambucina

Sambucus = plant genus; named after the elderberry (genus *Sambucus*), the fruits of which were used to make a blue dye; some members of this genus have blue flowers.

Cormous geophyte to 14 cm tall, with stems below ground. **Leaves** 14 cm long, ribbed, hairy and lanceolate. **Flowers** Fragrant, 6 cm wide, blue to purple with white markings, with bracts forked at the tips. Flowering in early spring. **Habitat** Fairly common on sandy loam soils in Transitional Shrubland and Grassy Fynbos. **Distribution** Very widespread from Bokkeveld (Nieuwoudtville) to Gqeberha (Port Elizabeth). **Notes** 2 subspecies: subsp. *sambucina* (described here) and subsp. *longibracteata*, which is endemic to the Nieuwoudtville area.

Kouenek, Kouga 25.09.2011

BOBARTIA

Rush iris, blombiesie

Named after Jacob Bobart, 1599–1680, a German botanist who cultivated a garden of over 1,600 medicinal herbs – it was the first of its kind in England.

Rhizomatous perennials with long cylindrical leaves in a tuft. Flowers star-like, yellow, with several on a long leafless stem. 15 species in South Africa; 2 species in Baviaanskloof.

Bobartia orientalis

orientalis = eastern; referring to the geographic distribution of the species in South Africa

Tufted perennial to 130 cm. **Leaves** Terete, 0.3 cm wide, yellowish green. **Flowers** In a dense head of green spathes, each flower 3 cm wide, yellow, star-shaped. Flowering in spring. **Habitat** Frequent on dry, stony slopes in Grassy Fynbos. **Distribution** Widespread from Villiersdorp to Makhanda.

Onder Kouga, Joubertina, 18.01.2009

CHASMANTHE

Cobra lily

khasme = wide open, gaping, *anthos* = flower; pertaining to the trumpet-shaped flowers

3 species in South Africa; 1 species in Baviaanskloof.

Chasmanthe aethiopica

aethiopica = from sub-Saharan Africa; in classical times this region had not yet been explored by Europeans.

Cormous geophyte to 60 cm tall, with papery corm tunics. **Leaves** Arranged in a fan, soft-textured, sword-shaped. **Flowers** In a single-ranked spike, orange. Fruit is bird dispersed, seeds pea sized, orange and slightly fleshy. Flowering in autumn and winter. **Habitat** Grassy Fynbos, usually in relatively damp situations at Forest margins. **Distribution** Widespread from Darling to East London, mostly on coastal forelands.

Langkloof, Joubertina 13.05.2021

DIETES

Wild iris, forest iris

dis = twice, *etes* = associated; alluding to the 2 genera, *Iris* and *Moraea*, to which this genus is most closely related

Popular in gardens around the world. 6 species in Africa and Australasia; 5 species in South Africa; 1 species in Baviaanskloof.

Dietes iridioides Small forest iris

Iris = plant genus, *oides* = resembling; referring to the resemblance to genus *Iris*

Rhizomatous, tufted perennial to 50 cm. **Leaves** In a loose fan, dark green, sword-shaped. **Flowers** 6 cm wide, white marked with yellow, with violet style branches that look like inner tepals. Flowering in spring and summer. **Habitat** Occurring in shade under Forest, often near streams. **Distribution** Very widespread from Riviersonderend to East Africa. **Other name** Imbotyi kaxam (X).

Poortjies, Cambria 22.01.2022

FREESIA

Named after Friedrich Heinrich Theodor Freese, 1795–1876, a German doctor and botanist who studied South African plants

This genus is popular in horticulture and hybrids have been developed. Cormous geophytes with leaves in a fan, flowering in spring. 17 species in southern and tropical Africa; 1 species in Baviaanskloof.

Freesia corymbosa Kammetjie

corymbosa = corymbose; pertaining to the corymbose inflorescence in this species

Cormous geophyte to 50 cm. **Leaves** 7–10, in a fan, lanceolate, with acute tips. **Flowers** Usually 6–10, 3 cm long, yellowish, pink or white, slightly scented. Flowering in spring. **Habitat** Occasional on loamy soils in Grassy Fynbos and grassy Transitional Shrubland. **Distribution** Widespread from Baviaanskloof and Langkloof to Makhanda. **Other name** Flissie (A).

Stompie se Nek, Baviaans 03.07.2003

GEISSORHIZA   Satinflower, sysie

geisson = tile, *rhiza* = root; pertaining to the neatly overlapping, woody corm tunics in some species

Spectacular and colourful flowers with short style branches. ± 100 species in South Africa; 4 species in Baviaanskloof.

Geissorhiza bracteata

bracteata = with bracts; referring to the prominent green bracts at the base of the flower

Cormous geophyte to 15 cm tall, with woody tunics. **Leaves** Sword-shaped, flat, less than 1 cm wide, usually prostrate. **Flowers** 1 to a few per stem, 1 cm wide, white, sometimes with pink markings outside, tube greenish. Flowering in spring. **Habitat** Shale in Transitional Shrubland. **Distribution** Widespread from Albertinia and the Little Karoo to Somerset East.

Kouenek, Kouga 26.09.2011

Geissorhiza roseoalba

roseo = rosy, *albus* = white; alluding to the white flowers with rosy pink and reddish markings

Cormous geophyte to 30 cm tall, with tunics woody and concentric. **Leaves** Lanceolate, 0.6 cm wide, with margins and midrib thickened. **Flowers** 5 cm wide, light pink with darker pink markings and reddish-purple nectar guides. Flowering in spring. **Habitat** Occasional on upper rocky slopes and ridges in Mesic and Grassy Fynbos. Nearly endemic from around De Rust to Kariega (Uitenhage).

Ridge west of Elandsvlakte, Baviaans 29.09.2011

GLADIOLUS

gladiolus = a small sword; referring to the sword-shaped leaves

± 260 species in Africa, Madagascar and Eurasia; 170 species in South Africa; 8 species in Baviaanskloof.

Gladiolus floribundus

floribundus = profusely flowering; alluding to the tendency to produce many flowers

Bergplaas, Combrink's Pass 23.09.2011

Cormous geophyte to 45 cm tall, stout, with papery tunics. **Leaves** In a fan, sword-shaped, leathery, 2 cm wide. **Flowers** Large, 6 cm wide, white with a dark streak, long tubed (7 cm long), lower tepals smaller. Flowering in spring. **Habitat** Fairly frequent in loamy soils in Grassy Fynbos and Transitional Shrubland. **Distribution** Widespread from Cederberg to Hlanganani.

Gladiolus geardii

Named after the landowner of the farm near Kariega (Uitenhage) where the plant was first recorded for science

Poortjies, Cambria 15.12.2019

Cormous geophyte to 150 cm tall, with papery tunics and branched stems. **Leaves** Sword-shaped, 1.2 cm wide. **Flowers** Funnel-shaped, 6 cm long, pink with markings on lower tepals. Flowering in summer. **Habitat** Uncommon on damp stream banks in sandy and peaty soils in Forest. **Distribution** Narrow distribution from Baviaanskloof to Kariega (Uitenhage). Easily seen at Poortjies. **Status** Near Threatened.

Gladiolus huttonii   Eastern Cape flame

Named after Henry Hutton, 1825–1896, amateur botanist who collected the type specimen near Makhanda with his wife Caroline

Suuranysberg, Langkloof 17.09.2017

Cormous geophyte to 60 cm with fibrous tunics. **Leaves** narrow, cross-shaped in section. **Flowers** With a narrow tube, 5 cm long, red to orange, tube with maroon markings, lower tepals sometimes yellow and smaller. Flowering in winter and early spring. **Habitat** Occurring in Grassy Fynbos and Sour Grassland. **Distribution** Widespread from Plettenberg Bay to Makhanda.

Gladiolus leptosiphon

lepto = thin, narrow, *siphon* = tube or pipe;
referring to the narrow floral tube of this species

Cormous geophyte to 50 cm, with fibrous
tunics. **Leaves** Narrow. **Flowers** Tube
narrow, 5 cm long, white with red-purple
lines on the 3 lower tepals. Flowering in
late spring. **Habitat** Uncommon on dry
lower slopes in Transitional Shrubland.
Distribution Widespread from Ladismith to
Kariega (Uitenhage).

Between Sewefontein and Beacosnek, Kouga 22.09.2010

Gladiolus maculatus Bruinafrikaner

macula = spot; pertaining to the speckled markings on the flower

Cormous geophyte to 60 cm with
fibrous tunics. **Leaves** Linear, leathery
and relatively short. **Flowers** Funnel-
shaped, tube long, tepal margins wavy,
usually cream-yellow with darker brown
spots, sweetly scented. Flowering in
autumn and winter. **Habitat** Found
on shale bands in Grassy Fynbos.
Distribution Widespread from the Cape
Peninsula to Makhanda. **Other name**
Kaneelpypie (A).

Joubertina 14.09.2011

Gladiolus patersoniae

Named after Florence Paterson, 1869–1936, who collected the type specimen in Baviaanskloof Mountains in 1910

Cormous geophyte to 50 cm, tunics
roughly fibrous. **Leaves** 4-grooved,
terete. **Flowers** Hanging, bell-like,
blue to pale grey with transverse
yellow markings on 3 lower tepals,
apple-scented. Flowering in spring.
Habitat Rocky slopes in Grassy
Fynbos. **Distribution**
Widespread from
Worcester to
Baviaanskloof and to
Kariega (Uitenhage).

Enkeldoring, Baviaans 21.09.2011

Gladiolus permeabilis subsp. edulis

permeabilis = allowing to pass through; referring to the
gap between the upper tepal and the lateral tepals,
giving the flower a 'window' if viewed from the side

Cormous geophyte to 50 cm tall, with fibrous
tunics. **Leaves** Narrow, 0.3 cm wide, with raised
midrib. **Flowers** 2 cm wide, cream to brown or pink
with yellow markings and wine-red lines. Flowering
in spring. **Habitat** Occurring mainly in Transitional
Shrubland but also found in Succulent Karoo,
Thicket and Grassy Fynbos. **Distribution** Very
widespread from Caledon to Zimbabwe. **Notes**
2 subspecies: subsp. *permeabilis* is endemic to the
Western Cape.

Zandvlakte, Baviaans 30.08.2010

Gladiolus stellatus

stellatus = starry; pertaining
to the star-shaped flowers

Cormous geophyte to 50 cm,
with fibrous tunics. **Leaves**
Narrow, 0.3 cm wide, with
raised midrib. **Flowers**
Star-shaped, 2 cm wide,
white to pale blue, scented.
Flowering in spring. **Habitat**
Lower slopes in Transitional
Shrubland and Grassy Fynbos.
Distribution Widespread from
Swellendam to Gqeberha (Port Elizabeth).

Elandsvlakte, Baviaans 03.11.2011

Gladiolus virescens   Kalkoentjie

virescens = greenish; the 3 lower tepals are
tinged with a yellowy-green band.

Cormous geophyte to 20 cm, with
papery tunics. **Leaves** Narrow,
0.4 cm wide, terete, ribbed. **Flowers**
With upper tepals erect, yellow to
pink with dark veins, scented, tube
to 1.5 cm long. Flowering in early
spring. **Habitat** Occurring on lower
slopes in Transitional Shrubland
and Grassy Fynbos. **Distribution**
Widespread from Ceres to Gqeberha
(Port Elizabeth).

Langkloof 28.09.2015

IXIA

Kalossie

Ixia = a name of Greek origin, coined by Linnaeus to indicate variation in flower colour. The Afrikaans name comes from the name for a headdress worn by early Cape Malay slaves in Cape Town, which the flowers are said to resemble.

The genus can be detected by looking at the base of the symmetrical flowers for 2 papery bracts that have 2- or 3-toothed tips. ± 50 species in South Africa; 1 species in Baviaanskloof.

Ixia orientalis

orientalis = eastern; referring to the eastern distribution of the species

Cormous geophyte to 50 cm. **Leaves** Linear, grass-like. **Flowers** 5–10 per spike, quite crowded, white or pink; floral tube narrowly funnel-shaped, 1 cm long. Flowering in spring. **Habitat** Occasional on damp, loamy slopes in Grassy Fynbos. **Distribution** Widespread from Villiersdorp to Port Alfred.

Toverwater, Kouga 25.11.2009

MORAEA

Uintjie, tulp

Named after Sara Elisabeth Moraea, 1716–1806, whom Linnaeus married in 1739

± 220 species in sub-Saharan Africa and Mediterranean to the Middle East; 180 species in South Africa; 6 species in Baviaanskloof.

Moraea algoensis

algoensis = from Algoa; where the species was first scientifically recorded, near Gqeberha (Algoa Bay)

Slender geophyte to 30 cm. **Leaves** Single, narrow, channelled. **Flowers** 3 cm wide, purple, with yellow or cream markings on the outer tepals. Flowering in early spring. **Habitat** Occurring on lower slopes in Transitional Shrubland. **Distribution** Widespread from Worcester to Gqeberha. **Notes** Easily confused with *M. tripetala*, but the flowers of *M. algoensis* are bigger and outer tepals are wider.

Onder Kouga road, Joubertina 24.08.2009

Moraea bipartita Blue tulp

bipartita = divided into two parts; referring to the inner tepal lobes that are split longitudinally, almost to the base

Cormous geophyte to 45 cm, usually smaller. **Leaves** Few, narrow, 0.4 cm wide. Flowering stems branched. **Flowers** 3 cm wide, bluish purple with yellow markings at the throat. Flowering in winter and spring. **Habitat** Succulent Karoo and Thicket. **Distribution** Widespread from Ladismith to Kariega (Uitenhage). **Other name** Bloutulp (A).

Bruintjieskraal, Grootrivierpoort 03.09.2020

Moraea exiliflora

exilis = weak, small, slender; referring to the small, flimsy flowers that open in the late afternoon only

Slender, cormous geophyte to 25 cm. **Leaves** Single, linear, furrowed. **Flowers** Small, 2.5 cm wide, pale blue with white and yellow nectar guides, opening in late afternoon. Flowering in spring. **Habitat** Uncommon on shallow soils in Transitional Shrubland and Thicket. **Distribution** Narrow distribution from the Little Karoo to Baviaanskloof.

Vlakkie, Zandvlakte 25.08.2010

Moraea falcifolia Patrysuintjie

falcifolia – sickle-shaped leaves; referring to the shape of the leaves

Stemless geophyte to 5 cm. **Leaves** Several, spreading, slightly twisted, with wavy margins. **Flowers** 2.5 cm wide, outer lobes white with yellow nectar guides, sometimes outlined in brown, inner lobes with purple markings. Flowering in autumn and winter. **Habitat** Localised on shale soils in Succulent Karoo and Transitional Shrubland. **Distribution** Widespread from southern Namibia through the Little Karoo to Hlanganani.

Langkloof 30.08.2009

Moraea fugacissima Clockflower

fugax = fleeting, not lasting long, *issima* = very; pertaining to the flowers that last for a few hours, from 10:30–16:00

Stemless geophyte to 6 cm. **Leaves** Numerous, narrow, linear to terete. **Flowers** bright yellow, 2 cm wide, cup-shaped, tepals similar, stigmas fringed, fragrant. Flowering late winter and early spring. **Habitat** uncommon on richer soils in Transitional Shrubland or Grassy Fynbos. **Distribution** Widespread from Namaqualand to Langkloof and Humansdorp.

Joubertina 17.08.2011

Moraea lewisiae Volstruisuintjie

Named after Dr Gwendoline Joyce Lewis, 1909–1967, a South African botanist who specialised in Iridaceae; she has 5 other plant species named after her.

Slender geophyte to 90 cm. **Leaves** 1–3, linear, twisted, channelled and trailing. **Flowers** Bright yellow, 3 cm wide, opening late afternoon, tepals all equal size, fragrant. Flowering in late spring and summer. **Habitat** Occurring in dry and exposed places in Transitional Shrubland and Grassy Fynbos. **Distribution** Widespread from Namaqualand to Jeffreys Bay.

Joubertina 27.11.2009

Moraea polyanthos Bloutulp

polyanthos = many-flowered; referring
to the plentiful inflorescence

Cormous geophyte to 60 cm. **Leaves** Several,
narrow, channelled. **Flowers**
Flowering stem branched,
each flower 3 cm wide, blue,
with small yellow nectar guides
at the throat. Flowering in
spring. **Habitat** Lower slopes
in Transitional Shrubland.
Distribution Widespread from
Worcester to Baviaanskloof.

Onder Kouga road, Joubertina 24.08.2009

Moraea ramosissima Vlei-uintjie

ramosissima = very much branched;
referring to the much-branched stems

Robust cormous geophyte to 120 cm, only flowering
in the first year after fire. **Leaves** 1 cm wide, several,
furrowed. **Flowers** Many on branched stems, 3.5 cm
wide, bright yellow. Flowering in late spring and early
summer. **Habitat** Occurring in localised stands in
damp places in Transitional Shrubland and Grassy
Fynbos. **Distribution** Widespread from Gifberg
to Humansdorp. **Notes** The corms are eaten and
dispersed by porcupines and baboons.

Louterwater 15.11.2009

Moraea spathulata Bergflap

spathulata = spatula-shaped; referring to the shape
of the broad, gradually tapering outer tepal lobes

Robust, cormous geophyte
to 1 m. **Leaves** Single,
long, 2 cm wide. **Flowers**
Large, 6 cm wide, yellow
with orange-yellow nectar
guides on the outer lobes,
scented. Flowering in
winter and spring. **Habitat**
Localised in rocky areas
on upper slopes in Grassy
Fynbos. **Distribution**
Very widespread from
Kammanassie to
Zimbabwe.

Kouenek, Kouga 20.09.2011

Moraea tricuspidata   Rietuintjie

tricuspidata = three-pointed; pertaining to the
shape of the inner tepals

Cormous geophyte to 60 cm. **Leaves** Single, long,
linear and channelled. **Flowers** White to cream,
outer tepals with dark speckles at the base, inner
tepals 3-pointed. Flowering in spring. **Habitat**
Transitional Shrubland and Grassy Fynbos.
Distribution Widespread from Cederberg to Cape
Peninsula and to Makhanda.

Langkloof 13.10.2009

Moraea unguiculata

unguiculata = contracted into a claw shape; referring to the inner
tepals that have 3 teeth, which are sometimes rolled inward

Cormous geophyte to 50 cm. **Leaf** Solitary, 0.5 cm wide,
furrowed. **Flowers** Inconspicuous, 1 cm wide, cream-
white to brownish, outer tepals with violet and lime-yellow
spotted nectar guides, inner tepals with 3 points and claw-
like. Flowering in spring and early summer.
Habitat Occurring on shale in Transitional Shrubland
and Grassy Fynbos. **Distribution** Widespread from
Namaqualand to Gqeberha.

Joubertina 17.10.2010

TRITONIA   Agretjie

triton = weathercock; referring to the elevated position of the anthers

± 28 species in eastern and southern Africa; 5 species in Baviaanskloof.

Tritonia linearifolia

linearifolia = linear-leaved;
referring to the shape of the leaves

Cormous geophyte to 60 cm.
Leaves Linear. **Flowers** Several
in a spike, all hanging to
1 side. **Flowers** Tube 1.5 cm
long, cream or yellow with
veins visible; bracts short and
acute. Flowering in spring.
Habitat Damp seepages on
south-facing slopes and slabs
in Mesic and Grassy Fynbos. **Distribution**
Fairly narrow distribution from George to
Baviaanskloof Mountains.

Kleinrivier, Kouga Wildernis 18.12.2011

TRITONIOPSIS
Rietpypie

Tritonia = plant genus, *iopsis* = resembling; the flowers superficially resemble those of *Tritonia*.

24 species in winter-rainfall parts of South Africa; 3 species in Baviaanskloof.

Tritoniopsis antholyza   Karkaarblom

antho = flower, *eyssa* = enraged; alluding to the mouth of the flower that is reminiscent of an angry, open-mouthed beast. *Antholyza* is an old genus similar in appearance to this species. The common name refers to the sound made when dry leaves are rubbed together.

Cormous geophyte to 90 cm. **Leaves** Several, lanceolate, relatively broad, 3–6-veined. **Flowers** Tube 3 cm long, pink to orange-red. Flowering in late spring, summer and autumn. **Habitat** Frequent but seldom abundant on middle to upper sandstone slopes in Grassy and Mesic Fynbos. **Distribution** Widespread from Bokkeveld (Nieuwoudtville) to Gqeberha (Port Elizabeth).

Quaggasvlakte, Kouga 13.02.2016

WATSONIA
Named after William Watson, 1715–1787, an English naturalist and scientist who studied electricity. He coined the word 'circuit'.

52 species in southern Africa; 4 species in Baviaanskloof.

Watsonia knysnana

knysnana = from Knysna; the species has since been found a considerable distance from the coastal village.

Robust, cormous perennial to 1.6 m. **Leaves** Sword-shaped. **Flowers** Tube 4 cm long, pink to purple. Capsules blunt. Flowering in summer. **Habitat** Sandstone slopes in Grassy Fynbos and Mesic Fynbos. **Distribution** Widespread from Mossel Bay to East London.

Scholtzberg, Baviaans 23.12.2011

Watsonia schlechteri

Named after Friedrich RR Schlechter, 1872–1925, a German botanist, explorer and plant collector who worked with Harry Bolus at the Bolus Herbarium in Cape Town

Cormous perennial to 1 m. **Leaves** Sword-shaped, 1.5 cm wide, with distinctly thickened margins and midrib. **Flowers** 4 cm wide × 6 cm long, bright red to orange. Flowering in summer. **Habitat** Occasional, uncommon on rocky slopes and ridges in Mesic Fynbos. **Distribution** Widespread from Citrusdal to the Kouga Mountains.

thickened margins

Kouenek, Kouga 02.12.2009

JUNCACEAE   Rushes

A family of rhizomatous plants that resemble grasses or sedges. *Juncus* is the largest and best-known genus. 8 genera and 464 species worldwide.

JUNCUS

jungere = to bind or tie together; referring to the old practice of using rushes to make ropes

± 250 species worldwide; 5 species in Baviaanskloof.

Juncus acutus

acutus = acute; referring to the pointed leaf tips or acute bracts

Robust, tough, neatly tufted perennial to 2 m. **Leaves** Cylindrical, dark green. **Flowers** In panicles, red to brown. Flowering in spring and summer. **Habitat** Slightly brackish marshes. **Distribution** Very widespread in temperate parts of the world.

Apieskloof, Baviaans 26.10.2018

Juncus capensis

capensis = from the Cape; referring to the geographical origin of the species

Soft, tufted perennial to 50 cm. **Leaves** Bright green, grass-like. **Flowers** In cymes, pale with dark keels. Flowering in summer and autumn. **Habitat** Found in sand in perennial freshwater streams. **Distribution** Widespread from Clanwilliam to Eastern Cape.

Zandvlaktekloof, Baviaans 18.02.2016

PRIONIUM

prionium = a saw blade; referring to the razor-sharp, saw-like edges of the leaf

1 species in the genus, endemic to South Africa.

Prionium serratum   Palmiet, bobbejaanstert

serrata = serrated; referring to the serrated edges of the leaf. Old leaves resemble baboon tails, thus the Afrikaans common name.

Robust shrub to 2 m. **Stems** 10 cm wide and covered in the dark, fibrous remains of the old leaves. **Leaves** In crowded tufts, with serrated margins, hairless. **Flowers** In tall panicles, very small, brown. Flowering mainly in summer. **Habitat** Forming dense stands on sandy boulder beds in perennial streams and marshes. **Distribution** Widespread from Gifberg to KwaZulu-Natal.

Kromrivier, Langkloof 13.05.2021

LANARIACEAE

Only 1 species in this family.

LANARIA

lana = wool, *aria* = connected with; referring to the woolly inflorescence

Lanaria lanata   Kapok-lily

lanata = hairy; referring to the overall hairiness

Rhizomatous, tufted perennial to 80 cm. **Leaves** Tough, 0.5 cm wide, channelled, with serrated margins. **Flowers** In white-woolly panicles, pale blue, honey-scented. Flowering in summer. **Habitat** Common in Grassy and Mesic Fynbos. **Distribution** Widespread from Bainskloof to Makhanda. **Other name** Perdekapok (A).

Suuranysberg, Kareedouw 21.01.2022

ORCHIDACEAE Orchids

A massive family with 763 genera and ± 26,000 species worldwide. Horticulturists have developed over 100,000 cultivars and hybrids. Orchid flowers are usually vividly attractive, often scented and they have highly specialised pollination systems. In South Africa there are almost 500 species in 54 genera; 24 species have been recorded in Baviaanskloof.

ACROLOPHIA

akros = at the tip, *lophos* = crest; alluding to the raised edges on the lip of the flower, perhaps also referring to the mountainous habitat

7 species in South Africa; 2 species in Baviaanskloof.

Acrolophia capensis

capensis = from the Cape; referring to the geographical origin of the species

Resprouting, robust, geophytic, to 80 cm. **Leaves** Arranged in a fan, linear, keeled and leathery, 30 cm long × 1.5 cm wide, lime-green. **Flowers** Several in a branched raceme, well-spaced, 2 cm wide, greenish to brown, with a creamy lip with 5–7 ridges. Flowering in summer. **Habitat** Occasional on rocky ridges in Mesic Fynbos. **Distribution** Widespread from Ceres to Malihanda.

Scholtzberg, Baviaans 28.12.2011

BARTHOLINA Spider orchid

Named after Thomas Bartholin, 1616–1680, Danish doctor who discovered the lymphatic system in humans

2 species in Namibia and South Africa; 1 species in Baviaanskloof.

Bartholina burmanniana Spinnekoporgidee, Spider orchid

Named after Johannes Burman, 1707–1780, Dutch botanist and doctor who was a friend of Linnaeus and collected specimens in the Cape

Tuberous geophyte to 20 cm. **Leaves** Single, heart-shaped, 1 cm wide, flat on the ground, often dry when flowering. **Flowers** Single on a hairy scape, white, 3 cm wide, lip with numerous thread-like lobes, spur 1 cm long. Flowering mostly after fire in spring. **Habitat** Found sporadically in Grassy Fynbos and Arid Fynbos. **Distribution** Widespread from Clanwilliam to Katberg Mountains.

Toverwater, Kouga 20.08.2011

BRACHYCORYTHIS

Helmet orchid

brachys = short, *korys* = helmet; pertaining to the perianth that sometimes resembles a helmet

About 32 species in Africa and Asia; 1 species in Baviaanskloof.

Brachycorythis macowaniana

Named after Peter MacOwan, 1830–1909, originally an English professor of chemistry, who moved to South Africa and collected many specimens. He was one of the first professors of botany at the University of Cape Town.

Tuberous geophyte to 30 cm. **Leaves** 5–15, neatly overlapping up the stem, narrowly lanceolate, to 6 cm long × 1.5 cm wide. **Flowers** Many in a raceme, brownish and green, lateral sepals 0.4 cm long, lip with a 0.2 cm-long spur. Flowering after fire in summer. **Habitat** Infrequent in Grassy and Mesic Fynbos. **Distribution** Widespread from Swellendam to Port Alfred.

Toverwater, Kouga 25.11.2009

DISA

dis = rich; pertaining to the colourful flowers or the relatively high number of species in the genus. May be named after the mythical Queen Disa of Sweden.

183 species, mostly in Africa; 143 species in South Africa; 6 species in Baviaanskloof.

Disa bifida

bi = two, *fidus* = divided; referring to the divided tips of the petals

Slender, tuberous geophyte to 30 cm, with a wiry stem. **Leaves** Basal, in a rosette, flat on the ground, egg-shaped, 2 cm long × 1 cm wide, median line paler green spotted with purple dots. **Flowers** Few, in a raceme, each 1.5 cm wide, pink, lip tip greenish to purple. Flowering in spring. **Habitat** Uncommon in Grassy and Mesic Fynbos. **Distribution** Widespread from Cederberg to Gqeberha (Port Elizabeth).

Scholtzberg, Baviaans 24.07.2011

Disa comosa

comosa = with a tuft of leaves or hairs; referring to the tufted habit

Tuberous geophyte, erect or slanted, to 30 cm. **Leaves** 2 or 3, basal, elliptic, up to 12 cm long, cauline leaves smaller and sheathing the stem. **Flowers** Several in a loose raceme, lime-green, sometimes tinged red, 2 cm long, spur cylindrical, lateral sepals 0.6 cm long. Flowering in spring and early summer, after fire. **Habitat** Uncommon on upper peaks and outcrops in Mesic and Subalpine Fynbos. **Distribution** Widespread from Cederberg to Kouga Mountains.

Toverwater, Kouga 02.10.2009

Disa cornuta

cornuta = horned; referring to the pointy, paired bracts around the flower, which turn black and resemble a pair of horns

Resprouting, robust geophyte to 1 m. **Leaves** Cauline, overlapping, lanceolate, often barred with red near the base. **Flowers** Numerous in a close-packed raceme, silvery-yellow and purple, median sepal deeply hooded, with a pointy tip, spur to 2 cm long. Flowering in spring and summer. **Habitat** Loam-rich soils with some clay, in Transitional Shrubland and Grassy Fynbos. **Distribution** Widespread from the West Coast to Zimbabwe.

Toverwater, Kouga 03.11.2009

Disa harveyana subsp. *harveyana*   Lilac disa

Named after William Henry Harvey, 1811–1866, an Irish botanist who visited the Cape in 1834. He co-authored the first 3 volumes of *Flora Capensis* (published between 1860 and 1865) with Otto Wilhelm Sonder.

Resprouting, slender geophyte to 60 cm. **Leaves** Strap-shaped, 20 cm long, dry at flowering. **Flowers** Few in a lax raceme, cream or mauve, with purple streaks, spur slender, to 9 cm long. Flowering in midsummer. **Habitat** Infrequent on damp, shaded slopes in Grassy and Mesic Fynbos. **Distribution** Widespread from the Cape Peninsula to Kouga Mountains. **Notes** Pollinated by long-proboscid flies. 2 subspecies: subsp. *longicalcarata* is found around the Cederberg area in the Western Cape and flowers in spring.

Kouga Wildernis 18.12.2011

Disa lugens var. *lugens* Blue bonnet disa, green-bearded disa

lugens = mourning; alluding to the deeply
dissected, beard-like frills hanging from
the lower petal, giving the flower a
drooping, weeping appearance

Resprouting, slender, reed-like geophyte to
1 m. **Leaves** Basal, linear, 6 cm long × 0.2 cm
wide, dry at flowering. **Flowers** Few in a lax
raceme, each 3 cm wide, cream-green, with
frilly lip. Flowering in spring. **Habitat** Seldom
seen in young veld in Grassy and Mesic
Fynbos. **Distribution** Widespread from the
Cape Peninsula to Makhanda. **Notes** Pollinator
unknown. 2 varieties: var. *nigrescens* has
purplish-black flowers and is endemic to
coastal plains around Humansdorp.

East of Kougakop 14.02.2012

Disa porrecta

porrecta = stretched out, protracted; alluding to the long, gradually tapering spur on the flower

Resprouting, reed-like geophyte to 60 cm. **Leaves** Basal, linear, 30 cm long, developing after flowering.
Flowers Up to 15 clustered in a dense raceme, pointing downwards, pink, orange or red, with
ascending spur to 4 cm long. Flowering in late summer. **Habitat** High altitude in Mesic and Subalpine
Fynbos. **Distribution** Widespread from Kammanassie Mountains to Eastern Cape and Lesotho. **Notes**
Easily confused with *D. ferruginea*, but that species is endemic to the Western Cape.

Peaks west of Scholtzberg, Baviaans 06.03.2012

Disa spathulata subsp. *tripartita*   Oupa-met-sy-pyp

spathulata = shaped like a spatula; referring to the shape of the strange long lower lip

Tuberous geophyte to 30 cm. **Leaves** Many, basal, linear and narrow, and a few sheathing the stem. **Flowers** 1–5 in a loose raceme, varying from maroon to pale lime or green and blue, spur club-shaped, to 0.3 cm long, lip (the 'grandfather's pipe' of the common name) spathulate and 3-lobed at the end, twisted and wavy. Flowering in spring. **Habitat** Localised in loamy and sandy soils in Transitional Shrubland and Grassy Fynbos. **Distribution** Widespread from Worcester to Langkloof. **Notes** Subsp. *spathulata* is found only between Namaqualand and Worcester and usually has a less divided and shorter-lobed lip. **Status** Endangered.

Louterwater, Langkloof 17.09.2011

DISPERIS

dis = twice, *pera* = bag, pouch; referring to the pouches in the 2 lateral sepals; these pouches do not contain a reward, but serve to protect the floral parts in bud.

Oil is secreted from the tip of the lip appendage, attracting oil-collecting bees that pollinate the flowers. 78 species, all in Africa except 1; 26 species in South Africa; 3 species in Baviaanskloof.

Disperis macowanii

Named after Peter MacOwan, 1830–1909, a professor of chemistry at Huddersfield College in West Yorkshire, England. He relocated to Cape Town where he became a productive botanist; 20 plants have been named after him.

Tiny tuberous geophyte to 10 cm, with hairy stem. **Leaves** 2, egg-shaped, alternate, spreading, 2 cm long, purple below, with hairy margins. **Flowers** 1 flower, less than 1 cm long, white or mauve, with magenta lip apex. Flowering in autumn and winter. **Habitat** Sporadic on shaded cliffs and ledges in Forest, often on moss banks. **Distribution** Widespread from Cape Peninsula to Drakensberg.

Nooitgedacht, Kouga 12.05.2021

EULOPHIA

Harlequin orchid

eulophia = good crest; referring to the crest on the lip of the flower

± 250 species in tropical and subtropical parts of the world; ± 27 species in South Africa; 2 species recorded in Baviaanskloof, although it is likely that there are more to be found.

Eulophia tenella

tenellus = delicate; referring to the slender and delicate habit

Rhizomatous geophyte to 20 cm, seldom taller. **Leaves** Linear-lanceolate, 30 × 0.5 cm. **Flowers** Few to many in a compact raceme, sepals dark green to brownish purple, petals pale straw outside and pale brown inside, lip with a bright yellow crest. Flowering in summer. **Habitat** Uncommon in grassy Transitional Shrubland and Grassy Fynbos. **Distribution** Widespread from Joubertina to Zimbabwe.

Joubertina 26.12.2012

Eulophia tuberculata

tuberculate = covered with wart-like projections; referring to the appearance of the yellow crest on the lip

Rhizomatous geophyte to 40 cm. **Leaves** Tough, leathery and succulent, only partly developed at flowering time, 20 × 2 cm, margins often tinged with maroon. **Flowers** Few to many in a loose raceme, sepals yellow-green with brown or purple, petals pale yellow with reddish veins inside, lip crest bright yellow. Flowering in spring and early summer. **Habitat** Exposed places and rocky outcrops in grassy Transitional Shrubland and Thicket. **Distribution** Widespread from Mossel Bay to tropical Africa.

Joubertina 25.11.2012

HABENARIA

habena = strap, reins, *aria* = possessing; pertaining to the long, narrow, strap-like divisions of the petals

839 species worldwide, 239 of which are in Africa; 30 species in South Africa; 1 species in Baviaanskloof.

Habenaria arenaria

arena = sand, *aria* = possessing; referring to the sandy-soiled habitat of this species

Slender, tuberous geophyte to 40 cm. **Leaves** Few, mostly basal, spreading almost flat, shiny, 17 cm long × 5 cm wide, sometimes mottled with grey. **Flowers** Several in a lax raceme, greenish, lip 3-lobed, spur 2 cm long. Flowering in autumn and early winter. **Habitat** Infrequent in well-shaded areas under Forest. **Distribution** Widespread from Riversdale to Mpumalanga.

Waterkloof, Kouga Wildernis 14.05.2021

HOLOTHRIX
Thread orchid

holos = entire, *thrix* = hair; alluding to the hairiness of most parts, entirely hairy

± 46 species in Africa, Madagascar and Arabia; 23 species in South Africa; 4 species recorded in Baviaanskloof.

Holothrix burchellii

Named after William John Burchell, 1781–1863, an English explorer who collected plants in southern Africa and Brazil. He amassed over 50,000 plant specimens in southern Africa and donated them to Kew Gardens, each record with meticulous, detailed notes on habit and habitat. He described his journey in a 2-volume work published in 1822 and 1824, *Travels in the Interior of Southern Africa*.

Resprouting, thin, tuberous geophyte to 50 cm. **Leaves** 2, basal, flat, paired, ovate, 6 cm wide. **Flowers** Numerous, in a dense raceme, each flower 1 cm long, greenish to cream, with petals and lip divided into many filiform lobes. Flowering in spring. **Habitat** Seldom seen in grassland in Transitional Shrubland. **Distribution** Widespread from Piketberg to Hlanganani.

Tulpieskraal, Kouga 19.10.2015

Holothrix mundii

Named after Johannes Ludwig Leopold Mund, 1791–1831, a Prussian pharmacist, botanist and land surveyor who was sent to South Africa in 1816. He collected specimens around Cape Town and as far east as Kariega.

Small, tuberous geophyte to 15 cm. **Leaves** 2, roundish, flat on the ground, 2 cm wide, fleshy, margins hairy. **Flowers** Few to several in a congested raceme on a hairy stalk, small, white, lip with 7 broad lobes. Flowering in spring. **Habitat** Infrequent in Grassy Fynbos and grassy Transitional Shrubland. **Distribution** Widespread from Piketberg and the Cape Peninsula to Gqeberha (Port Elizabeth).

Joubertina 07.10.2017

Holothrix pilosa

pilosa = hairy; pertaining to the fine hairs on the leaves, flower stalk and sepals

Tuberous geophyte to 50 cm. **Leaves** Single, sometimes 2, round, flat, 1.4 cm wide, margins and lower surface hairy, usually dry at flowering. **Flowers** Many, mostly 1-sided, stalk and sepals softly hairy, petals creamy white with a greenish central stripe, with up to 7 short lobes. Flowering from spring to autumn. **Habitat** Localised in dry, rocky places in Grassy Fynbos and Transitional Shrubland. **Distribution** Widespread from Riviersonderend to Gqeberha (Port Elizabeth).

Joubertina 03.01.2011

Holothrix schlechteriana

Named after Friedrich Richard Rudolf Schlechter, 1872–1925, a German botanist who worked with Harry Bolus in the Bolus Herbarium from 1891–1897. He also explored other parts of the world and described over 1,000 species of orchid. He published over 300 scientific papers.

Tuberous geophyte to 30 cm. **Leaves** 2, round, flat, to 10 cm wide. **Flowers** Profuse, mostly 1-sided, on a bracteate stalk covered in short, velvety hairs, sepals green, petals and lip green or yellow, divided into many filiform lobes. Flowering in spring and summer. **Habitat** Localised in the shade of shrubs or rocks in Succulent Karoo and Transitional Shrubland. **Distribution** Widespread from Namaqualand to Makhanda.

Gamtoos Valley, Patensie 14.05.2006

Satyr orchid, trewwa

Satyrion = the two-horned creature from Greek mythology, who is half man–half goat; pertaining to the pair of spurs on the flower, this was the name used for an orchid by Dioscorides and Pliny.

± 91 species, mostly in Africa and a few in Asia; 41 species in South Africa; 5 species recorded in Baviaanskloof.

Satyrium bracteatum   Bracket satyre

bracteatum = with bracts; pertaining to the leaves on the stem that look like bracts below the flowers

Resprouting, small, tuberous geophyte to 15 cm. **Leaves** 3–8, reducing in size up the stem, lanceolate. **Flowers** Several in a fairly dense raceme to 6 cm long, yellow with reddish-brown stripes, margins often tinged reddish; with pouched spurs and spreading bracts above. Flowering in spring. **Habitat** Sporadic in Grassy Fynbos, usually on marshy or peaty ledges. **Distribution** Widespread from the Cederberg and Cape Peninsula to Limpopo. **Notes** Pollinated by flies.

Nooitgedacht, Kouga 28.09.2011

Satyrium jacottetiae

Named after Hélène A. Jacottet, c. 1867–?, a Swiss botanist who collected in and around Lesotho; she collected the type specimen.

Tuberous geophyte to 50 cm. **Leaves** 2, broadly egg-shaped to elliptic, to 12 cm long, sheaths on the stem dry and membranous. **Flowers** Several to many, petals white to pale pink, fringed with thick hairs, spurs 2.5 cm long, bracts becoming dry and deflexed. Flowering in spring. **Habitat** Occasional on open slopes in Grassy Fynbos. **Distribution** Widespread from Swellendam to Matatiele. **Notes** This species was once included with *S. membranaceum*, but further study showed it to be a distinct species.

Joubertina 02.10.2009

Satyrium longicolle

longi = long, *collum* = neck; referring to the long, spreading 'neck' of the flower, comprising the spur and sepals

Resprouting, slender, stout, tuberous geophyte to 30 cm. **Leaves** 2, flat on the ground, ovate to roundish, to 10 cm long. **Flowers** Numerous in a lax raceme, pale pink to white with purple markings, anthers dark purple, spurs to 3 cm long, with deflexed bracts. Flowering in spring and early summer. **Habitat** Uncommon on damp slopes in Grassy and Mesic Fynbos. **Distribution** Widespread from Riversdale to Makhanda.

Koud Nek's Ruggens, Kouga 20.12.2011

POACEAE

Grasses

The grasses are the fifth-largest plant family in the world with ± 780 genera and ± 12,000 species. These were the first plants to be cultivated, about 10,000 years ago. Primarily cereal crops, the grasses include species such as maize, wheat, rice, barley, millet and bamboo. Grasses fuel the frequent fires that happen in savanna or grassland ecosystems. All the grasses described here, except the annuals, can resprout after fire. Grasses are most common and abundant in the first few years after fire. They utilise the habitat exposed by fires, as well as the nutrients that are released in the soil. As soon as the surrounding shrub species grow taller and begin to cast a shadow, the shaded grasses tend to become moribund and less prevalent in the veld. Some scientists believe that the evolution of bipedalism in apes was partly an adaptation to outrun the fires that burnt through grassy ecosystems; without grass, hominids may never have evolved.

There are ± 200 genera and 1,000 species of grass in southern Africa; ± 70 species of indigenous grass and over 10 exotic species in Baviaanskloof.

A unique set of terms is used in the description of grasses; the basic ones are illustrated here. Grass can be described as tufted or creeping, with hollow stems and long, narrow leaves. The base of the leaves clasp the stem to form a leaf sheath, and there is a ligule at the point where the sheath ends and the blade begins. Ligules are useful parts to help with grass identification, because they are absent on restios and sedges (both of which are easily confused with grasses). The texture and hairiness of the ligule varies between grass species, so they are another useful diagnostic feature.

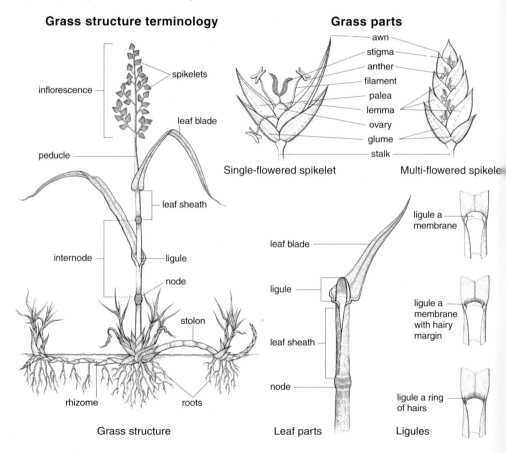

Grass structure terminology

- inflorescence
- spikelets
- peducle
- leaf blade
- leaf sheath
- internode
- ligule
- node
- stolon
- rhizome
- roots

Grass structure

Grass parts

- awn
- stigma
- anther
- filament
- palea
- lemma
- ovary
- glume
- stalk

Single-flowered spikelet

Multi-flowered spikele

- leaf blade
- ligule
- leaf sheath
- node

Leaf parts

- ligule a membrane
- ligule a membrane with hairy margin
- ligule a ring of hairs

Ligules

Both the structure and the shape of a grass inflorescence help with identification. There are four main kinds of grass inflorescences, illustrated below: unbranched (spike or raceme), panicle, digitate panicle and false panicle. Spikelets are the flower-bearing parts of a grass. Their size, arrangement on the inflorescence, and hairiness, together with the presence of awns, are key features to look at when identifying grasses.

Inflorescence types

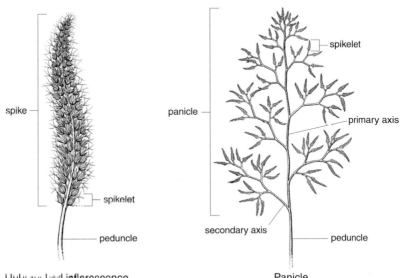

Unbranched inflorescence

Panicle

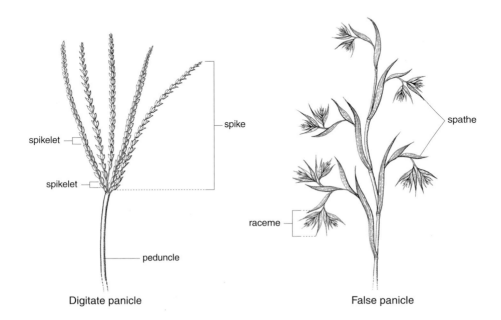

Digitate panicle

False panicle

ARISTIDA

Three-awn wiregrass, steekgras

arista = awn or bristle; referring to the 3 long awns on the lemmas of this genus

Cosmopolitan genus with over 300 species. These are unpalatable grasses that become more prevalent when the veld is overgrazed and/or burnt too frequently. The 3-awned seed is an easy aid to identification. 4 species in South Africa; all 4 species found in Baviaanskloof.

Aristida congesta Katstertsteekgras

congesta = congested; referring to the compact inflorescence that looks like a cat's tail, hence the Afrikaans common name

Small, slender, short-lived, densely tufted grass to 60 cm. **Leaves** Linear, rolled and wiry, with a few long white hairs near the ligule; ligule a fringed membrane. **Inflorescence** Congested panicle. Spikelets to 3 cm long including the awns. Flowering from summer to autumn. **Habitat** Common and abundant on lower, open, barren, overgrazed and/or recently burnt slopes. **Distribution** Widespread throughout southern Africa and in tropical Africa.

Uitspan, Baviaans 19.02.2016

Aristida vestita Large woolly three-awn

vestio = to clothe; referring to the white-woolly leaf sheaths and culms near the ligule

Densely tufted to 80 cm. **Leaves** 25 cm long × 0.4 cm wide. Culm internodes woolly-hairy; leaf-sheath hairs woolly; ligule a fringe of hairs. **Inflorescence** Open panicle, 15—30 cm long. Spikelets to 5 cm long including the awns. Flowering mostly from summer to autumn. **Habitat** Open, dry, rocky slopes in Transitional Shrubland and Thicket. **Distribution** Widespread throughout southern Africa to Tanzania. **Other name** Harde steekgras.

Rooikloof, Kouga 24.04.2018

ARUNDINELLA

arundo = reed, *nella* = diminutive; referring to the resemblance to a small thin reed

55 species in tropical and subtropical parts of the world; 1 species in South Africa.

Arundinella nepalensis   Riviergras, beesgras

nepalensis = from Nepal; the type specimen was collected in Nepal.

Perennial, upright grass with unbranched culms to 150 cm, with creeping rhizomes. **Leaves** 50 cm long × 0.5 cm wide, usually hairy, especially the sheath. **Inflorescence** Contracted panicle with the main stem much thicker than the branches. Spikelets 0.5 cm long, awn 0.5 cm long, pale green to straw-coloured. Flowering in late summer. **Habitat** Forming dense stands in marshes and riverbeds. **Distribution** Very widespread from Kouga Mountains through Africa to Asia, China and Japan. **Notes** Livestock may graze early growth. Also useful for stabilising riverbanks.

Kleinrivier, Kouga Wildernis 14.05.2021

ARUNDO*

arundo = Latin name for a reed

Robust, reed-like grasses with 6 species native to Europe, North Africa and Asia; 1 invasive weed in Baviaanskloof.

*Arundo donax**   Spanish reed, giant reed

donax = Greek name for a reed

Robust perennial, rhizomatous, reed-like, to 6 m. **Leaves** Cauline, drooping, 80 cm long × 7 cm wide, lobed at the base; ligule a fringed membrane. **Inflorescence** Fluffy panicle to 60 cm long. Spikelets 1.5 cm long. Flowering all year round but mostly late summer and autumn. **Habitat** Invasive. Declared weed along watercourses in southern Africa. **Distribution** Native to West and Central Asia and temperate East Asia; now invasive in many parts of the world. **Notes** The hollow culms are used for fences and ceilings. The plant is also used medicinally.

leaf lobed at base

clumps of Spanish reed

Kouga Kliphuis, Kouga River 13.05.2021

CAPEOCHLOA (= *Merxmuellera*)

Capeo = Cape, *chloa* = a grass; restricted to the Cape flora; the name refers to the geographical origin of the genus.

3 species in the Cape; 2 species in Baviaanskloof.

Capeochloa arundinacea Olifantsgras

arundo = reed, *acea* = resembling; referring to the reed-like nature of this species

Robust, tufted, reed-like perennial to 1.3 m tall, sometimes forming dense colonies. **Leaves** Tough, 60 cm long × 7 cm wide, pale green, with razor-sharp margins; ligule a fringe of hairs. **Inflorescence** Large, in a dense panicle, 25 cm long, straw-coloured. Spikelets to 1.6 cm long; lemmas and glumes covered with white hairs. Flowering in spring. **Habitat** Common on open slopes in Grassy and Arid Fynbos. More prevalent on slightly deeper soils on the gently sloping rocky ridges. **Distribution** Widespread from Bokkeveld (Nieuwoudtville) to Gqeberha (Port Elizabeth). **Notes** This is the only large, robust grass on dry Fynbos slopes in the area. It could be confused with *C. cincta*, but that species is larger and restricted to stream banks.

Scholtzberg, Baviaans 23.12.2011

CENCHRUS (= *Pennisetum*)   Sandbur

kenchros = millet; the unbranched, spike-like inflorescence of this species resembles that of millet grass, most likely pearl millet, *C. americanus* (= *Pennisetum glaucum*).

92 species in tropical and warm, temperate regions around the world; 8 native species and 2 exotics in South Africa; 2 native and 2 exotic species in Baviaanskloof.

Cenchrus caudatus (= *Pennisetum macrourum*)   Riverbed grass

caudatus = tailed; referring to the large tail-like inflorescence

Robust, densely tufted perennial to 1.5 m. **Leaves** Drooping, rough, 60 cm long × 1 cm wide, with rolled tips; ligule a fringe of hairs. **Inflorescence** Dense, long spike to 30 cm. Spikelets 0.6 cm long, surrounded by bristles. Flowering in summer and autumn. **Habitat** Marshes and stream banks. **Distribution** Widespread throughout southern and tropical Africa. **Notes** Useful for stabilising eroding stream banks.

Mistkraal, Baviaanskloof River 02.02.2020

Cenchrus ciliaris   Foxtail buffalo grass

ciliaris = like an eyelash, short hairs; referring to the fine wavy bristles surrounding the spikelets

Densely tufted perennial with much-branched culms, to 1 m. **Leaves** 25 cm long × 0.8 cm wide, blue-green, palatable. **Inflorescence** Spike-like, bristly, 12 cm long, purple to strawcoloured. Spikelet 0.5 cm long, surrounded by numerous wavy, scabrid or plumose bristles. Flowering from spring to autumn. **Habitat** Sandstone on open, hot, dry, lower slopes and valley floor in Thicket and Transitional Shrubland. **Distribution** Very

Vlakkie, Zandvlakte 27.02.2010

widespread throughout southern Africa and India. **Notes** A useful fodder grass for livestock. It has been used for restoring overgrazed Spekboomveld in Baviaanskloof. Fast growing, becoming abundant after good rains, but disappears during droughts. **Other names** Bloubuffel, buffelsgras (A).

Cenchrus clandestinum* (= *Pennisetum clandestinum*)   Kikuyu

clandestinum = hidden; the inflorescence is partly hidden in the upper leaf sheath. The common name was given by a tribe in Kenya where this species was first collected for propagation.

Mat-forming, leafy perennial to 30 cm. **Leaves** Leaf sheaths overlapping, leaves linear-oblong, 20 cm long × 0.7 cm wide, slightly hairy, usually folded; ligule a fringe of hairs **Inflorescence** Obscure, enclosed in upper leaf sheaths. Spikelets to 2 cm long. Flowering from spring to autumn. **Habitat** Riverbanks and irrigated areas near homesteads, also on road verges. **Distribution** Originally from the East African highlands, now a cosmopolitan weed. **Notes** Used as a garden lawn and for pasture. It forms monospecific stands where it can become invasive, crowding out the natural vegetation.

23.06.2020

Cenchrus setaceus* (= *Pennisetum setaceum*)   Fountain grass

setaceus = bristled; referring to the purplish bristles on the inflorescence

Densely and neatly tufted perennial to 1 m. **Leaves** Rolled and wiry, 40 cm long × 0.2 cm wide, unpalatable; ligule a fringed membrane. **Inflorescence** Dense spike, purple, bristly-hairy. Spikelets 0.6 cm long. Flowering in summer, autumn and winter. **Habitat** Invading roadsides in dry, rocky areas. **Distribution** Native to North Africa. Introduced as an ornamental grass in many parts of the world where it has become a troublesome weed. Spreading fast in southern Africa; first recorded in Baviaanskloof in 2018, where it is rapidly becoming a problematic invader.

Keerom, Baviaans 26.10.2018

CYMBOPOGON

Turpentine grass

kymbe = boat, *pogon* = beard; referring to the hairy spikelets that protrude from the boat-shaped spathes

The popular culinary herb, lemongrass, is from the tropical Asian species *C. citratus*.
± 40 species in Africa and Asia; 6 species in southern Africa; 3 species in Baviaanskloof.

Cymbopogon pospischilii (= *C. plurinodis*) Bitter turpentine grass

Named after a certain A. Pospischil, who collected in East Africa

Densely tufted perennial to 1 m. **Leaves** 35 cm long × 0.4 cm wide, blue-green, often folded, releasing a turpentine smell when crushed; ligule a fringed membrane. **Inflorescence** Many clustered, rather disorganised paired racemes, to 30 cm long, in purple-rust-coloured leafy spathes. Spikelets 0.6 cm long, with hairy pedicels. Flowering from spring to autumn. **Habitat** Common on the lower, rocky sandstone slopes in Thicket and Transitional Shrubland. **Distribution** Widespread in southern Africa and tropical Africa. **Notes** Easily confused with *C. marginatus*, usually found higher up, growing in Fynbos; its leaves are not folded.

Rooikloof, Kouga 24.04.2018

CYNODON

Quick grass

kuon = dog, *odon* = tooth; derivation unresolved

8 species in tropics and subtropics; 6 species in South Africa; 1 species in Baviaanskloof.

Cynodon dactylon Couch grass

dactylos = a finger; alluding to the finger-like inflorescence branches

Mat-forming perennial with rhizomes and stolons, to 30 cm tall, but usually flat on the ground and mostly only a few centimetres tall. **Leaves** 12 cm long × 0.4 cm wide; ligule a fringe of hairs. **Inflorescence** 4 or 5 racemes in a single whorl. Numerous awnless spikelets, each 0.2 cm long, neatly arranged on the underside of the raceme, purplish. Flowering mostly in late summer. **Habitat** Forming colonies in disturbed areas, mostly on the valley floor. It also forms grazing lawns in flat areas in the mountains. These are frequented by game. **Distribution** Widespread throughout Africa and a cosmopolitan weed. **Notes** An extremely useful grass that can endure heavy grazing. It is also used for urban and suburban lawns, land restoration, and in traditional medicine. **Other name** Kweekgras (A).

Misgunst, Antoniesberg 23.04.2018

DIGITARIA
Finger grass

digitus = finger, *aria* = resembling; referring to the paniculate inflorescence, which appears to be finger-like

± 200 species, tropical and subtropical; 25 species in South Africa; 1 species in Baviaanskloof.

Digitaria eriantha Woolly finger grass

erion = wool, *anthos* = flower; alluding to the slightly woolly spikelets

Tufted, palatable, perennial grass to 1 m tall, sometimes with long-hairy stolons. **Leaves** Linear to lanceolate, relatively soft, 40 cm long × 1 cm wide. **Inflorescence** 3–15 digitately arranged racemes on the end of tall spathes. Spikelets 3 cm long, without awns, finely woolly, alternately arranged along the ribbed racemes. Flowering in late summer and autumn. **Habitat** Fairly common in most habitats at mid- to lower altitudes. Most common in Transitional Shrubland. **Distribution** Widespread in southern Africa. **Notes** A valuable fodder grass and highly regarded as natural or cultivated pasture.

Rooikloof, Kouga 24.04.2018

EHRHARTA

Named after Jakob Friedrich Ehrhart, 1742–1795, a Swiss-born botanist who worked in Germany. He was a pupil of Carl Linnaeus and the first person to use the term 'subspecies' in botanical literature.

± 36 species in the southern hemisphere; ± 24 species in South Africa; 5 species in Baviaanskloof.

Ehrharta calycina Rooigras, polgras

kalyx = cup, *ina* = indicating possession; the glumes are equal in length to or longer than the lemma; together they form a cup shape.

Loosely tufted, highly palatable, perennial grass to 60 cm. **Leaves** Mostly basal, linear-lanceolate, soft, 30 cm long × 0.7 cm wide, with margins often undulate, especially near the base; ligule a fringed membrane. **Inflorescence** An open panicle, with flimsy secondary branches. Spikelets 0.6 cm long, green becoming magenta, hanging down and waving in the wind; glume as long as spikelet and villous lemma. Flowering mainly in spring. **Habitat** Most habitats on lower slopes and in the bottom of valleys. **Distribution** Widespread from Namaqualand to KwaZulu-Natal.

Rooikloof, Kouga 24.04.2018

Ehrharta dura Brandgras

dura = hard; pertaining to the tough, hard leaf blades

Tufted, rhizomatous, tough perennial to 80 cm. **Leaves** Flat, linear-lanceolate, 30 cm long × 1 cm wide. **Inflorescence** A verticillate panicle on a long, bendy spathe. Spikelets 2.5 cm long, including the long-awned lemma, green, hanging down. Flowering in summer. **Habitat** Restricted to south-facing slopes in Mesic and Subalpine Fynbos. **Distribution** Widespread from Tulbagh to Kariega (Uitenhage).

29.01.2019

Ehrharta erecta Shade ehrharta

erecta = held upright; referring to the inflorescence and spikelets, which are both erect

Delicate, sprawling or tufted, annual or perennial to 50 cm. **Leaves** Flat and soft, 15 cm long × 1 cm wide, with wavy margins; ligule a fraying membrane. **Inflorescence** A simple spike with spikelets held erect. Spikelets small, 0.4 cm long, with glumes shorter than lemmas, awns lacking. Flowering mostly in spring. **Habitat** Damp, shaded areas, often in the narrow ravines under Forest. **Distribution** The most widespread species of *Ehrharta*; widespread in southern Africa and East Africa. It has also spread to other continents where it has become naturalised and weedy. **Other name** Skadugras (A).

Waterkloof, Bokloof 26.04.2018

Ehrharta ramosa Pypsteelgras

ramus = branch, *osa* = abundance; alluding to the much-branched culms

Much-branched, rigid, shrub-like perennial to 1 m with olive-green culms. **Leaves** Seemingly absent, sheath-like, 6 cm long × 0.5 cm wide, fading white. **Inflorescence** A raceme-like panicle. Spikelets on ascending pedicels, glabrous, 0.7 cm long, with glumes usually shorter than lemmas, sometimes ending in a sharp point. Flowering in spring and summer. **Habitat** Common on rocky slopes in all types of Fynbos and Transitional Shrubland. **Distribution** Widespread from Cederberg to Kariega (Uitenhage).

Bosrug, Baviaans 27.10.2018

ENNEAPOGON

ennea = nine, *pogon* = beard; referring to the lemma that has 9 plumose hairs

Suurgras

± 30 species in pantropics and subtropics; 6 species in South Africa; 2 species in Baviaanskloof.

Enneapogon scoparius   Bottlebrush grass

scopae = several twigs or a broom, *aria* = pertaining to; referring to the contracted inflorescence that appears broom-like

Densely tufted, unpalatable, hardy perennial to 60 cm. **Leaves** Narrow, rolled, slightly hairy, 20 cm long × 0.3 cm wide; ligule a fringe of hairs. **Inflorescence** A spike-like contracted panicle to 12 cm long. Spikelets 0.4 cm long, lemma with 9 hairy awns. Flowering mostly in summer. **Habitat** Common on disturbed rocky, lower slopes in Transitional Shrubland, Thicket and Succulent Karoo. **Distribution** Very widespread in southern Africa. **Notes** Useful for stabilising loose soil on barren, overgrazed slopes. A common feature in certain parts of Baviaanskloof. **Other name** Borselgras (A).

Rooikloof, Kouga 24.04.2018

ERAGROSTIS

eros = love, *agrostis* = grass; alluding to the heart-shaped spikelets of some species

Love grass

Easily recognised by the laterally compressed spikelets with several closely overlapping, glabrous and awnless lemmas. ± 300 species, cosmopolitan; ± 65 species in South Africa; 6 species in Baviaanskloof.

Eragrostis capensis   Hartjiegras

capensis = from the Cape; referring to the geographical origin of the species

Tufted, moderately palatable perennial to 50 cm. **Leaves** Mostly basal, linear, open to rolled, 30 cm long × 0.5 cm wide, blue-green, slightly hairy near the sheath; ligule a fringed membrane. **Inflorescence** A raceme, 10 cm long, with relatively few spikelets on unbranched, bendy culms. Spikelets large, 1.5 cm long × 0.5 cm wide, greenish turning gold, heart-shaped. Flowering from spring to autumn. **Habitat** Common but seldom abundant, found in all fire-prone vegetation, mostly Grassy Fynbos. **Distribution** Widespread in the Cape, throughout southern tropical Africa and Madagascar. **Notes** An attractive grass in the garden. Available at some retail nurseries.

Rooikloof, Kouga 24.04.2018

Eragrostis curvula Weeping love grass

curvus = bent, *ula* = diminutive; referring to the curved leaf blades

Densely tufted, robust, well-rooted, moderately palatable, leafy perennial to 1 m. **Leaves** Basal sheaths densely hairy, open to rolled, hanging over, with tapering filiform tips, 40 cm long × 0.4 cm wide; ligule a fringe of hairs. **Inflorescence** An open to contracted panicle to 20 cm long. Spikelets linear to oblong, 0.5 cm long × 0.1 cm wide, olive-green, appressed to the branches. Flowering mostly in summer. **Habitat** Very common in all habitats, but most abundant in disturbed areas where the soil is deeper. **Distribution** Very widespread throughout southern and tropical Africa. **Other names** Berg-soetgras, blousaadgras, oulandsgras (A).

Misgunst, Antoniesberg 23.04.2018

Eragrostis obtusa Dew grass

obtusa = blunt; referring to the lemmas, which are obtuse

Tufted, moderately palatable perennial to 40 cm, with lower nodes of culms bent in a knee-like fashion. **Leaves** 15 cm long × 0.4 cm wide, with rolled margins and collared sheaths; ligule a line of short hairs. **Inflorescence** A lax to slightly contracted panicle, to 12 cm long. Spikelets heart-shaped, 0.5 cm long × 0.4 cm wide, with lower lemma obtuse to rounded. Flowering all year round except in June. **Habitat** Occurring mainly on the flattish plateaus in Baviaanskloof, in Transitional Shrubland where there is some clay or loam in the soil. **Distribution** Widespread in South Africa. **Other name** Kwaggakweek (A).

Engelandkop, Kouga 13.07.2011

Eragrostis racemosa Narrow heart love grass

racemus = bunch or cluster of grapes, *osa* = abundance; referring to the spikelets in contracted panicles

Densely tufted, moderately palatable perennial to 60 cm. **Leaves** Mostly basal, linear, 10 cm long × 0.4 cm wide, with basal sheaths sometimes slightly hairy; ligule a fringe of hairs. **Inflorescence** An open or contracted, sparsely branched panicle, 12 cm long, on sturdy upright culms. Spikelets elongated heart-shaped, 1 cm long × 0.4 cm wide, ascending, on short stout pedicels. Flowering from spring to autumn. **Habitat** Fairly common in shallow, stony soil in Grassy Fynbos and Transitional Shrubland. **Distribution** Very widespread throughout southeast Africa, Sudan and Madagascar. Populations in the Western Cape are naturalised weeds. **Other name** Smalhartjiegras (A).

31.12.2016

EUSTACHYS

eu = well, *stachys* = spike, resembling that of wheat; alluding to the inflorescence, which is like that of wheat – 1-sided, with 2 rows of spikelets; the Afrikaans common name refers to the inflorescence, which resembles the tracks left by chicken feet.

12 species in the Americas and Africa; 1 species in South Africa.

Eustachys paspaloides   Red Rhodes grass

Paspalum = plant genus, *oides* = resembling; the inflorescence resembles that of the inflorescence of genus *Paspalum* (crowngrass).

Tufted, highly palatable, short-rhizomatous, geniculately ascending perennial to 80 cm. **Leaves** Flat or folded, 15 cm long × 0.5 cm wide, blue-green, with compressed basal sheaths tinged with purple; ligule a fringe of hairs. **Inflorescence** Digitate, with 4–7 sturdy, spike-like fingers. Spikelets 0.3 cm long, golden brown, in 2 neat rows to 1 side of the rachis. Flowering from spring to autumn. **Habitat** Occurring mostly in grasslands on flattish plateaus in Transitional Shrubland and Grassy Fynbos. **Distribution** Very widespread from Worcester to the Arabian Peninsula. It has become a naturalised weed in the southern parts of the USA and in Brazil. **Notes** It may be confused with *Chloris gayana*, but that species is more erect, leafier and it has awned spikelets. **Other name** Bruinhoenderspoor (A).

Grasnek, Kouga 17.04.2012

FINGERHUTHIA

Named after Carl Anton Fingerhuth, 1798–1876, a German botanist and physician who described numerous botanical taxa

2 species in Africa; 1 species in Baviaanskloof.

Fingerhuthia africana   Thimble grass

africana = from Africa; referring to the geographical origin of the plant

Tufted, moderately palatable, smallish perennial to 90 cm. **Leaves** Open or folded, 35 cm long × 0.5 cm wide; ligule a fringe of hairs. **Inflorescence** A dense, spike-like raceme that coincidentally resembles a thimble. Spikelets drop from the top down, exposing the stalk and leaving hollows, reinforcing the thimble comparison. Spikelets 0.5 cm long, have glumes with recurved awns and hairy keels, and lemmas with rounded apices. Flowering from spring to autumn. **Habitat** Open, stony soil in Transitional Shrubland, Thicket and Succulent Karoo. **Distribution** Very widespread in southern Africa, tropical Africa, the Arabian Peninsula and Afghanistan.

Tchnuganoo, Kouga 30.04.2018

GEOCHLOA (= *Merxmuellera*)

geo = earth, *chloa* = a grass; referring to the geophytic habit of the genus

This genus of 3 species from the Western and Eastern Cape was newly described in 2010, based on molecular data. 1 species in Baviaanskloof.

Geochloa decora

decora = elegance; referring to the neat, tufted habit and the woolly base

Tufted perennial to 60 cm tall, with swollen white-woolly bases. **Leaves** Curling and twisting, tough, 20 cm long × 0.4 cm wide; ligule a fringe of hairs. **Inflorescence** A contracted panicle, 10 cm long. Spikelets to 2.5 cm long, containing several woolly-hairy florets. Flowering mostly after fire, in spring and summer. **Habitat** Occurring mostly on south-facing slopes in Grassy and Mesic Fynbos. **Distribution** From the Cape Peninsula to Cockscomb Peak in Baviaanskloof.

Cockscomb, Groot Winterhoek 17.02.2016

HETEROPOGON   Pylgras

hetero = different, *pogon* = beard; referring to the difference between the male and female spikelets; the female spikelet is awned and the male spikelet awnless.

6 species in Africa and southern Europe; 1 species in Baviaanskloof.

Heteropogon contortus

contortus = twisted; referring to the twisted and tangled spikelets

A moderately palatable tufted perennial to 1 m. **Leaves** Folded, with compressed sheaths, 30 cm long × 0.8 cm wide; ligule a fringed membrane. **Inflorescence** A spike-like raceme. Spikelets 1.2 cm long (excluding awns). Florets have dark, twisted awns to 8 cm long, which cluster together when dry. Flowering from spring to autumn. **Habitat** Transitional Shrubland and Grassy Fynbos. **Distribution** Widespread in tropical and subtropical regions.

Rooikloof, Kouga 24.04.2018

MELICA

melica = old Italian name for sorghum

70 species worldwide; 2 species in South Africa; 1 species in Baviaanskloof.

Melica racemosa   Haakgras, dronkgras

racemus = bunch or cluster of grapes, *osa* = abundance;
referring to the clustered appearance of the raceme

Tufted perennial, soft and weak
stemmed, to 1 m. **Leaves** Flat or
rolled, often scabrous, to 30 cm long
× 0.5 cm wide; ligule an unfringed
membrane. **Inflorescence** A single
raceme, sometimes paniculate.
Spikelets 0.8 cm long, silvery,
without awns; lemmas hairy on
margins. Flowering from spring to
autumn. **Habitat** Thicket, Fynbos
Woodland and Grassy Fynbos,
on rocky ridges and ledges, and
hanging off cliffs in shaded ravines.
Distribution Widespread from
Namaqualand to Lesotho.

Misgund, Willowmore 16.01.2022

MELINIS

meline = Greek name for millet, a general term used for a wide variety of small, seeded grasses used for
making bread

± 26 species, mostly around the Indian Ocean; 8 species in South Africa; 2 species in
Baviaanskloof.

*Melinis repens**   Natal red top

repens = sudden or unexpected; alluding to the plant's
tendency to grow rapidly in disturbed areas

Annual or weak perennial, growing in tuft or tussocks
to 1.2 m. **Leaves** Linear, 20 cm long × 1 cm wide, with
leaf sheaths not overlapping. **Inflorescence** Paniculate,
to 20 cm long. Spikelets 0.3 cm long, covered with long
and velvety red, pink-and-white hairs that shimmer in the
sun. Flowering from spring to autumn. **Habitat** Common
in disturbed areas on lower slopes and flats, often on
roadsides. **Distribution** Originally from southeast Africa, it
has spread throughout southern Africa and to most other
parts of the world. **Notes** Low grazing value, but useful for
stabilising disturbed ground quickly.

Rooikloof, Kouga 24.04.2018

MICROCHLOA

micro = small, *chloa* = a grass; referring to the small habit of this genus

6 species, 5 in southern Africa; 1 cosmopolitan species, described here.

Microchloa kunthii Pincushion grass
Named after Carl Sigismund Kunth, 1788–1850, a German botanist

Rus en Vrede, Kouga 25.04.2018

Densely tufted, perennial dwarf grass to 40 cm, with mostly basal leaves. **Leaves** Filiform and twisted, to 8 cm long × 0.1 cm wide; ligule a fringed membrane. **Inflorescence** A thin curved spike with all spikelets in a single row. Spikelets 0.3 cm long. Flowering from spring to autumn. **Habitat** Dry, stony open areas with shallow soil in Succulent Karoo and Thicket. **Distribution** Very widespread all around the world. **Other name** Elsgras (A).

MISCANTHUS   Dabagrass, ruigtegras

mischos = pedicel, *anthos* = flower; referring to the spikelets on short pedicels and the inflorescence on a long stalk

± 7 species in the tropics and subtropics; 2 species in South Africa; 1 species in Baviaanskloof.

Miscanthus ecklonii (= *M. capensis*)
Named after Christian Frederick Ecklon, 1795–1868, a Danish apothecary who became interested in botany while travelling in the Cape of Good Hope. He amassed a large collection of plants in South Africa, which he took back to Germany for further study.

Large, densely tufted, perennial grass to 2.5 m. **Leaves** Hard, drooping, to 100 cm long × 1.6 cm wide; ligule a fringed membrane. **Inflorescence** A contracted panicle, to 45 cm long, hairy. Spikelets to 0.6 cm long, with pilose glumes, florets with 1 awn. Flowering in late summer. **Habitat** Banks of perennial streams. **Distribution** Widespread from the Kouga and Baviaanskloof rivers to Eswatini.

Miskraal, Baviaanskloof River 23.08.2010

PANICUM Panic grass

panicum = millet; a general term used for a wide variety of small, seeded grasses used for making bread

Several species are grown for millet. Over 600 species in tropical and warm temperate regions; ± 33 species in South Africa; 3 species in Baviaanskloof.

Panicum maximum Guinea grass

maximum = greatest; alluding to the extremely tall culms

Loosely to densely tufted, leafy, highly palatable perennial grass to 2 m, which spreads by rooting at the nodes. **Leaves** Lanceolate, soft, 40 cm long × 2 cm wide, with leaf sheaths usually hairy. Ligule a fringed membrane. **Inflorescence** An open panicle with flimsy secondary branches and whorled lower branches. Spikelets 0.3 cm long, green to purple; fertile lemma pale and transversely rugose, a characteristic that distinguishes it from other *Panicum* species. Flowering from spring to autumn. **Habitat** Occurring mostly on lower slopes and the valley floor, in partial shade of Thicket or sweet thorn (*Vachellia karroo*). **Distribution** Widespread in the Western and Eastern Cape and tropical Africa. **Notes** One of the best grazing grasses. **Other name** Blousaad-soetgras (A).

Rus en Vrede 4X4, Kouga 25.04.2018

PASPALUM

paspalos = millet; a general term used for a wide variety of small, seeded grasses used for making bread

± 250 species, mainly growing in the tropics and subtropics; 3 indigenous species in South Africa; 3 exotic species in Baviaanskloof.

*Paspalum dilatatum** Dallis grass

dilatatus = broadened, expanded, widened; referring to the broad spikelets, and the spreading habit

Loosely tufted, short-rhizomatous, mat-forming perennial to 40 cm. **Leaves** Flat, soft, 15 cm long × 1 cm wide. Ligule a membrane to 0.4 cm long. **Inflorescence** Drooping, digitately branched, with a tuft of white hairs where the branch meets the main stem. Spikelets in 4 rows on each branch, each spikelet 0.4 cm long, roundish, with silky-hairy margins, and dark purple anthers and stigmas. Flowering in summer. **Habitat** Moist banks next to perennial rivers. **Distribution** Exotic weed originating from South America, now all over the world.

02.05.2021

Paspalum distichum* Water couch

distichum = in 2 rows; referring to the spikelets, neatly arranged in 2 rows

Mat-forming, creeping grass that grows to 30 cm in waterlogged soil. **Leaves** Linear, 20 cm long × 0.8 cm wide. Ligule a hairy membrane. **Inflorescence** Digitate with 2 arms, curving horizontally or downwards when mature. Spikelets glabrous, 0.3 cm long, arranged in 2 rows. Flowering in summer. **Habitat** Growing in or near water on riverbanks. **Distribution** Originally from tropical Africa and the Americas, now in warmer parts of the world. **Notes** A problematic weed in many river systems and well-watered, cultivated lands around the world. However, it has a robust root system that holds strong in rivers exposed to periodic scouring floods, and thus plays a stabilising role in these ecosystems. **Other name** Bankrotkweek (A).

Voetpadskloof, Baviaanskloof River 27.04.2018

Paspalum urvillei* Giant paspalum

Named after Jules Sébastien César Dumont d'Urville, 1790–1842, a French explorer who first collected this species in Brazil in 1822

Tufted, upright perennial to 2 m, without rhizomes. **Leaves** 50 cm long × 1 cm wide, with rough margins and basal sheath covered with stiff hairs. Ligule a pale brown membrane. **Inflorescence** 10–20 hanging racemes, up to 30 cm long. Spikelets 0.2 cm long, covered with fluffy white hairs. Flowering in summer and autumn. **Habitat** Damp areas on stream banks. **Distribution** Native to South America; now a problematic weed in many parts of the world. **Other name** Langbeen-paspalum (A).

Poortjies, Cambria 18.05.2021

PENTAMERIS (= *Pentaschistis*)

penta = five, *meros* = a part; referring to the lemmas that are 5-veined in some species

± 83 species in Africa and Madagascar; 62 species in South Africa; 11 species in Baviaanskloof.

Pentameris distichophylla (= *P. dregeana*)

distichous = 2 rows, *phyllon* = leaf; referring to the symmetrical arrangement of leaves in 2 rows on the stem

Tufted, curly-leaved, hardy, bushy perennial to 90 cm, with stems branching near the base. **Leaves** Densely leafy, with blades curling, rolled and hairy, 30 cm long × 0.2 cm wide, and leaf sheaths woolly. **Inflorescence** A loosely contracted panicle, to 10 cm long. Spikelets 2-flowered, less than 2 cm long. Flowering in spring and early summer. **Habitat** Rocky upper slopes of the mountains in Grassy, Mesic and Arid Fynbos. **Distribution** Widespread from Cederberg to Baviaanskloof. **Notes** Traditionally used as bedding in caves and mountain huts.

Bosrug, Baviaans 27.10.2018

Pentameris eriostoma

erion = wool, *stoma* = mouth; referring to the white-woolly leaf sheath and lower part of the culm

Tufted, tussock-forming, tough, unpalatable perennial to 90 cm. **Leaves** Basal, mostly rolled, 40 cm long × 1.5 cm wide, with white-woolly sheaths and sheath mouth. Ligule a dense fringe of hairs. **Inflorescence** An open panicle to 20 cm. Spikelets 1.2 cm long, on thin pedicels, with awned lemmas. Flowering in spring. **Habitat** Common and abundant on rocky slopes, this is the most frequently encountered grass in Baviaanskloof, in Arid and Grassy Fynbos. **Distribution** Widespread from Namaqualand to Eastern Cape. **Notes** The densely woolly sheaths make identification easy.

Ridge west of Elandsvlakte, Baviaans 29.09.2011

Pentameris macrocalycina

makros = large, *kalyx* = cup, *ina* = indicating possession; alluding to the large glumes

Tufted, branched, unpalatable perennial to 1 m. **Leaves** Erect, rigid, tightly rolled, filiform, 35 cm long × 0.2 cm wide. Leaf sheaths bearded at the mouth. **Inflorescence** A contracted panicle, 12 cm long. Spikelets to 2.5 cm long, 2-flowered. Flowering in spring and early summer. **Habitat** Common on upper slopes in Mesic and Subalpine Fynbos. **Distribution** Widespread from Cederberg to Kariega (Uitenhage).

Kouenek, Kouga 25.09.2011

Pentameris pallida

pallida = pale; pertaining to the spikelets

Small, tufted perennial to 40 cm. **Leaves** Usually rolled, sometimes with stalked glands, 20 cm long × 0.4 cm wide. **Inflorescence** A contracted panicle. Spikelets numerous, 0.5 cm long, awn 0.6 cm long. Flowering in spring. **Habitat** Mesic and Grassy Fynbos. **Distribution** Widespread from Namaqualand to Eastern Cape.

Voetpadskloof, Baviaans 27.04.2018

Pentameris rigidissima

rigidus = stiff, *issimus* = most; referring to the rolled and rigid leaves

A tough, hard-leaved grass to 30 cm. **Leaves** Small, rigid, 10 cm long × 0.1 cm wide. **Inflorescence** A contracted panicle. Spikelets 0.8 cm long, lemmas awned. Flowering in spring and summer. **Habitat** Occurring in rock crevices on the ridge lines of mountain ranges in Mesic Fynbos. **Distribution** Widespread from Cederberg to Baviaanskloof.

West of Bosrug, Baviaans 14.03.2012

PHRAGMITES

phragma = a hedge; referring to the tall habit and the numerous dense stems arising from a rhizome

3 species, cosmopolitan; 1 species in Baviaanskloof.

Phragmites australis   Common reed

australis = of the south; curiously, it has a worldwide distribution.

Robust, rhizomatous, reed-like perennial to 3 m. Culms thick and erect, unbranched. **Leaves** 45 cm long × 4 cm wide, flat, with persistent leaf sheaths. Ligule long, with a fringe of hairs. **Inflorescence** A plumose panicle to 40 cm long. Spikelets to 1.8 cm long, covered with silky hairs. Flowering in autumn. **Habitat** Forming dense stands on riverbanks, especially muddy rivers such as the Groot River. **Distribution** One of the most widely distributed plants in the world. **Notes** Used as thatching and for making mats and baskets. Can also be used in the chemical industry and to make paper. **Other name** Fluitjiesriet (A).

Blue Hill Escape, Kouga 21.06.2020

SETARIA
Bristle grass

seta = bristle, *aria* = possessing; referring to the bristles surrounding the spikelets

± 140 species, in pantropical and warm temperate areas; ± 14 species in South Africa; 1 species in Baviaanskloof.

Setaria sphacelata var. *torta*   Golden Timothy

sphacelata = speckled with brown or black; referring to the purple to black-tinged spikelets

Tufted or creeping perennial, moderately palatable, to 50 cm. **Leaves** Basal, compressed and overlapping, 30 cm long × 0.5 cm wide, pale blue-green, with dry leaves curling. Ligule a fringe of hairs. **Inflorescence** A spike-like panicle, 5 cm long, golden yellow, with persisting bristles surrounding the spikelets. Spikelets 0.3 cm long. Flowering from spring to autumn. **Habitat** Occurring sporadically in most habitats, but mostly in Transitional Shrubland and Grassy Fynbos. **Distribution** From the Cape to tropical Africa. **Notes** An extremely variable grass with 9 different taxa recognised within the species. **Other name** Borselgras (A).

Engelandkop, Kouga 13.07.2011

STIPA

stuppa = the coarse part of flax; the name refers to the long feathery awns of some species.

± 300 species, mostly in temperate to subtropical areas; 2 species in South Africa; 1 species in Baviaanskloof.

Stipa dregeana

Named after Johann Franz Drège, 1794–1881, a German-born botanical collector and explorer who was based in South Africa

Tufted perennial to 1 m. **Leaves** Flat, lanceolate, striated, 60 cm long × 1 cm wide, tapering to a fine point. Ligule an unfringed membrane. **Inflorescence** An open panicle, 20 cm long, with slender branches, drooping. Spikelets to 2 cm long, including the 1.5 cm-long awns. Flowering mostly in summer. **Habitat** Common in shady openings under Thicket or Forest, or under thicket clumps in Transitional Shrubland and Grassy Fynbos. **Distribution** Widespread from the Cape to the highlands of East Africa.

Rus en Vrede 4X4, Kouga 28.07.2011

TENAXIA (= *Merxmuellera*)

tenax = holding fast, tenacious, strong; referring to the well-rooted habit, and the tough, persistent, wiry foliage of most species

8 species in Afromontane regions, reaching the Himalayas; 5 species in South Africa; 2 species in Baviaanskloof.

Tenaxia stricta (= *Merxmuellera stricta*) **Bokbaardgras**

stricta = erect; alluding to the branches of the inflorescence, which are held erect

Tufted perennial, tough, glabrous, to 80 cm. **Leaves** Stiff, filiform, to 45 cm long × less than 0.1 cm wide. Ligule a fringe of hairs. **Inflorescence** An open or contracted panicle to 13 cm long. Spikelets 2 cm long, 3-awned. Flowering in spring and early summer. **Habitat** Common and abundant on sandy loam soils in Transitional Shrubland and Fynbos habitats; most prevalent in Grassy Fynbos. **Distribution** From Nieuwoudtville to the Cape Peninsula and KwaZulu-Natal.

Cockscomb, Groot Winterhoek 16.02.2016

THEMEDA

Rooigras

thamada = depression filled with water after rain (Arabic); derivation unresolved

18 species worldwide; 1 species in southern Africa.

Themeda triandra

tri = three, *andra* = man; referring to the 3 male spikelets in the clustered inflorescence

Highly variable, very palatable, tufted perennial to 75 cm tall, turning red when dry. **Leaves** 30 cm long × less than 1 cm wide, with compressed leaf sheaths. Ligule a notched membrane. **Inflorescence** Several clusters of spikelets hanging off thin branches, partially nestled in a spathe. Spikelets 0.7 cm long excluding awns; awns dark, twisted and up to 4 cm long. Flowering from spring to autumn. **Habitat** Common in Transitional Shrubland and Grassy Fynbos. **Distribution** Widespread throughout southern Africa, tropical Africa and Asia. The rusty red colour of grassy slopes and mountains in winter is largely due to the colour of the dry foliage of this species. **Other names** Umsinde (X), seboko (S), insinde (Z).

dark twisted awns

Nuwekloof Pass, Baviaans 08.10.2016

TRAGUS Kousklits

tragos = goat; alluding to the bristles on the spikelet, which are similar to a goat's beard; these burrs are often found in goats' fur.

1 species cosmopolitan; 6 species in Africa; the cosmopolitan species is found in Baviaanskloof.

*Tragus berteronianus** Small carrot-seed grass, burr grass

Named after Luigi Carlo Giuseppe Bertero, 1789–1831, an Italian naturalist, botanist and doctor. He was lost at sea, presumably in a shipwreck, while sailing from Tahiti to Chile.

hooked hairs

Loosely tufted, moderately palatable annual to 60 cm. **Leaves** Flat, short, 6 cm long × 0.5 cm wide, with wavy margins. Ligule a fringe of hairs. **Inflorescence** A narrow false spike to 15 cm long. Spikelets 0.3 cm long, awnless, burr-like, with rigid, hooked glume hairs. Flowering mostly in summer and autumn. **Habitat** Very common in disturbed areas in sandy and loamy soils in all habitats; more commonly found in non-Fynbos vegetation. **Distribution** Originally from the Americas and now a weed in many parts of the world. **Notes** Tenacious burrs often stick to socks and animal fur, a testament to the efficacy of the prickles as a dispersal mechanism.

Rooikloof, Kouga 24.04.2018

TRIBOLIUM

tri = three, *bolus* = fiery, arrow-shaped meteor; alluding to the spikelets with 3 florets

The spikelets are coarsely hairy, sometimes with awns. 14 species in southern Africa; 2 species in Baviaanskloof.

Tribolium hispidum

hispidum = hispid, roughly hairy; referring to the roughly hairy spikelets

Tufted, moderately palatable perennial to 40 cm. **Leaves** Convoluted, 15 cm long × 0.4 cm wide. Ligule a fringe of hairs. **Inflorescence** A contracted panicle, spike-like, 7 cm long. Spikelets 0.4 cm long, glumes very bristly-hairy, awnless. Flowering in spring and summer. **Habitat** Loamy soils. Most common in Transitional Shrubland and Grassy Fynbos, but also found in other habitats. **Distribution** Widespread from Namaqualand to Eastern Cape.

bristly hairs

Engelandkop, Kouga 19.04.2012

TRIRAPHIS

Needlegrass

tri = three, *raphis* = needle; referring to the spikelets, which in some species have 3 sharp awns

10 species in Africa and Australia; 4 species in South Africa; 1 species in Baviaanskloof.

Triraphis andropogonoides   Broom needlegrass

Andropogon = plant genus, *oides* = resembling; some parts resemble the
bearded or hairy male spikelets of the grass genus *Andropogon*.

Relatively hard, unpalatable, rhizomatous
perennial to 1.2 m. **Leaves** Linear, rolled, 30 cm
long × 1.2 cm wide. Ligule a fringe of hairs.
Inflorescence A dense, hairy panicle, to 30 cm
long, brown. Spikelets with 5–15 florets, 1 cm
long, hairy, shortly awned. Flowering from spring to
autumn. **Habitat** Infrequent and never abundant
in stony or sandy soil in Grassy Fynbos and
Transitional Shrubland. **Distribution** Widespread
from Potberg to Botswana and Limpopo. **Other
names** Koperdraadgras, besemgras (A).

Rooikloof, Kouga 24.04.2018

TRISTACHYA

Trident grass

tri = three, *stachys* = spike; referring to the arrangement of the spikelets in groups of 3

± 20 species in Africa, Madagascar and America; 3 species in South Africa; 1 species in
Baviaanskloof.

Tristachya leucothrix   Rooisaadgras

leucos = white, *thrix* = hair; referring to the long white hairs covering the plant

Densely tufted, moderately
palatable perennial to 90 cm.
Leaves Linear, 40 cm long
× 0.6 cm wide, silvery-hairy.
Basal leaf sheaths with dense
brown hairs. Ligule a fringe
of hairs. **Inflorescence** A
panicle with few spikelets.
Spikelets relatively large and
hairy, to 4 cm long, in clusters
of 3, central awn 5–10 cm
long. Flowering from spring
to autumn. **Habitat** Nutrient-
poor, sandy soils in Mesic and
Grassy Fynbos. **Distribution**
Widespread from Swellendam
to Mpumalanga.

Ridge west of Elandsvlakte, Baviaans 03.11.2011

UROCHLOA (= *Brachiaria*)

oura = tail, *chloa* = grass; pertaining to the lemma that is usually awned (tailed) or ending in a sharp point

± 12 species worldwide in tropics and subtropics; 8 species in southern Africa; 1 species in the Baviaanskloof.

Urochloa serrata (= *Brachiaria serrata*)　　　　Red-topped signal grass

serrata = serrated; referring to the serrated margins of the leaf

anthers

Rus en Vrede 4X4, Kouga 25.04.2018

Tufted, clump-forming, low-growing, rhizomatous perennial to 50 cm. **Leaves** Lanceolate, relatively short and wide, to 15 cm long × 1 cm wide at the base, crinkly-serrated. Ligule a fringe of hairs. **Inflorescence** A hairy raceme with short branches, to 2.5 cm long. Spikelets to 0.4 cm long, covered with purple-red hairs. Flowering from spring to autumn. **Habitat** Occurring mostly in Transitional Shrubland and Grassy Fynbos, in sandy-loam soils. **Distribution** Widespread from Bredasdorp to tropical Africa. **Other name** Ferweelgras (A).

RESTIONACEAE　　　　　　　　　　　　　　　　　　　Restios

A family of evergreen reeds mostly confined to the Fynbos biome. + 500 species of Restionaceae in the world, mostly in the southern hemisphere. Australia hosts ± 150 species from this family, ± 350 (almost 70 per cent) of which are from the Cape Floristic Region.

Restios have tufted, photosynthetic stems (culms) and the leaves are reduced to leaf sheaths. Flowers are small, contained in brownish spikelets found at the end of the culms. All members of the family are dioecious, with males and females on separate plants. The flowers are wind pollinated and the fruit is a nut or capsule.

The common presence of members of this family in South African vegetation is a most reliable indicator of Fynbos. Furthermore, restio species are also useful indicators of the different Fynbos communities or habitat types. Common names include restios, Cape reeds, biesies and dekriet. The latter, *Thamnochortus insignis*, is most famous as a thatching material, hence its Afrikaans common name 'dekriet', which means covering reed. Some species have recently become popular in suburban gardens.

19 genera and ± 350 species in southern Africa, mostly confined to the Western Cape; 9 genera and 29 species in Baviaanskloof.

CANNOMOIS

kanna = cane, *omois* = resembling; suggesting that the plants look like cane

This genus has relatively large nuts, flattened on 1 side, with elaiosomes for ant dispersal. A Cape Fynbos endemic with 13 species mostly in Western Cape; 2 species in Baviaanskloof.

Cannomois scirpoides Kouga cannomois

scirpus = rush, *oides* = resembling; the plant appears rush-like.

Resprouting, shortly rhizomatous, clump-forming, stiffly erect to 1 m. **Culms** Relatively thick and usually unbranched. Sheaths dark brown to almost red. **Inflorescence** Pale or straw-coloured in male. Female bracts hard, persistent, and dark purple. The membranous tepals comprise about one-tenth of the 1 cm-long black nut. Nut with a prominent white elaiosome, to 0.2 cm long, dispersed by ants. Flowering in autumn, winter and spring. **Habitat** Most common on well-drained, rocky and/or stony slopes, in Arid and Mesic Fynbos. **Distribution** Widespread from Touws River along the inland mountains to Cockscomb. **Other name** Dekriet (A).

Bosrug, Baviaans 14.03.2012

Cannomois virgata Bergbamboe, olifantsriet, besemriet

virgate = branching; referring to the branching habit

Robust, erect, rhizomatous, resprouting, mat-forming, to 1.5 m. **Culms** Hollow, much branched above. **Inflorescence** Female spikelets solitary, occasionally paired, to 3 cm long, acute. Tepals as long as black nut, 1 cm long, with a very small white elaiosome, dispersed by ants. Flowering in spring and summer, best after fire. **Habitat** Forming dense stands on upper, south-facing slopes, usually on shale, in Mesic Fynbos. **Distribution** Widespread from Porterville to Cape Peninsula and Kariega (Uitenhage). **Notes** Used to make brooms. The possibility of an undescribed species of *Cannomois* on the higher peaks of the Kouga and Baviaanskloof mountains needs further investigation.

Kouenek, Kouga 20.07.2011

ELEGIA (= *Chondropetalum*)

elegeia = song of lamentation (elegy); derivation unresolved

This genus is easily recognised by the loosely convoluted sheaths that often fall off early, leaving dark abscission rings at the nodes along the culm. Cape Fynbos endemic with ± 50 species; 5 species in Baviaanskloof.

Elegia capensis   Horsetail restio

capensis = from the Cape; referring to the geographical origin of the species

Spectacular, big plants with strong rhizomes, resprouting, forming dense stands to 3 m. **Culms** Branching, with whorls of thin, sterile culms at each node, thus the horsetail analogy. Nut brown, triangular in cross section, 0.2 cm long × 0.05 cm wide, without elaiosome. Flowering in spring or autumn. **Habitat** Seepage areas and along stream banks where there is groundwater. **Distribution** Widespread from Cederberg to Gqeberha (Port Elizabeth). **Notes** One of the more popular garden plants in this family. **Other names** Fonteinriet, katstert, kanet (A)

29.09.2020

Elegia filacea

filum = a thread; pertaining to the thin and bendy culms

Reseeding, tufted, neat, to 0.5 m, without spreading rhizomes. **Culms** Unbranched, thin and flimsy, bending easily in the wind. Spathes light brown, usually acute. Nut small, triangular in cross section, to 0.15 cm long, brown, without an elaiosome. Flowering mostly in summer. **Habitat** Uncommon in mountains in Baviaanskloof, but it can form large stands where it occurs. Only found in Mesic and Subalpine Fynbos, mostly on the upper, well-drained slopes of the mountains. **Distribution** Widespread from Cederberg and Cape Peninsula to Gqeberha (Port Elizabeth).

26.05.2021

Elegia juncea Vlerkiesriet

junceus = rush-like, made of rushes; suggesting
that this plant looks like rushes

Resprouting, clumped or mat forming,
with spreading rhizomes, to 1 m. **Culms**
Unbranched, slightly thicker and stiffer than
E. filacea. **Inflorescence** Spathes sometimes
with wavy margins and tending to fold over
backwards when flowering is over. Nut 0.2 cm
long, brown, without an elaiosome. Flowering in
summer. **Habitat** Mesic and Subalpine Fynbos,
mostly on the upper, well-drained, south-facing
slopes of mountains with higher rainfall in Mesic
Fynbos. **Distribution** Widespread from the
Kouebokkeveld Mountains to Cockscomb.

Cockscomb, Groot Winterhoek 16.02.2016

Elegia vaginulata

vagina = sheath, scabbard; pertains to the shape of the leaf sheaths

Resprouting, mat-forming plant with spreading rhizomes,
which distinguishes it from from tufted *E. filacea*. **Culms**
Unbranched, to 0.5 m tall. Nut 0.2 cm long, brown,
without an elaiosome. **Habitat** Damper sites in the
mountains, but not marshy areas. It also seems to prefer
the finer-textured soils of shale bands and cooler, south-
facing slopes in Mesic Fynbos. **Distribution** Widespread
from Clanwilliam to Makhanda. **Notes** Culms shorter and
thinner than those of *E. juncea*.

Cockscomb, Groot Winterhoek 17.02.2016

HYPODISCUS

Pineapple reed

hypo = beneath, *diskos* = disc; pertaining to the toothed or lobed disc that crowns the ovary and nut in some species, possibly for ants to grip during transportation to their underground nest. The inflorescences of some species resemble tiny pineapples.

Females have a single round nut per inflorescence, with elaiosome. The arrangement of the rows of teeth above the cylindrical nuts, and the surface texture of the nuts, are useful for identification of some species. 15 species from Namaqualand to Eastern Cape; 5 species in Baviaanskloof.

Hypodiscus aristatus

aristatus = hair; pertaining to the acuminate and awned bracts on the inflorescences

Resprouting, forming neat and dense tussocks up to 0.8 m. **Culms** Unbranched. **Inflorescence** Pineapple-like. Bracts awned, prickly/bristly, becoming tanned, white, shimmering in the sun. Tepals papery, same length as the 1 cm, oblong, smooth nut, with 0.3 cm-long elaiosome. Flowering in winter and spring. **Habitat** Seldom abundant in Baviaanskloof, preferring the upper northern slopes in Mesic Fynbos, where the soil is relatively deep, stony and well drained. Not found in seepages or areas that can get damp. **Distribution** Widely distributed and ubiquitous in the Cape Floristic Region from Clanwilliam to Baviaanskloof. **Notes** Can be mistaken for *H. albo-aristatus*, which has a single node per culm or no nodes at all.

Above Coleskeplaas, Baviaans 06.04.2012

Hypodiscus striatus

stria = furrow or groove; pertaining to the striate or vertically lined culms, which can be seen by holding the culms up against the light (best seen under magnification)

Resprouting, dense, clump forming with spreading rhizomes, with unbranched culms to 0.6 m. **Culms** Striated, dark green or olivaceous, always with a blue-grey appearance. Nut smooth-capped, shiny and torpedo-shaped, to 1 cm long, black, with a 0.3 cm-long white elaiosome. Flowering is from late summer through to winter. **Habitat** Very common in Baviaanskloof, becoming extremely abundant on the dry and rocky, well-drained, north-facing slopes in all Fynbos habitats, but especially in Arid Fynbos. An arid-tolerant species. Often the first restio that one encounters when moving from non-Fynbos into Fynbos habitats, especially on the north-facing slopes. Able to tolerate the finer-textured soils, therefore ubiquitous throughout Grassy Fynbos. **Distribution** Very widespread in the Cape mountains, from Kamiesberg to Kariega (Uitenhage). **Notes** The dense clumps are important, both for binding and trapping soil on eroding slopes, especially after fire.

nuts

elaiosome

Tafelberg, Rooihoek 10.03.2012

Hypodiscus synchroolepis

syn = with, together, *lepis* = scales; referring to the
interesting nuts, which have several rows of teeth
(or scales) on the surface

Reseeding, slender culmed, neatly tussocked,
to 0.7 m. **Culms** Relatively flexible and light
green in colour. Bracts awned, sheaths dark
brown in the middle and pale straw-coloured
on the margins. Nut to 0.5 cm long, with
several rows of ridges or teeth, bottom half
dark brown, or black, and the 2 top rows of
teeth lime-green. This nut lacks an elaiosome;
perhaps the lime-green teeth are fake
elaiosomes. Flowering in autumn. **Habitat**
Occurring sporadically on the upper slopes in
Mesic Fynbos. Usually absent from drier parts
of the mountains. **Distribution** Widespread on
mountains from Caledon to Makhanda.

Quaggasvlakte, Kouga 13.02.2016

MASTERSIELLA

Named after Maxwell Tylden Masters, 1833–1907, an English botanist and physician, and a fellow of the
Linnean Society of London and the Royal Society; he became a restio specialist at Kew Gardens.

There are 3 species in the genus, 2 of which occur in Baviaanskloof. Plants do not resprout after
fire. They have much-branched culms that tend to be rather weak or decumbent relative to those
of other members of the family. Male spikelets are somewhat reflexed; female spikelets are quite
different from the male and have a single flower. The seeds with elaiosomes are dispersed by ants.

Mastersiella purpurea

purpurea = purple; pertaining to the colour at the base of the plants and sometimes on the inflorescence

Reseeding, tufted, without spreading rhizomes, to 0.8 m. **Culms** Branched and erect. **Inflorescence**
Male racemose with up to 10 spreading to somewhat reflexed spikelets. Female has up to 10 spikelets,
with 1 flower per spikelet. Nut smooth, brown,
to 0.6 cm long, with a white elaiosome 0.3 cm
long. Flowering in autumn. **Habitat** Mesic Fynbos.
Distribution Fairly widespread on mountains from
around Montagu to Cockscomb.

Quaggasvlakte, Kouga 12.02.2016

Mastersiella spathulata

spathula = spoon; the reference is not clear, could allude to the slightly flattened culms near the tips, and the female inflorescence, which is spoon-shaped when in fruit.

Reseeding, tufted and decumbent, to 0.6 m. Plants do not have spreading rhizomes. **Culms** Branched, rather weak and drooping, pale blue-green. **Inflorescence** Male racemose with up to 10 spikelets, spreading or reflexed. Female with 1 to a few spikelets that are obtuse or obovate. Nut smooth, brown, to 0.3 cm long, elaiosome white and as long as the nut. Flowering in spring or autumn. **Habitat** Found on the upper slopes of the higher peaks, mostly south-facing slopes, in Mesic and Subalpine Fynbos. **Distribution** Widespread from the Agulhas plain to Cockscomb.

smooth nuts

Quaggasvlakte, Kouga 13.02.2016

PLATYCAULOS

platys = broad, flat, *kaulos* = stem; pertaining to the flattened culms of this genus

Very few genera have compressed culms like this genus. The variability in growth form within this genus is remarkable: some species form cushions similar to *Anthochortus*, while others form large erect tussocks. Usually associated with wet seepages and stream banks in the mountains. 12 species in Africa and Madagascar; 2 species in Baviaanskloof.

Platycaulos callistachyus

calli = beautiful, *stachyus* = spike; referring to the attractive inflorescence

Erect, up to 2.5 m tall, but with top parts drooping. Unique in the genus for its large size. **Culms** Much branched, especially upper ends, 0.7 cm wide at the base, and very thin (0.05 cm) at the apex. Tufted or clumped, with strong rhizomes for standing in flood waters and to support the large structure. **Inflorescence** Male racemose, with up to 10 spikelets, each 0.8 cm long. Female with 1 spikelet, to 2.5 cm long, with 5–15 flowers each with 3 white feathery styles. Nut round, 0.14 cm long, white, without an elaiosome. Flowering in autumn. **Habitat** Only found on stream sides in Fynbos. **Distribution** Widespread from Ceres to Gqeberha (Port Elizabeth).

Erasmuskloof River, Cockscomb 17.02.2016

RESTIO

restio = a ropemaker; the culms or rhizomes resemble ropes.

The largest and most diverse genus in the Restionaceae. The family was re-classified in 2010; genera such as *Calopsis* and *Ischyrolepis* have been included in *Restio*; 7 species are common and widespread, while *R. vallis-simius* is endemic. Relative to other genera, the male and female plants look fairly similar. A distinctive feature is the laterally compressed flowers and the villous keels of the outer lateral tepals. ± 167 species, 8 of which occur in Baviaanskloof.

Restio andreaeanus

Named after Hans Karl Christian Andreae, 1884–1966, a German-born scientist and chemist, who worked in Stellenbosch as an assistant to Dr HW Rudolf Marloth, helping with botanical work and collecting plants. His main interest was beetles and he spent many years as curator of Coleoptera for the South African Museum, Cape Town.

Quaggasvlakte, Kouga 13.02.2016

Reseeding, compact, tufted, without spreading rhizomes, to 1 m tall; mostly less than 0.6 m in Baviaanskloof. **Culms** Branched, light or lime-green in colour, with sheaths and spathes pale brown or orange. **Inflorescence** Male racemose or paniculate, with up to 10 spikelets. Female paniculate and up to 7 cm long, with up to 10 spikelets. Nut smooth, to 0.25 cm long, brown, without an elaiosome. Flowering in autumn. **Habitat** Arid-tolerant. Found on hot, north-facing slopes at the dry limit of Arid Fynbos. **Distribution** Not a very widespread species, confined to mountains around the Little Karoo and as far east as Baviaanskloof.

Restio capensis

capensis = from the Cape; referring to the geographical origin of the species

Resprouting, tufted, and without spreading rhizomes, to 0.6 m. **Culms** Sparsely branched, dark olive-green; membranous margins on the sheaths soon decay and fray. Vegetative or sterile growth present on young culms, especially near the base. **Inflorescence** Male and female spikelets with tiny, recurved awns on the bracts are a typical feature. **Habitat** Favouring finer-textured soils, common on shale bands in Grassy Fynbos. Tolerant of relatively dry conditions; can be found at the edges of Fynbos vegetation. **Distribution** Widespread throughout the Fynbos biome from Southern Namaqualand to Gqeberha (Port Elizabeth).

Nuwekloof Pass, Baviaans 13.10.2016

Restio gaudichaudianus

Named after Charles Gaudichaud-Beaupré, 1789–1854, a French naturalist. Despite extensive travels, he never worked in the Cape.

Resprouting, rather scruffy and tangled, up to 1.2 m. **Culms** Shiny, smooth and branching. **Inflorescence** Male racemose, to 5 cm long, with up to 50 spikelets. Male spikelets thin and tending to curve. Female inflorescence to 3.5 cm long, with up to 10 spikelets, each to 0.8 cm long. Nut elliptical in shape, silvery or brown, 0.24 cm long, with surface pitted, lacking elaiosome. Flowering in autumn. **Habitat** Rocky slopes and ledges at the arid end of Fynbos, often on the cooler, south-facing slopes. **Distribution** A very widespread species from the northern Cederberg to Cockscomb.

Uitspan, Baviaans 19.02.2016

Restio hystrix

hystrix = a genus of porcupines; referring to the prickly female spikelets

Reseeding, robust, compact, tufted, up to 1 m. **Culms** Smooth, sparsely branched, up to 0.4 cm wide at the base. **Inflorescence** Male to 15 cm long, with as many as 50 spikelets, each straight, thin, 2.5 cm long × 0.3 cm wide. Female inflorescence to 20 cm long, also with numerous spikelets, each 2.5 cm long × 1.2 cm wide. Female bracts awned. Nut 0.3 cm long, white, elliptical in side view, lacking an elaiosome. Flowering in winter and spring. **Habitat** Infrequently found on the upper slopes of the mountains in Mesic Fynbos. **Distribution** Common in the mountains around the Little Karoo, extending eastwards onto the Kouga and Baviaanskloof mountains.

Quaggasvlakte, Kouga 13.02.2016

Restio paniculatus

panicle = branching inflorescence

Plants killed by fire, sometimes resprouting. A grand plant to 2.5 m tall, forming large clumps on perennial riverbanks. Rhizomes spreading, to 2 cm wide. **Culms** Much branched, with numerous fine and sterile branches off the main culm. Sheaths closely convoluted, with membranous margins that soon decay and fray. **Inflorescence** Male paniculate, to 35 cm long, and with up to 500 spikelets. Female much the same, but the spikelets become wider as the seed sets. Male and female spikelets small, 0.5 cm long, with membranous margins on the bracts. Nut obovate in side view, to 0.15 cm long, black, without an elaiosome. Flowering in late autumn. **Habitat** Stream banks under Forest. **Distribution** Very widespread from Gifberg to Cape Peninsula and extending into KwaZulu-Natal.

Poortjies, Cambria 26.10.2018

Restio sejunctus

sejunctus = a separation; may be referring to the taxonomic process whereby this taxon was separated from a closely related species

Reseeding, with spreading rhizomes and much-branched culms, to 0.6 m. **Culm** Surface smooth or finely warty, olivaceous. **Inflorescence** Male with up to 10 sessile spikelets in a raceme, each spikelet 1 cm long, with yellow anthers. Female inflorescence similar but fewer spikelets, with white stigmas. Flowering in autumn. **Habitat** South-facing slopes in Grassy and Mesic Fynbos. **Distribution** Very widespread from Cape Peninsula to Lesotho.

Suuranysberg, Kareedouw 11.05.2021

Restio triticeus

triticum = wheat; a misnomer as this plant does not resemble wheat, other than being grass-like and tufted

Resprouting, tufted, to 60 cm, often with sterile growth at the base. **Culms** Branched, culm surface rough to the touch and often white-dotted, olive-green, sheaths closely convoluted, brown, with a short mucron tip, and membranous shoulders soon decaying. **Inflorescences** Few to many spikelets in a raceme. Male spikelets smaller and more numerous. Female spikelets usually slightly curved, flower keel villous. **Habitat** Common on rocky or stony, well-drained and loamy soils, in Grassy Fynbos and Sour Grassland. **Distribution** Widespread from Malmesbury to Makhanda.

Suuranysberg, Kareedouw 11.05.2021

Restio vallis-simius

vallis = valley, *simius* = monkey or baboon; referring to the Baviaanskloof where the plant was first recorded

Resprouting, rather diffuse plants, with single isolated shoots, to just over 1 m. **Culms** Branched, quite thick at the base (0.5 cm) and thin at the apex (0.05 cm). A distinctive feature is the way the culm curves upwards from the base, so the first 10–20 cm of culm is always curved. **Inflorescence** Male with up to 20 spikelets, each spikelet 1.5 cm long, erect and straight. Female inflorescence with up to 10 spikelets, each spikelet 2 cm long. Spathes and sheaths typically pale brown. Nut smooth, 0.2 cm long, without an elaiosome. **Habitat** The only known member of this family that is endemic to the area. Found in Mesic Fynbos on the upper, south-facing slopes of the Baviaanskloof range from Scholtzberg Peak across to Cockscomb (± 1,000 m above sea level). **Status** Rare.

Ridge west of Elandsvlakte, Baviaans 16.02.2016

RHODOCOMA

rhodon = rosy, *kome* = hair of the head; referring to the red or rosy styles of *R. capensis*, and the much-frayed sheaths

Similar to *Thamnochortus* as it also has pendulous male spikelets, but the female bracts in *Rhodocoma* are shorter. Endemic to South Africa; 3 of the 8 species occur in Baviaanskloof.

Rhodocoma capensis Katstertkanet

capensis = from the Cape; referring to the geographical origin of the species

Resprouting, forming colonies, spreading with rhizomes, to 1 m tall, but in Baviaanskloof around 0.6 m. **Culms** Distinctive short, thin branches in whorls at nodes on the main culm, giving a bushy appearance. **Inflorescence** Males and females look remarkably similar and are only told apart by close examination. Spikelets to 0.7 cm long. Flowering in spring and early summer. **Habitat** Found sporadically on shallow soil over rock slabs on the upper mountain slopes in Mesic Fynbos. **Distribution** Widespread on Cape mountains from Cederberg to Makhanda.

Kouenek, Kouga 20.01.2022

sheath

Rhodocoma fruticosa

fruticosus = bushy or shrubby; unsuitable description as the plant has unbranched culms and is not bushy

Resprouting, with spreading rhizomes, to 1 m. **Culms** Sheaths to 5 cm long, with membranous margins that decay and fray. **Inflorescence** Male has numerous pendulous spikelets that wave around in the wind on flimsy pedicels. Female inflorescence erect with up to 50 spikelets, each 0.8 cm long. Nut elliptical, colliculate or rugose, less than 0.2 cm long, brown or grey, and without an elaiosome. Flowering in autumn and winter. **Habitat** Common in all the Fynbos habitats, never becoming very abundant or dominant in the vegetation, and more common on the drier, rocky slopes. **Distribution** Very widespread from Cape Peninsula to KwaZulu-Natal.

Guerna, Kouga 16.02.2016

THAMNOCHORTUS

thamnos = bush, shrub, *chortus* = green herbage, grass, fodder; referring to the sterile vegetative growth found on the culms, giving the plant the appearance of being more palatable

The culms are mostly unbranched, and the plants form neat tufts with culms fanning outwards. The sheaths are tightly rolled on the culm, but with the upper membranous half soon decaying. Male spikelets are pendulous. Female spikelets are stiffly erect, with bracts taller than the flowers. A relatively large genus of ± 33 species, 2 of which occur in Baviaanskloof.

Thamnochortus rigidus

rigidus = rigid; pertaining to the firm tufts of sterile shoots, a feature that differentiates it from *T. cinereus*, which has long and soft sterile growth

Resprouting, tufted, culms unbranched, to 0.8 m. **Culms** Velvety pubescent, with tufts of sterile growth at the nodes. Rhizome short, keeping the culms well aggregated at the base. **Inflorescence** Male with up to 100 pendulous spikelets. Female with up to 50 spikelets, each to 4 cm long and stiffly erect. Flowering in autumn to spring. Nut elliptical, 0.2 cm long, brown, without an elaiosome. **Habitat** Common in most kinds of Fynbos, but most prevalent on stony sand, especially on the ridges, in Arid Fynbos. **Distribution** Widespread from Cederberg to Baviaanskloof, along the inland mountains.

Uitspan, Baviaans 19.02.2016

WILLDENOWIA

Named after Karl Ludwig Willdenow, 1765–1812, a German botanist, plant taxonomist, pharmacist, physician, naturalist, and professor of botany at the University of Berlin. He was the founder of phytogeography, which is the study of plant distribution.

12 species endemic to Cape Fynbos; 1 species in Baviaanskloof.

Willdenowia teres

teres = round in cross section; referring to the round culms and also the nut

Resprouting, with spreading rhizomes, to 50 cm. **Culms** Branched, smooth, without striations. **Inflorescence** Spikelets greyish. Nut visible, smooth, less than 1 cm long, with an elaiosome. Flowering in winter and spring. **Habitat** Occasional on well-drained stony slopes in Grassy and Mesic Fynbos. **Distribution** Widespread from Gifberg to Cape Peninsula and Baviaanskloof.

Kouenek, Kouga 20.01.2022

RUSCACEAE

A relatively small family with ± 6 genera and 200 species worldwide. Genus *Ruscus* is mostly from Europe and has 6 species; the foliage of some is used in the flower trade. Some species of *Eriospermum* are popular pot plants, with unusual leaves. Mother-in-law's-tongue (*Sansevieria hyacinthoides*) is common in gardens.

ERIOSPERMUM
Cottonseed

erion = wool, *sperma* = seed; alluding to the seeds that are covered with long, fine, woolly hair

Winter-growing, tuberous geophytes with leaves dry at flowering. The shape of the tuber and the leaf is useful for species identification. Flowers on thin racemes are not showy. The wind-dispersed woolly seeds are more noticeable in autumn. ± 102 species in sub-Saharan Africa; 4 species in Baviaanskloof.

Eriospermum capense

capense = from the Cape; referring to the geographical origin of the species

Tuberous geophyte to 30 cm when flowering. **Leaves** Heart-shaped, spreading, almost flat on the ground, glabrous, 7 cm long, dark green above and reddish below, without enations. **Flowers** White to pale yellow, on long pedicels. Flowering in summer and autumn. **Habitat** Occasional on heavier soils in Grassy Fynbos and Transitional Shrubland. **Distribution** Widespread from Namaqualand to Makhanda.

Rus en Vrede 4X4, Kouga 25.04.2018

Eriospermum ciliatum

ciliatum = ciliated, hairy; referring to the hairy leaf margins

Tuberous geophyte to 25 cm. **Leaves** Leaf flat on ground, ovate to heart-shaped, with hairy margins. **Flowers** Held upright, star-shaped, bright yellow. Flowering in autumn. **Habitat** Infrequent on sandstone slopes in Grassy and Mesic Fynbos. **Distribution** Narrow distribution from Baviaanskloof to Gqeberha (Port Elizabeth).

Ridge west of Elandsvlakte, Baviaans 16.02.2016

Eriospermum paradoxum   Haasklossie

paradoxum = paradox; referring to the leaf, which has a strange woolly, plumose appendage

Tuberous geophyte to 10 cm. **Leaves** Leaf blade small, with an erect, plumose appendage, partly covered with white-woolly hairs. **Flowers** Crowded, star-shaped, white, scented. Flowering in autumn. **Habitat** Infrequent on rocky slopes in Grassy Fynbos. **Distribution** Widespread from Namaqualand to Makhanda. **Notes** Photograph shows leaf appendage; the actual leaf is hidden below.

leaf appendage

Rus en Vrede 4X4, Kouga 25.04.2018

Eriospermum zeyheri

Named after Karl Ludwig Philipp Zeyher, 1799–1858, German botanist who collected plants and insects in South Africa for 22 years. Many of his specimens were lost in disastrous events. Best remembered as co-author of *Enumeratio Plantarum Africae Australis*, a catalogue of South African plants.

Tuberous geophyte to 50 cm. **Leaves** Prostrate, heart-shaped, new at flowering. **Flowers** 1 cm wide, crowded in spikes at the end of a long stem, white with green keels, fragrant. Flowering in late summer and autumn. **Habitat** Frequent on clay soils in Transitional Shrubland. **Distribution** Widespread from McGregor to Makhanda.

Bergplaas, Combrink's Pass 02.02.2020

SANSEVIERIA

Erroneously named by Thunberg after Raimondo di Sangro, Prince of San Severo, 1710–1771. The intention had been to name the plant after Pietro Antonio Sanseverino, 1724–1771, Duke of Chiaromonte, in honour of a garden of exotic plants that he established in Italy.

± 60 species, mostly in tropical Africa; 6 species in South Africa; 2 species in Baviaanskloof.

Sansevieria aethiopica   Wildewortel

aethiopica = from sub-Saharan Africa; in classical times this region had not yet been explored by Europeans.

Rhizomatous, stemless succulent to 50 cm. **Leaves** Several, sword-shaped, thickly succulent and tough, U-shaped in cross section, mottled with grey, with red-and-white margins. **Flowers** Whitish on elongated racemes. Fruit is a yellow-orange berry to 1 cm wide. **Habitat** Common in the understorey of dense Thicket. **Distribution** Very widespread

Vlakkie, Zandvlakte 09.12.2011

from Baviaanskloof to tropical Africa. **Notes** The roots and leaves are used medicinally. The other species, *S. hyacinthoides*, is well known by its common name 'mother-in-law's-tongue'. It has longer, flatter, less-succulent leaves and is only found eastwards of Kariega (Uitenhage).

TECOPHILAEACEAE   Cyanellas

A small family of cormous geophytes with 9 genera and 27 species in the world. Noticeable for the arrangement of 5 short anthers and 1 long anther. Pollinated by bees.

CYANELLA

cyaneus = greenish blue, *ella* = diminutive; pertaining to the colour of the flowers in some species

8 species in South Africa and Namibia; 1 species in Baviaanskloof.

Cyanella lutea subsp. *lutea*   Five fingers, lady's hand

lutea = yellow; referring to the yellow flowers

Cormous geophyte to 25 cm. **Leaves** 4–6 in a basal rosette, linear-lanceolate, ribbed, with wavy margins. **Flowers** In a branched raceme, each 1.5 cm wide, yellow. Flowering in spring. **Habitat** Frequent on richer soils in Succulent Karoo, Thicket and Transitional Shrubland. **Distribution** Widespread from Roggeveld to Kouga Mountains. **Notes** Corms edible and eaten by porcupines and francolins. *C. lutea* subsp. *rosea* has pinkish flowers and is found from southern Namibia through the arid interior to East London. **Other name** Geelraaptol (A).

Bosrugkloof, Baviaans 22.09.2006

DICOTYLEDONS

Dicots

The dicotyledons – or dicots – are the largest of the two major groups of flowering plants. A cotyledon is the first leaf that appears when a seed germinates. The dicots group is also referred to as the eudicots or 'true dicots' because the scientific world only recently established, through DNA studies, that several families assumed to be dicots actually belong in the palaeodicots (basal angiosperms).

Dicots have net-veined leaves. Their flower parts are always in multiples of four or five. Dicot pollen has three or more pores. The vascular bundles (growth tissue) inside the stems of dicots are arranged in rings. Dicots have strong taproots that can develop secondary roots.

Although the group includes annual herbs and shrubs, over 50 per cent of the family members are woody plants. Well-known and utilised members of this group include proteas, roses, sunflowers, mustard, beans, mint and parsley. Relative to other plant groups, this group is probably the most diverse in terms of growth forms and flower structures.

There are ± 175,000 species of dicots in the world in over 300 families, making up an impressive three-quarters of all flowering plants.

ACANTHACEAE

Acanthus, pistol bush

A large family of 350 genera and 4,350 species throughout the world, some of which are popular garden plants. Also known as the pistol bush family after *Justicia adhatodoides* (= *Duvernoia adhatodoides*) from the eastern parts of southern Africa, because the seed capsules burst open with an audible bang to eject the seeds. The plants in this family have club-shaped pods with two seed chambers. Hook-like growths on the walls of the pods aid in the ejection and dispersal of the seeds when the pods burst open. Other diagnostic features include opposite leaves, the absence of stipules, large, showy and often spiny bracts around the flowers, irregular flowers that are 1- or 2-lipped (a character that divides the family into two groups), and cystoliths (enlarged cells containing crystals of calcium carbonate, thought to serve as protection from leaf-eaters) that appear as white streaks on the leaves of some species. The ecologically important mangroves (genus *Avicennia*) are members of this family. 8 genera and 12 species in Baviaanskloof.

BARLERIA

Named after Jacques Barrelier, 1606–1673, a Dominican monk and French botanist who worked in Paris

Small shrubs, sometimes spiny, recognised by the 4-lobed calyx, with the outer-lobe pair larger and leaf-like. ± 250 species in the world, mostly tropical; ± 60 species in South Africa; 2 species in Baviaanskloof.

Barleria obtusa Bush violet

obtusa = obtuse; referring to the rounded shape of the leaves

Scrambling, weak-stemmed, leafy shrub, using other plants for support, to 2 m. **Leaves** Roundish, 3 cm long × 2 cm wide, dark green, with translucent hairs. **Flowers** Funnel-shaped, 2–3 cm wide, mauve to purple, with a long slender stigma and 2 anthers. Flowering mostly in autumn, but also after summer rain.

Rus en Vrede 4X4, Kouga 25.05.2010

Habitat Fairly common in open, rocky areas in Thicket. **Distribution** Widespread from the Little Karoo to Zimbabwe. **Notes** Heavily grazed by stock and game, sometimes pruned down to a dense, twiggy, well-rounded shrublet. **Other names** Bosviooltjie (A), inzinziniba (X), idololenkonyane (Z).

Barleria pungens

pungens = pungent, piercing, ending in a rigid and sharp point; alluding to the very spiny leaves

Spiny shrub to 40 cm. **Leaves** Spine-tipped, 2.5 cm long, fresh green, margins and spines whitish. **Flowers** In axillary clusters, 1.5 cm wide, blue-purple, sepals ovate, with spinous teeth on the margin, spine-tipped. Flowering from spring to autumn. **Habitat** Valleys in Succulent Karoo. **Distribution** Widespread from Swellendam to Makhanda.

Zandvlakte, Kouga 27.03.2012

BLEPHARIS

blepharis = eyelash; pertaining to the fringe of hairs on the anthers

Generally small shrublets with spine-toothed leaves and 4 distinctive hard and protruding stamens that appear flat above, with a fringe of hairs. ± 125 species from Africa to India; 50 species in South Africa; 2 species in Baviaanskloof.

Blepharis capensis

capensis = from the Cape; referring to the geographical origin of the species

Resprouting, erect shrub, often low and gnarled, to 1 m. **Leaves** Very spiny, dimorphic, 3 cm long, margins revolute, with spinous teeth. **Flowers** In terminal spikes, up to 2 cm long, whitish with greenish-brown lines; bracts oblanceolate, spinous toothed and spine-tipped, white on veins. Flowering in summer and autumn. **Habitat** Various habitats including Grassy Fynbos, Transitional Shrubland, Thicket and Succulent Karoo. **Distribution** Widespread from southern Namibia, through the Great and Little Karoo to Port Alfred. **Other names** Ubuhlungu besigcawu, unomatshinotshino (X).

spiny margins

Grasnek, Kouga 17.04.2012

Blepharis integrifolia

integer = whole, entire, *folia* = leaf; referring to the simple leaves

Resprouting, dwarf, prostrate, shortly-hairy shrublet, to 10 cm. **Leaves** Simple, opposite, elliptic, 2 cm long. **Flowers** Solitary in the axils, with bracts in 4 or 5 pairs, ending with 5–7 purple-red, recurved bristles. Corolla sparsely hairy, to 2 cm long, pale blue, mauve or purple with darker veins. Flowering in spring, summer and autumn. **Habitat** Disturbed grassland, often around the edges of thicket clumps. **Distribution** Widespread from Riversdale, through Africa to India.

recurved bristle

Patensie 25.10.2018

DICLIPTERA

di = two, *kleio* = to shut, *ptera* = wings; referring to the 2-chambered capsules that open to release the seeds

± 170 species in tropical areas; 20 species in South Africa; 1 species in Baviaanskloof.

Dicliptera cernua

cernua = nodding; pertaining to the lower petal of the flower that nods when alighted on by a pollinator

Sprawling, weak-branched, herbaceous shrublet to 50 cm. **Leaves** Ovate, opposite, 2 cm long, fresh green. **Flowers** On cymes in the axils, up to 2 cm long, magenta, 2-lipped, with 2 bracts. Fruit a 2-chambered capsule that does not burst open. Flowering in winter and spring. **Habitat** Occurring at bush clump margins in Thicket. **Distribution** Widespread from Grootrivierpoort to KwaZulu-Natal. **Other name** Umhlolowane (X).

Bruintjieskraal, Grootrivierpoort 26.10.2018

DYSCHORISTE (= *Chaetacanthus*)

dyschortas = inhospitable, with little grass or feed; referring to the small stature of most species, rendering them of little grazing value

± 170 species in warm and tropical areas; 11 species in South Africa; 1 species in Baviaanskloof.

Dyschoriste setigera

seta = bristle, *gero* = carry or bear; referring to the
numerous bracts around the flowers, which are thin
and bristle-like and remain dry on the stems

Resprouting from a woody base, small, diffuse
shrublet to 30 cm. **Leaves** Opposite, obovate,
obtuse, 1 cm long × 0.5 cm wide, young leaves
glandular-hairy. **Flowers** Axillary, narrowly funnel-
shaped, 1.5 cm long, white or blue. Flowering
in spring and summer. **Habitat** Uncommon and
sporadic, easily overlooked, in Grassy Fynbos and Transitional
Shrubland where clay or loam is present in the soil. **Distribution**
Widespread from Mossel Bay to Mpumalanga.

Elandsvlakte, Baviaans 01.11.2011

HYPOESTES

hypo = beneath, *estia* = house; pertaining to the calyx, which is covered or 'housed' by the bracts

± 40 species, mostly in Australia; 3 species in South Africa; 2 species in Baviaanskloof.

Hypoestes aristata   Ribbon bush

aristata = with bristles; referring to the bristly bracts

Upright shrub to 1.5 m. **Leaves** Opposite, ovate,
hairy, 10 cm long, dark green. **Flowers** 3 cm long,
mauve, in dense bristly spikes in the axils. Flowering
in summer and autumn. **Habitat** Found at the margins
of Forest and Thicket. **Distribution** Very widespread
from De Hoop to tropical Africa. **Other names**
Seeroogblommetjie (A), uhlololwane (X),
idolo-lenkonyane-elimhlophe, uhlonyane (Z).

14.12.2020

Hypoestes forskaolii

Named after Swedish botanist Pehr Forskaol, 1736–1768

Reseeding, erect shrub to 1 m. **Leaves** Opposite, ovate,
softish, petiolate, 5 cm long × 2 cm wide, becoming
smaller below the flowering spike. **Flowers** Inflorescence
a false spike at the end of the branches, with numerous
clusters of short spikes arranged around the stem, calyx
glandular-hairy. Each flower narrowly funnel-shaped,
2 cm long, white or mauve with purplish markings
on upper side, 3-lobed petals, lower petal entire.
Flowering in late summer and autumn. **Habitat** Shaded
understorey of Forest. **Distribution** Very widespread
from Mossel Bay to Somalia. Easily seen at Poortjies.

Poortjies, Cambria 11.07.2016

ISOGLOSSA

isos = equal, *glossa* = tongue; referring to the tongue-like lower petal

± 50 species, mostly in tropical parts of Africa and Asia; 15 species in South Africa; 1 species in Baviaanskloof.

Isoglossa ciliata

ciliata = fine hairs; referring to the glandular-hairy bracts that also have non-glandular hairs

Sprawling subshrub to 50 cm. **Leaves** Opposite, ovate, 6 cm long × 3 cm wide, hairy on midvein and margin. **Flowers** In dense spikes, 1 cm wide, pink or white, with upper lip hooded, and lower petal upraised with herringbone markings; bracts glandular-hairy. Flowering in autumn, winter and spring. **Habitat** Uncommon in Baviaanskloof in damp, shaded areas in Forest. **Distribution** Widespread from Knysna to Mozambique.

Poortjies, Cambria 18.05.2021

JUSTICIA

Named after Sir James Justice, 1698–1763, a Scottish horticulturist

The lobes of the lower petals spread sideways. ± 420 species worldwide, mostly in the tropics; 23 species in South Africa; 3 species in Baviaanskloof.

Justicia cuneata Bloubos

cuneatus = wedge-shaped; referring to the shape of the leaves that lack petioles

Erect shrub, with whitish stems, to 60 cm. **Leaves** Opposite, elliptic, sessile, ascending, up to 1 cm long, pale green. **Flowers** Single, axillary, 1 cm long, with a short peduncle, cream with pink markings. Flowering in spring and early summer. **Habitat** Fairly common in valleys in open Thicket and Succulent Karoo in sandy-rocky areas and dry riverbeds. **Distribution** Widespread from Namaqualand to Addo.

Doringkloof 27.08.2010

Justicia tubulosa **subsp.** *late-ovata* (= *J. leptantha; Siphonoglossa leptantha*)

tubus = tube, *osa* = well developed; referring to the tubular flower, with the tube being longer than the petal lobes; *lata* = broad, *ovum* = egg, *ata* = possessing; with broadly ovate leaves

Sprawling, soft-stemmed, leafy shrub to 50 cm. **Leaves** Opposite, broadly ovate, hairy, 2.5 cm long × 2 cm wide, dark green, petiolate. **Flowers** Solitary, in axils near tips of the branches, hairy, 1.8 cm long, white to lilac, with narrow tube. Flowering from summer to early winter. **Habitat** Fairly common in protected and shady ravines under Forest. **Distribution** Widespread from Knysna to KwaZulu-Natal.

Gannalandkloof, Baviaans 30.12.2017

MONECHMA

mono = one, *echma* = obstacle, support, holdfast; referring to the well-rooted nature of some species

The lobes of the lower petal are short and recurved. ± 40 species in Africa and India; 14 species in South Africa; 1 species in Baviaanskloof.

Monechma spartioides Maklikbreekbos

Spartium = plant genus, *oides* = resembling; *Spartium* is a genus in the pea family, of which Spanish broom is a member; unclear where the resemblance lies.

Upright, brittle, white-stemmed, non-hairy shrub to 1 m. **Leaves** Opposite, linear to ovate, 2 cm long, yellowish green. **Flowers** Solitary, in the axils, 1 cm long, white, cream, mauve or blue, with purple markings on the lower lip. Flowering after rain. **Habitat** In valleys in Succulent Karoo and Thicket. **Distribution** Widespread from Namibia to Graaff-Reinet and Baviaanskloof. **Notes** Often heavily browsed by stock and game.

Rus en Vrede 4X4, Kouga 28.07.2011

ACHARIACEAE

± 33 genera and 155 species globally. Mostly tropical herbs, shrubs and trees. Chaulmoogra oil, extracted from *Hydnocarpus kurzii* (from India and Myanmar), has antimycobacterial properties and was previously used to treat leprosy. Characteristic features include more petals than sepals, and ornamented fruits. 6 genera and 7 species in southern Africa; 2 species in Baviaanskloof.

CERATIOSICYOS

keration = little horn, *sikyos* = cucumber; referring to the elongated, cylindrical fruits

1 species endemic to South Africa.

Ceratiosicyos laevis

laevis = smooth; referring to the smooth, hairless leaves and stems

Herbaceous monoecious climber. **Leaves** Soft, palmately 5–7-lobed, 14 cm long × 12 cm wide, with serrated margins. **Flowers** Axillary, 0.5 cm long, yellowish green, solitary in females, racemes in males. Fruit elongated, cylindrical, ribbed, 8 cm long × 0.8 cm wide. Flowering through winter. **Habitat** Occasional in Forests and Forest margins in sheltered ravines. **Distribution** Widespread from Wilderness to Modjadjiskloof (Duiwelskloof) in Limpopo.

Kamassiekloof, Zandvlakte 21.04.2012

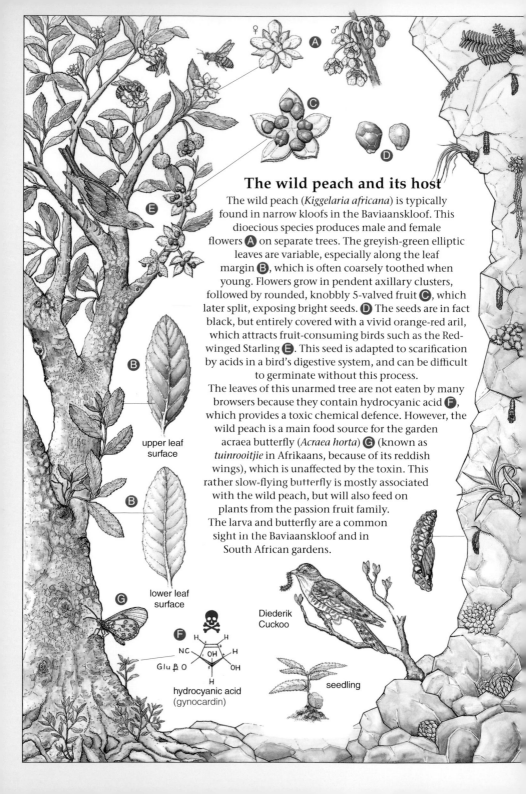

The wild peach and its host

The wild peach (*Kiggelaria africana*) is typically found in narrow kloofs in the Baviaanskloof. This dioecious species produces male and female flowers **A** on separate trees. The greyish-green elliptic leaves are variable, especially along the leaf margin **B**, which is often coarsely toothed when young. Flowers grow in pendent axillary clusters, followed by rounded, knobbly 5-valved fruit **C**, which later split, exposing bright seeds. **D** The seeds are in fact black, but entirely covered with a vivid orange-red aril, which attracts fruit-consuming birds such as the Red-winged Starling **E**. This seed is adapted to scarification by acids in a bird's digestive system, and can be difficult to germinate without this process.

The leaves of this unarmed tree are not eaten by many browsers because they contain hydrocyanic acid **F**, which provides a toxic chemical defence. However, the wild peach is a main food source for the garden acraea butterfly (*Acraea horta*) **G** (known as *tuinrooitjie* in Afrikaans, because of its reddish wings), which is unaffected by the toxin. This rather slow-flying butterfly is mostly associated with the wild peach, but will also feed on plants from the passion fruit family. The larva and butterfly are a common sight in the Baviaanskloof and in South African gardens.

upper leaf surface

lower leaf surface

hydrocyanic acid (gynocardin)

Diederik Cuckoo

seedling

Life cycle of the garden acraea butterfly

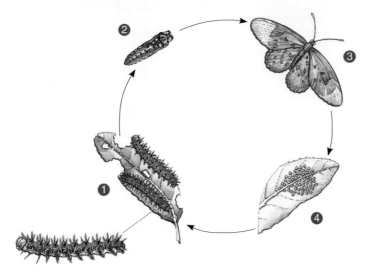

Life cycle of the garden acraea butterfly

Acraea horta belongs to a group of insects known as Lepidoptera, which comprise mainly moths and butterflies, with ± 180,000 species; 32 species are known in the genus *Acraea*.

1. The larvae of *Acraea horta* are often seen in the vicinity of wild peach trees. These caterpillars are known to defoliate the wild peach completely before they form a pupa.

2. The yellowish-black pupa is left well alone by most insect-consuming birds because it contains hydrocyanic acid, a quick-acting poison, to which the garden acraea is immune.

3. A butterfly emerges from the pupa. Males have a wingspan of 4.5–5 cm, females 4.9–5.3 cm. Wings are bright orange with black spots. Although the adults are on wing year-round, they are more commonly found in spring and summer (October–April).

4. The female lays tiny yellowish-cream-coloured eggs in neat patches on the lower surface of the leaves of the wild peach.

5. Soon little caterpillars appear. These newly-hatched larvae gorge on the leaves of the wild peach. (The larvae store the hydrocyanic acid toxin in their spines.) As the caterpillars mature, they spin a silken mat around themselves with their posterior prolegs, and attach to an upright surface – often a vertical cliff.

6. Some birds, such as the Diederik Cuckoo (*Chrysococcyx caprius*), are adapted to feed on the garden acraea larvae. The cuckoo does not swallow the entire caterpillar, but rather beats it open on a branch and consumes the innards, discarding the spiny and poisonous skin.

KIGGELARIA

Named after Franz Kiggelaar, 1648–1722, a botanist, medicine maker and curator of Simon von Beaumont's garden in The Hague. His published catalogue of this garden included many plants from the Cape of Good Hope.

1 species in tropical and southern Africa.

Kiggelaria africana Wild peach

africana = from Africa

Uitspankloof, Baviaans 19.02.2016

A semideciduous, dioecious tree, also a sun-loving forest pioneer, to 20 m, with smooth grey bark. **Leaves** Elliptic, 10 cm long × 5 cm wide, often light green, discolorous, usually toothed, with pockets of hairs in the lower vein axils. **Flowers** Female flowers solitary, male flowers in cymes, yellowish green and inconspicuous. Fruit round, greyish green and roughly textured, splitting open to reveal several seeds covered with red flesh, dispersed by birds. Flowering from late summer to winter. **Habitat** Common in the narrow ravines in Forest. **Distribution** Very widespread from Namaqualand to tropical Africa. **Notes** One of the first species to appear after a fire. A fast-growing tree that will provide good shade in domestic gardens. **Other names** Speekhout, wildeperske (A), idungamuzi, umduma, umhlandela, umhlizinyathi, umkhokohkho, umvethi (X), lekhatsi (S), imfeyenkomo, isikali, umhlabahlungulu (Z).

AIZOACEAE (= MESEMBRYANTHEMACEAE) Vygies, mesembs, ice plants

± 135 genera and 1,800 species in southern Africa, Australia and the central Pacific area. Mesembs range from dwarf stone plants to small trees reaching to 3.5 m; most are leaf succulents and most of the shrubby species have woody stems.

The popular and endearing Afrikaans name 'vygie' — meaning 'little fig' — refers to the fruiting structure. The English common name 'mesemb' derives from the old family name Mesembryanthemaceae. 'Ice plant' is most commonly used in North America for *Carpobrotus edulis*, which has become invasive there. This familiar name derives from the plants' bladder cells that glisten icily in the sun, acting as tiny reservoirs of water.

The greatest diversity of mesembs is found in the winter-rainfall Succulent Karoo parts of South Africa. The Baviaanskloof to the east of this area also plays host to various mesemb species. Plants here vary from the dwarf stone plants (*Conophytum* spp.) found in crevices, and the tiny *Delosperma esterhuyseniae* confined to cliffs, to larger shrubs such as the donkievy (*Mestoklema arboriforme*) and the witvygie (*Delosperma laxipetalum*), which grows up to a metre high. Most species however are procumbent, mat-forming opportunists (*Drosanthemum* and *Malephora* spp.), often seen along roadways in disturbed soil, especially when flowering en masse. Others, such as *Machairophyllum*, are small, tufted succulents growing on exposed sandstone outcrops.

Vygies are popular garden plants, with some species highly sought after by succulent collectors. They grow easily from cuttings: horticulturists recommend that they are left to 'cure' in a cool, dry place for about two weeks before planting. Sour fig fruits (*Carpobrotus* spp.) are edible and can be made into a delicious preserve. The leaves of some edible species are used in traditional cures for a sore throat and fungal infections, while kougoed (*Sceletium* spp.) have a mildly stimulating and hypnotic effect and have been traditionally used to manage depression and anxiety. Certain mesembs have also been used to make soap. In California, *Carpobrotus edulis* is used in fire breaks.

This is South Africa's largest succulent plant family, with bright, showy and familiar flowers. The Succulent Karoo contains ± 96 per cent of the family's diversity. ± 17 genera and 60 species in Baviaanskloof.

The vygie capsule

A small miracle of nature, the vygie seed capsule is a complex and sophisticated structure – a biologically engineered phenomenon. A knowledge of the shape of the capsule is key when identifying vygies, as these structures vary from species to species. The capsules are hygrochastic – they open only when wet. Those interested in propagating mesembs use hot water to facilitate speedy opening of the capsules. (They open in seconds in hot water.)

Most vygies in Baviaanskloof (with the exception of *Carpobrotus* spp. with their edible fruit) have woody capsules, which provide an easy way to recognise genera and species in this family. The opportunistic, rapid-growing species (*Mesembryanthemum aitonis* and *Drosanthemum hispidum*) are densely flowered, with capsules produced at – or just above – the leaf canopy. The longer-lived species tend to flower more sparsely, producing capsules that are not always above the leaf canopy.

Vygie fruiting capsules are usually top-shaped, with 5 or more locules (chambers containing the tiny seed, which is usually pear-shaped). These cavities are capped with 5 or more valves (depending on the species), attached to narrow, expanding keels. (The valves sometimes have wings.) The keels are part of the plant's hygrochastic mechanism. Water expands the keels, which push open the valves, exposing the locules. Sometimes the locules have translucent covering membranes. When present, these remain intact while the valves are open. They perform a different function: when a raindrop hits the surface, pressure builds up inside the water-filled locule, causing the seeds to shoot out through distal openings at the top. As a vygie capsule dries, the keels contract, which closes the valves in preparation for the next bout of moisture. The capsule protects the seed from predation, keeping it safe during dry seasons.

The capsule structures of 8 vygie genera are illustrated overleaf, showing what they look like when closed, and then opened, after exposure to moisture.

The locules in *Conophytum truncatum* – from the western Baviaanskloof – are more basic, not covered by membranes. In their case, when it rains, the seeds are splashed or washed out by raindrops and locally dispersed. *Delosperma* and *Trichodiadema* have similar kinds of capsules; all three of these genera have funicular hairs, which hold the seeds in tension so that they are not all dispersed in a single rainfall event.

There is significant variation in seed size, colour and surface texture within and between vygie genera, but sometimes the capsule type does not differ much between taxa. It is therefore helpful to look at all features when trying to identify a species.

Lampranthus coralliflorus, Zandvlakte bergpad, Kouga 08.08.2010

Characteristics of select vygie capsules

GENUS	LOCULES	VALVE WINGS	EXPANDING KEELS	MEMBRANES
Aizoon	2–5	absent	present or lacking	absent
Carpobrotus	10	absent (capsule closed)	absent	absent
Conophytum	3–8	narrow	parallel	translucent
Delosperma	4–6	broad	parallel	vestigial
Drosanthemum	5	broad	divergent	translucent
Glottiphyllum	8	folded downwards	divergent at tips	small, acute, concave
Lampranthus	5–10	large, not always present	divergent	rigid, convex, rimmed
Lithops	4–9	broad	fused, divergent at tips	reduced or absent
Machairophyllum	5–15	narrow	with serrated margins	present
Malephora	8–12	present	fused, divergent at tips	small, elliptical
Mesembryanthemum	4–6	usually folded inwards	parallel	vestigial
Mestoklema	5	recurved	divergent, with narrow, pointed wings	reduced
Rhombophyllum	5	very narrow, pointy	divergent, edge serrated	stiff
Ruschia	5	absent	divergent, reddish	convex, rimmed, closing rodlets
Tetragonia	4	absent (capsule closed)	absent	absent
Trichodiadema	4–7	present	present	present

Distinctive open and closed capsules of select vygie genera

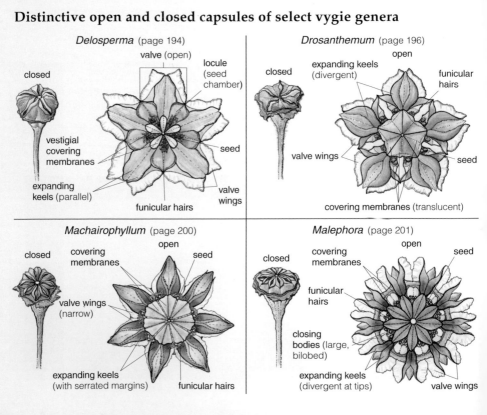

Delosperma (page 194)

valve (open)

locule (seed chamber)

closed

vestigial covering membranes

seed

expanding keels (parallel)

valve wings

funicular hairs

Drosanthemum (page 196)

open

closed

expanding keels (divergent)

funicular hairs

valve wings

seed

covering membranes (translucent)

Machairophyllum (page 200)

open

closed

covering membranes

seed

valve wings (narrow)

expanding keels (with serrated margins)

funicular hairs

Malephora (page 201)

open

closed

covering membranes

seed

funicular hairs

closing bodies (large, bilobed)

expanding keels (divergent at tips)

valve wings

CLOSING BODIES	CAPSULE TEXTURE	SIMILAR GENERA	SEED
absent	pale, woody		small
absent	leathery, pulpy inside		small, pear-shaped, embedded in pulp
absent	soft, pale	*Delosperma*	pear-shaped
absent	soft, pale	*Conophytum, Trichodiadema*	pale brown, globose
absent	pale, short lived, long stalks	*Mestoklema*	small, pear-shaped
large, whitish	pale, soft, falling off		large
absent	woody		D- or pear-shaped, dark, medium–large
absent	woody		variable
present	woody		egg-shaped, golden brown
large, whitish, bilobed	woody		flattish, rough-textured
absent	corky, pale		small
absent	small	*Drosanthemum*	smooth, brown
large, bilobed	woody, long-stalked		brown, dark brown
rod or hook-shaped	woody		small, brown to black
absent	spongy, pale, winged		small
absent	spongy	*Delosperma*	small, pear-shaped

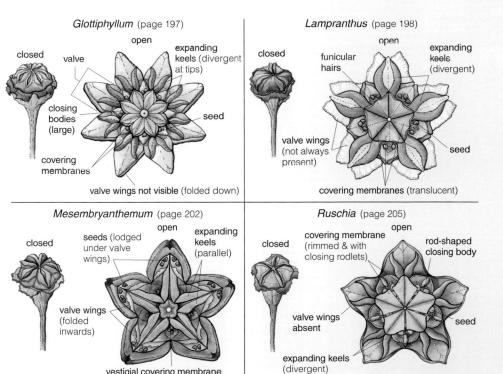

Glottiphyllum (page 197)

closed
open
valve
expanding keels (divergent at tips)
closing bodies (large)
seed
covering membranes
valve wings not visible (folded down)

Lampranthus (page 198)

closed
open
funicular hairs
expanding keels (divergent)
valve wings (not always present)
seed
covering membranes (translucent)

Mesembryanthemum (page 202)

closed
open
seeds (lodged under valve wings)
expanding keels (parallel)
valve wings (folded inwards)
vestigial covering membrane

Ruschia (page 205)

closed
open
covering membrane (rimmed & with closing rodlets)
rod-shaped closing body
valve wings absent
seed
expanding keels (divergent)

AIZOON

Spekvygie

aei = always, *zoos* = alive; referring to the ability of the plants to survive drought

Genus *Galenia* was incorporated into *Aizoon* in 2017, including an additional 27 species, making a total of 43 species in *Aizoon*. Mostly in southern Africa but also in North Africa, India and Afghanistan; 7 species in Baviaanskloof.

Aizoon africanum (= *Galenia africana*) Kraalbos, geelbos

africanum = from Africa; referring to the geographical origin; the common name refers to its abundance in kraals.

Upright, yellow, rounded shrub to 1 m. **Leaves** 2 cm long × 0.5 cm wide, yellowish. **Flowers** Many on a spreading inflorescence, tiny, to 2 cm long, creamy yellow. Flowering in summer and autumn. **Habitat** Common in disturbed or barren areas, on loam or sand, on the valley floor in Succulent Karoo and Thicket. **Distribution** Very widespread from Namibia to Baviaanskloof. Not browsed by most animals. **Other name** Iqina (X).

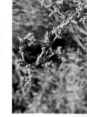

Verlorenrivier, Baviaans 19.05.2021

Aizoon glinoides Spekvygie

Glinus = plant genus, *oides* = resembling; resembling *Glinus*, a genus in Molluginaceae

Prostrate, sprawling perennial, 5 cm. **Leaves** Usually obovate, tip pointed, silky-hairy, to 2 cm long, silvery-grey. **Flowers** 1 cm wide, yellow. Flowering in spring and summer. **Habitat** Open, disturbed ground on stony soils with some clay in Thicket or Succulent Karoo. **Distribution** Widespread from Langkloof to KwaZulu-Natal.

Zandvlakte 27.09.2010

Aizoon portulacaceum (= *Galenia portulacacea*) Brakopslag

Portulaca = plant genus; referring to the superficial similarity to some species of *Portulaca*

Prostrate, sprawling perennial to 10 cm. **Leaves** Narrowly obovate, almost hairless, to 3 cm long × 0.5 cm wide, fresh or yellow-green. **Flowers** Tiny, to 0.5 cm wide, pink. Flowering in spring. **Habitat** Common in disturbed open areas on sand or loam in valleys in Succulent Karoo and Thicket. **Distribution** Widespread from Namaqualand through Karoo to Baviaanskloof.

Vlakkie, Zandvlakte 25.08.2010

Aizoon procumbens (= *Galenia procumbens*)

procumbens = procumbent; referring to the procumbent growth form (spreading along the ground without sending out roots)

Rounded, yellowish shrub to 1 m tall, with drooping branches. **Leaves** Opposite, oblanceolate, recurved, to 1 cm long × 0.5 cm wide, yellowish. **Flowers** In small cymes, tiny, to 0.3 cm wide, whitish. Flowering in spring. **Habitat** Found on the valley floor in stony, alluvial fans in the western part of the Baviaanskloof. **Distribution** Widespread from Namaqualand to Free State.

Speekhout, Baviaans 14.02.2016

CARPOBROTUS

Suurvy, sour fig

karpos = fruit, *brota* = edible; pertaining to the edible fruits

The juice from the leaves can be used to treat fungal infections, skin wounds, itchy bites and sunburn. The leaf juice is mixed with water for a gargle to treat mouth and throat infections. The species has been introduced in various parts of the world; 13 species from the Western Cape through to KwaZulu-Natal, also in Chile and Australia; 3 species in Baviaanskloof.

Carpobrotus deliciosus Ghoukum

deliciosus = tasty; referring to the edible fruit

Succulent perennial with trailing stems to 2 m wide. **Leaves** Triangular, straight, almost 2 cm wide × 7 cm long, bright green. **Flowers** Large and showy, 8 cm wide, magenta, pink or white. Fruit narrowing abruptly into pedicel. Flowering in winter and spring. **Habitat** All kinds of Fynbos. Often associated with dung middens. **Distribution** Widespread from Langeberg to KwaZulu-Natal. **Other names** Ghouna-vy (A), igcuthuma (X), ikhambilamabulawo (Z).

Duiwekloof, Baviaans 05.09.2015

Carpobrotus edulis Sour fig, perdevy

edulis = edible; pertaining to the edible fruit

Succulent perennial with trailing stems to 2 m wide. **Leaves** Triangular, occasionally curving, to 2 cm wide × 7 cm long, fresh green sometimes tinged reddish. **Flowers** Large, 6 cm wide, yellow and fading to pink with time. Fruit tapering gradually into pedicel. Flowering in spring. **Habitat** Disturbed, sandy soils in Fynbos and Transitional Shrubland. **Distribution** Widespread from Namaqualand to Makhanda. **Other names** Unomatyumtyum, igcuthuma (X).

Beacosnek, Baviaans 03.11.2011

CONOPHYTUM

Toontjies, cone plant

konos = cone, *phyton* = plant; referring to the cone-shaped leaves of most species

86 species from southern Namibia to Baviaanskloof; 1 species in Baviaanskloof.

Conophytum truncatum

truncatum = truncate; referring to the short, truncated leaves

Succulent perennial forming domes to 15 cm wide. **Leaves** Leaf bodies obconic to cylindrical, truncate, to 3 cm long × 1 cm wide, top surface with clear or reddish dots. **Flowers** To 1.5 cm long, usually yellow but also white or greyish. Fruit with 4–7 locules. Flowering at night in autumn. **Habitat** Rocky outcrops and cliffs in the western part of Baviaanskloof. **Distribution** Widespread from Robertson to Steytlerville.

Nuwekloof Pass, Baviaans 05.05.2006

DELOSPERMA

Skaapvygie, intelezi (X)

delos = visible, *sperma* = seed; the capsules have no covering membranes, so the seeds are easily seen once the capsule is open.

± 160 species from Western Cape and Namibia to East Africa; 5 species in Baviaanskloof, with several others close by.

Delosperma echinatum Pickle plant

echinos = hedgehog or sea urchin; referring to the tough bristles on the leaves

See p. 190 for labels

Hankey 25.10.2018

Sprawling succulent shrub to 15 cm. **Leaves** Opposite, barrel-shaped, 2.5 cm long, fresh green, covered with white spots and bristles. **Flowers** 2 cm wide at the branch tips, yellow. Flowering in winter and spring. **Habitat** Rocky ground in disturbed areas in Thicket and Transitional Shrubland. Fairly localised from Patensie to Addo. **Notes** Popular with gardeners as a pot plant. **Other name** Krimpvarkvygie (A).

Delosperma esterhuyseniae Klipvygie

Named after Elsie Esterhuysen, 1912–2006, a botanist and prolific
collector of plants from the mountainous areas of South Africa,
especially Fynbos from the Western Cape. She made the first
scientific collection of ± 150 taxa, and has 2 genera and
± 34 species named after her.

Usually compact to cushion-like, sometimes hanging. **Leaves**
Subclavate with rounded apex, to 3 cm long, often tinged
reddish in dry conditions. **Flowers** To 3.5 cm wide, white
with yellow centre. Fruit with tall rims on valves. Flowering in
summer and autumn. **Habitat** A dweller of sandstone cliffs,
easily seen at Raaskrans in Nuwekloof Pass. **Distribution**
Endemic to the Kouga and Baviaanskloof mountains.

Nuwekloof Pass, Baviaans 08.10.2016

Delosperma multiflorum Klein-kopervygie

multi = many, *florum* = flower; referring to the many-flowered cymes

Perennial, weak-stemmed, velvety-leaved and -stemmed
shrublet to 30 cm. **Leaves** Spreading, subterete to terete, finely
papillate, 2 cm long, fresh green sometimes grey-green. **Flowers**
Many in cymes, 1.2 cm wide, in various shades of copper to
dark pink, with petals drying to black. Flowering in autumn
and winter. **Habitat** Clay soils in Transitional Shrubland and
Succulent Karoo. **Distribution** Widespread from Calitzdorp to
Makhanda and inland from Loxton to Cradock.

Kleinpoort 06.03.2018

Delosperma prasinum Blinkvygie

prasinus = leek green; pertaining to the light green leaves

Small creeping succulent to 20 cm. **Leaves** Opposite, slightly
fused at the base, oblong, triangular, keeled, tip pointed, to
2 cm long, fresh green. **Flowers** Solitary, 1.5 cm wide, white
or pink. Flowering in summer. **Habitat** Thicket. **Distribution**
Narrow range from Baviaanskloof to Gqeberha (Port Elizabeth).

Onverwacht, Cambria 04.02.2020

Delosperma stenandrum Kleinblomvygie

stenos = narrow, *andrum* = male; referring to the thin anthers

Cushion forming or spreading succulent to 15 cm. **Leaves**
Trigonous, 2 cm long × 0.8 cm wide; fresh leaves finely
papillate and hairy, with slightly downcurved tips. **Flowers** 1 cm
wide, bright pink. Flowering in spring, summer and autumn.
Habitat Shaded cliffs in Thicket. **Distribution** Widespread from
Baviaanskloof to the Wild Coast, mostly along the seaboard.

Witruggens, Kouga 08.03.2012

DROSANTHEMUM

drosos = dew, *anthos* = flower; referring to the glistening
bladder cells in the leaves, reminiscent of dewdrops

110 species from Namibia to the Eastern Cape;
5 species in Baviaanskloof.

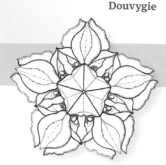

Drosanthemum hispidum Muisvygie

hispidum = coarsely hairy, bristly; referring to the bristly
hairs that cover the stems and leaves

See p. 190 for labels

Sprawling succulent shrub with thin, hairy branchlets, to
50 cm. **Leaves** Densely papillate, terete and obtuse, 1 cm
long. **Flowers** Solitary, purple, 3 cm wide. Flowering during and after rain in spring and summer. **Habitat**
Useful pioneer for covering open ground in Succulent Karoo and Thicket. **Distribution** Widespread
throughout the drier parts of southern Africa.

hairy stems

Gannaland, Zandvlakte 10.09.2011

Drosanthemum lique Doublaarvygie

liqueo = 'I am liquid', fluid, transparent; referring to the
succulent leaves and leaf surface with transparent bladder cells

Erect, twiggy shrub, older branches
shiny and smooth. **Leaves** Opposite,
spreading, widely spaced on the upper
stems, cylindrical, obtuse and papillate,
1 cm long. **Flowers** Solitary, stalked,
2 cm wide, magenta. Fruit 5-locular.
Flowering in spring and summer.
Habitat Pioneer of open ground on
the valley floor in Succulent Karoo
and Thicket. **Distribution** Widespread
from the Little Karoo to Cradock and
Gqeberha (Port Elizabeth).

Uitspan, Kouga 02.05.2018

Drosanthemum wittebergense Wittebergvygie

Named after the Witteberg Mountains south of Matjiesfontein, where it was first recorded

Neatly erect shrub with ascending branches. **Leaves** Up to 2 cm long × 0.3 cm wide, dark green. **Flowers** Large and showy, 4 cm wide, magenta with white centres. Flowering in spring, especially October. **Habitat** Occasional on disturbed ground on shale soils at biome boundaries in Succulent Karoo and Transitional Shrubland. **Distribution** Fairly widespread from Matjiesfontein to Willowmore. Easily seen on roadside at Elandspoort farm south of Antoniesberg.

Elandspoort, Antoniesberg 08.10.2016

GLOTTIPHYLLUM Skilpadkos

glottis = mouth of the human windpipe, *phyllon* = leaf; referring to the tongue-shaped leaves

16 species in karroid parts of Western and Eastern Cape; 2 species in Baviaanskloof.

Glottiphyllum depressum

depressus = pressed down, flattened; referring to the form of the leaves

See p. 191 for labels

Low growing and clump forming to 10 cm. **Leaves** Opposite, 8 cm long, green, some leaves with a hooked point at the tip.
Flowers Nestled between the leaves, 3 cm wide, yellow. Fruit with a spongy top, valves thickened and with low rims. Flowering in winter, especially after rain. **Habitat** Usually in the shade of other shrubs in karroid parts of Succulent Karoo and Transitional Shrubland. **Distribution** Widespread from Ceres to Humansdorp. **Notes** The other species that might be found here, *G. difforme*, has leaves with irregular protuberances that resemble the uvula above the human throat.

Above Nuwekloof Pass 08.05.2006

LAMPRANTHUS

Unomatyumtyuma (X)

lampros = bright, *anthos* = flower; referring to the large
showy flowers of many species

An under-collected and poorly studied genus with
many spectacular species that are popular in gardens.
The capsules have 5 compartments with no closing
bodies, and valve wings with transparent membranes.
Likely over 200 species from Namibia to KwaZulu-
Natal, especially in the Western Cape; possibly more
than the 5 species recorded in Baviaanskloof.

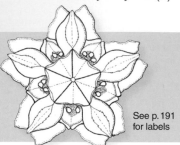

See p.191
for labels

Lampranthus affinis Dassievygie

affinis = similar to, allied to; referring to the similarity with a closely related species

Decumbent or sprawling succulent
shrub to 30 cm. **Leaves** Ascending
and in bunches, usually held upright
on the trailing or hanging stems, to
4 cm long × 0.3 cm wide, greyish
green with pink tips. **Flowers** At the
branch tips, to 4 cm wide, white,
pale mauve or magenta. Capsule
less than 1 cm wide. Flowering in
late spring and summer. **Habitat**
Sandstone cliffs and rocky outcrops,
often in well-shaded habitat of narrow
kloofs. Occurring in Arid Fynbos,
Fynbos Woodland and Transitional
Shrubland. **Distribution** Narrow, from
Oudtshoorn to Baviaanskloof. **Notes**
Grows well in pots.

Uitspan, Baviaans 28.10.2018

Lampranthus coralliflorus Koraalvygie

coralli = coral, *florus* = flower; referring
to the coral-bright flowers

Sprawling succulent shrub to 30 cm.
Leaves 4 cm long × 0.3 cm wide,
pale green. **Flowers** At the branch
tips, 4 cm wide, shining bright pink.
Flowering in spring. **Habitat** Exposed
rocky slopes in Spekboomveld and
Transitional Shrubland. **Distribution**
Widespread from Montagu to
Baviaanskloof. **Notes** Regarded by
some as the same species as *L. affinis*.

Rus en Vrede 4X4, Kouga 08.08.2010

Lampranthus haworthii   Blinkvygie

Named after Adrian Hardy Haworth, 1767–1833, an
English entomologist, botanist and crustacean expert. He
described several new taxa, including 22 genera of moths.

Relatively robust, erect, well-branched shrub to
80 cm. **Leaves** Opposite and fused at the base,
spreading from the stems, to 4 cm long, greyish
green. **Flowers** Large, to 7 cm wide, shiny magenta,
sometimes with white centres. Flowering in spring.
Habitat Shales in Transitional Shrubland, especially
after fire. **Distribution** Widespread from the
Cederberg, through the Little Karoo to Baviaanskloof.

Duiwekloof, Baviaans 28.10.2018

Lampranthus spectabilis   Duikervygie

spectabilis = remarkable or showy; referring to
the spectacularly colourful flowers

Fairly robust, spreading shrub to 25 cm
tall, rooting at the nodes. **Leaves**
Trigonous, 2.5 cm long × 0.4 cm wide.
Flowers To 6 cm wide, rose-magenta,
on relatively thick stalks. Capsules to
1.8 cm wide. Flowering in spring. **Habitat**
Flattish low plateaus with shallow,
reddish soil, usually containing some
clay, in Transitional Shrubland (grasslands)
or Grassy Fynbos. **Distribution** From
Baviaanskloof eastwards to East London.

Guerna, Kouga 09.02.2012

Lampranthus stayneri   Ruigtevygie

Named after Frank James Stayner,
1907–1981, a succulent specialist
and curator of the Karoo National
Botanical Garden at Worcester

Decumbent to erect shrub to
35 cm. **Leaves** Acuminate, to
2.5 cm long, with recurved tips.
Flowers 3.5 cm wide, magenta,
on short stalks. Capsules 0.8 cm
wide. Flowering in spring and
summer. **Habitat** Shale or clay-
rich soils in grassy Transitional
Shrubland. **Distribution** Eastern
distribution from Baviaanskloof
to Addo.

Elandsvlakte, Baviaans 01.11.2011

LITHOPS
Living stones

lithos = stone, *ops* = face; referring to the roundish leaves, flat on the ground, resembling stones or pebbles

36 species from dry parts of southern Africa; 1 species in Steytlerville Karoo.

Lithops localis Beeskloutjie

localis = of a place; referring to the plant's localised habitat of gravel flats on heavy clays

Stemless, geophytic succulent, usually found in clumps, to 3 cm. **Leaves** Paired, 2 cm wide, separated by the crack from which the flower arises. Leaf surface smooth, slightly convex, regularly dotted, greyish brown. **Flowers** 2.5 cm wide, yellow. Flowering in autumn. **Habitat** Occasional on black shales and gravel flats in Succulent Karoo. **Distribution** Widespread from Western Karoo to Steytlerville.

Steytlerville Karoo 19.05.2006

MACHAIROPHYLLUM

machaira = sword, dagger, *phyllon* = leaf; referring to the large, sharply tapering, sword-shaped leaves

5 species, mainly in the Little Karoo; 2 species in Baviaanskloof.

Machairophyllum bijliae Sabelvygie

Named after the genus *Bijlia*, the Prince Albert vygie, with leaves and growth form scarcely like this plant. Lady Johanna Susanna Faure (van der Bijl), 1851–1928, had a succulent collection near Prince Albert, and sent specimens to NE Brown, who named genus *Bijlia* after her.

See p. 190 for labels

Tufted, forming clumps, to 20 cm. **Leaves** Trigonous, smooth, and sometimes seeming waxen, 10 cm long, pale green. **Flowers** Many-petalled, to 6 cm wide, golden yellow with red reverse. Flowering at night in late spring. **Habitat** Rocky outcrops in Arid and Grassy Fynbos. **Distribution** Fairly widespread from George and the Swartberg to the Zuurberg, Eastern Cape. **Notes** Another species, *M. albidum*, might occur in the extreme west. It has 2-angled pedicels that are not as long.

Uitspan, Baviaans 19.02.2016

MALEPHORA

male = armhole, *pherein* = to bear; referring to the membranous coverings on the capsules with divergent keel arms

17 species from Namibia to Eastern Cape;
2 species in Baviaanskloof.

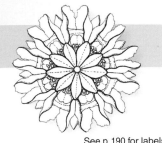

Malephora uitenhagensis Springbokvygie

Named after the town of Uitenhage (now Kariega)
in the Eastern Cape

See p. 190 for labels

Trailing succulent-forming mats to 2 m wide. **Leaves** Terete, smooth, 2.5 cm long × 0.5 cm wide, bright green. **Flowers** 3.5 cm wide, yellow or copper-red, with 4 sepals. Capsules 8–12-locular. Flowering in early spring. **Habitat** Deep alluvial soils in Doringveld and Thicket. **Distribution** Widespread from Calitzdorp to Graaff-Reinet and to Gqeberha (Port Elizabeth). **Notes** The other species that may be found in Baviaanskloof, *M. lutea*, is similar but has longer, waxy, blue-green leaves.

Doringkloof 04.09.2006

MESEMBRYANTHEMUM

mesembria = middle of the day, *anthemon* = flower;
alternatively: *mesos* = centre, *embryon* = pistil or embryo; 2 derivations
are possible – the opening of flowers at midday or referring to the
flower with an embryo in the centre.

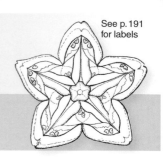

See p. 191
for labels

This genus now includes *Aptenia, Aridaria, Brownanthus,*
Phyllobolus, Prenia, Psilocaulon, Sceletium and *Sphalmanthus.*
103 species in southern Africa; 7 species in Baviaanskloof.

Mesembryanthemum aitonis   Sea spinach, angled ice plant

Named after William Aiton, 1731–1793, a Scottish botanist who worked on
Hortus Kewensis, a catalogue of plants cultivated at Kew Gardens

Annual or biennial with trailing
stems and angled branches.
Leaves With distinct bladder
cells, flat, ovate or spathulate,
to 4 cm long × 3 cm wide, often
tinged red. **Flowers** 2 cm wide,
white to pinkish. Fruit oblong,
with perianth lobes reflexed at
maturity. Flowering in spring and
summer. **Habitat** Pioneer of open
or disturbed ground in Thicket
and Succulent Karoo. **Distribution**
Widespread from Caledon to East
London. **Notes** A weed on coastal
plains of southwestern Australia.
Other names Brakslaai, slaaibos
(A), iqina (X).

Nuwekloof Pass, Baviaans 01.09.2011

Mesembryanthemum articulatum   (= *Psilocaulon articulatum*)   Asbosvygie, lidjiesvygie

articulatum = jointed, articulated; referring to the jointed stems

Decumbent perennial with stems
finely rough and hairy, forming dense
mats to 30 cm. **Leaves** Subterete,
hairy, 2 cm long, present briefly after
rain. **Flowers** 1 cm wide, white to
pink, bracts with twisted and pointy
tips. Flowering mostly in spring and
summer, also after rain. **Habitat**
Sandy and stony soils in dry, exposed
or disturbed places in Thicket and
Succulent Karoo. **Distribution**
Widespread from Namibia to
Eastern Cape.

hairy
stems

Tchnuganoo, Baviaans 22.10.2011

Mesembryanthemum bicorne   Loogbossie

bicorne = two horns; alternatively
Bicorne is a genus of intestinal
worms; the persistent dried leaves
are horn-like and the articulated
stems are worm-like.

Perennial with many trailing,
smooth, articulated, round, pale
green, succulent stems and small,
slender leaves. **Leaves** Subterete,
2 cm long. **Flowers** 2 cm wide,
white to pale pink. Flowering in
spring. **Habitat** Disturbed or open sites in more arid
parts of the valley floor. **Distribution** Widespread from
Worcester to Gqeberha (Port Elizabeth).

01.11.2021

Mesembryanthemum cordifolium (= *Aptenia cordifolia*)   Brakvygie

cordi = heart, *folium* = leaf; referring to the heart-shaped
leaves; *apten* = wingless; referring to the unwinged
valves of the capsules

Perennial, mat-forming or climbing succulent with
4-angled stems. **Leaves** Opposite, heart-shaped,
flat, 4 cm long × 2.5 cm wide, fresh green. **Flowers**
Usually single at the branch tips, 1.5 cm wide, white
through to magenta. Flowering in spring, summer and
autumn. **Habitat** Moist, shaded habitat in Thicket and
Forest. **Distribution** Originally possibly a narrow coastal
distribution from the Baviaanskloof to East London, but
today extends into the Western Cape and to Limpopo.
Notes Easy to grow and popular in gardens, but
sometimes regarded as weedy. Also used medicinally.

Grasnek, Kouga 08.03.2012

Mesembryanthemum granulicaule   Groenblomvygie

granuli = granule, *caule* = stem; referring to the rough, granular stem

Annual or biennial, decumbent or erect, the
rough articulated branches with dome-shaped
bladder cells. **Leaves** 1.5 cm long, pale green,
sometimes tinged purplish. **Flowers** To 1 cm wide,
pale yellow or cream-coloured. Flowering in late
spring. **Habitat** Disturbed open areas in arid parts
of Succulent Karoo and Thicket. **Distribution**
Widespread from Namibia through the Little Karoo
and Eastern Cape to Free State. Naturalised in
parts of southern Australia.

rough granular
stems

Drinkwaterskloof, Kouga 09.11.2011

Mesembryanthemum guerichianum   Ice plant

Named after Georg Julius Ernst Guerich, 1859–1938, a German
geologist, palaeontologist, botanist, university teacher and plant
collector in South West Africa (now Namibia)

Prostrate annual, spreading over the ground. **Leaves** Large,
thickly fleshy, with bladder cells glistening, 15 cm long × 5 cm
wide, green tinged red, margins wavy. **Flowers** To 5 cm wide,
white. Flowering in spring and summer. **Habitat** Disturbed,
dry, sandy areas in Succulent Karoo and Thicket. **Distribution**
Widespread from Angola to Cape Peninsula and to Cradock and
Addo, Eastern Cape. **Other names** Brakslaai, soutslaai (A).

Rus en Vrede 4X4, Kouga 05.11.2011

Mesembryanthemum splendens   Bosvygie, donkievygie

splendens = shiny or bright; referring to the showy
white flowers that shine at night

Perennial shrub that climbs through and
over neighbouring bushes, to 1 m. **Leaves**
Cylindrical, 5 cm long × 0.5 cm wide, grey-green
tinged pinkish. **Flowers** To 4 cm wide, white.
Flowering all year round but especially after
good rain. **Habitat** Common in Succulent Karoo.
Distribution Widespread in karroid areas from
Worcester to Cradock and Makhanda.

Elandspoort, Antoniesberg 08.10.2016

MESTOKLEMA   Donkievygie, kopervygie, igabushe (X)

mestos = full, *klema* = small branch, twig; referring to the much-branched, shrubby habit

6 species from southern Namibia to Eastern Cape; 1 species in Baviaanskloof.

Mestoklema tuberosum   Copper mesemb

tuberosum = with a tuber; referring to the tuberous roots

Much-branched shrub to 70 cm,
with tuberous roots. **Leaves**
Trigonous, finely papillate, to 1.5 cm
long, grey-green, with recurved tips.
Flowers In terminal cymes, 0.8 cm
wide, copper-coloured. Flowering in
autumn. **Habitat** Succulent Karoo.
Distribution Widespread from
southern Namibia to Graaff-Reinet
and Addo. **Notes** Popular in pots
with tuberous roots exposed. **Other
name** Kopervygie (A).

Vaalwater, Antoniesberg 12.02.2016

RHOMBOPHYLLUM
Bylvygie

rhombus = rhombus, lozenge, *phyllon* = leaf; referring to the slightly rhomboid leaves

5 species in the Eastern Cape; 1 species in Baviaanskloof.

Rhombophyllum dolabriforme

dolabriforme = shaped like an axe head or hatchet; referring to a tooth-like protuberance on the leaves, giving it an axe shape, with the base of the leaf forming the axe 'handle'

Perennial subshrub, densely branched, forming solitary tufts to 30 cm. **Leaves** Shaped like elk horns or hatchets, to 3 cm long, dull green, covered with translucent dots. **Flowers** 4 cm wide, bright yellow, sometimes tinged red on back, opening in the afternoon or evening. Capsule small, less than 1 cm wide when open, well stalked. Flowering in spring and early summer. **Habitat** Common on stony shales in flattish, overgrazed and open areas in Thicket. **Distribution** Eastern, from Baviaanskloof to Gqeberha (Port Elizabeth) and Graaff-Reinet. **Notes** Not eaten by stock. Popular in pots.

Zandvlakte, Baviaans 26.10.2018

RUSCHIA
Igcakriya (X)

Named after Ernst Julius Rusch, 1867–1957, a German who settled and farmed at Lichtenstein in Namibia, where, together with his son, he collected and grew succulents

Largest genus in the mesemb family with 220 species from dry parts of southern Africa; 12 species in Baviaanskloof.

See p. 191 for labels

Ruschia cradockensis Doringvygie

Named after Cradock, the town where it was first recorded

Rounded shrublet to 30 cm, with sharp, paired spines derived from the dried inflorescence. **Leaves** Triangular in cross section, rough, 0.5 cm long, dark green. **Flowers** Few in terminal dichasia, each flower to 1 cm wide, magenta, hardening and becoming spiny when the capsules are mature. Flowering in winter and early spring. **Habitat** Common in overgrazed veld in Succulent Karoo or Thicket. **Distribution** Widespread from Namaqualand, through the arid interior, to Richmond and Cookhouse. **Notes** Easy to identify by the spines, since the other spiny *Ruschia* spp. do not occur in Baviaanskloof.

Bosrugkloof, Baviaans 02.04.2012

Ruschia inclusa Vlaktevygie

inclusa = included; the derivation is unresolved.

Dense, mound-forming, succulent shrub to 15 cm. **Leaves** Fused basally, 0.5 cm long, grey-green, tinged pink or red. **Flowers** Small, 0.7 cm wide, pink. Flowering in spring. **Habitat** Infrequently encountered on stony, clay flats in Succulent Karoo. **Distribution** Widespread from Ceres to Baviaanskloof.

Damsedrif, Kouga 11.10.2016

Ruschia knysnana Knysnavygie

Named after the town of Knysna

Erect shrublet to 40 cm. **Leaves** Apiculate, 3 cm long × 0.5 cm wide, pale green, shortly fused on the stem. **Flowers** 1 to a few on cymes, 2.5 cm wide, pink, on pedicels 2.5 cm long. Capsule 6-locular. Flowering in summer and autumn. **Habitat** Rocky slopes in Grassy Fynbos and Transitional Shrubland. **Distribution** Fairly widespread from Knysna to Makhanda.

Engelandkop, Kouga 13.07.2011

Ruschia multiflora Witvygie

multi = many, *flora* = flowers; referring to the numerous flowers

Robust shrub to 1 m. **Leaves** 2 cm long × 0.5 cm wide, fresh green, fused to form a sheath around the stem, tip recurved and mucronate. **Flowers** In cymes, 3 cm wide, white. Flowering in spring and summer. **Habitat** Succulent Karoo and Thicket. **Distribution** Widespread from Worcester to Baviaanskloof.

Rus en Vrede 4X4, Kouga 02.11.2011

Ruschia perfoliata   Vetstingelvygie

per = through, *foliata* = leaves; referring to
the leaves that are fused around the stem

Densely branched, well rounded, mound forming
to 50 cm. **Leaves** Short, 1 cm long, dull green,
fused around the stem, recurved, each tip with 2
small downward-pointing teeth. **Flowers** Solitary
at branch tips, 2 cm wide, pink. Flowering
in summer. **Habitat** Deep alluvial sands in
Transitional Shrubland and Thicket. **Distribution**
Widespread from Laingsburg to Graaff-Reinet
and Baviaanskloof. Easily seen in the vicinity of
Speekhout farm.

Speekhout, Baviaans 24.12.2017

Ruschia pungens   Stekelvygie

pungens = pungent, sharply pointed; referring
to the old inflorescences that become woody
and hard, forming blunt spines

Robust shrub to 1 m. **Leaves** To 5 cm long
× 0.5 cm wide, greyish blue-green with
reddish tips. **Flowers** Several in a branched
inflorescence that later becomes woody,
1.5 cm wide, purple. Flowering in spring.
Habitat Succulent Karoo and Thicket.
Distribution Widespread from Montagu
to Baviaanskloof.

Damsedrif, Kouga 11.10.2016

Ruschia tenella   Ribbokvygie

tenella = tender or dainty; referring to
the relatively soft and slender petals

Much-branched, erect
shrublet to 30 cm.
Leaves Fused at the
stem, recurved, 1 cm
long, fresh green with
white translucent dots.
Flowers Small, 1.5 cm
wide, white. Flowering
in early summer or
autumn. **Habitat**
Clay-rich soils in grassy
Transitional Shrubland.
Distribution Widespread from Botrivier to
Gqeberha (Port Elizabeth).

Jeffreys Bay 28.05.2014

TETRAGONIA

Kinkelbos, klapperbrak, utyuthu (X)

tetra = four, *gonia* = an angle; referring to the 4-angled shape of the fruits

60 species in Africa, South America and Australia; 21 species in South Africa; 3 species in Baviaanskloof.

Tetragonia echinata Klappiesbrak

echinatus = prickly; pertaining to the fruit that is covered with spiny ridges and horns

Prostrate, trailing annual to 30 cm. **Leaves** Ovate to orbicular, subsucculent, 2 cm long. **Flowers** Small, 0.3 cm wide, greenish, with 4 anthers and 4 sepals. Fruit globose, covered in spiny ridges and horns. Flowering in winter, spring and summer. **Habitat** Disturbed or overgrazed ground in Thicket. **Distribution** Widespread from Namibia to Makhanda.

Vlakkie, Zandvlakte 21.12.2011

Tetragonia fruticosa Klimopkinkelbossie

fruticosa = shrubby or bushy; pertaining to the bushy habit

Sprawling shrub to 1 m with white woody stems, clambering over other plants. **Leaves** Lanceolate-elliptic, soft, 4 cm long × 1 cm wide, with recurved margins. **Flowers** In terminal racemes, 0.5 cm wide, yellow. Fruit with 4 wings, and turning black with maturity. Flowering in spring. **Habitat** Common in non-Fynbos habitat in sandy or loamy soils. **Distribution** Widespread from Namaqualand to Gqeberha (Port Elizabeth).

Coleskeplaas, Baviaans 23.08.2010

Tetragonia spicata Opslagkinkelbossie

spicata = spike; referring to the spike-like inflorescence

Similar to *T. fruticosa*, but leaves are rhomboid-lanceolate. **Flowers** In terminal racemes or 1 to a few in the leaf axils. Each flower to 0.4 cm wide. Fruit has 4 broad wings with knobs in-between. Flowering in winter and spring. **Habitat** The valley floor in Doringveld and Thicket. **Distribution** Widespread from Namaqualand to Makhanda.

Geelhoutbos, Kouga 20.04.2012

TRICHODIADEMA

Diadem vygie

trichos = hair, *diadema* = band or crown; referring to the tips of the leaves, which have a characteristic tuft of hairs or bristles, a reliable feature for identifying the genus

Shapes of the expanding keels together with the covering membranes on the capsules are used to identify the species. 34 species in South Africa and Namibia; 3 species in Baviaanskloof, and several additional species near Baviaanskloof, especially around Kariega (Uitenhage).

Trichodiadema barbatum Kareemoer

barbatum = bearded; referring to the diadema
(the bristles at the tips of the leaves)

Low, dense succulent to 10 cm tall, with tuberous roots.
Leaves To 1.2 cm long × 0.4 cm wide, with noticeable bladder cells. **Flowers** Solitary in axils, 3 cm wide, pinkish. Capsules 5-locular. Flowering in autumn, winter and spring. **Habitat** Shallow soil in rocky areas in Transitional Shrubland and Grassy Fynbos. **Distribution** Widespread from the Little Karoo to Kariega (Uitenhage).

bristles form a diadema

Bosrug, Baviaans 07.05.2006

Trichodiadema densum Doringkroonvygie

densum = dense; pertaining to the compact and low growth form, with small leaves packed closely together

Low and compact, to 8 cm tall, with fleshy roots. **Leaves** Crowded together in dense clumps, each diadema 0.5 cm long, white, usually with a yellowish centre; with 15–20 bristles. **Flowers** To 5 cm wide, bright pink with white-and-yellow centres. Flowering in winter and spring. **Habitat** Uncommon in rocky areas with shallow soil in Transitional Shrubland and Arid Fynbos. **Distribution** Widespread from Great Karoo to Baviaanskloof. Easily seen at the top of the Nuwekloof Pass.

Nuwekloof Pass, Baviaans 08.10.2016

Trichodiadema mirabile Doringblaarvygie

mirabile = admirable; referring to the diadema

Erect shrublet, 10 cm tall, with white-hairy stems. **Leaves** Papillate, often with dark tips; diadema with 12–26 bristles, 0.5 cm long, bright green. **Flowers** 4 cm wide, white. Capsules 6-chambered. Flowering in summer. **Habitat** Infrequent on stony slopes and rocky outcrops in Transitional Shrubland. **Distribution** Widespread from Laingsburg to Kariega (Uitenhage).

dark-tipped leaves

Uitspan, Baviaans 30.01.2020

AMARANTHACEAE

Amaranths

Now including members of the family Chenopodiaceae, this cosmopolitan family grows in the tropics and cool temperate parts of the world and includes members such as spinach (*Spinacia oleracea*), beetroot and quinoa. Many species have the capacity to tolerate very salty conditions, allowing them to dominate in saline habitats. ± 165 genera and 2,040 species; 15 genera and ± 32 species in the Western and Eastern Cape; 4 genera and 6 species have been recorded in Baviaanskloof, although more species are likely to be found.

ACHYRANTHES

Chaff flower

achyron = chaffy, husk, *anthos* = flower; the flowers are chaffy.

Medicinal and ornamental plants. 13 species from tropical and warm temperate climates; 1 species in Baviaanskloof.

Achyranthes aspera   Devil's horsewhip

aspera = hard, bitter; referring to the chaffy flowers

Sprawling, herbaceous perennial to 50 cm. **Leaves** Elliptic, usually attenuate, 14 cm long, pale below. **Flowers** Many in 15 cm-long spikes, each flower to 0.5 cm long, whitish. Flowering in summer and autumn. **Habitat** Partial shade of Thicket and Forest. **Distribution** Widespread from Knysna to tropical Africa and Asia; a weed in other parts of the world. **Notes** Used in East Africa to treat wounds and ringworm. The plant has antibacterial, antifungal and anthelmintic properties. **Other names** Knapsekêrel (A), isinama sebhokhwe, isanama sokugabha (X), bohomane (S), ibundlubundlu, ulimilwengwe, usibambangubo (Z).

Zandvlaktekloof, Baviaans 04.01.2003

ATRIPLEX*

Saltbush, orach

Atriplex was a name applied by Pliny to the edible orachs

± 250 species from warm-temperate and subtropical parts of the world; 3 species in Baviaanskloof.

Atriplex lindleyi subsp. *inflata**   Lindley's saltbush

inflata = inflated; named after John Lindley, 1799–1865, a British botanist who collected in Australia. In 1838 he saved the Royal Botanic Gardens, Kew from destruction in the interests of urban expansion.

Annual or perennial to 30 cm tall, grey-mealy. **Leaves** Rhomboid, to 3 cm long, toothed. **Flowers** Clustered in the axils, tiny, 0.2 cm wide, pale yellow. Bracts of the fruit are fused into an inflated bladder, just over 1 cm long. Flowering in autumn, winter and spring. **Habitat** A weed from Australia that colonises disturbed or overgrazed, dry, stony flats in arid areas throughout southern Africa. **Other names** Soutbos, klappiesbrak (A).

Doringkloof, Kouga 12.09.2015

*Atriplex nummularia** Old man's saltbush

nummus = coin, *alaria* = similar to; pertaining to the circular shape of the leaves

Dioecious or monoecious, large, hardy, erect or sprawling, grey, perennial shrub to 3 m. **Leaves** Irregularly shaped, with a scaly feel, to 5 cm long, silvery-grey, sometimes toothed. **Flowers** With tiny flowers in clusters at stem tips and in leaf axils. Fruit to 0.6 cm long. Flowering in winter or under favourable conditions. **Habitat** Disturbed or overgrazed areas in semi-arid regions throughout southern Africa. **Distribution** A weed from Australia, often grown as fodder. **Notes** Sheep that have fed on this plant produce meat that contains high levels of Vitamin E. This not only has health benefits for the sheep, but also extends the shelf life of the meat, enabling it to keep a fresh red colour for longer. **Other name** Soutbos (A).

Gannaland, Zandvlakte 09.12.2011

*Atriplex semibaccata** Creeping saltbush

semi = partly, *baccatus* = berry-like, pulpy; referring to the strange-looking fruit that becomes red and fleshy

Prostrate, spreading, perennial shrub to 30 cm tall × 2 m wide, with a deep taproot. **Leaves** Elliptic-obovate, glabrescent above, hairy below, to 3 cm long × 1 cm wide, grey-mealy occasionally tinged pink on the margins, sometimes coarsely toothed. **Flowers** Minute; female flowers clustered in the leaf axils, male flowers at the branch tips. Fruit 0.5 cm long, red and fleshy. Flowering mainly in spring and summer. **Habitat** Disturbed or overgrazed areas in semi-arid ground throughout southern Africa. **Distribution** A weed from Australia, useful as fodder for stock, also for erosion control.

Zandvlakte 19.01.2022

CYATHULA   Wolweklits

kyathos = cup, *ula* = diminutive; the stamens are united at their base to form a small cup.

± 25 species, mostly in tropical Africa; 1 species in Baviaanskloof.

Cyathula uncinulata Burweed

uncinatus = hooked, barbed; referring to the barbed fruit

Erect or sprawling shrublet to 1 m tall, covered with yellowish hairs. **Leaves** Petiolate, ovate, hairy below, 10 cm long. **Flowers** In dense rounded heads, 2 cm wide, green to whitish. Fruit burr-like, 4 cm wide. Flowering in summer and autumn. **Habitat** Sporadic along pathways in rocky areas in sheltered ravines in Thicket and Forest. **Distribution** Widespread from Oudtshoorn to tropical Africa. **Other names** Klits (A), isinama (X), bohome (S).

Zandvlaktekloof, Baviaans 01.03.2010

DYSPHANIA* (= *Chenopodium*) Goosefoot

dys = poorly, *phaner* = visible, easily seen; referring to the nondescript, unshowy appearance of many species, rendering them easily overlooked; derivation of the former name: *cheno* = goose, *pod* = foot; the leaves of some species resemble goose feet.

A large genus of ± 43 species. Quinoa (*Chenopodium quinoa*) is a popular gluten-free food source, relatively high in protein. There is evidence of *Dysphania* spp. being used as a food source since 4,000 BC. Several naturalised weedy species in South Africa and in Baviaanskloof; 4 species are indigenous in South Africa, but none of these occur in Baviaanskloof.

*Dysphania carinata** (= *Chenopodium carinatum*) Keeled goosefoot

carinatum = keeled; referring to the wing-like, hairy keel on each perianth segment

Semi-decumbent, spreading, aromatic, annual herb to 60 cm. **Leaves** Alternate, ovate-elliptic, 3 cm long, with 2–6 lobe-like blunt teeth. **Flowers** In dense axillary clusters, minute, greenish, glandular. Flowering in summer and autumn. **Habitat** A naturalised weed from Australia occurring in disturbed areas such as roadsides and exposed riverbanks. **Distribution** Widespread in southern Africa, common in Baviaanskloof.

Rooikloof, Kouga 24.04.2018

MAIREANA*

Named after Joseph Francois Maire, 1780–1867, a little-known naturalist and amateur botanist who curated an important herbarium in Paris

± 57 species of shrubs and herbs endemic to Australia; 1 species in Baviaanskloof.

*Maireana brevifolia** Australian bluebush

brevi = short, *folia* = leaf; alluding to the relatively short leaves

Erect, perennial shrub to 1 m. **Leaves** Alternate, succulent, cylindrical or flat, 0.5 cm long. **Flowers** Tiny, 0.3 cm wide. Fruiting body white or pink, 5-winged and wind dispersed. Flowering from late spring to early winter. **Habitat** Common in overgrazed kraals and other barren areas in Thicket and Succulent Karoo. **Distribution** Naturalised in South Africa, Canary Islands and the Middle East. **Notes** Used as fodder for stock, spreading in the valley west of Zandvlakte.

Zandvlakte, Baviaans 21.12.2011

PUPALIA

Pupali = eastern name for the genus; derivation unresolved

4 species; 1 species in Baviaanskloof.

Pupalia lappacea   Burweed

lappacea = burr-like; referring to the burred seed heads that attach to animal fur and passers-by for dispersal

Sprawling perennial to 50 cm. **Leaves** Ovate, attenuate, petiolate, to 7 cm long, paler below. **Flowers** In dense, elongated spikes, becoming longer in fruit. Fruit spiny and forming burr-like clusters, which adhere to animals and other passers-by. Flowering in summer and autumn. **Habitat** Occurring along path-sides in Forest and Thicket. **Distribution** Widespread from George to tropical Africa and Asia. **Notes** A plant with many medicinal uses throughout its range. **Other names** Bosklits (A), esibomvu (Z).

Damsedrif, Baviaans
01.05.2018

SUAEDA   Seepweeds, sea-blites

suaed = silty, Arabic word; referring to the habitat of very fine silty soils

Succulent shrubs and bushes that usually grow in saline habitats near the sea. 110 species worldwide; 4 species in South Africa.

Suaeda plumosa   Brakbos

plumosa = feathery; referring to the numerous tiny flowers on the stems that appear feathery when in flower. Often confused with *S. fruticosa*, but that species grows only on the Arabian Peninsula.

Sprawling succulent to 1 m, with weak stems that root at the nodes. **Leaves** Ovoid-ellipsoid, fleshy, glabrous, 0.6 cm long, grey-blue. **Flowers** Numerous in axillary clusters, tiny, each 0.2 cm long, yellowish. Flowering in spring and summer. **Habitat** Saline and muddy inland pans, riverbanks and estuaries. **Distribution** Widespread from southern Namibia to Gqeberha (Port Elizabeth).

Gamtoos Valley, Patensie 17.05.2021

ANACAMPSEROTACEAE

Small family of succulents. 3 genera and 36 species worldwide.

ANACAMPSEROS

anakampto = to cause the return of, *eros* = love; having the power to restore love

± 30 species in Africa; 3 species in Baviaanskloof.

Anacampseros arachnoides

arachnoides = spider-like; referring to the growth form and hairiness that are reminiscent of a baboon spider

Dwarf succulent to 8 cm. **Leaves** Tightly packed, ovoid, 0.5 cm long, dull brownish green, with recurved tips, woolly in axils. **Flowers** 3–5 on a branched peduncle, petals elliptic and pointed, 2 cm wide, pink, lasting only one afternoon. Flowering in summer. **Habitat** Rock crevices on stony slopes in Grassy Fynbos and Transitional Shrubland. **Distribution** Widespread from Namaqualand to Worcester and to Makhanda and Tarkastad.

Scholtzberg, Baviaans 23.12.2011

Anacampseros telephiastrum   Gemsboksuring

telephium = succulent, *astrum* = star; referring to the succulent leaves and star-shaped flowers. *Telephiastrum* was an old genus name for *Anacampseros*, suggesting that this species is typical of the genus or one of the first described species in the genus.

Low succulent to 15 cm tall when flowering. **Leaves** Crowded, ovate, to 3 cm long, tip pointed, surface finely dotted and sometimes with transverse wrinkles, with basal bristles shorter than leaves or inconspicuous. **Flowers** On stout peduncles, 2 cm wide, pink, with 30–45 anthers. **Habitat** Sporadic in loamy soils on rocky outcrops in Grassy Fynbos and Transitional Shrubland. **Distribution** Widespread from Worcester to Somerset East. **Notes** Popular in pots.

Nooitgedacht, Kouga 12.05.2021

ANACARDIACEAE

Cashews, sumac

A family of trees and shrubs including several economically important plants such as cashews, pistachios, mangoes, marula and sumac. The resin or milky sap in some members can be poisonous, e.g. poison ivy. The edible nuts from this family are said to support good mental health. ± 83 genera and 860 species worldwide; 14 genera and 133 species in southern Africa; 2 genera and 13 species plus 1 exotic species in Baviaanskloof.

LOXOSTYLIS

Teerhout

loxos = oblique, *stylis* = style; pertaining to the styles that arise laterally from the ovary, also referring to the oblique fruits

1 species in South Africa, including Baviaanskloof.

Loxostylis alata Teerhout

alatus = winged; referring to the winged leaf midrib

Evergreen dioecious tree, to 6 m tall, often multi-stemmed. **Leaves** Compound, imparipinnate, 2–5 pairs of leaflets and a terminal leaflet, each leaf 10 cm long, with a winged rachis. **Leaves** Red when young, becoming pale green with maturity, and sometimes with a silvery-waxy shine. **Flowers** In dense terminal panicles, white, with petals falling early but bracts persisting and turning pink-red around the developing fruit. Flowering in summer. **Habitat** Grassy Fynbos, Sour Grassland and Fynbos Woodland, more common on south-facing slopes. **Distribution** Widespread from Baviaanskloof to KwaZulu-Natal. **Notes** Fruit contains a sweet, dark, sticky substance favoured by baboons and birds. In KwaZulu-Natal, the bark is harvested for medicinal use. **Other names** Wild pepper tree (E), breekhout (A), isibara (Z, X).

Keerom, Baviaans 21.12.2011

SCHINUS*

Schinos was the name for the mastic tree; some species of *Schinus* produce similar resins.

Trees and shrubs native to South America with ± 8 species; 1 invasive species in Baviaanskloof.

Schinus molle Pepper tree

Mulli was the name of this tree in the Peruvian highlands.

Aromatic dioecious tree to 10 m, with drooping branches. **Leaves** Pinnately compound with 20–40 alternate leaflets, 20 cm long × 8 cm wide. **Flowers** Numerous in long panicles hanging from drooping branch tips, small, white. Fruit a drupe, 0.5 cm wide, with woody seed turning from green to red, pink or purplish. Flowering all year round. **Habitat** Planted for shade, it has invaded watercourses

Uitspan, Baviaans 19.02.2016

and roadsides throughout southern Africa. **Distribution** Native to the Peruvian Andes. Now an invasive weed in various parts of the world. **Notes** Incas used oil from the leaves to preserve their dead. Pink peppercorns are used in the food industry. **Other names** Umngcunube (X), umpelempele (Z).

SEARSIA (= *Rhus*)
Karee, korentebos, taaibos, intlolokotshane (X)

Named after American ecologist and botanist Paul Bigelow Sears, 1891–1990, professor of conservation at Yale University, among other achievements

Ecologically important, the plants in this large genus are dioecious (separate male and female plants) and easy to recognise by the trifoliate leaves, with the central leaflet usually slightly bigger than the 2 lateral leaflets. The foliage produces a resinous smell when crushed. Most species can resprout after fire. ± 111 species in southern Africa; 13 species in Baviaanskloof.

Searsia dentata   Nanabessie

dentata = serrated; referring to the margins of the leaves that are deeply toothed, dentate

Deciduous shrub to 2 m tall, becoming taller in higher rainfall, further north. **Leaves** Trifoliate, obovate, sometimes partly folded and curving downward, bright green above, paler below, margins deeply toothed, central leaflet 3.5 cm long × 2 cm wide; some leaves turn yellow-orange in autumn. **Flowers** With styles slightly fused at the base. Drupe round, shiny, 0.4 cm wide, pale to dark brown. Flowering in spring. **Habitat** Rocky outcrops in Grassy Fynbos at the margins of Forest. **Distribution** Widespread from Storms River to Limpopo. **Other names** Nana-berry (E), intlolokotshane yedobo (X), mabelebele (S).

Suuranysberg, Kareedouw 11.05.2021

Searsia incisa var. *effusa*   Rub-rub berry

incisa = incised; referring to the teeth in the margins of the leaf, which are more pronounced in *S. incisa* var. *incisa*

Multi-stemmed bush to 2 m. **Leaves** Trifoliate, discolorous (white-felted below), with well-lobed and/or dentate margins, central leaflet 3 cm long × 1.2 cm wide. **Flowers** Yellowish green. Flowering in winter. Drupe ellipsoid, rough-hairy (thus common names), to 1 cm wide, becoming rust-coloured. **Habitat** Occasional on rocky mountain slopes in Grassy Fynbos and Transitional Shrubland. **Distribution** Widespread from Richtersveld to Makhanda. **Other names** Baardbessie (A), unongqutu (X).

MK

Zandvlakte bergpad, Kouga 18.10.2011

Searsia lancea   Karee

lancea = lance-shaped; referring to the lance-shaped leaves

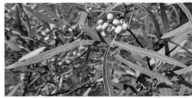

Medium-sized tree to 7 m tall, usually with several stems from the base. **Leaves** Trifoliate, sessile, linear-lanceolate, olive-green, 7 cm long × 1 cm wide. **Flowers** Greenish yellow. Drupe roundish, slightly flattened, hairless, shiny,

Baviaanskloof River, Vero's 28.10.2018

0.5 cm wide, becoming yellow and brown. Flowering in autumn and winter. **Habitat** Occurring along rivers and streams in Succulent Karoo and Thicket. **Distribution** Widespread from Ceres to Zimbabwe. **Notes** The tree house at Speekhout is in a grand old karee. The edible fruit of the karee has been used as an ingredient in mead or honey beer; 'karee' is the original Khoisan name for mead. The wood is hard and has been used for constructing wagons, fence posts, tool handles, bowls and tobacco pipes. **Other names** Iqunguwe, iqwela (X), mosotlhoana (S).

Searsia longispina Besemkraaibessie, doringtaaibos

longi = long, *spina* = spines; referring to the long, woody spines

Multi-stemmed, tough, armed, bushy tree to 4 m. The thorns are specially developed side branches, very tough, but not too sharp. **Leaves** Clustered, trifoliate, hairless, sometimes irregularly notched, olive-green above, slightly paler below, central leaflet 2.5 cm long × 1 cm wide. **Flowers** Pale yellow. Flowering in autumn and winter. Drupe lens-shaped, shiny, 0.5 cm wide, brown. **Habitat** Common in Thicket and Transitional Shrubland. **Distribution** Widespread from Namaqualand to southern KwaZulu-Natal. **Notes** Adapted to withstand the onslaught of browsing mega-herbivores. **Other name** Spiny currant (E).

Rus en Vrede, Kouga 25.04.2018

Searsia lucida Blinktaaibos

lucida = shining; pertaining to the glossy leaves

Multi-stemmed, neatly erect bush to 3 m. **Leaves** Trifoliate, sessile, obovate to spathulate, hairless, shiny, central leaflet 3.5 cm long × 1.5 cm wide, petiole 1 cm long, sometimes narrowly winged. **Flowers** Yellow. Drupe shiny, round, to 0.5 cm wide. Flowering in spring. **Habitat** Fynbos and Transitional Shrubland. **Distribution** Widespread from Citrusdal to Zimbabwe. **Other names** Intlokoshane, intlakoshanebomvu, umchane, amapozi (X).

Nooitgedacht, Kouga 20.01.2022

Searsia pallens Bleekkoeniebos

pallidus = pale; referring to the pale green leaves and
veins, especially when held up to the sun

Multi-stemmed, erect bush, with distinctly ribbed
young branchlets, to 3 m. **Leaves** Trifoliate,
sessile, oblanceolate, hairless, leaflets 4 cm long
× 1 cm wide, bright green to olive-green. **Flowers**
Yellow. Drupe elliptic, shiny, 0.5 cm wide, chestnut-
brown. Flowering in autumn and winter. **Habitat**
Rocky slopes in a wide range of vegetation types
except Mesic Fynbos. **Distribution** Widespread

Zandvlakte bergpad, Kouga 08.08.2010

from Robertson to KwaZulu-Natal. Common along roadside fence lines. **Notes** Easily confused with
S. undulata, which does not occur in Baviaanskloof. **Other name** Ribbed kuni-bush (E).

Searsia pterota Pendoringtaaibos

pterota = winged; referring to the winged petiole

Tough, spiny shrub to 2 m, with leaves in clusters.
Leaves Trifoliate, obovate, rigid, leathery, with
margins rolled under, central leaflet 3 cm long ×
1 cm wide, with winged petiole 2 cm long. **Flowers**
Pale yellow. Drupe ellipsoid, shiny, 0.5 cm wide,
dark reddish brown, drying black. Flowering in late
summer and autumn. **Habitat** Shale in Thicket.
Distribution Close to the coast from Saldanha
Bay to East London. **Notes** Often confused with *S.
longispina*, but differs in that it does not form into a
large rounded tree, the petiole is distinctly winged,
and drupe is a different colour. **Other name** Winged
currant (E).

Andrieskraal Road, Patensie 05.02.2020

Searsia pyroides Gewonetaaibos

pyros = fire; referring to the burning sensation felt when injured by the spines

Multi-stemmed tree to 7 m tall, sometimes with
sharp spines. **Leaves** Trifoliate, obovate, largest
leaflet 7 cm long × 3 cm wide, pubescent
above and slightly hairy below, dull olive-green.
Flowers Pale yellow. Drupe small, round and
smooth, to 5 cm wide, dull yellow to reddish.
Flowering in spring and summer. **Habitat**
Occurring along watercourses in Succulent
Karoo and Thicket. **Distribution** Widespread
from Namibia to Baviaanskloof and through
Free State to Limpopo. **Other names** Common
currant (E), incakotshi (X).

Suuranysberg, Kareedouw 11.05.2021

Searsia rehmanniana Suurtaaibos

Named after Anton Rehmann, 1840–1917, a Polish
botanist, geographer and explorer who visited
South Africa twice between 1875 and 1880, collecting
over 9,000 specimens

Deciduous, spreading shrub to 4 m. **Leaves** Trifoliate,
obovate, slightly hairy, sparsely toothed and very
variable in number and size of teeth on the same
plant, olive-green, central leaflet 5 cm long × 3 cm
wide. **Flowers** Greenish yellow. Drupe round and
smooth, to 0.4 cm wide, yellowish. Flowering in
summer and autumn. **Habitat** Stream and riverbanks
in Thicket. **Distribution** Widespread from Piketberg
to Limpopo. **Other names** Blunt-leaved currant (E), stompblaartaaibos
(A), umhlakothi, umhlanghoti, intlolokotshane, intlokotshane ephakathi,
imbokodi, incakotshi (X).

Skuinspadkloof,
Baviaans 01.02.2020

Searsia rosmarinifolia Roosmaryntaaibos

rosmarini = rosemary, *folia* = leaf; referring to the
leaflets resembling those of rosemary

Resprouting, relatively thin-stemmed, evergreen dwarf
shrub to 1 m. **Leaves** Trifoliate, sessile, discolorous,
linear, 3 cm long × 0.5 cm wide. **Flowers** Cream-
coloured. Drupe hairy, elliptic, 0.8 cm wide. Flowering
in autumn and winter. **Habitat** Grassy Fynbos and
Transitional Shrubland. **Distribution** Widespread from
Clanwilliam to Gqeberha (Port Elizabeth).

Engelandkop, Kouga 13.07.2011

Searsia tomentosa Korentebos

tomentosa = hairy; referring to the hairy undersides of the leaves and the drupes

Resprouting, multi-stemmed, evergreen
shrub to 2 m. **Leaves** Trifoliate, obovate,
discolorous, white-felted below, leaflets 4 cm
long × 2 cm wide. **Flowers** Cream-coloured.
Drupe ellipsoid, finely hairy, 0.5 cm wide.
Flowering in winter and early spring. **Habitat**
Occasional on rocky slopes in Grassy Fynbos
and Transitional Shrubland, and along
drainage lines. **Distribution** Widespread from
Clanwilliam to Zimbabwe. **Notes** The bark and
wood are rich in tannic acid and have been
used for tanning processes in the past.
Other names Bicoloured currant (E),
rosyntjiebos (A).

Witrivier, Cambria 04.02.2020

APIACEAE (= UMBELLIFERAE)   Celery, carrots and parsley

A large family containing more than 434 genera and over 3,700 species worldwide, including herbs such as anise, coriander, cumin, dill and fennel. The flowers of this family are quite distinctive and almost always in terminal umbels, explaining why the family is also known as the Umbelliferae or umbellifers. Many species produce phototoxic substances that sensitise human skin to sunlight. ± 40 genera and ± 165 native species in southern Africa; 13 genera with 19 species and 2 exotic species in Baviaanskloof.

ALEPIDEA   Kalmoes
a = without, *lepis* = a scale; referring to the fused involucral bracts

Several species are used in traditional medicine and spiritual practices. ± 40 species in southern Africa; 2 species in Baviaanskloof.

Alepidea capensis

capensis = from the Cape; referring to the geographical origin of the species

Resprouting, perennial herb to 40 cm tall when in flower. **Leaves** Crowded in a basal rosette, oblong, with long incurving bristles on the margins; each leaf 10 cm long, long-petiolate. **Flowers** In simple 15 cm-wide umbels, each flower to 1 cm, white or pink. Flowering in summer. **Habitat** Damp seepages on shale in Grassy and Mesic Fynbos. **Distribution** Widespread from Swellendam to KwaZulu-Natal. **Other names** Inkatazo, iqwili (X).

Cockscomb, Groot Winterhoek 17.02.2016

Alepidea delicatula

delicatula = small and dainty; referring to the small habit

Resprouting, perennial herb to 20 cm tall when in flower. **Leaves** In a basal rosette, with spreading bristles on the toothed margins, inter-tooth bristles incurved; each leaf 4 cm long × 1.5 cm wide. **Flowers** In simple umbels, to 1 cm wide, white or pink. Flowering in late summer. **Habitat** Damp situations in Mesic Fynbos. **Distribution** Outeniqua, Swartberg, Kammanassie and Kouga mountains. **Status** Rare.

Quaggasvlakte, Kouga 13.02.2016

ANGINON
Wildeseldery

Name for hemlock (*Conium maculatum*) given by Dioscorides, probably alluding to the poisonous nature of this species

13 species in Western and Eastern Cape; 1 species in Baviaanskloof.

Anginon difforme Common needle-leaf

difforme = unlike the usual; referring to the tall, rigid habit that is unusual in the genus and family

Resprouting, rigid, stiff-leaved, sparsely branched shrub to 3 m. **Leaves** In axillary tufts, terete, undivided, 15 cm long × 0.3 cm wide, lime-green. **Flowers** In compound umbels, tiny, yellowish. Flowering in summer and autumn. **Habitat** Grassy Fynbos and Transitional Shrubland. **Distribution** Widespread from Richtersveld to Grabouw and to Zuurberg.

Inflorescence visited by the balbyter ant (*Camponotus fulvopilosus*)

Nuwekloof Pass, Baviaans 19.12.2011

ARCTOPUS
Platdoring, sieketroos

arktos = a bear, *pous* = a foot; Linnaeus is responsible for naming this genus for its resemblance to a bear's paw.

3 species in the Western Cape, 1 species in Baviaanskloof.

Arctopus echinatus

echinatus = prickly; referring to the spines on the leaves and fruit

Resprouting, dioecious, stemless perennial to 5 cm tall, with thickened roots. **Leaves** Flat on the ground in a basal rosette, glossy, lobed to fringed, with bristly margins and spines on the surface, 15 cm long × 10 cm wide. **Flower** Male flowers white to pink on short stalks, female flowers basal and very spiny, especially when in fruit. Flowering in winter and spring. **Habitat** Flat areas with shallow soil in Grassy Fynbos and Transitional Shrubland. **Distribution** Widespread from Namaqualand to Makhanda. **Notes** The roots have been used in traditional medicine for treating venereal diseases.

The Heights, Langkloof 09.07.2019

CENTELLA

Pennywort, varkoortjies

centella = derivation unresolved

45 species in southern Africa; 1 species (*Centella asiatica*) pantropical; 2 species in Baviaanskloof.

Centella asiatica Waternael

asiatica = from Asia; a widespread species first known from Asia, but which has a worldwide distribution

Prostrate, trailing perennial herb, rooting at the nodes. **Leaves** Kidney-shaped, petiolate, sometimes slightly hairy, 2 cm wide, with crenate margins. **Flowers** Inconspicuous, on short stalks, 0.3 cm wide, reddish. Fruit obovate. Flowering in summer and autumn. **Habitat** Damp areas usually under Forest. **Distribution** Very widespread from Cape Peninsula to tropical Africa and other continents.

Waterkloof, Kouga Wildernis 15.05.2021

Notes Harvested and eaten raw in salads, cooked meals and beverages in Asia. Also used medicinally for various ailments. Easily confused with *Hydrocotyle verticillata*, which has the petiole meeting the leaf in the centre rather than at the base. **Other names** Pennywort (E), bolila-balinku (S), icudwane (Z).

Centella virgata Wortelgras

virgata = twiggy or branching; pertaining to the well-branched habit

Much-branched, upright or sprawling perennial to 60 cm. **Leaves** Linear, needle-like, to 5 cm long, sometimes woolly. **Flowers** Tiny, yellowish. Fruit slightly ribbed, terminal, ovate, longer than bracts, to 0.4 cm wide. Flowering in spring and summer. **Habitat** Occurring sporadically on rocky slopes and cliffs in Grassy Fynbos and Transitional Shrubland. **Distribution** Widespread from Kogelberg to Kariega (Uitenhage).

Smitskraal, Kouga 09.03.2012

HERMAS

hermas = a person's name; derivation unresolved, but possibly an individual that Linnaeus knew

9 species in Cape flora; 1 species in Baviaanskloof.

Hermas ciliata Doily tinder-leaf

ciliata = fringed with hairs; referring to
the fringe of hairs on the leaf margins

Resprouting, prostrate herb with leaves in a basal
rosette, to 60 cm tall when in flower. **Leaves** Elliptic,
base cuneate, glabrous above, white-felted below,
to 8 cm long, with hairy margins. **Flowers** Clustered
in a roundish, dense, compound umbel, white.
Flowering in summer and autumn. **Habitat** Damp,
partly shaded situations on upper slopes in Mesic
and Subalpine Fynbos. **Distribution** Widespread
from Du Toitskloof to Van Staden's Mountains.

Quaggasvlakte, Kouga 13.02.2016

HETEROMORPHA

hetero = dissimilar, *morpha* = forms; referring to the different varieties, or perhaps the variability in the
leaves on an individual plant

7 species from sub-Saharan Africa and the Arabian Peninsula; 1 species in Baviaanskloof.

Heteromorpha arborescens Parsley tree

arborescens = attaining the size or character of a tree; this species grows into a small tree.

Small tree with smooth, flaking bark, to 9 m. **Leaves** Either palmately
compound, trifoliate or simple, petiolate, elliptic-lanceolate, 7 cm long, pale
green or grey-green, with crenulate margins. **Flowers** In compound umbels
10 cm wide, each flower 0.3 cm wide, greenish white. Flowering in summer.
Habitat Occasional on
stream banks in or near
Forest. **Distribution**
Widespread from the
Cape to the Arabian
Peninsula. **Notes**
Used medicinally for
its antibacterial and
antifungal properties.
Other names
Wildepietersieliebos (A),
umbangeza, iyeza-
lempambano (X),
umbangadlala (Z).

Waterkloof, Bokloof 26.04.2018

LICHTENSTEINIA

Kalmoes, intlwathi (X)

Named after Professor Hinrich von Lichtenstein, 1780–1857, a German physician, explorer, botanist and zoologist who travelled around the Cape from 1803–1806 and was the director of the Berlin Zoological Museum

7 species in southern Africa; 1 species in Baviaanskloof.

Lichtensteinia latifolia

lati = broad, *folia* = leaf; referring to the broad leaves

Resprouting, prostrate, with basal leaves, to 1.4 m when in flower. **Leaves** Rotund to ovate, sometimes lobed to trifid, more than 30 cm long × 20 cm wide, with dentate margins. **Flowers** In compound umbels on long, hollow peduncles, each flower 0.5 cm wide, yellow. Flowering in summer. **Habitat** Middle to lower slopes in shallow, loamy soils in Grassy Fynbos. **Distribution** Widespread from Robertson to Kariega (Uitenhage).

Kouga Valley, Suuranysberg 01.11.2011

NANOBUBON

nano = northern, *Bubon* = plant genus; *Bubon* was a genus that was divided into new northern and southern genera, *Nanobubon* and *Notobubon* respectively.

3 species in Western and Eastern Cape; 1 species in Baviaanskloof.

Nanobubon capillaceum

capillaceum = slender; referring to the thin, linear leaves and stems

Resprouting, tufted perennial to 60 cm, with thick woody roots. **Leaves** Compound, needle-like, 15 cm long, petiole much longer than the blade, rachis and leaflets bent backwards. **Flowers** In a flat-topped umbel 10 cm wide, each flower 0.5 cm wide, cream. Flowering in summer. Fruit prominently ribbed, more than 1 cm long. **Habitat** Loam-rich soils in Grassy Fynbos. **Distribution** Widespread from Hottentots Holland Mountains to Kariega (Uitenhage).

Geelhoutbos Ruggens, Kouga 07.02.2012

NOTOBUBON

noto = southern, *Bubon* = plant genus; *Bubon* was a genus that was divided into new northern and southern genera, *Nanobubon* and *Notobubon* respectively.

12 species in southern Africa; 3 species in Baviaanskloof.

Notobubon ferulaceum

ferulaceus = reed-like; pertaining to the thin, rigid green stems that give the plant a reed-like habit

Resprouting, erect, simple-stemmed shrub to 1 m. **Leaves** Finely dissected, with visible venation, 10 cm long, same colour green on both sides, spaced in bunches up the stem. **Flowers** In flat-topped compound umbels, each flower small, 0.3 cm wide, yellow. Flowering in spring and summer. **Habitat** Rocky slopes in Mesic and Grassy Fynbos. **Distribution** Widespread from the Kouebokkeveld Mountains to Gqeberha (Port Elizabeth).

Cockscomb, Groot Winterhoek 17.02.2016

Notobubon gummiferum

gummiferum = producing gum; referring to the gum produced by this plant

Fairly robust shrub or small tree to 5 m. **Leaves** Compound, large, discolorous, glaucous and silvery below. **Flowers** Many in large, orbicular, compound umbels, to 15 cm wide. Each flower 0.7 cm wide, yellow. Flowering in spring and summer. **Habitat** Occurring sporadically along stream beds in forested kloofs. **Distribution** Widespread from Swellendam to Humansdorp.

Kaseykloof, Baviaans 28.10.2018

APOCYNACEAE
Dogbanes, milkweed, ubungxani, isiqaji (X)

Now including Asclepiadaceae. A large family of 366 genera and over 5,100 species, mostly in warm tropical parts of the world, especially Africa. Most produce a milky sap which is often poisonous and, in some cases, lethal if eaten. The type genus *Apocynum* means 'poisonous to dogs'. 38 genera and over 120 species in South Africa; 23 genera and 36 species in Baviaanskloof.

ACOKANTHERA
Gifboom

akoke = point, cutting edge, *anthera* = anther; pertaining to the pointed anthers

15 species in Africa; 1 species in Baviaanskloof.

Acokanthera oppositifolia Intlungunyembe

oppositi = opposite, *folia* = leaf; referring to the opposite leaves

Resprouting, woody shrub or tree to 4 m tall, with milky sap. **Leaves** Opposite, ovate, tough and leathery, glossy, with a distinct spine-like tip held upright, 6 cm long × 3 cm wide. **Flowers** In axillary clusters, fragrant, to 1 cm long, white tinged with pink. Flowering in autumn, winter, spring and early summer. **Habitat** Thicket and Forest. **Distribution** Widespread from Mossel Bay to tropical Africa. **Notes** Highly toxic, used to make a lethal poison applied to arrows by Khoisan hunters and warriors in the past. Also used in traditional medicine. **Other names** Bushman's poison bush (E), gifboom, boesmansgifpyl (A), ubuhlungu, intlungunyembe, inxinene (X), inhlungunyembe (Z).

Rooikloof, Baviaans 12.07.2016

CARISSA

Carissa = an Indian name ascribed to plants in this genus; the poisonous substance in the bark is called carrisin.

7 species, mostly subtropical; 4 species in South Africa; 1 species in Baviaanskloof.

Carissa bispinosa (= *C. haematocarpa*) **Num-num**

bi = two, *spinosa* = spined; referring to the paired spines

Resprouting, densely twiggy, rounded shrub to 5 m tall, with Y-shaped spines and milky sap. **Leaves** Opposite, ovate, glossy above, with a spine-like tip at the apex, 4 cm long × 2 cm wide. **Flowers** In dense clusters at the branch tips, each flower 1 cm wide, white, sweet-smelling. Fruit to 1.5 cm long, red, edible. Flowering in spring and summer. **Habitat** Succulent Karoo and Thicket. **Distribution** Widespread from Namibia to Gqeberha (Port Elizabeth) and to tropical Africa. **Notes** Some experts believe that *C. haematocarpa* is a distinct species that should be reinstated. **Other names** Noemnoem (A), isibetha nkunzi (X).

Waterkloof, Bokloof 26.04.2018

CYNANCHUM Bokhoring

kyon = dog, *agchein* = to kill; referring to the toxic nature of the species; the common name pertains to the shape of the fruiting body.

100 species, cosmopolitan; 13 species in South Africa; 3 species in Baviaanskloof.

Cynanchum ellipticum   Monkey rope, melktou

ellipticum = elliptic; pertaining to the shape of the leaves

corona

Climber with milky sap. **Leaves** Opposite, elliptic to ovate, 4 cm long × 2 cm wide, glossy green above, paler below. **Flowers** To 1 cm wide, with green or brown petals forming a recurved star; corona white and cup-shaped, to 0.5 cm wide, in clusters in the axils. Flowering in summer. **Habitat** Thicket and Forest. **Distribution** Widespread from George to Mozambique. **Notes** Poisonous.

Ysrivier, Cambria 18.05.2021

Cynanchum gerrardii

Named after William Tyrer Gerrard, 1831–1866, an English naturalist who collected in KwaZulu-Natal and Madagascar in the 1860s and discovered many species new to science. He died at an early age from yellow fever.

Zandvlakte 19.01.2022

Succulent climber to 1 m with finely striated stems and milky sap. **Leaves** Opposite, rudimentary and falling early. **Flowers** Small, 0.5 cm wide, with green reflexed petals and a white corona. Flowering all year round except late summer. **Habitat** Uncommon in Transitional Shrubland and Thicket. **Distribution** Very widespread from Baviaanskloof through Africa to the Arabian Peninsula and Madagascar.

Cynanchum viminale (= *Sarcostemma viminale*)   Melktou, giftou

vimen = long shoots that are easily bent; referring to the relatively thin pliant stems that tend to intertwine; derivation of the former name: *sarx* = flesh, *stemma* = garland, wreath; referring to the long succulent stems that could be rolled into a wreath

Resprouting, clambering or climbing, leafless stem succulent, to 3 m long, with milky sap. **Stems** 0.5 cm wide, grey-green, sometimes tinged purplish, with leaves reduced to minute scales. **Flowers** In dense umbels, fragrant, 1 cm wide, yellow and white. Flowering all year round, especially after rain. **Habitat** Common in all lowland habitats, especially Thicket. **Distribution** Very widespread from Clanwilliam through to India and Australia. **Notes** The sap has been used in traditional medicine and can cause burning of the skin. Some individuals are more toxic than others. **Other names** Iphozi, umbelebele (X), ntlalamela (S), ingotshwa (Z).

Zandvlakte 03.04.2012

DUVALIA
Gortjie

Named after Henri Auguste Duval, 1774–1814, a French botanist and physician with a special interest in succulents; he was the person who described the genera *Gasteria* and *Haworthia*.

16 species in Africa and the Arabian Peninsula; 3 species in Baviaanskloof.

Duvalia caespitosa subsp. *caespitosa*

caespitosa = tufted; pertaining to the tufted habit

Leafless dwarf succulent. **Stems** 4- or 5-angled, to 2 cm wide, with non-milky sap. **Flowers** 2.5 cm wide, dark reddish brown to almost black, with narrow petals and deflexed margins. Flowering in autumn, winter and spring. **Habitat** Shady areas in Thicket and Succulent Karoo. **Distribution** Widespread from Namibia through Karoo to Kariega (Uitenhage). **Notes** 2 other subspecies are not found in Baviaanskloof.

Kariega 28.04.2018

GOMPHOCARPUS
Katoenbos

gomphos = a club, *karpos* = fruit; referring to the club-shaped fruit of some species

Shrubs with milky sap and large inflated fruits; the sap of some species is poisonous.
± 30 species in Africa and the Arabian Peninsula; 3 species in Baviaanskloof.

Gomphocarpus cancellatus Gansiebos

cancellatus = latticed; referring to the flowerheads that appear lattice-like

Resprouting, rigid, hairy shrub to 1.5 m tall, with milky sap. **Leaves** Opposite, elliptic, leathery, rounded at the base, silvery-hairy, 5 cm long × 3 cm wide. **Flowers** In dense heads, to 5 cm wide, each flower to 1 cm wide, cream-coloured. Flowering in autumn, through winter and spring, to early summer. **Habitat** Occurring sporadically on rocky ground in various habitats. **Distribution** Very widespread from southern Namibia to Cape Peninsula and to Graaff-Reinet and Makhanda. **Other name** Bergmelkbos (A).

Langkloof 10.07.2020

Gomphocarpus fruticosus

fruticosus = shrubby; referring to the shrubby habit

Shrub with milky sap to 2 m. **Leaves** Linear, opposite, 7 cm long × 1 cm wide. **Flowers** Several in hanging umbels, 1 cm wide, white tinged with purple. Fruit inflated, ovoid-acute, covered with fleshy protuberances. Flowering in spring and summer. **Habitat** Disturbed areas in most habitats in the lowlands. **Distribution** Widespread throughout southern Africa and almost cosmopolitan. **Notes** Easily confused with *G. physocarpus*, which has round fruit and is a weedy, roadside species from tropical Africa. **Other names** Ukakhayi, igwada (X), lebejana, moethimolo (S), usingalwesalukazi (Z).

Zandvlakte, Baviaans 06.02.2016

GONIOMA

Kamassie

gonia = an angle, *ma* = result of an action; pertaining to the fruits that grow at right angles to the stalk

1 species in South Africa.

Gonioma kamassi Boxwood

kamassi = Khoisan name for this species

Neat, usually single-stemmed, upright tree to 6 m tall, with clear sap. **Leaves** Usually in whorls of 3 or 4, oblanceolate, smooth and hairless, glossy above, paler below, 8 cm long × 2 cm wide. **Flowers** In small clusters at the branch tips, pleasantly scented, 0.7 cm long, white. Fruit 2 horn-like carpels, green maturing to brown and dry, splitting to release papery, winged seeds. Flowering in spring and summer. **Habitat** An understorey tree in tall Forest. **Distribution** Mostly in southern and Eastern Cape with scattered populations in KwaZulu-Natal and Mpumalanga. **Notes** The sap is poisonous. **Other name** Igalagala (X).

Ysrivier, Cambria 18.05.2021

HUERNIA

Named after Justus Heurnius, 1587–1652, a Dutch missionary, doctor and first recorded collector of plants at the Cape. The genus name was misspelled by Robert Brown, who published the genus in 1810.

± 50 species in Africa and the Arabian Peninsula; 25 species in South Africa; 2 species in Baviaanskloof.

Huernia echidnopsioides (= *H. longii* subsp. *echidnopsioides*)

Echidnopsis = plant genus, *oides* = resembling; named after another genus in this family that has a remarkably similar growth form

Stoloniferous succulent to 6 cm tall, with well-spaced stems. **Stems** Cylindrical, 6 cm long × 1 cm wide, furrowed and with short bristles or lacking bristles. **Leaves** Absent. **Flowers** On peduncles to 0.7 cm long, flowers star-like, 3 cm wide, yellowish with small red to brown spots. Flowering in autumn. **Habitat** Rare on exposed, rocky conglomerate cliffs and steep slopes in Transitional Shrubland. **Distribution** Local endemic near Kouga Dam and Kariega (Uitenhage). **Status** Rare.

Kouga Dam 01.06.2014

Huernia thuretii

Named after Gustav Adolphe Thuret, 1817–1875, a French botanist who specialised in marine algae reproduction. He established a famous botanical garden in southern France to study how tropical plants adapted to the climate there; today the garden contains over 3,000 species.

Clump-forming stem succulent. **Stems** To 20 cm long × 2 cm wide, grey-green, sometimes mottled purple, with deltoid teeth. **Flowers** Campanulate, 3 cm wide, cream, sometimes covered with red to brown spots, smooth or with conical papillae. Flowering in summer and autumn. **Habitat** Shade of other shrubs in Thicket and Succulent Karoo. **Distribution** Eastern distribution from Willowmore to Makhanda.

Tafelberg, Rooihoek 10.03.2012

MICROLOMA
Bokhorings

mikros = very small, *loma* = fringe, edge; referring to the minute hairs in the flower tube and the tiny opening at the tip of the flower tube

10 species in southern Africa; 1 species in Baviaanskloof.

Microloma tenuifolium Kannetjies

tenui = slender, thin, *folium* = leaf; pertaining to the thin, linear, thread-like leaves

Climber to 50 cm, with swollen roots and clear sap. **Leaves** Opposite, drooping, up to 5 cm long × 0.3 cm wide. **Flowers** Several held upright in a tight bunch 2 cm wide, shiny; each flower 1 cm long, pinkish red. Flowering in autumn, winter and spring. **Habitat** South-facing slopes in Grassy Fynbos. **Distribution** Widespread from Gifberg to Cape Peninsula and to Gqeberha (Port Elizabeth).

Boplaats, Kouga 09.11.2021

NERIUM*

Nerion = the classical name for the oleander plant, derived from the Greek word for water; alluding to the preferred habitat of the genus along rivers and streams

Only 1 species worldwide; unclear where it originated, but most likely in the Mediterranean.

*Nerium oleander** Oleander, nerium

ollyo = I kill, *andros* = man; alluding to the toxicity of oleander

Multi-stemmed shrub or small tree to 6 m. **Leaves** Opposite or in whorls of 3, 10 cm long × 3 cm wide, dull green, veins distinct with a strong midrib and many parallel cross-veins. **Flowers** Clustered at the branch tips, 4 cm wide, pink, red or white. Fruit finger-like, 15 cm long, splitting to release many downy, wind-dispersed seeds. Flowering all year round but not in winter. **Habitat** Invasive weed in South Africa and other parts of the world. **Notes** Clearing operations have brought it under control along the Baviaanskloof River where it was once common. Poisonous. **Other name** Selonsroos (A).

08.02.2021

PACHYCARPUS

pachys = thick, *karpos* = fruit; pertaining to the large, leathery fruits

30 species in Africa; 25 species in South Africa; 2 species in Baviaanskloof.

Pachycarpus dealbatus Tongued thick-fruit

dealbatus = whitened, whitewashed; the fine white hairs that sometimes cover the leaves and stems make the plant appear whitewashed.

Resprouting, stout-stemmed, to 50 cm, with milky sap. **Leaves** Opposite, spreading, undulating, 8 cm long × 3 cm wide. **Flowers** Globose, to 1.5 cm wide, with petals and sepals recurved, pale green with purple markings. Flowering in late summer. **Habitat** Occurring sporadically in Grassy Fynbos. **Distribution** Widespread from Riversdale to KwaZulu-Natal. **Notes** The other species that may occur in Baviaanskloof in grasslands, *P. grandiflorus*, has globose flowers, yellow with purple-brown spots. **Other names** Igqobulenja (X), ishongwe, ukhatimuthi (Z).

10.01.2013

PACHYPODIUM

Dikvoet

pachys = thick, *podion* = foot of a vase, or *pous* = foot; alluding to the swollen, succulent stems or roots that are sometimes partly exposed

20 species in Africa and Madagascar; 5 species in South Africa; 2 species in Baviaanskloof.

Pachypodium bispinosum Dikvoet

bi = two, *spinosum* = spines; pertaining to the paired spines on the branches

Tough succulent shrub with partly exposed, swollen roots, and spiny branches to 60 cm. **Leaves** Lanceolate, glabrous above, hairy below, 1.5 cm long × 0.5 cm wide. **Flowers** Tube funnel-shaped, to 1 cm wide, lobes 0.5 cm wide, pink to purple, sometimes with white. Flowering in spring and early summer. **Habitat** Succulent Karoo, Thicket and Transitional Shrubland. **Distribution** From Calitzdorp to Makhanda. **Notes** An important and reliable food source for porcupines. Popular potplant subject. **Other names** Halfmens, hobbejaankos, kragman (A).

spines

Zandvlakte bergpad, Kouga 02.11.2011

Pachypodium succulentum Dikvoet

succulentum = juicy; alluding to the succulent stems and roots

Remarkably like *P. bispinosum.* **Leaves** Lanceolate, glabrous above, hairy below, 1.5 cm long × 0.5 cm wide. **Flowers** White with a dark reddish-pink line down the centre of the lobes, tube 0.4 cm wide. Flowering in spring and summer. **Habitat** Sunny, rocky slopes in Thicket, Succulent Karoo and Transitional Shrubland. **Distribution** Widespread from Ladismith and the Great Karoo to the Fish River. **Other name** Ystervarkkambroo (A).

Zandvlakte bergpad, Kouga 03.11.2011

SECAMONE

secamone = derived from an Arabic name, squamouna, for *S. aegyptiaca*

± 80 species in Africa, Madagascar and India, to Australia; 5 species in South Africa; 2 species in Baviaanskloof.

Secamone alpini   Monkey rope

Named after Prospero Alpino, 1553–1617, an Italian botanist. He drew the earliest known European illustration of the Arabian coffee plant (*Coffea arabica*).

Climbing or scrambling vine with milky sap. **Leaves** Opposite, shiny, held nearly perpendicular to the stem and in branched axillary clusters in some areas, 0.4 cm wide, finely hairy. **Flowers** With corona lobes erect and curving in over anthers, greenish yellow or white. Flowering in spring and summer. **Habitat** Occasional in Forest. **Distribution** Very widespread from Clanwilliam to tropical Africa. **Notes** Protuberances sometimes seen on the leaves are galls, abnormal growths formed in response to the presence of psyllids — tiny, sap-sucking lice. **Other names** Iyeza lentloko, ityholo (X).

galls formed by psyllids

13.04.2019

Secamone filiformis

fili = thread, *formis* = form; alluding to the narrow leaves or the thin climbing stems

Slender, slightly woody climber to 3 m, with milky sap. **Leaves** Opposite, linear-lanceolate, 3 cm long × 0.4 cm wide, pale green. **Flowers** Clustered in the leaf axils, small, 0.2 cm wide, yellowish, on short pedicels less than 1 cm long. Flowering in summer. **Habitat** Thicket. **Distribution** Widespread from Humansdorp to Zimbabwe. **Other names** Ubuka, ikhubalo, ikhubalo elimnyama, imbijela (X).

Ysrivier, Cambria 22.01.2022

STAPELIA
Carrion flower, aasblom

Named by Linnaeus in honour of Johannes Bodaeus von Stapel, 1602–1636, a Dutch physician from Amsterdam with an interest in botany, who worked on the Latin version of Theophrastus's *Historia Plantarum*. This work was originally written some time from c. 350–287 BC in ten volumes, of which nine survive. The Afrikaans common name aasblom means 'bait flower', and alludes to the bad smell.

Leafless stem succulents with foul-smelling flowers. 30 species mainly in southern Africa; 4 species in Baviaanskloof.

Stapelia grandiflora Makghaap

grandi = large, *flora* = flowers; alluding to the large flowers

Leafless, clump-forming, erect stem succulent. **Stems** 4-angled, 30 cm long × 3 cm wide, with erect whitish teeth on the angles. **Flowers** On the ground, large, 10–20 cm wide, purple-brown with cream, often silky-hairy, smelling of rotting fish. Flowering in autumn. **Habitat** Rocky sandstone soil at the edge of thicket clumps in Thicket. **Distribution** Widespread from Calitzdorp to Addo and to Cradock and the Free State. **Other name** Slangghaap (A).

Rooihoek, Kouga Valley 10.03.2012

Stapelia obducta

obducta = clouded, gloomy; referring to the dark flowers covered with fine hairs

Leafless, clump-forming, semi-erect stem succulent. **Stems** 4-angled, to 25 cm. **Flowers** Lobes strongly recurved, covered with fine hairs, to 4 cm wide, purple-brown, held on 6 cm-long peduncles. Flowering from autumn to spring. **Habitat** Uncommon on shaded, stony slopes and cliffs in Fynbos Woodland. **Distribution** Local endemic from Baviaanskloof Mountains to Cockscomb.

Uitspankloof, Baviaans 19.02.2016

Stapelia paniculata subsp. *kougabergensis*

paniculata = paniculate, *kougabergensis* = from the Kouga Mountains; flowers clustered into a short, paniculate inflorescence

Leafless, clump-forming, erect stem succulent. **Stems** 4-angled, 10 cm long × only 0.5 cm wide. **Flowers** To 3 cm wide, purple-brown, sometimes with fine white hairs. Flowering in autumn. **Habitat** North-facing rocky outcrops and ridges on hard sandstones in Arid or Grassy Fynbos. **Distribution** Local endemic restricted to the Kouga and Baviaanskloof mountains. **Notes** This subspecies was first recorded in the Kouga Mountains. 2 other subspecies, subsp. *paniculata* and subsp. *scitula*, occur on the West Coast near Doringbaai, and inland around Robertson.

ZandVlaktekloof, Baviaans 01.03.2010

TROMOTRICHE

tromo = trembling, *trichos* = hair; referring to the habit of hanging down from cracks on cliff faces, also alluding to the soft hairs on the flowers

11 species in Namibia and South Africa; 1 species in Baviaanskloof.

Tromotriche baylissii

Named after Roy Douglas Abbot Bayliss, 1909–1994, an English military officer and businessman who moved to Africa in 1947. He collected mainly succulents from various countries in Africa and has ± 7 succulent species named after him.

23.01.2003

Leafless stem succulent. **Stems** Pendulous or creeping to 3 m long × 1 cm wide, roughly 4-angled, dull green with grey blotches. **Flowers** Tubular-campanulate, 1.5 cm long, brown, with transverse corrugations in the tube. Flowering in summer and autumn. **Habitat** Cliffs in narrow, shaded kloofs in Thicket. **Distribution** Local endemic from Kouga Mountains to Cockscomb. Easily seen at the rock pool in Geelhoutboskloof.

XYSMALOBIUM

xysma = lint, covering, *lobos* = pod; referring to the pods of some species, which are covered with hairs

+ 43 species in Africa; 21 species in South Africa; 2 species in Baviaanskloof.

Xysmalobium gomphocarpoides

Gomphocarpus = plant genus, *oides* = resembling; alluding to the similarity to some species of *Gomphocarpus*

Resprouting, leafy shrub with few stems and a milky sap to 50 cm. **Leaves** Opposite, linear, wavy, 10 cm long × 0.5 cm wide, margins crisped, rough to the touch. **Flowers** In small umbels in the leaf axils, 2cm wide, pale green tinged purplish, lobes pointed and curving upwards, glabrous. Fruit to 10 cm long, covered in soft, greenish spikes. Flowering in spring, summer and autumn. **Habitat** Uncommon on shale soils in Succulent Karoo and Transitional Shrubland. **Distribution** Widespread from Bokkeveld and Karoo to KwaZulu-Natal.

Willowmore 16.01.2022

AQUIFOLIACEAE

Holly

Comprising 2 genera of over 400 species of trees and shrubs in the world. Best known for the familiar holly tree (*Ilex aquifolium*) that is used in many Christmas decorations in Europe. Only 1 species in southern Africa.

ILEX

ilex = based on the species name for holm oak (*Quercus ilex*) as the leaves look similar

1 species in Baviaanskloof.

Ilex mitis African holly

mitis = unarmed; referring to leaves that are entire, since most other members of the genus have toothed leaves with sharp points

Large tree to 30 m tall, dioecious, with greyish bark and reddish-purple twigs. **Leaves** Alternate, elliptic, 8 cm long × 3 cm wide, glossy green, with wavy margins; petiole sometimes reddish, midrib and petiole grooved above. **Flowers** In axillary clusters, fragrant, to 3 cm long, 0.3 cm wide, white. Fruit becoming bright red, small, to 1 cm wide, and round with a sharp point. Flowering in spring and summer. **Habitat** Stream banks in Forest in the narrow kloofs. **Distribution** Widespread from the Cape Peninsula to tropical Africa. **Notes** A soapy lather for washing can be produced by rubbing the leaves together. The Dendrological Society identified the depicted tree as big tree number 397 in November 2022. **Other names** Without, waterboom (A), umduma, isidumo, ububhubhu (X), iphuphuma (Z).

Uitspankloof, Baviaans 02.05.2018

ARALIACEAE

Ginseng

Mostly trees and shrubs with large compound leaves. Closely related to Apiaceae. Containing ± 700 species mostly in Southeast Asia and tropical America. Probably best known for the rice-paper plant (*Tetrapanax papyriferum*) which is used to make paper.

CUSSONIA

Cabbage tree, kiepersol

Named after Pierre Cusson, 1727–1783, who was a qualified medical doctor and became a professor of botany at the University of Montpellier, France

25 species in Africa and Madagascar; 8 species in South Africa; 3 species in Baviaanskloof.

Cussonia gamtoosensis Gamtoos cabbage tree

Gamtoos = a river and valley in the Eastern Cape, *ensis* = from;
referring to the Gamtoos Valley where this species is found

Relatively slender-stemmed, sometimes multi-stemmed, tree
to 4 m. **Leaves** 2-digitate, relatively small, 30 cm wide, dull
grey-green. Leaflets simple or with 1 or 2 teeth. **Flowers** Usually
in spikes of 4 forming an umbellate inflorescence at the
branch tips, parts of tree crowded in spikes to 10 cm
long, each flower 0.5 cm wide, greenish yellow. Flowering
in autumn. **Habitat** Thicket. **Distribution** Localised
endemic of the Gamtoos Valley around Hankey and
Patensie. **Status** Rare.

Hankey 25.10.2018

Cussonia paniculata Karoo cabbage tree

paniculata = paniculate; pertaining to the branched inflorescences

Thickset tree to 5 m. **Leaves** Compound, to 45 cm wide,
with 7–9 leaflets, shallowly to deeply lobed, pale blue-
grey-green (almost whitish). **Flowers** In robust panicles at
the branch tips, crowded on spikes to 13 cm long, each
flower 0.6 cm wide, green. Flowering in autumn. **Habitat**
Dry, rocky, sandstone slopes in Arid and Grassy Fynbos,
and Transitional Shrubland. **Distribution** Widespread
from Swartberg and Baviaanskloof to Mpumalanga.
Easily seen in Nuwekloof Pass. **Notes** The roots can be
peeled and eaten as a source of water. **Other names**
Nooiensboom, bergkiepersol (A), umsenge, umngqokhwe
(X), umsengembuzi (Z).

Nuwekloof Pass, Baviaans 24.05.2010

Cussonia spicata Cabbage tree

spicata = spike-like; referring to the spike-like inflorescences

Thick-stemmed tree to 10 m. **Leaves** Twice compound, to
70 cm wide, with 5–9 leaflets that are 2-digitate, leaves dark
or bluish green, bunched at branch tips. **Flowers** Parts of tree
crowded in spikes of 8–12 florets, forming a robust umbellate
head, with spikes up to 15 cm long × 4 cm wide, each floret
0.5 cm wide, greenish yellow. Flowering in summer and
autumn. **Habitat** Common in Thicket and Forest, especially in
fire-safe sites. **Distribution** Very widespread from Swartberg
to tropical Africa. **Notes** The most commonly seen cabbage
tree in Baviaanskloof, with spectacular tall specimens in
the slot-kloofs (local name for narrow ravines). The roots
are edible and have also been used for treating malaria. Its
soft wood was used to make break blocks for wagons. **Other
names** Nooiensboom, gewone kiepersol (A), umsenge (X).

Rus en Vrede 4X4, Kouga 01.12.2015

ASTERACEAE

Daisies, ubulawu (X)

1,911 genera with ± 33,000 species globally. This is probably the most diverse plant family in the world, with the orchid family in close competition. *Aster* means star and refers to the star-like inflorescence.

In southern Africa, there are ± 200 native genera and 2,270 species. Exotic species that have become naturalised or invasive include an additional 59 genera and 157 species. Approximately 42 genera and 139 species are cultivated for everyday foodstuffs. The common sunflower (*Helianthus annuus*) is cultivated for the oil extracted from its fruits and lettuce (*Lactuca sativa*) is cultivated as a popular salad ingredient. Chamomile, chicory and artichokes are other familiar members. Many species are popular in gardens and succulent species are sought-after by container-gardening enthusiasts. Several species are used medicinally and some have great potential as essential oils.

123 genera and ± 1,078 species of daisies make this the largest family in the Fynbos biome. There are over 53 genera and 200 species, and more than 10 exotic weed species, in Baviaanskloof.

Parts of a daisy flower

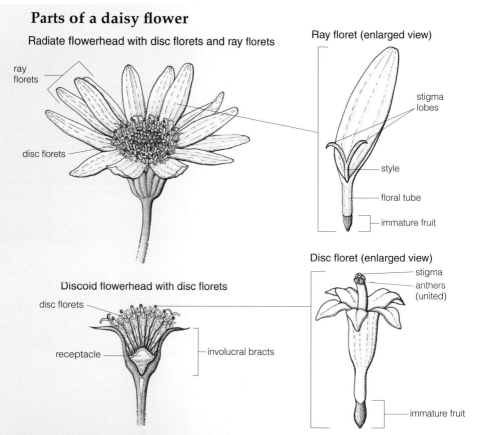

Radiate flowerhead with disc florets and ray florets

Ray floret (enlarged view)

ray florets

disc florets

stigma lobes

style

floral tube

immature fruit

Discoid flowerhead with disc florets

disc florets

receptacle

involucral bracts

Disc floret (enlarged view)

stigma

anthers (united)

immature fruit

All daisy plants have many tiny flowers or florets grouped into a flowerhead. Some species have a single flowerhead on a stalk, while others have compound inflorescences (a 'synflorescence'), consisting of a few to many flowerheads clustered together. The central ('button') part of a daisy comprises disc florets, and some taxa have an outer row of ray florets ('petals'). Disc and ray florets together on the same daisy make a radiate flowerhead. Some taxa have only disc florets, no ray florets – this is a discoid flowerhead. An essential first step when identifying daisies is to determine whether the specimen has a discoid or a radiate flowerhead.

Involucral bracts and fruits of the daisy family

A knowledge of the structure of the involucral bracts and fruits of daisies is helpful when trying to identify a member of the family. Dried flowerheads may persist on a plant long after flowering, making it possible to identify it by closely examining the bracts and fruit. The illustrations that follow highlight some of the more distinctive bracts and fruits of select genera.

Involucral bracts surround the base of the disc and ray florets. *Senecio* species have near-identical bracts lying next to each other, not overlapping, with calycles (small tooth-like leaves) at the base of the bract. *Cullumia* and *Berkheya* have prickly bracts with sharp awns. *Helichrysum*, *Achyranthemum* and *Syncarpha* never have ray florets, and their bracts have a papery texture. *Othonna* species have a single row of bracts, smooth and green, partly fused to form a cup, without calycles.

Daisy fruits are consistent across all or most members of a genus, and can be used to identify genera without looking at any other part of the plant. Noteworthy examples are *Ursinia* species with their umbrella-like fruits, *Cullumia* species with nut-like fruits, and *Eriocephalus* species with woolly fruits. The colour and nature of the pappus (structures attached to the fruits and aiding dispersal) can also help with identification. In *Senecio* and *Othonna*, the pappus is white, soft and flimsy, but in *Pteronia* the pappus is rigid, straw-coloured, and sometimes with bristles. In some genera, the pappus is lacking and the fruit is a simple nut (*Cullumia*) or a winged nut (*Dimorphotheca*). Most daisies have wind-dispersed fruit, but a noteworthy exception is bitou (*Osteospermum monoliferum*), which has berry-like fruit that is dispersed by birds, and some osteospermums are dispersed by ants.

Distinctive involucral bracts and fruits of select genera

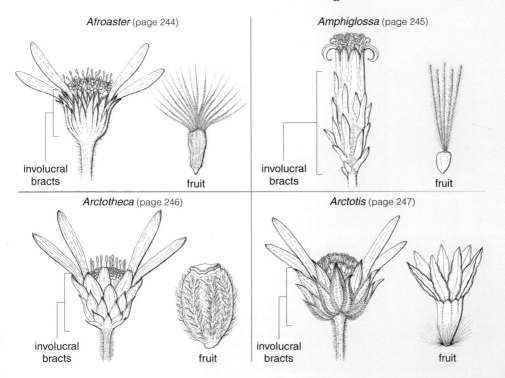

Afroaster (page 244)
involucral bracts
fruit

Amphiglossa (page 245)
involucral bracts
fruit

Arctotheca (page 246)
involucral bracts
fruit

Arctotis (page 247)
involucral bracts
fruit

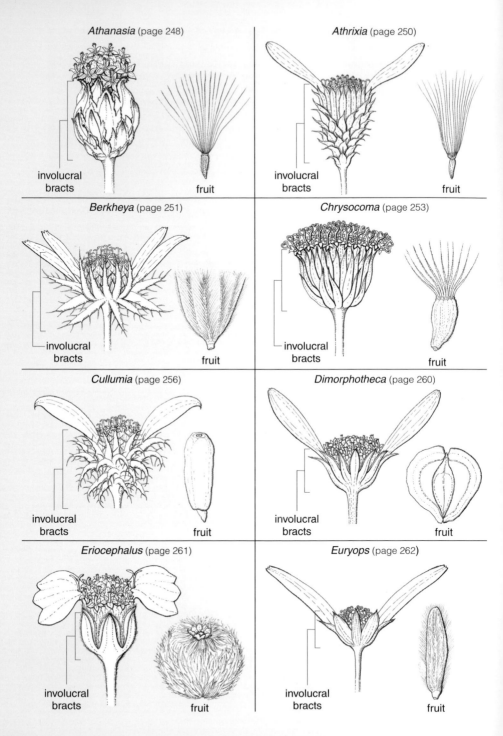

Athanasia (page 248)

involucral
bracts

fruit

Athrixia (page 250)

involucral
bracts

fruit

Berkheya (page 251)

involucral
bracts

fruit

Chrysocoma (page 253)

involucral
bracts

fruit

Cullumia (page 256)

involucral
bracts

fruit

Dimorphotheca (page 260)

involucral
bracts

fruit

Eriocephalus (page 261)

involucral
bracts

fruit

Euryops (page 262)

involucral
bracts

fruit

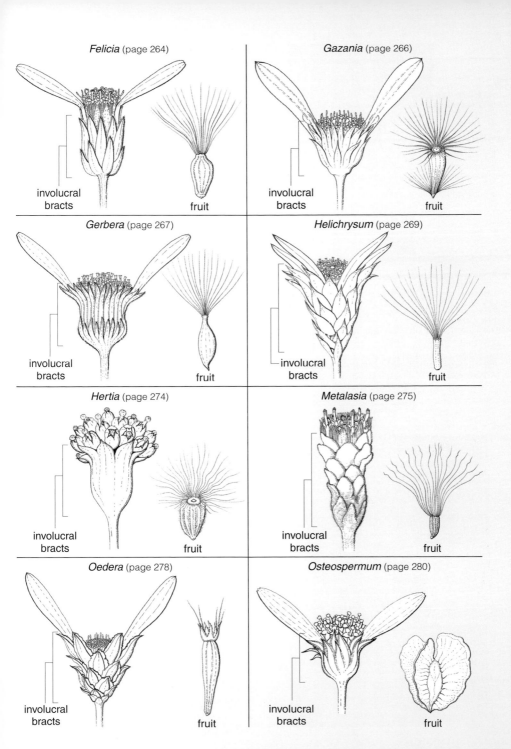

Felicia (page 264)

involucral bracts

fruit

Gazania (page 266)

involucral bracts

fruit

Gerbera (page 267)

involucral bracts

fruit

Helichrysum (page 269)

involucral bracts

fruit

Hertia (page 274)

involucral bracts

fruit

Metalasia (page 275)

involucral bracts

fruit

Oedera (page 278)

involucral bracts

fruit

Osteospermum (page 280)

involucral bracts

fruit

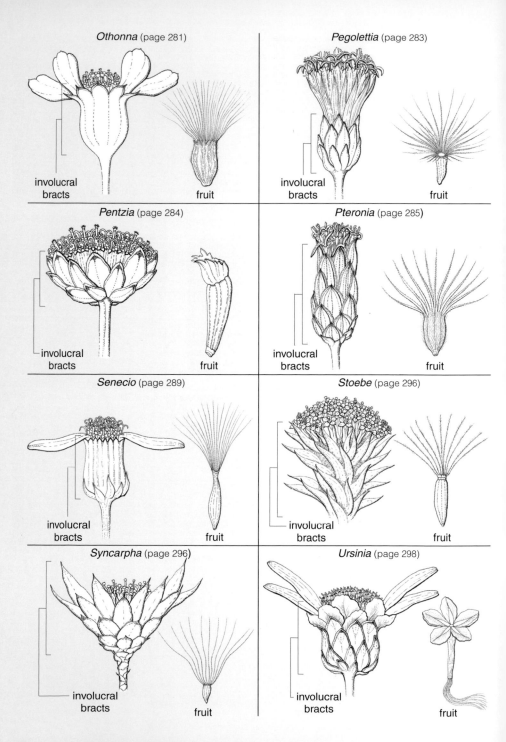

Othonna (page 281)

involucral bracts

fruit

Pegolettia (page 283)

involucral bracts

fruit

Pentzia (page 284)

involucral bracts

fruit

Pteronia (page 285)

involucral bracts

fruit

Senecio (page 289)

involucral bracts

fruit

Stoebe (page 296)

involucral bracts

fruit

Syncarpha (page 296)

involucral bracts

fruit

Ursinia (page 298)

involucral bracts

fruit

ACANTHOSPERMUM* Starburr

akanthos = thorn, *sperma* = seed; pertaining to the spiny fruits or burrs

3 species native to South America, all of which have become weeds on other continents. The burrs contaminate and downgrade sheep's wool. 3 species in southern Africa; 1 species in Baviaanskloof.

*Acanthospermum australe** Paraguayan starburr, prostrate starburr, kruipsterklits

australe = southern; referring to its geographical origin in South America

Prostrate, creeping, mat-forming, short-lived perennial herb. **Leaves** Rhombic-ovate to triangular, 3 cm long × 2 cm wide, with irregularly serrated margins; the surface is faintly gland-dotted. **Flowerheads** Tiny heads, 0.5 cm wide, with a few white-yellow ray and disc florets. Fruit is a single-seeded, prickly burr; several are clustered together in a bunch. Flowering all year round. **Habitat** Invading roadsides and other disturbed areas. **Distribution** Native to South America, now a weed in North America, southern Africa and Australia.

Suuranysborg, Kareedouw 11.05.2021

ACHYRANTHEMUM Chaff flower

achyron = chaff, husk, straw, *anthemon* = flower; referring to the papery bracts of the flowerheads

A genus recently instated (2019) for a distinct group of syncarphas. 7 species in South Africa; 2 species in Baviaanskloof.

Achyranthemum paniculatum (= *Syncarpha paniculata*) Sewejaartjie

paniculata = paniculate; referring to the paniculate inflorescence

Reseeding, upright, silvery-felted shrublet to 50 cm. **Leaves** Ascending, linear, apiculate, 1.5 cm long × 0.5 cm wide. **Flowerheads** Few, in a branched, terminal synflorescence, each flowerhead 1 cm wide, tinged pink when immature, maturing to white, with acute and papery bracts. Flowering in spring, summer and autumn. **Habitat** Most Fynbos habitats. **Distribution** Widespread from Gifberg to Makhanda.

Oubos, Krakeel 06.10.2011

Achyranthemum striatum (= *Syncarpha striata*)

striata = striated; referring to the 3-veined leaves that
appear striated with grooves or lines

Reseeding, upright, well-rounded, leafy shrub
to 60 cm. **Leaves** Linear, ascending, 3-veined,
hairy, sometimes slightly curved. **Flowerheads**
Few to several, in loose clusters at the tips of the
branches, each 1 cm wide, white, with acute and
papery bracts. Flowering in spring, summer and
autumn. **Habitat** Grassy Transitional Shrubland
and Grassy Fynbos. **Distribution** Widespread from
Humansdorp and Patensie to East London. The
other species are found closer to the coast, and
mostly to the east of this study area.

Hankey 25.10.2018

AFROASTER (= *Aster*)

afro = from Africa, *aster* = star; referring to its star-like shape

Distinguished by the pointed glandular hairs covering the fruit, and the barbed
bristles and row of scales on the fruit. Easily confused with *Felicia*. ± 18 species
in Africa; 3 species in South Africa; 2 species in Baviaanskloof.

fruit

Afroaster hispida (= *Aster bakerianus*)   Udlatshana (Z)

hispida = bristly or rough; alluding to the leaves that feel
like sandpaper and are covered in small tough hairs

Tough, resprouting, with a few erect stems to 70 cm, but usually only
30 cm. **Leaves** Few, held upright, lanceolate, 6 cm long × 1 cm wide,
with revolute margins with slight teeth, 3–5 veined from the base.
Flowerheads Solitary, 3.5 cm wide, with yellow centres and white or
blue rays, at the tips of 20 cm-long scabrid peduncles. Flowering mostly
after fire in summer and autumn. **Habitat** Fairly common in rocky areas
in Sour Grassland or Grassy Fynbos. Seldom seen in flower. **Distribution**
Widespread from Outeniqua
Mountains to East African
montane grasslands. **Notes**
A decoction of the roots has
been used medicinally and as
a purgative. The other species
in the area, *A. laevigatus*, has
been recorded on Cockscomb
Mountain, previously known to
occur only on the Van Staden's
Mountains. **Other names**
Noxgxekana, umthekisana (X),
umhlungwana (Z).

Rus en Vrede 4X4, Kouga 25.04.2018

AMELLUS

Astertjie

Amellus = name of a flower first used in a poem by the Roman poet Virgil, in 29 BC; describing a daisy with blue flowerheads

Easily confused with *Felicia*, but *Amellus* fruit have bristles and scales. 12 species in southern Africa; 1 species in Baviaanskloof.

Amellus tridactylus

tri = three, *dactylus* = finger, toe or claw; referring to the 3 teeth on some of the leaves

Annual to 10 cm. **Leaves** Oblanceolate and hairy, 1.5 cm long. **Flowerheads** Radiate, single, 2 cm wide, yellow with bright blue-purple rays. Fruit with a pappus of scales and bristles. Flowering from winter through to spring. **Habitat** Occurring in disturbed areas where more nutrients are available, often areas visited frequently by stock or game in Transitional Shrubland. **Distribution** Widespread from Namibia to Karoo and Baviaanskloof. **Notes** The other species occurring in Baviaanskloof, *A. strigosus*, has leaf hairs adpressed, while this species has spreading hairs.

Nooitgedacht, Kouga 26.09.2011

AMPHIGLOSSA

Kopseerbossie

amphis = both ways, *glossus* = tongue; referring to the stigma with 2 tongues or the ray florets

Recognised by the small flowerheads, a chaffy involucre, and papillose fruit that has a finely bristly and plumose pappus. 11 species in South Africa and Namibia, mostly in the western parts of the Karoo; 1 species in Baviaanskloof.

fruit

Amphiglossa callunoides

Calluna = a genus of heather from Europe and Asia, *oides* = resembling; resembling the many-flowered, spiked inflorescences of the genus *Calluna*

inner bracts mucronate

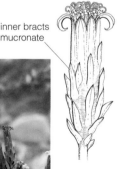

Very tough, sprawling, with branches arching over and rooting at the nodes. Thinly cobwebby shrublet to 50 cm. **Leaves** Narrowly lanceolate, adpressed, 0.2 cm long. **Flowerheads** Radiate, small, 0.6 cm long, solitary on tips of short, erect, shoots at right angles to the arching stem. Rays small, 4–5 per head. Flowering in late summer. **Habitat** Growing in boulder beds on open sand banks. **Distribution** Endemic to the Kouga and Baviaanskloof riverbeds. **Notes** The root system helps to stabilise recently formed boulder beds following scouring events. **Status** Vulnerable.

Rooihoek, Kouga River 10.03.2012

ARCTOTHECA

arctos = brown bear, *thecus* = capsule or case; the fruit is woolly like a bear.

Annual or perennial herbs. The inner bracts have broad membranous margins; outer bracts have a fleshy apical appendage. 5 species in southern Africa; 2 species in Baviaanskloof.

fruit

Arctotheca calendula   Cape weed, botterblom

calendae = little calendar; acknowledging that the plant is a reliable indicator of the arrival of spring as the flowers always bloom then

Tufted to sprawling annual to 20 cm. **Leaves** Mostly basal, hairy, dissected, woolly below, 16 cm long × 4 cm wide, dying back early. **Flowerheads** With rays, singular 5 cm-wide heads, on rough scapes, with short, reflexed outer bracts; achenes woolly, pappus chaffy. Flowering mostly in spring. **Habitat** Common on roadsides and disturbed areas, mainly in the valley floors. **Distribution** Widespread from Namaqualand to the Cape Peninsula and to Humansdorp, near the coast. **Other name** Isiqwashumbe (X).

Bosrugkloof, Baviaans 04.09.2006

Arctotheca prostrata

prostrata = prostrate; referring to the prostrate growth habit over the soil surface

Perennial herb, prostrate, sprawling, rooting at nodes. **Leaves** Auriculate at base, with soft hairs, 15 cm long × 4 cm wide. **Flowerheads** 4 cm wide, yellow with white-tipped involucral bracts and silky achenes. Flowering mainly in spring. **Habitat** Occurring in seasonally damp riverbeds and roadside ditches. **Distribution** Widespread from Cederberg to KwaZulu-Natal. **Notes** An easy-to-grow perennial. Useful for stabilising damp riverbeds that dry up periodically.

Nuwekloof Pass, Baviaans 08.10.2016

ARCTOTIS

Arctotis, gousblom

arco = wedge in, pack, tighten, compress, *totus* = all together; alluding to the flowerheads that are packed tightly together

Bracts in this genus are similar to those of *Arctotheca*. Fruits are often with a tuft of hairs at the base and a pappus of scales. ± 60 species in southern Africa; 2 species in Baviaanskloof.

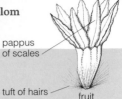

pappus of scales

tuft of hairs

fruit

Arctotis acaulis Renostergousblom

acaulis = without a stem; referring to the low and prostrate habit

Perennial to 20 cm, with a woody rootstock. **Leaves** In a rosette on the ground, usually lyrate-pinnatifid to oblanceolate, often tinged purplish, rough-hairy above, grey-felted below. **Flowerheads** 5 cm wide, with black-and-yellow or orange centres, ray petals white or yellow, outer bracts with a thin woolly tip. Flowering in spring. **Habitat** Occurring in loam-rich soils in Transitional Shrubland and Grassy Fynbos. **Distribution** Widespread from Bokkeveld Mountains to Cape Peninsula and to Langkloof.

Baviaans 10.09.2011

Arctotis arctotoides (= *Venidium arctotoides*) Bitterguusblom

arctotoides = resembling the genus *Arctotis*

Perennial, creeping, mat-forming. **Leaves** Tufted, convex, to 15 cm long, lobed and toothed, usually hairless above, white-felted below. **Flowerheads** Radiate, up to 4 cm wide, yellow. Flowering in spring, summer and autumn. **Habitat** Occurring in damp stream banks or drainage ditches in lowland habitats in Succulent Karoo. **Distribution** Widespread through southeastern and central South Africa. In Baviaanskloof it can be seen around the campsite at the entrance to Uitspankloof. **Notes** Used in traditional medicine for treating epilepsy, indigestion and other stomach problems. Juice from the leaves, or a paste of the leaves, is proven to have antibacterial properties and has been used to treat sores. The plant makes a useful groundcover for stabilising eroding dongas. It is recommended for domestic gardens and is easily grown from cuttings or seed. **Other names** Botterblom (A), isigwamba, ubushwa (X), putswa-pududu (S).

Uitspankloof, Baviaans 19.02.2016

Arctotis linearis

linearis = linear; referring to the shape of the leaves, which seem to grow more linear with age; younger leaves are oblanceolate.

Tufted, grey-felted perennial to 40 cm. **Leaves** Toothed, 10 cm long, pale grey. **Flowerheads** 5 cm wide, solitary on long spathes that hang over in bud, erect when fertile. Disc black, rays yellow, bracts obscurely tailed; achenes with 2 elongate cavities. Flowering all year round. **Habitat** Fairly limited. Most common in first few years after fire in Grassy Fynbos. It resprouts in some situations and regenerates well from seed after fire. **Distribution** From the Langeberg to Patensie, mostly in the Kouga Mountains.

Zandvlakte, Kouga 04.12.2015

ARTEMISIA
Named after Artemis, the Greek goddess of hunting

Most have aromatic leaves and a very bitter taste. Several species are used in traditional medicine. A drug derived from *A. annua* is a preferred treatment for malaria. ± 400 species, mostly in the northern hemisphere; 2 species in southern Africa, also in Baviaanskloof.

Artemisia afra   Wormwood, wilde-als

afra = from Africa; native to Africa

Resprouting, much branched from the base, erect, soft, aromatic, greyish shrub to 1.5 m. **Leaves** Bipinnatifid, 6 cm long × 3 cm wide, grey-green above, finely white-hairy below. **Flowerheads** Discoid, crowded in a false spike at the branch tips, 0.4 cm wide, yellowish. Flowering in autumn. **Habitat** Damp sites on shale in Grassy Fynbos or on stream banks and Forest margins. **Distribution** Widespread from the Cape to tropical Africa. **Notes** Used in traditional medicine. Dispensed to treat malaria, colic, fever, headache, coughs, colds, earache and intestinal worms. A traditional way to clear the sinuses is

Ysrivier, Cambria 04.02.2020

to roll the fresh leaves into a ball and insert it into the nostrils. It is also believed to act as an appetite suppressant. Some recipes for medicinal hot toddies call for the addition of a leaf of wormwood. **Other names** Umhlonyane (X, Z), lengana (S).

ATHANASIA   Klaaslouwbos
a = without, *thanatos* = death; pertaining to the persistent dry involucral bracts, and the tendency for many species to survive through droughts

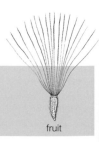

For easy identification, look for the corymbose inflorescence with many heads (never with ray florets), the scale-like bracts below the flowers, and the distinctive leaves with minute branched hairs (only visible with a hand lens). ± 40 species in southern Africa; 6 species recorded in Baviaanskloof.

fruit

Athanasia linifolia

lini = line, thread, *folia* = leaf; referring to the linear shape of the leaves; the midrib has a slightly raised line on the outside surface.

Erect, slender, reseeding shrub to 2 m. **Leaves** Linear, ascending close to the stem, hairless, 3 cm long × 0.2 cm wide. **Flowerheads** Discoid, several in terminal corymbs (synflorescence) to 5 cm wide, each flowerhead 1 cm wide, yellow. Bracts glabrous, shiny, pale yellow, with a dark black tip, margins serrate, with scales. Flowering from spring through to summer. **Habitat** Upper slopes of Kouga Mountains in Grassy and Mesic Fynbos. **Distribution** Widespread in Cape Mountains from Cederberg to Kouga mountains.

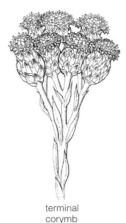

terminal corymb

Engelandkop, Kouga 25.09.2011

Athanasia pinnata

pinnata = pinnate; referring to the leaves that appear to be pinnate

Handsome, erect, grey-velvety, sparsely branched, reseeding shrub to 2 m. **Leaves** Dense, ascending, tightly packed (especially during dry periods), pinnatisect with linear to filiform lobes, 4 cm long, covered with fine silver-grey hairs. **Flowerheads** Discoid, many very densely packed in terminal corymbs 5 cm wide, each head to 0.5 cm wide, yellow, with scales, bracts woolly-hairy. Flowering in early summer. **Habitat** Occurring sporadically on rocky slopes in Arid and Grassy Fynbos. **Distribution** Eastern Cape from Kouga Mountains to Makhanda. Easy to see in Nuwekloof Pass (R332).

Koud Nek's Ruggens, Kouga 07.11.2011

Athanasia tomentosa

tomentum = wool or hair; referring to the leaves that are covered with silvery-grey hairs

Erect, reseeding, velvety grey shrub to 2 m. **Leaves** Simple, entire, oblanceolate, sometimes with a few teeth, covered with silvery-grey hairs, 3 cm long × 0.4 cm wide. **Flowerheads** Discoid, many densely packed in terminal corymbs 5 cm wide, each head very narrow, 0.3 cm wide, pale yellow. Involucre covered with fine velvety hairs. Flowering in summer. **Habitat** Rocky slopes in Arid Fynbos, sometimes in semi-shaded, open areas in Fynbos Woodland and Passerina Veld (a kind of Transitional Shrubland), and on sloping sandstone ledges with exposed bedrock. **Distribution** Nearly endemic from mountains in the Little Karoo to the Baviaanskloof and Kouga mountains.

Nuwekloof Pass, Baviaans 04.12.2011

Athanasia vestita

vestitus = clothed; the needle-shaped leaves covering most of the stems and branches are symmetrically arranged, making this a 'well-dressed' plant.

An attractive, resprouting, much-branched, compact, well-leaved, rounded shrub to 1.5 m. **Leaves** Linear, to 2 cm long, needle-like (0.1 cm wide), spreading. **Flowerheads** Many in a flattish-topped, corymbose synflorescence, 5 cm wide; individual heads small, 0.3 cm wide, with fewer than 10 florets. Flowering in spring and summer. **Habitat** Occurring in Grassy Fynbos and Transitional Shrubland. **Distribution** Narrow from Robertson to the Kouga Mountains. **Notes** A plant with good potential for domestic garden spaces.

Zandvlakte bergpad, Kouga 04.12.2015

ATHRIXIA

thrix = hair; referring to the fine point at the end of the involucral bract

Leaves are narrow and woolly or felted beneath. Bracts are also woolly, with bristly tips. The fruit is ribbed, hairy and with a pappus of bristles and/or scales. An important medicinal plant. 14 species in Africa and Madagascar; 10 species in South Africa; 1 species known from Baviaanskloof.

fruit

Athrixia heterophylla   Iyeza logezo (X)

hetero = diverse, variable, *phyllon* = leaf; alluding to the leaves that are diverse in shape and size, and are also discolorous

Small shrub, 10–45 cm, resprouting from a woody rootstock. **Leaves** Simple, ovate to linear, mucronate, decurrent, sandpapery, 2 cm long, green above, felted beneath, with transparent bristly hairs above, and involute margins; lower leaves broader. **Flowerheads** Solitary, 2–3 cm wide, on short stalks 3 cm long. Bracts acuminate, lower bracts curling back, felted white. Rays pink or magenta with yellow centre. Flowering in spring and summer. **Habitat** Occurring in Grassy and Mesic Fynbos, often on shale bands or where the soil is loamier. **Distribution** Widespread from Cape Peninsula to Gqeberha (Port Elizabeth). **Notes** This is not the sought-after species *A. phylicoides*, which is used as a sore throat and cough remedy, but it probably holds the same potential.

Zandvlakte, Kouga 30.10.2011

side view of solitary flowerhead with overlapping involucral bracts; here showing only 2 rays

BERKHEYA　　　　　　　　Wild thistle, wildedissel

Named after Dutch botanist, Johannes Lefrancq van Berkhey, 1729–1812, best known for his illustrations of natural subjects

All species have spines on the leaves and involucral bracts, often also on the stems. Heads can be with or without ray florets. ± 75 species in southern and tropical Africa; ± 25 species in the Cape; 5 species in Baviaanskloof.

fruit

Berkheya angustifolia

angusti = narrow, *folia* = leaf; referring to the relatively narrow leaves

Attractive, dense, rounded shrub to 60 cm, resprouting and much branched from a woody base. **Leaves** Narrowly lanceolate, spiny, leathery, white-felted underneath, 2 cm long × 0.5 cm wide, discolorous. **Flowerheads** Radiate, singular at branch tips, large, to 5 cm wide, yellow, with spiny bracts. Flowering from winter to spring. **Habitat** Uncommon in Baviaanskloof, found in Grassy Fynbos. **Distribution** Widespread from Ladismith to Makhanda.

radiate flowerhead

Engelandkop, Kouga 30.10.2011

Berkheya carduoides　　Ikhakhakhaka (X)

carduus = thistle, *oides* = resembling; referring to the similarity between this and other thistle-like daisies of the genus *Carduus*

Erect perennial to 60 cm tall that sprouts after fire. **Leaves** Basal leaves tufted and rosulate, oblanceolate, pinnatifid, with lobes rounded, toothed and spiny, 15 cm long × 3 cm wide, with revolute margins; decurrent on stems and appearing wing-like. **Flowerheads** Few in a corymb at the end of a single stem, discoid, to 2 cm wide, yellow. Flowering in summer. **Habitat** Occurring in Grassy Fynbos. **Distribution** From Swellendam to Bathurst.

Guerna, Kouga 06.02.2012

Berkheya cruciata　　Disseldoring

cruciata = cross-shaped; alluding to the spines at the tips of the leaves and bracts that look like crosses

Glabrous shrub to over 1 m tall, killed by fire, regenerating from seed. **Leaves** Rigid, auriculate-ovate, toothed and spiny, spreading, 5 cm long × 2 cm wide, grey-green, streaked with white veins. **Flowerheads** In loose corymbs at the end of the branches, with short rays, 3 cm wide, yellow, with very spiny involucral bracts. Flowering in summer. **Habitat** Occurring in most kinds of Fynbos, but mostly rocky slopes in Arid Fynbos. **Distribution** Near endemic, with narrow distribution from the Swartberg to Kouga and Baviaanskloof mountains.

Verberg, Kouga 12.02.2016

BRACHYLAENA

brachus = short, *klaina* = a cloak; referring to the short bracts on the flowerheads

Dioecious trees and shrubs with alternate leaves. Coastal silver oak (*B. discolor*) is fast-growing and popular in gardens and car parks. 15 species in Africa, Madagascar and the Mascarene Islands; 9 species in southern Africa; 2 species in Baviaanskloof.

Brachylaena glabra   Malbar

glabra = smooth and hairless; referring to the glabrous upper surfaces of the leaves

Tall, dioecious tree to 15 m tall, with young leaves and buds felted with rusty brown hairs. **Leaves** Obovate to broadly lanceolate, strongly petioled, 15 cm long, with a few blunt teeth near the apex. **Flowerheads** Many clustered in a terminal, branched synflorescence, each to 1 cm wide, white. Seeds wind-dispersed. Flowering in autumn, winter and spring. **Habitat** Occurring in damp Forest, usually near mountain streams. **Distribution** Widespread from Humansdorp to KwaZulu-Natal. **Other names** Silver-oak (E), amacirha (X).

Ysrivier, Cambria 03.02.2020

Brachylaena ilicifolia   Small bitterleaf

Ilex = Cape holly genus, *folia* = leaf; the leaves are reminiscent of those of the Cape holly.

Tough, rigid shrub or small tree to 3 m. **Leaves** 4 cm long, bright glossy green above, densely felted below, and margins with small pointy teeth. **Flowerheads** Few in axils, 0.6 cm wide, creamy yellow. Flowering in winter and spring. **Habitat** Occasional in dry Thicket. **Distribution** Widespread from Baviaanskloof to Mozambique. **Other name** Hulsbitterblaar (A).

Grootrivierpoort 03.02.2020

CAPUTIA   (= *Senecio*)

Genus *Caputia* was created in 2012 for some succulent members of *Senecio*. *Caput bonae spei* was the original Latin name for the Cape of Good Hope. The name alludes to the geographical origin of the genus, as this name was applied to the wider area of southern Africa, not only the Cape Peninsula.

The involucre of *Caputia* is identical to that of *Senecio*. 4 species in South Africa; 2 species in Baviaanskloof.

Caputia pyramidata

pyramidata = pyramid-shaped; referring to the pyramid-like clusters of leaves up the stem, as well as the inflorescence

Thick-stemmed, succulent, white-woolly shrub to 1 m. **Leaves** Cylindric, 12 cm long × 1 cm wide, crowded at the branch tips. **Flowerheads** With rays, yellow, 3 cm wide, on a long white-woolly peduncle. Flowering in winter and spring. **Habitat** Thicket. **Distribution** From the eastern end of Baviaanskloof to Makhanda.

Hankey 16.10.2012

Caputia scaposa

scaposa = with a scape; referring to the elongated peduncle or scape

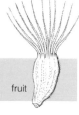

Dwarfed, tufted, silver-leaved succulent, to 40 cm. **Leaves** Lanceolate, usually spathulate, fleshy, 10 cm long × 1 cm wide. **Flowerheads** Few, 3.5 cm wide, on 45 cm-long branched peduncles. Flowering in summer. **Habitat** Rocky slopes and cliffs in Thicket. **Distribution** from Baviaanskloof to the Kei River.

Gert Smitskloof, Kleinpoort 04.03.2003

CHRYSOCOMA   Beesbos

chrysos = golden, *kome* = hair, locks; alluding to the yellow terminal heads

Similar in appearance to *Felicia*, but this genus lacks ray florets. 23 species in Africa and Australia; 21 species in South Africa; 2 species in Baviaanskloof.

fruit

Chrysocoma ciliata Bitterbos

ciliata = hairy; referring to the minute tuft of hairs on the tip of the achene

Single-stemmed, erect, shrublet to 60 cm; young stems bright green. **Leaves** Linear, ascending, 0.2–1.4 cm long, bright green. **Flowerheads** Singular at the tips, discoid, relatively small, to 1 cm wide, yellow. Flowering from spring to summer, mostly after rain. **Habitat** Common in disturbed areas in Transitional Shrubland, Thicket and Succulent Karoo. **Distribution** Widespread from Namaqualand to Mpumalanga. **Notes** Bitterbos is a miracle plant that can grow when all else fails. Sometimes regarded as weedy, it has potential for restoration of overgrazed and barren ground. This plant is toxic and can result in 'kraalsiekte' and even death of some stock. Yet, traditionally a useful plant for treating stomach ailments. **Other names** Bitterkaroo (A), ibhosisi (X), sehalahala (S).

Grasnek, Kouga 17.04.2012

Chrysocoma valida

validus = strong; a hardy plant able to survive drought

Single-stemmed, slender, glabrous shrublet to
30 cm. **Leaves** Linear, ascending, up to 1.5 cm
long. **Flowerheads** Discoid, solitary, 1 cm wide,
yellow. Flowering in spring. **Habitat** Sporadic in Arid
and Grassy Fynbos. **Distribution** Widespread on
mountains from Namaqualand to Baviaanskloof.

Nuwekloof Pass, Baviaans 08.10.2016

discoid
flowerhead

CICHORIUM*   Chicory

kichore = chicory in Greek; derived from the classical name for chicory

All florets in this genus have rays. 10 species in Africa and Europe; 1 species in Baviaanskloof.

Cichorium intybus*   Blue dandelion, chicory, sigorei

tybi = January; the plant is said to be ready to be harvested and eaten in January.

Bokloof, Baviaans 09.12.2011

Perennial weedy herb to 1 m tall, with a long fleshy taproot. **Leaves** Tufted,
lanceolate, entire, large, about 20 cm long × 5 cm wide, green, exuding a
milky sap, sometimes toothed. **Flowerheads** In axillary clusters subtended by
smaller leaves, several dozen well-spaced up the stem, flowers 2–4 cm wide,
blue. Flowering from late winter to spring. **Habitat** A weed of fallow lands and
roadsides. **Distribution** Originally from Eurasia, then introduced for cultivation;
it has become naturalised in many parts of the world. **Notes** The leaves are very bitter, but once cooked
are edible. Roasted roots are a supplement or substitute for coffee. Cultivated since 300 BCE, this is a
useful plant that has been dispensed medicinally for generations. The roots and leaves, when cooked,
have been used to treat gout, arthritis, liver problems, diabetes and many other ailments.

CINERARIA

cinerareus = of ashes; pertaining to the greyish wool on the leaves of some species

A close relative of *Senecio*, with similar involucral bracts. Leaves may be broad and shaped like
those of ivy. 35 species in Africa and Madagascar; 32 species in South Africa; 3 species recorded
in Baviaanskloof.

Cineraria lobata subsp. *platyptera*

lobatus = lobed; referring to the lobed leaves of this species

Annual, soft, flimsy, almost glabrous, upright to 60 cm. **Leaves** Lobed, glabrous,
6 cm long × 3 cm wide, bright green. **Flowerheads** In terminal corymbs, 1.5 cm
wide, each capitulum with 5 yellow rays. Flowering in spring and summer.
Habitat On the valley floor under sweet thorn trees or Thicket. **Distribution**
Found from Baviaanskloof to Gqeberha (Port Elizabeth). **Notes** 4 subspecies,
but only this one occurs in Baviaanskloof. **Status** Near Threatened.

Kamassiekloof, Zandvlakte 18.09.

Cineraria saxifraga

saxifraga = of the rocks; alluding to the rocky habitat
where the plants usually grow

Soft, diffuse perennial to 30 cm.**Leaves** Lobed, palmate,
1.5 cm wide, with cuneate base; auricles lacking at the base of
the petiole. **Flowerheads** 1 cm wide in a few-headed corymb,
with yellow rays and glabrous bracts. Flowering from spring
to early summer. **Habitat** Shaded rocky places in various veld
types. **Habitat** Spreading over rocks in shaded areas; often in
kloofs and wooded areas or Spekboomveld. **Distribution** Fairly
widespread from Baviaanskloof to the Wild Coast.

Zandvlakte, Baviaans 15.07.2011

CORYMBIUM

Heuningbos, plampers

korumbos = a cluster; the inflorescence is a tightly packed cluster of heads.

Leaves are distinctive and look like monocot leaves, strap-shaped with parallel veins. 9 species in
Western and Eastern Cape; 2 species in Baviaanskloof.

Corymbium africanum Plampers

africana = from Africa

Resprouting strongly from a stout, silky-hairy
rhizome. **Leaves** Tufted, tough, filiform to
linear, scabrid, to 30 cm. **Flowerheads** In
dense panicles, producing a relatively dense,
flat-topped synflorescence, mostly bright pink,
but also purple and white. Involucral bracts scabrid. Flowering
mostly in early summer, but also other times depending on fire
and rainfall. Only flowering in the first few years after fire. **Habitat**
Common in Grassy Fynbos. **Distribution** Widespread from the
Cederberg and Cape Peninsula to Makhanda.

Zandvlakte, Baviaans 29.04.2018

Corymbium glabrum

glabrum = hairless, smooth; referring to the
smooth leaves and involucral bracts

Resprouting, tufted, to 60 cm tall,
silky-hairy at the base. **Leaves** Flat,
lanceolate, leathery, with veins
sometimes prominent, 30 cm long ×
0.6 cm wide, lime-green. **Flowerheads** In loose corymbs, discoid,
to 1 cm long, white or pink. Involucral bracts smooth. Flowering
in summer, only after fire. **Habitat** Uncommon in Baviaanskloof
on rocky upper slopes in Grassy and Mesic Fynbos. **Distribution**
Widespread from Cederberg and Cape Peninsula to Makhanda.

Sipoonkop, Kouga 30.04.2018

CRASSOTHONNA

crass = thick, *Othonna* = plant genus; a recently formed genus that placed all those species of *Othonna* with succulent leaves into genus *Crassothonna*

Involucral bracts are in a single row, glaucous, and fused together along their length. 12 species in South Africa; 2 species in Baviaanskloof.

Crassothonna cacalioides (= *Othonna carnosa*)

Cacalia = plant genus, *oides* = resembling; the growth form resembles an old European genus, *Cacalia*.

Succulent with sprawling branches to 30 cm. **Leaves** Cylindrical, fleshy, 4 cm long × 0.5 cm wide, grey-green, sometimes tinged purple at the tips. **Flowerheads** Few on long, flimsy peduncles, 1.5 cm wide, rays yellow. Flowering mainly from autumn to late spring. **Habitat** Rocky outcrops mostly in Grassy Fynbos. **Distribution** Widespread from Namibia to Makhanda, Eastern Cape. **Notes** Easy to grow and popular in gardens and nurseries.

Damsedrift, Baviaans 17.01.2022

Crassothonna capensis Little pickles

capensis = from the Cape; referring to the geographical origin of the species

Trailing, rooting at nodes, perennial succulent herb to 10 cm, with leaves resembling small gherkins (pickled cucumbers). **Leaves** Cylindrical, 2 cm long, grey-green, woolly in axils. Under stress, the leaves turn shades of lavender and red. **Flowerheads** Radiate, single or paired on short, 5 cm-long peduncles, yellow. Flowering from autumn to spring. **Habitat** Rocky areas in Succulent Karoo. **Distribution** Widespread from the Little Karoo to the Eastern Cape.

Elandspoort, Antoniesberg 23.04.2018

CULLUMIA

fruit

Named after the Cullum brothers: Sir John Cullum, 1733–1785, historian, and his brother, Sir Thomas Gery Cullum, 1741–1831, a doctor and well-regarded writer on science and botany

Visually similar to *Berkheya*, but the stems are leafy up to the flowerheads and the fruit is simple. 15 species in parts of South Africa that receive winter rainfall; 4 species in Baviaanskloof.

Cullumia bisulca

bi = two, *sulcus* = furrow; referring to the underside of the leaves

Resprouting, prickly, stiff, upright shrub to 1.5 m. **Leaves** With spines and bristles, 3 cm long × 0.5 cm wide, bright green, with

Bo Kouga road, Haarlem 03.10.2009

strongly recurved margins. **Flowerheads** Radiate, 2.5 cm wide, yellow, with yellowish involucral bracts. Flowering in spring and early summer. **Habitat** Dry, rocky, north-facing slopes in Grassy and Arid Fynbos. **Distribution** Widespread from Calvinia to the Kouga Mountains.

Cullumia decurrens

decurrens = running down; referring to the leaf base running down the stem

Reseeding, sprawling, spiny shrublet to 90 cm. **Leaves** Recurved, tightly packed around the stem, glabrous and glossy, 1 cm long × 0.5 cm wide, margins strongly revolute with several short spines. **Flowerheads** Radiate, 2.5 cm wide, yellow. Flowering in spring and summer. **Habitat** Occasional in Grassy and Mesic Fynbos. **Distribution** Fairly widespread from the Swartberg and Outeniquas to Gqeberha (Port Elizabeth).

Bosrug, Baviaans 27.10.2018

CURIO (= *Senecio, Kleinia, Cacalia*)
curiosus = curious; pertaining to these intriguing succulents

The involucral bracts of this genus are like those of *Senecio*. ± 20 species in South Africa; 6 species in Baviaanskloof.

Curio acaulis

a = without, *caulis* = a stem; the plant is stemless.

Perennial succulent to 30 cm. **Leaves** Tufted, cylindrical, curving upwards, to 10 cm long × 0.5 cm wide, greyish. **Flowerheads** Discoid, 1 cm wide, white. Flowering in spring. **Habitat** Rocky areas in Transitional Shrubland and Succulent Karoo. **Distribution** From Ceres to the Baviaanskloof.

Bosrug, Baviaans 22.07.2011

Curio articulatus (= *Kleinia articulatus; Senecio articulatus*) **Worsies**

articulata = divided into distinct parts, articulate; referring to the articulated stems that resemble small sausages, hence the Afrikaans common name

Perennial succulent, sprawling, with thick, sausage-like stems, 1.5 cm wide, grey with purple markings. **Leaves** Deciduous, often absent in dry periods, 5 cm long, soft, fleshy, irregularly notched, pale green above, tinged purple below. **Flowerheads** Discoid, 0.5 cm wide, white. Flowering in autumn and winter. **Habitat** Mainly in Spekboomveld, creeping under the shade of other plants. **Distribution** Widespread from Montagu to Kariega (Uitenhage). **Notes** Leaves contain a prebiotic fibre beneficial for human digestion. It is also found in many other plants, including asparagus, bananas, onions and garlic.

Klipfontein, Baviaans 19.05.2021

Curio crassulifolius   Blue fingers, blouvingertjies

crassulifolius = crassula-leaved; with leaves resembling those of a *Crassula*

Perennial succulent, tufted, forming colonies on sandstone rocks, to 30 cm tall. **Leaves** Tufted, fusiform-terete, faintly striated, sometimes flattened, to 10 cm long, blue-grey-green, sometimes tinged purple, tip pointed. **Flowerheads** Discoid, usually solitary on long peduncles, to 1 cm long, white or yellow. Flowering in spring and summer. **Habitat** Found on rocky outcrops and ledges, often on summit ridges, in Mesic Fynbos. **Distribution** Widespread from Cederberg to Kariega (Uitenhage).

Elandsvlakte, Baviaans 29.09.2011

Curio ficoides   Blue chalk-sticks

ficoide = ice plant or sour fig of the mesemb family; derived from *ficus* for fig, referring to the succulent leaves

Erect, succulent shrub to 1 m. **Leaves** Slightly flattened, 10 cm long × 1.5 cm wide, greyish, ascending from thick, erect stem. **Flowerheads** Several in few branched corymbs, without rays, 0.7 cm wide, creamy white; very variable depending on habitat. Flowering in autumn and winter. **Habitat** In Baviaanskloof it is commonly found on rocky slopes, usually south-facing, in Spekboomveld and Fynbos Woodland. **Distribution** Fairly narrow distribution from the Swartberg (Little Karoo) to Zuurberg, Eastern Cape. **Notes** Popular in gardens and easy to grow from cuttings. Leaves contain inulin, a prebiotic fibre that is beneficial for the human digestive system. It is contained in many other plants, including ground chicory root, asparagus, bananas, fresh herbs, yams, leeks, onions and garlic.

Zandvlakte, Baviaans 15.07.2011

Curio radicans   Bobbejaantoontjies

radico = rooting; pertaining to the trailing stems that root at the nodes

Trailing, perennial, succulent herb. **Leaves** Facing upwards off the horizontal stem, cylindrical, pointed at the tip, 1 cm long × 0.7 cm wide, with a narrow translucent line on 1 side. **Flowerheads** Usually solitary, discoid, at the end of long 10 cm stalks, 0.5 cm wide, white or mauve, involucral bracts sometimes covered with short glandular hairs. Flowering from autumn to spring. **Habitat** In rocky areas trailing under other shrubs in Spekboomveld and Succulent Karoo. **Distribution** Widespread in arid areas from Namibia through the Great and Little Karoo to Makhanda. **Other names** String-of-bananas (E), vingertjies (A).

Rus en Vrede 4x4, Kouga 19.09.2010

DICEROTHAMNUS (= *Elytropappus*)   **Renosterbos**

dicero = rhino, two horns, *thamnus* = bush; direct translation of renosterbos (rhinoceros bush); *elytron* = a sheath, *pappos* = down or fluff; refers to the sheath around the base of the pappus bristles

10 species in Western and Eastern Cape; 2 species in Baviaanskloof.

Dicerothamnus rhinocerotis (= *Elytropappus rhinocerotis*)   **Renosterbos**

rhino = nose, *ceras* = horn, horn-like projection; referring to the plant's association with the rhinoceros, probably the black rhino, which is thought to have eaten the plant in the past

Single-stemmed, grey-woolly shrub to 2 m tall, with many branches, regenerating from seed. **Leaves** Minute, to 0.2 cm long, scale-like, adpressed. **Flowerheads** Few at tips of branches, mostly 3-flowered, without rays, 0.4 cm long, purple. Flowering in autumn. **Habitat** Transitional Shrubland and Grassy Fynbos, especially after disturbance and/or fire. **Distribution** Widespread from southern Namibia and Karoo to Eastern Cape. **Notes** Rarely browsed, although reportedly eaten by rhinoceros, but this has not been confirmed. Used in traditional medicine, for both humans and stock. Contains a chemical that has significant anti-inflammatory effects. **Other name** Ibhobhosi (X).

Rus en Vrede 4X4, Kouga 25.04.2018

DICOMA   **Wildekarmedik**

di= two, *koma* = tuft of hair; referring to the double row of pappus bristles on the fruit

Easily confused with *Macledium*, but this genus does not have a spiny involucre. *D. anomala* and other species are used in traditional medicine for various ailments. ± 30 species in Africa, Madagascar and India; 16 species in southern Africa; 1 species in Baviaanskloof.

Dicoma picta   Knoppiesdoringbossie

picta = brilliantly marked, referring to the spectacular flowers

Woody shrublet to 75 cm, with stems striated grey-and-green. **Leaves** Opposite, spathulate, 3 cm long × 0.5 cm wide, grey-mealy, dropping off readily. **Flowerheads** 3.5 cm wide, purple-pink, rays becoming white with age, bracts pointy. Flowering after rain in spring and summer. **Habitat** Uncommon on dry stony slopes in Succulent Karoo. **Distribution** Widespread from Cederberg to Kariega.

Krakeel, Langkloof 03.12.2009

DIMORPHOTHECA

Marguerite, magriet

dimorphe = of two forms, *theca* = a case; referring to the 2 forms of fruit in some species

Easily confused with *Osteospermum*, but the seeds of *Dimorphotheca* are flattened and the disc florets are fertile. 15 species in southern Africa; 4 species in Baviaanskloof.

fruit

Dimorphotheca cuneata Bosmagriet, jakkalsbos

cuneata = wedge-shaped; referring to the shape of the leaves

Rounded, erect, perennial shrub, mostly to 50 cm. **Leaves** Toothed to almost lobed, sticky, glandular with a varnished appearance, 2.5 cm long × 1 cm wide, aromatic when crushed. **Flowerheads** With rays, to 4 cm wide, rays usually white and darker on reverse, with yellow centre. Flowering mostly in spring, but also after good rains. **Habitat** In Baviaanskloof it is confined to clay-rich, shale-derived soils in Transitional Shrubland and Succulent Karoo. **Distribution** Widespread and abundant in arid, rocky areas from Namaqualand to Kariega (Uitenhage). **Other name** Witgousblommetjie (A).

Beacosnek, Kouga 08.08.2010

Dimorphotheca ecklonis Vanstaden's daisy

Named after Christian Friedrich Ecklon, 1795–1868, a Danish botanist who collected extensively in South Africa; the herbarium that he collated became the basis for *Flora Capensis*. Several other species have been named after him, including a genus of kelp (*Ecklonia*).

Softly woody, sprawling, perennial shrub to 1 m tall, using other shrubs for support. **Leaves** Elliptic, toothed, glabrous or slightly scabrid, and sometimes petiolate, 10 cm long × 4 cm wide. **Flowerheads** Solitary on long peduncles, large, to 8 cm wide, rays white with darker reverse, disc violet-blue. Achenes trigonous, ridged, rugose. Flowering from autumn to spring. **Habitat** Lower bushy slopes and valleys in Grassy Fynbos. **Distribution** Endemic with narrow distribution from the Langkloof to Kariega (Uitenhage). **Notes** Rarely seen in the wild, more commonly seen in gardens and nurseries. A naturalised weed in California and Australia. **Status** The red list status of this species should be reviewed since the habitat is threatened by invasive plants and agriculture, and several of the original populations may have become extinct.

Nooitgedacht, Kouga 05.11.2011

DISPARAGO

Basterslangbos

dispar = unlike or dissimilar, *ago* = resemblance; referring to the different kinds of florets in the flowerheads

Easily confused with *Stoebe*, but *Disparago* has ray florets. 9 species in South Africa; 2 species in Baviaanskloof.

Disparago tortilis

tortilis = twisted; referring to the twisted leaves

Upright, profusely branched shrub to 1 m tall,
regenerating from seed, not browsed. **Leaves**
Linear, twisted and curling, packed tightly
around the stem, less than 1 cm long × 0.2 cm
wide, tipped with a bristle. **Flowerheads** Packed
into a compact, roundish head, 1.3 cm wide, in various shades of
pink, also white. Pappus bristles plumose, cypsela surface woolly.
Flowering from autumn to spring. **Habitat** Favouring disturbed
areas and overgrazed grasslands. Frequently encountered in
Grassy Fynbos on the mountains. **Distribution** Widely distributed
from Grabouw to Port Shepstone. **Notes** Some people may have
called this species *D. ericoides*, but that only occurs in the Western
Cape, and it has 2 florets per head. In *D. tortilis*, the heads are
1-flowered, with either a single ray or a single disc, with different
types of florets mixed together in the rounded synflorescence.

Bo Kouga road, Langkloof 20.08.2011

ERIOCEPHALUS   Kapokbossie, wild rosemary, Cape snow bush

erion = wool, *kephale* = head; referring to the woolly fruiting flowerhead, which is also
the derivation for the common name

Used in traditional medicine and as an essential oil. A genus with distinctive
woolly fruit. ± 34 species in southern Africa; 3 species in Baviaanskloof.

fruit

Eriocephalus africanus   Wild rosemary, Cape snow bush

africanus = from Africa; the origin of the plant

Reseeding, twiggy shrub to 1 m. **Leaves** Arranged in tufts on the
branches, linear, woolly, 1.5 cm long × 0.2 cm wide, silvery-grey.
Flowerheads In small umbels at branch tips, radiate, 0.5 cm wide, rays
white, disc purplish. Flowering mainly from midsummer to midwinter.
Habitat Often associated with shale bands or loam-rich soils in Grassy
Fynbos and Transitional Shrubland. **Distribution** Widespread from
southern Namaqualand to Eastern Cape. **Other name** Kapokbossie (A).

Zandvlakte bergpad, Kouga 11.04.2012

Eriocephalus capitellatus

capitellatus = with a small head or heads;
referring to the tiny flowerheads

Reseeding, slender, silvery-silky shrub, to 1.6 m.
Leaves In tufts, linear or trifid, 0.7 cm long ×
0.05 cm wide. **Flowerheads** in axillary glomerules
arranged in spikes, each flowerhead 0.5 cm
wide, with small, white rays. Flowering in summer
and autumn. **Habitat** Transitional Shrubland on dry, rocky slopes.
Distribution Widespread from the Cederberg to Baviaanskloof.

Duiwekloof, Baviaans 14.02.2016

Eriocephalus ericoides   Gewone kapokbossie

ericoides = like plants in genus *Erica*; the leaves are erica-like.

Twiggy shrub to 1 m. **Leaves** In axillary tufts, opposite, linear,
minute, 0.05 cm long. **Flowerheads** Solitary in axils of upper
leaves, 0.4 cm wide, appearing discoid but with small rays.
Flowering from winter to early summer. **Habitat** Mainly in
Succulent Karoo habitats and sometimes in Transitional
Shrubland in the west. **Distribution** Widespread in drier parts of
southern Africa.

Zandvlakte bergpad, Kouga 28.10.2011

EURYOPS   Harpuisbos

eurys = large, *ops* = eye; referring to the showy flowerheads

Leaves are usually lobed or forked and needle-like, arranged neatly around the stems.
Flowerheads are always borne from axillary peduncles that branch from within the
topmost leaves of the flowering shoot. A single row of smooth bracts is joined at the
base. Pappus bristles are deciduous. Several species are popular in gardens. 97
species in Africa and southern parts of the Arabian Peninsula; ± 91 species in South
Africa; 11 species in Baviaanskloof.

fruit

Euryops euryopoides

euryopoides = like *Euryops*; this species is typical of the genus.

Erect, reseeding, sparsely branched,
single-stemmed, leafy shrub to 1.5 m.
Leaves Mostly 3-forked and needle-
like, to 3 cm long. **Flowerheads**
Radiate, 2 cm wide, yellow. The fruit
is hairless and ribbed. Flowering
most of the time. **Habitat** Lower
slopes in Grassy and Arid Fynbos.
Distribution Eastern distribution from
Baviaanskloof to Makhanda.

Engelandkop, Kouga 19.04.2012

Euryops lateriflorus Soetharpuisbos

lateri = at the side, *florus* = flowers; referring to the flowers
that are borne on the side of the branches, near the tips

Resprouting, stiffly erect shrub to 1.5 m. **Leaves**
Oblanceolate to obovate, closely leafy ascending the stem,
3-veined from base, smooth and leathery,
3 cm long × 1.5 cm wide, grey-green.
Flowerheads Single, on peduncles to 5 cm
long, crowded in leaf axils near the branch
tips, radiate, 1.5 cm wide, yellow. Flowering
in winter and early spring. **Habitat** Loamy
soils in Grassy Fynbos and Transitional
Shrubland. **Distribution** Very widespread
from Namibia to Baviaanskloof.

Rus en Vrede 4X4, Kouga 21.05.2010

Euryops munitus Umsola (X)

munitus = fortified; referring to the stiff,
tightly packed, ascending leaves that
appear to be protecting the stems; the leaves
are also tiny ('minute').

Erect, reseeding, wiry-twiggy shrublet
to 30 cm. **Leaves** Pinnatisect, tightly
packed, stiff, ascending, curving in
towards the stem, 0.5 cm long, with
mucronate lobes. **Flowerheads** On long
thin peduncles 10 cm long, each floret
1 cm wide, yellow. Fruit almost smooth.
Flowering in autumn. **Habitat** Common
in Grassy and Arid Fynbos. **Distribution**
Nearly endemic with a narrow
distribution from the Baviaanskloof to
Gqeberha (Port Elizabeth).

Bosrug, Baviaans 22.07.2011

Euryops pinnatipartitus

pinnati = pinnate, *partitus* = divided;
referring to the divided, pinnatisect leaves

Upright, slender, reseeding, single-stemmed, sparsely
branched shrub to less than 1 m. **Leaves** Pinnatisect,
tightly packed, ascending, to 3 cm long, lobes oblong to
terete. **Flowerheads** On short peduncles, radiate, 1 cm
wide, yellow. Pappus lacking. Fruit with 10 warty ridges.
Flowering mainly in spring and summer. **Habitat** Upper
slopes in Mesic and Subalpine Fynbos. **Distribution** On
high mountains from Swellendam to Humansdorp.

Quaggasvlakte, Kouga 13.02.2016

Euryops rehmannii

Named after Anton Rehmann, 1840–1917, a Polish botanist, geographer and explorer who visited South Africa twice between 1875 and 1880, and collected over 9,000 specimens

Upright, reseeding, much-branched, rounded shrub to 2 m, waving in the wind. **Leaves** Needle-like, smooth, neatly arranged, 5 cm long × 0.1 cm wide. **Flowerheads** Borne above the leaves, on stalks 10 cm long, each flower 1.5 cm wide, yellow. Fruit shortly-hairy. Flowering in spring. **Habitat** Relatively dry, rocky habitat in Grassy and Arid Fynbos, and sometimes Transitional Shrubland. **Distribution** Widespread from Cederberg to Kariega (Uitenhage).

Bosrug, Baviaans 27.07.20

Euryops spathaceus   Harpuisbos

spathaceus = resembling a spathe; referring to the flat, spathe-like leaves

Erect, reseeding, single-stemmed, slender shrub to 1.5 m. **Leaves** Flat, spreading, oblanceolate, up to 5 cm long. **Flowerheads** Solitary in upper axils, radiate, 2 cm wide, yellow; bracts fused initially, later splitting irregularly. Flowering mostly late summer and autumn. **Habitat** Rocky areas and ridges in Grassy Fynbos. **Distribution** Widespread from Swartberg to Komani (Queenstown), Eastern Cape. **Other names** Iyeza lehlaba (X), ulwapesi (X).

Grasnek, Kouga 17.04.2012

FELICIA   Astertjie

felix = happy, or the genus for cats, *Felix*; the derivation is unresolved but perhaps the charming flowers inspire happiness, or the fruits and other parts are hairy or have whiskers like a cat, or it is named after a woman.

Easily recognised by the blue-mauve rays (rarely white), and a flattened, elliptical fruit with pappus bristles. ± 85 species from southern Africa to the Arabian Peninsula; ± 56 species in South Africa; 7 species recorded in Baviaanskloof.

fruit

Felicia douglasii

Named after the author, Douglas Euston-Brown, who found and recorded the plant on the Kouga Mountains in 2011

Erect, single-stemmed, twiggy shrub to 1.5 m. **Leaves** Spreading, apiculate-uncinate, to 2 cm long × 0.4 cm wide, apple-green above, paler below, with recurved and scabrid margins. **Flowerheads** Arranged in a compound synflorescence of up to 20 radiate flowerheads, each 2.5 cm wide, with white or pale blue rays, reverse pale blue or turquoise; disc florets with purple throats and yellow lobes. Fruit glabrous, pappus bristles deciduous. **Habitat** On or at the base of rock slabs that would form seepages after good rain, in Mesic or Grassy Fynbos, or Fynbos Woodland. **Distribution** Rare endemic on steep, south-facing ledges of the Baviaanskloof and Kouga mountain ranges.

Kouenek, Kouga 26.09.2011

Felicia filifolia Draaibossie

filum = thread, *folia* = leaves; referring to the thin linear leaves

Common, frequently encountered, floriferous and variable shrub that has 4 recognised subspecies. Densely branched, upright, to 1 m. **Leaves** Needle-like, in tufts, gland-dotted, to 1.5 cm long × 0.2 mm wide. **Flowerheads** To 2 cm wide, rays in various shades of blue or pink or rarely white, centre yellow. Flowering mainly in spring and summer. **Habitat** Rocky, sunny and exposed areas in most vegetation types. **Distribution** Widespread from Namibia to Mpumalanga. **Other names** Igabu (X), igangasi (X).

Bosrug, Baviaans 22.07.2011

Felicia joubertinae

Named after the town of Joubertina in the Langkloof, near to where the plant was discovered as new to science

Softly woody, semi-erect shrublet to 45 cm. **Leaves** Oblong, 2 cm long × 1 cm wide, with revolute margins, leaves and stems rough-hairy. **Flowerheads** On long, hairy peduncles, radiate, 3 cm wide, with pale blue rays and yellow centres. Flowering in spring. **Habitat** It has only been found on rocky ridges at the crest of mountains in Mesic Fynbos. **Distribution** Rare endemic of the Kouga and Baviaanskloof mountains.

Enkeldoring, Baviaans 21.09.2011

Felicia linifolia

linum = flax, *folia* = leaves; referring to the linear leaves that look like those of *Linum* spp. (flax)

Diffuse, twiggy shrublet to 45 cm. **Leaves** Opposite, spreading, linear-oblong, glabrous, 2 cm long × 0.3 cm wide. **Flowerheads** Solitary, on long peduncles, radiate, 2 cm wide, rays pale mauve and centres yellow. Flowering mainly autumn, winter and spring. **Habitat** Rocky areas on south-facing slopes in Grassy Fynbos and Transitional Shrubland. **Distribution** From Ladismith to Kariega (Uitenhage).

Uitspan, Baviaans 19.02.2016

Felicia microcephala

micro = small, *cephala* = head; referring to the small flowerheads

Soft, herbaceous, diffuse shrublet to 30 cm. **Leaves** Petiolate, obovate, single-veined, hairy, 1.5 cm long × 0.8 cm wide. **Flowerheads** 1 to several at branch tips, small, to 1.5 cm wide, yellow with pale blue rays. Flowering in spring and summer. **Habitat** Narrow rocky ravines in Thicket and Forest. **Distribution** Near endemic from Outeniqua Mountains to Humansdorp.

Keerom, Baviaans 22.12.2011

Felicia muricata Taai-astertjie, bloublommetjie

muricata = rough; referring to the surface texture of some parts

Upright, single-stemmed, much-branched, thinly-twiggy, glabrous shrublet to 50 cm. **Leaves** Linear, ascending, to 1.5 cm long × 0.2 cm wide. **Flowerheads** 2 cm wide, with blue rays and yellow centre. Flowering in spring and summer. **Habitat** Common on richer soils with more clay, or where animals or stock have enriched the soil with dung, in Succulent Karoo and Transitional Shrubland. **Distribution** Widespread from Oudtshoorn to tropical Africa.

Joubertina 24.08.2009

Felicia ovata Grootbloublommetjie

ovata = egg-shaped; pertaining to the egg-shaped or ovate leaves

Upright, softly-twiggy, conspicuously (yet sparsely) hairy shrublet, to 50 cm. **Leaves** Flat, narrowly ovate, 1.5 cm long × 0.3 cm wide, with long white hairs; tip recurved and acute. **Flowerheads** Borne on long stalks to 6 cm, 2.5 cm wide, with blue rays and yellow centres; ray petals often untidily arranged, twisted or curling down. Flowering from autumn to early summer. **Habitat** Occurring sporadically in most non-Fynbos habitats on the lower slopes. **Distribution** Widespread from Calvinia to the southern Drakensberg.

Modderfontein, Zandvlakte 18.09.2010

GAZANIA   Gazania, gousblom

gaza = riches; referring to the gold spots on the petals of some species, or the spectacular colours; alternatively, it was named after Theodorus Gaza, 1415–1475, who translated the botanical work of Theophrastus into Latin.

fruit

All species are radiate, with involucral bracts fused into a tough cup, often flattened at the base. ± 17 species in Africa; 12 species in the Cape; 2 species in Baviaanskloof.

Gazania krebsiana Common gazania, botterblom, rooigazania

Named after German naturalist Georg Ludwig Englehardt Krebs, 1792–1844, who spent his entire career collecting natural history specimens in South Africa

Perennial, tufted, to 20 cm tall, with leaves in a rosette close to the ground. **Leaves** Either simple or divided into narrow segments, 10 cm long × 1 cm wide, white-felted below, with margins rolled under. **Flowerheads** Up to 5 cm wide, on peduncles to 10 cm long, rays

variable in colour, yellow, orange or reddish, sometimes with dark marks at the base that attract monkey beetles. Flowering in spring and summer. **Habitat** Lower slopes and flats in disturbed, open, sunny areas in most habitats, often on roadsides. **Distribution** Very widespread throughout southern Africa to Tanzania. **Other names** Isaphepha (X), unongwe (X).

Above Nuwekloof Pass 22.09.2006

Gazania lichtensteinii Geelgazania, kougoed

Named after Martin Hinrich Lichtenstein, 1780–1857, a German physician and naturalist who explored southern Africa from 1803–1806. Not only has an Apiaceae genus been named after him, but also a species of sandgrouse, hartebeest and snake.

Tufted annual or perennial to 30 cm. **Leaves** Oblanceolate, 5 cm long × 1 cm wide, woolly below, margins toothed. **Flowerheads** 4 cm wide, bright yellow or orange with dark markings near the base of the ray petals, involucral bracts smooth. Flowering in spring and summer after rain. **Habitat** Occurring in disturbed or open places in Succulent Karoo. **Distribution** Widespread from Namibia through the Karoo to the western end of Baviaanskloof. **Notes** The flowering stems have a buttery taste, thus the common name.

Willowmore 16.01.2022

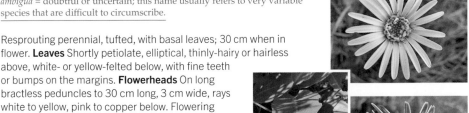

GERBERA   African daisy

Named after German naturalist, Traugott Gerber, 1709–1743, a plant collector

Tufted perennial herbs with basal leaves, only producing flowers in the first year after fire. ± 30 species in Africa and Asia; 13 species in South Africa; 8 species in the Cape flora; 4 species in Baviaanskloof.

fruit

Gerbera ambigua Isichwe (X)

ambigua = doubtful or uncertain; this name usually refers to very variable species that are difficult to circumscribe.

Resprouting perennial, tufted, with basal leaves; 30 cm when in flower. **Leaves** Shortly petiolate, elliptical, thinly-hairy or hairless above, white- or yellow-felted below, with fine teeth or bumps on the margins. **Flowerheads** On long bractless peduncles to 30 cm long, 3 cm wide, rays white to yellow, pink to copper below. Flowering after fire in spring and summer. **Habitat** Occasional in Grassy and Mesic Fynbos, where soil is deeper and with more loam. **Distribution** Widespread in grasslands from George to tropical Africa. **Notes** Inhaling smoke of the burning leaves is a treatment for head colds.

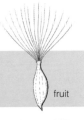

Engelandkop, Kouga 19.04.2012

Gazania • Gerbera | DICOTYLEDONS **267**

Gerbera ovata

ovata = egg-shaped; referring to the shape of the leaves

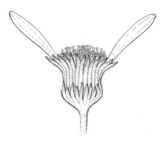

Resprouting, tufted perennial, with basal, decumbent to prostrate leaves. **Leaves** Petioles 8–20 cm long, blade to 8 cm long × 4 cm wide, dark green above, white-felted below. **Flowerheads** On long, felted peduncles to 40 cm long, each flowerhead 3.5 cm wide, rays white above, pink or mauve below, disc yellow; marked with small furry bracts. Flowering after fire in spring and summer. **Habitat** Upper south-facing slopes of the mountains in Grassy and Mesic Fynbos. **Distribution** Fairly narrow distribution from Baviaanskloof to Vanstadensberg. **Notes** A recently described species (2015) that was previously included in *G. tomentosa*, but that species is restricted to the Western Cape and has dentate leaf margins.

Zandvlakte, Baviaans 29.04.2018

Gerbera piloselloides   Swarttee

piloselloides = hawkweed-like; the plant resembles hawkweed, daisies in the genera *Hieracium* and *Pilosella*, with more than a thousand species, mostly in the northern hemisphere.

Resprouting, tufted perennial, with leaves basal in a rosette. **Leaves** Obovate, tapering at the base, 10 cm long × 3 cm wide, softly hairy or cobwebby, hairs yellowy to light brown. **Flowerheads** 3 cm wide, on furry, brown peduncles, involucre also hairy, rays white, pink, red or yellow. Flowering in winter and through summer. **Habitat** On shale bands in Grassy Fynbos or under Forest. **Distribution** Widespread from the Cape to the tropics. **Notes** The plant has been used in traditional medicine to treat head colds and earache. **Other names** Umgwashu, ubulawu, iyeza lamasi (X).

East of Bosrug, Baviaans 12.03.2012

HAPLOCARPHA   Bastergousblom

aplo = soft, *karphos* = scale or *karpos* = fruit; the most likely derivation is that the fruits are covered in silky hairs.

Involucral bracts of this species are similar to those of *Arctotis*. 9 species in Africa; 6 species in South Africa; 2 species in Baviaanskloof.

Haplocarpha lyrata

lyrata = lyre-shaped; derivation unclear; the space between the divided leaves or the curvature of the leaves may resemble the curved wood of the musical instrument.

Resprouting perennial, stemless, tufted. **Leaves** Variable, lyrate-pinnatisect, glabrous above, densely white-woolly below, 20 cm long × 6 cm wide. **Flowerheads** Single on short peduncles 10–20 cm long, each flowerhead 6 cm wide, with pale yellow rays. Fruit silky-hairy. Flowering in spring and summer. **Habitat** Grassy Fynbos and Transitional Shrubland. **Distribution** Widespread from George to Gauteng. **Notes** Possibly toxic to stock.

Engelandkop, Kouga 19.04.2012

Haplocarpha scaposa   False gerbera

scaposa = with scapes; referring to the long, leafless scapes or peduncles

Resprouting, tufted, stemless, with leaves mostly flat on the ground in a rosette. **Leaves** Paddle-shaped, 15 cm long × 7 cm wide, with venation prominent above and below, white-felted below, with margins toothed and slightly undulate. **Flowerheads** Borne on 40 cm-long, cobwebby spathes, flowerheads to 6 cm wide, yellow. Involucral bracts recurved and white-hairy. Fruit warty, with long hairs and pappus scales. Flowering through summer. **Habitat** Shale bands in Grassy Fynbos and Sour Grassland on Cockscomb Mountain. **Distribution** Essentially a grassland species found throughout eastern and tropical Africa; Baviaanskloof is its southwestern distribution limit. **Notes** Crushed leaves and roots have been used traditionally for treating menstrual disorders and female infertility. Sotho sangomas have used the plant when divining with bones. **Other names** Common haplocarpha (E), bietou (A), isikhali, umkhanzi (X), khutsana (S).

Cockscomb 17.02.2016

HELICHRYSUM   Strooiblom, strawflower

helios = sun or *helix* = spiral, *khrusos* = gold; referring to the golden or yellow flowerheads that track the sun; bracts are arranged in spirals, overlapping (like roof tiles).

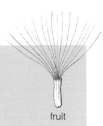

Several rows of dry, chaffy or papery involucral bracts are the main, unifying character of strawflowers. Heads are never radiate; leaves are never divided. The crushed leaves have a distinct aroma in many species. Several species are used in traditional medicine. ± 500 species mainly in Africa; ± 250 species in South Africa; ± 22 species in Baviaanskloof.

fruit

Helichrysum anomalum

anomalos = extraordinary, diverging from the normal; referring to the small flowerheads clustered together in terminal cymes

Dwarf, grey-woolly, reseeding, short-lived, twiggy shrublet to 30 cm. **Leaves** Ascending, overlapping, linear, 1 cm long × 0.2 cm wide, grey-felted, with margins rolled under. **Flowerheads** Crowded together in flat-topped cymes, small, 0.5 cm wide, ray florets lacking, outer bracts brown, inner bracts bright yellow. Flowering in spring, summer and autumn. **Habitat** Common in Grassy Fynbos. **Distribution** Widespread from Outeniquas to Drakensberg.

Kouga Wildernis 15.05.2021

Helichrysum cymosum   Impepho (Z)

cymosum = with cymes; referring to the flowerheads clustering into flat-topped cymes

Reseeding, softly-twiggy, straggling, thinly grey-woolly shrublet to 1 m;
mostly less than 50 cm tall, tending to become wider than it is tall.
Leaves Linear to elliptic, with 3 grooved veins above, 1.5 cm long ×
to 0.5 cm wide, silky-white to grey-felted below, margins slightly rolled
under. **Flowerhead** Flat-topped cyme cluster 5 cm wide, each flowerhead
0.3 cm wide, yellow. Flowering from spring to autumn. **Habitat** Common
in all Fynbos habitats, but more on south-facing slopes in Grassy Fynbos.
Distribution Widespread from the West Coast to Mpumalanga. **Notes**
Quick growing, underutilised in landscaping. The inhaled smoke from
dried leaves is believed to provide pain relief. One of several species
called 'impepho' and used as a purifying smudge in rituals. Also said to
cure infections. **Other name** Everlasting (E).

Guerna, Kouga 16.02.2016

Helichrysum excisum

excisum = collapsed, escaped or cut off; referring to the leaves, which appear
cut off, or the tendency for the inflorescence to fall off when in fruit

Reseeding, erect, densely twiggy, closely leafy, grey-felted shrub
to 50 cm. **Leaves** Spathulate, imbricate, slightly folded up, curving
downwards, hooked at tip, less than 1 cm long × 0.4 cm wide, margins
felted rusty brown; smelling spicy when crushed. **Flowerhead** Dense
cymes 5 cm wide, narrowly top-shaped, each flowerhead 0.5 cm wide,
bright yellow, with bracts deflexed at the tip. Flowering in late spring
and summer. **Habitat** Rocky sandstone in Transitional Shrubland at the
western end of Baviaanskloof. **Distribution** Widespread from Agulhas to
Baviaanskloof.

Uniondale 09.11.2020

Helichrysum felinum

felinum = cat; a certain Mr Hermann is believed to have written,
in 1737, that the cluster of flowerheads resembled a cat's padded paw.

Reseeding, short-lived, upright, soft, slender, sparsely branched shrublet
to 60 cm. **Leaves** Spreading or deflexed, ovate-lanceolate, 1.5 cm long
× 0.5 cm wide, green on top, white-felted
below. **Flowerhead** Spectacular, forming
a rounded synflorescence, often nodding
in bud, to 3 cm wide. Each flowerhead to
1 cm wide, white, sometimes with red.
Flowering in spring and early summer.
Habitat Frequent on upper mountain
slopes in Grassy, Mesic and Subalpine
Fynbos. **Distribution** Widespread from
Cape Peninsula to KwaZulu-Natal.

Krakeel, Langkloof 18.12.2011

Helichrysum herbaceum   Monkey-tail everlasting

herbaceum = herbaceous; referring to the growth form; alternatively, it could allude to the grass-green or yellow-green colour of the leaves.

Resprouting, erect, densely leaved perennial herb to 40 cm. **Leaves** Linear-lanceolate, ascending, becoming smaller and narrower up the stem, 2 cm long × 0.3 cm wide, greyish or grass-green, white-felted below. **Flowerheads** Single at the branch tips, 2.5 cm wide, with striking golden brown or yellow bracts. Flowering in spring, summer and autumn. **Habitat** Loamy soils in grassy vegetation in Transitional Shrubland and Grassy Fynbos. **Distribution** Very widespread from Langkloof to tropical Africa. **Other name** Impepho-yamakhosi (Z).

Nooitgedacht, Kouga 19.12.2011

Helichrysum lancifolium   Sewejaartjie

lanci = lance, *folium* = leaf; having lance-shaped leaves

Resprouting, sprawling, densely leafy, silvery-grey shrublet to 50 cm. **Leaves** Imbricate, oblong, folded, hooked, to 1.5 cm long × 0.4 cm wide, silvery silky. **Flowerheads** Solitary, to 5 cm wide, on scaly-leaved peduncles distinct from the leafy shoots, lower bracts brownish, upper bracts spreading and white, centre yellow. Flowering in spring and summer. **Habitat** Occurring in Grassy and Mesic Fynbos. **Distribution** Widespread from Ceres to the Baviaanskloof. **Other name** Everlasting (E)

Beacosnek, Kouga 22.09.2010

Helichrysum nudifolium var. *nudifolium*

nudi = naked, *folium* = leaf; referring to the glabrous leaves that are hairy but not felted below

Resprouting from a deep taproot, tufted, with leaves basal in a rosette, to 70 cm when in flower. **Leaves** Oblanceolate, glabrous or thinly woolly above, with 3–7 veins from the base, 25–50 cm long × 10cm wide, with margins and veins rough-hairy. **Flowerheads** Many in a compact, flat-topped, branched corymb 8 cm wide, each flowerhead 0.5 cm wide, yellow, with pale brown or yellow involucral bracts. Flowering in summer and autumn. **Habitat** Frequently encountered in Grassy Fynbos where the soils are less rocky and richer in loam. **Distribution** Very widespread from the Cape through Africa to the Middle East. **Notes** Similar to var. *oxyphyllum*, but that variety has white involucral bracts. The leaves have several medicinal uses, with younger leaves preferred. They are consumed as a treatment for colds — brewed and sipped in an infusion for coughs; also applied as a wound dressing. A decoction of the root is used for relief from stomach problems. **Other names** Cholocholo (X, Z), letapiso, mohlomela-tsie-oa-thaba (S), imphepho, isidwaba-somkhovu, umagada-emthini (Z).

Engelandkop, Kouga 06.11.2011

Helichrysum nudifolium var. *oxyphyllum* Uzandokwa (X)

oxy = sharp, *phyllon* = leaf; referring to the sharply pointed leaves

Similar to *H. nudifolium* var. *nudifolium*. Resprouting from a deep taproot, tufted, with leaves basal in a rosette, to 70 cm when in flower. **Leaves** Oblanceolate, with 3–7 veins from the base, 25–50 cm long × 10 cm wide, with margins and veins rough-hairy; darker green on top and white-felted below. **Flowerheads** Many in a compact, flat-topped, branched corymb 8 cm wide, each flowerhead 0.5 cm wide, white-and-pink or red, spathes white-felted. Flowering in spring and summer. **Habitat** Less frequently encountered in Grassy Fynbos. **Distribution** Very widespread from Caledon to tropical Africa. **Notes** The leaves have several medicinal uses, with younger leaves preferred. They are consumed as a treatment for colds – brewed and sipped in an infusion for coughs; also applied as a wound dressing. A decoction of the root is used for relief from stomach problems. **Other names** Cholocholo (X, Z), letapiso, mohlomela-tsie-oa-thaba (S), imphepho, isidwaba-somkhovu, umagada-emthini (Z).

Drinkwaterskloof, Kouga 09.11.2011

Helichrysum odoratissimum Kooigoed

odoratissimus = most fragrant; alluding to the strong smell of the leaves

Straggling, aromatic, reseeding, thinly white-woolly shrub to 60 cm, with winged stems. **Leaves** Decurrent, linear to spathulate, clasping the stem, 4 cm long × 1.5 cm wide, with undulate margins. **Flowerheads** In dense terminal cymes to 5 cm wide, each flowerhead 0.3 cm wide, yellow. Flowering in spring and summer. **Habitat** Grassy and Mesic Fynbos, usually on south-facing slopes associated with damp sites or seepages. **Distribution** Widespread from Gifberg to tropical Africa. **Notes** One of several species called 'impepho' and used as a purifying smudge in rituals. The leaves have been used in perfume and for insect-repellent ointment. Essential oil from the flowers is reputed to have many beneficial effects. **Other names** Impepho (X, Z, S), phefo-ea-setlolo, tooane (S).

leaf clasping the stem

Cockscomb, Groot Winterhoek 17.02.2016

Helichrysum rosum Bergankerkaroo

rosum = like a rose; the bracts of the flowerhead are vaguely rose-like; the Afrikaans name refers to the plants' ability to root at the nodes.

Reseeding, sprawling, softly twiggy, grey-woolly shrub to 1 m. **Leaves** Linear, in axillary tufts, 1.5 cm long × 0.3 cm wide, woolly below. **Flowerheads** Arranged in a dense terminal corymb to 2 cm wide, each flowerhead 0.4 cm wide, with rounded white bracts. Flowering in spring and summer. **Habitat** Open stony areas in most vegetation types except the upper mountain slopes. **Distribution** Widespread from Robertson to Addo and to Komani (Queenstown). **Other name** Strawflower (E).

Drinkwaterskloof, Kouga 09.11.2011

Helichrysum rugulosum

rugulosum = wrinkled; referring to the crisped bracts

Resprouting from a creeping rootstock, with several unbranched erect stems from the base, to 30 cm. **Leaves** Oblong-lanceolate, 1.5 cm long × 0.3 cm wide, margins rolled under, white-felted below, thinly felted above. **Flowerheads** In compact terminal corymbs 4 cm wide, each flowerhead 0.4 cm wide, creamy, with bracts crisped and curving down. Flowering in summer and autumn. **Habitat** Transitional Shrubland, Grassy Fynbos and Sour Grassland, often in grassy veld on shale bands. **Distribution** Widespread from Langeberg to Gauteng. **Other names** Marotole, motlosa-ngoaka, motoantoanyane, pulumo-tseou (S).

Bosrug, Baviaans 14.03.2012

Helichrysum teretifolium

tereti = terete, cylindrical, *folium* = leaf; referring to the narrow, linear leaves

Resprouting, many-branched from the base, upright shrub to 40 cm. **Leaves** Neatly arranged around the stem, spreading or upright against the stem, narrow, 1.5 cm long × 0.2 cm wide, woolly below and slightly twisted. **Flowerheads** Several in a tightly packed, flattish-topped synflorescence, each flowerhead 0.5 cm wide, with white bracts curving back slightly. Flowering in spring and summer. **Habitat** Growing in Grassy Fynbos and Transitional Shrubland. **Distribution** Widespread from Piketberg to KwaZulu-Natal.

Langkloof 02.11.2008

Helichrysum zeyheri Vaalbergkaroo

Named after Karl Ludwig Philipp Zeyher, 1799–1858, a botanist who collected extensively in South Africa. He was the co-author of *Enumeratio Plantarum Africae Australis, 1835–1837,* a descriptive catalogue of South African plants.

Greyish-white-woolly perennial, twiggy, upright, reseeding (sometimes resprouting), to 70 cm. **Leaves** Undulate or crisped, oblong to obovate, 2 cm long × 0.5 cm wide, mucronate, sometimes with golden hairs on the margins. **Flowerheads** In compact terminal cymes 2–3 cm wide, each pointy flowerhead 0.3 cm wide, white. Flowering in summer and autumn. **Habitat** Transitional Shrubland and Grassy Fynbos, usually on drier north-facing slopes. **Distribution** Widespread throughout drier parts of southern Africa.

Zandvlakte, Baviaans 19.01.2022

HERTIA

Named after Johann Casimir Hertius, 1679–1748, who wrote several books in Latin about the medicinal use of plants, most notably on *Pimpinella saxifraga*

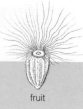

Easily confused with *Othonna*, but the soft pappus of *Hertia* falls off early. ± 10 species from southern Africa to Southwest Asia; 5 species in South Africa; 2 species in Baviaanskloof.

fruit

Hertia kraussii Springbokbos

Named after Christian FF Krauss, 1812–1890, a German scientist, traveller and collector. He collected specimens in the Cape and Natal between 1838 and 1840.

Reseeding, short-lived, erect, shrub to 1 m. **Leaves** Oblanceolate, smooth, fleshy and leathery, 2 cm long × 1 cm wide. **Flowerheads** Terminal or axillary, discoid, on 5 cm-long peduncles, 1 cm wide, yellow. Flowering in winter. **Habitat** Most common in the first few years following fire, or in disturbed open areas in Grassy and Arid Fynbos and Transitional Shrubland. **Distribution** Fairly narrow distribution from Kammanassie Mountains to Gqeberha (Port Elizabeth). **Notes** Not browsed, possibly toxic to stock.

Engelandkop, Kouga 13.07.2011

HILLIARDIELLA (= *Vernonia*)

Named after Dr Olive May Hilliard, 1926–2017, a botanist and taxonomist from Durban; this is a recently named genus comprising 8 species that were previously classified in *Vernonia*.

A genus of perennials with annual stems from a woody rootstock. The stems have glandular T-shaped hairs. 1 species in the Western and Eastern Cape, also in Baviaanskloof.

Hilliardiella pinifolia Blounaaldeteebossie

Pinus = a coniferous genus, *folium* = leaf; the needle-like leaves look like those of pine trees.

Resprouting strongly from a woody base, multi-stemmed, erect, perennial shrub to 50 cm. **Leaves** Linear, with margins rolled under, 6 cm long × 0.4 cm wide, silvery-hairy below. **Flowerheads** In dense corymbs 10–20 cm wide, flat-topped, discoid, each flowerhead 1 cm wide, purple. Flowering in summer. **Habitat** Occurring in Grassy Fynbos, especially on grassy slopes on shale or sandstone where loamy soils are better developed. **Distribution** Widespread from George to tropical Africa. **Other names** Inkanga, isiqaji (X).

Nooitgedacht, Kouga 20.01.2022

METALASIA

Blombos

meta = after, beyond, *lasios* = woolly; referring to the leaves that are woolly below

A genus of mostly shrubs with discoid flowerheads, sweetly scented, attracting bees. The involucre is papery and bright, and/or thin and translucent. 52 species in South Africa, mainly in Western Cape; 7 species in Baviaanskloof.

fruit

Metalasia acuta

acuta = acute; referring to the pointed tip of the leaves

Reseeding, single-stemmed, much-branched shrub to 1.5 m. **Leaves** Twisted, with axillary tufts, to 1.5 cm long × 0.1 cm wide, having a sharp, hard point. **Flowerheads** Several in dense terminal clusters, 3 cm wide, each flowerhead 0.7 cm long × 0.2 cm wide, white, with erect bracts. Flowering in summer and autumn. **Habitat** Grassy and Arid Fynbos. **Distribution** Widespread from Kogelberg to Kouga Mountains.

Quaggasvlakte, Kouga 13.02.2016

Metalasia aurea Geelblombos

aurea = golden; referring to the flowers

Reseeding, single-stemmed, much-branched shrub to 1.5 m. **Leaves** Spreading, twisted, with axillary tufts, 1.5 cm long × 0.1 cm wide, with sharp tips. **Flowerheads** Many in numerous terminal clusters, to 3.5 cm wide, each flowerhead 0.7 cm long × 0.2 cm wide, yellow. Flowering in autumn. **Habitat** Transitional Shrubland and Grassy Fynbos, usually where the soil is better developed. **Distribution** Widespread from Potberg to Gqeberha (Port Elizabeth).

Langkloof 17.06.2010

Metalasia massonii

Named after Francis Masson, 1741–1805, a Scottish botanist and gardener – Kew's original 'plant hunter'. He collected in South Africa between 1772 and 1775, and discovered many species new to science.

Reseeding, single-stemmed, much-branched, rounded shrub to 2 m tall, but only 50 cm tall on upper slopes. **Leaves** Twisted, closely clustered, 1 cm long × 0.2 cm wide, cream-woolly below. **Flowerheads** Several in terminal clusters, white to brown, with reflexed bracts. Flowering in spring. **Habitat** Common and abundant on upper slopes in Grassy, Mesic and Subalpine Fynbos. **Distribution** From the Little Karoo to Makhanda. **Notes** At the top of the high peaks there is a stunted form, with much shorter leaves and a neat, rounded growth form.

reflexed bracts

Bosrug, Baviaans 12.07.2011

Metalasia pallida

pallida = pale; referring to the white flowers; the leaves also tend to be pale green.

Reseeding, single-stemmed, upright, much-branched, grandiose shrub to 2.5 m. **Leaves** Not twisted, to 1.5 cm long × 0.2 cm wide, white-woolly below. **Flowerheads** In terminal clusters to 3 cm wide, each flowerhead to 0.5 cm wide, with white-and-brown bracts. Flowering in spring. **Habitat** Rocky slopes in Arid and Grassy Fynbos. **Distribution** Fairly narrow from the Little Karoo to Humansdorp.

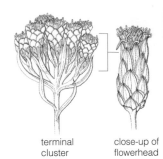

terminal cluster

close-up of flowerhead

Witruggens, Kouga 03.11.2011

Metalasia pungens

pungens = sharp, acrid or piercing; referring to the sharply pointed leaf tips

Reseeding, single-stemmed, erect, much-branched shrub to 2 m. **Leaves** Variable, not twisted, ascending, closely packed, with pointy tip curving outwards, 1 cm long × 0.2 cm wide. **Flowerheads** Many crowded in terminal clusters to 3 cm wide, each flowerhead 0.2 cm wide, pointy in bud, often tinged pink or red in bud, becoming white. Flowering mainly in autumn. **Habitat** Grassy and Mesic Fynbos, more common on upper slopes and rocky ridges. **Distribution** Widespread from Bredasdorp to Makhanda.

Rus en Vrede 4X4, Kouga 22.05.2010

Metalasia trivialis

trivialis = commonplace, ordinary; referring to the typical appearance of this species

Reseeding, single-stemmed, erect, much-branched shrub to 3 m, with white-woolly young branches. **Leaves** Twisted, spreading, to 1 cm long × 0.2 cm wide, bright green-glossy above, white-felted below, with axillary tufts. **Flowerheads** In terminal clusters 2–3.5 cm wide, each flowerhead 0.3 cm wide, white; involucral bracts erect, brown, cobwebby-woolly. Flowering in spring and summer. **Habitat** Rocky south-facing slopes in Grassy Fynbos, often in ravines adjacent to bush or forest. **Distribution** Widespread from mountains in the Little Karoo to Makhanda and Hogsback.

Suuranysberg, Kareedouw 11.05.2021

MICROGLOSSA

mikros = small, *glossa* = tongue; referring to the short, tongue-shaped ray florets

Scrambling bushes or climbers with flat-topped flowerheads and short rays. ± 10 species in Africa; 2 species in South Africa and Baviaanskloof.

Microglossa mespilifolia   Trailing daisy

Mespilus = plant genus, *folia* = leaf; alluding to the leaves that look like some species of *Mespilus* (medlar)

Fast-growing, scrambling climber to 4 m. **Leaves** Thinly textured, triangular, on narrow petioles, coarsely toothed, 4 cm long × 2.5 cm wide, bright green. **Flowerheads** Several in dense, flat-topped, clusters to 3 cm wide, each flowerhead 0.5 cm wide, with rays white and short, disc yellow. Flowering all year round. **Habitat** Forest margins. **Distribution** Widespread from Baviaanskloof along the coast to KwaZulu-Natal. Easily seen from the road at Poortjies. **Other names** Ithyolo (X), ikhambi-lentwala, ikhambi-lesiduli, indlondlo (Z).

Poortjies, Cambria 26.10.2018

MIKANIA

Named after Joseph Gottfried Mikan, 1743–1814, an Austrian-Czech botanist who was a professor of botany in Prague

Mostly climbers with opposite angular leaves, capitulum discoid and 4-flowered, pappus scabrid-bristly. Many species with medicinal uses. ± 430 species mainly in America; 2 species in southern Africa; 1 species in Baviaanskloof.

Mikania chenopodiifolia (= *M. capensis*)

chenopod = goose foot, *folia* = leaf; having leaves that look like goose feet

Scrambling climber to 3 m. **Leaves** Sagittate, sparsely toothed, 6 cm long × 3 cm wide. **Flowerheads** Many bunched together in a 4 cm-wide synflorescence, each flowerhead 0.3 cm wide, discoid, each head 4-flowered, white. Flowering in spring. **Habitat** Thicket and Forest. **Distribution** Very widespread from Knysna to tropical Africa.

Groendal, Kariega 03.09.2020

NIDORELLA   Vleikruid

nidor = strong smell; alluding to the aromatic leaves of some of the species; previously partly in genus *Conyza*

Herbs with alternate leaves. Many capitula in corymbs with very short rays; pappus barbed. ± 23 species in southern and tropical Africa; 8 species in South Africa; 2 species in Baviaanskloof.

Nidorella ivifolia Bakbesembossie, oondbos

ivifolia = ivy-like leaves; referring to the toothed leaves

Reseeding, slender shrub to 1.5 m. **Leaves** Lanceolate, petiolate, 8 cm long × 2 cm wide, with serrate-toothed margins. **Flowerheads** Discoid, many in a dense corymb 3 cm wide, flowerhead 0.5 cm wide, cream to pale yellow. Flowering mainly in summer and autumn. **Habitat** Damp south-facing slopes in Grassy Fynbos or on stream banks. **Distribution** Widespread from the Cape to Zimbabwe. **Notes** Leaves and roots have antibacterial properties and are used as herbal medicines for a wide variety of ailments. **Other names** Isavu, umfazi unengxolo, unonkangana, ingcethe, icegceya, iyeza lomoya (X).

Nuwekloof Pass 16.01.2022

OEDERA Perdekaroo

Named after George Christian Oeder, 1728–1791, German botanist, medical doctor, economist and social reformer; he was the author of *Flora Danica*.

All have dry, thin involucral bracts in rows, overlapping like roof tiles. Flowerheads always radiate, flowers yellow, rays often brown streaked or orange below. *Relhania* and *Rosenia* have been incorporated into *Oedera* (2018). 18 species in South Africa, mostly in the Cape; 7 species in Baviaanskloof.

fruit

Oedera calycina (= *Relhania calycina*)

calycinus = with a well-developed calyx; referring to the relatively large papery-brown involucral bracts

Resprouting, erect, sparsely branched shrub to 1 m. **Leaves** Ascending, lanceolate, pungent, with 3–9 veins, 1 cm long × 0.5 cm wide. **Flowerheads** Solitary, radiate, 2 cm wide, rays golden yellow, brownish below. Flowering in spring and summer. **Habitat** Grassy Fynbos and Transitional Shrubland. **Distribution** Widespread from Cederberg to Gqeberha (Port Elizabeth).

Hankey 25.10.2018

Oedera decussata (= *Relhania decussata*)

decussata = decussate; each pair of opposite leaves arranged around the stem at right angles to the one below

Resprouting, erect, stiff, divaricately branched, glabrous shrublet to 50 cm. **Leaves** Linear, ascending, tightly packed, decussate, 0.6 cm long × 0.1 cm wide. **Flowerheads** Showy, 2 cm wide, rays orange-yellow and often reddish below, acute brown bracts with pale margins. Flowering in spring. **Habitat** Occurring sporadically in Grassy and Mesic Fynbos, often on rocky ridges. **Distribution** Near endemic from Swartberg to Kouga and Baviaanskloof mountains. **Status** Rare.

flowerhead showing 2 rays

Enkeldoring, Baviaans 21.09.2011

Oedera genistifolia Kleinperdekaroo

Genista = genus name, *folia* = leaf; the leaves look similar to those of some species of *Genista*, a genus in the legume family that is commonly known as broom.

Reseeding, much-branched, twiggy shrub to 1 m. **Leaves** Densely packed around the stem, sometimes gummy on top, 1.5 cm long × 0.3 cm wide, grey-green or green. **Flowerheads** Few clustered in umbels, 0.5 cm wide, yellow, bracts straw-coloured. Flowering in spring. **Habitat** Clay-rich soils on shale bands in Transitional Shrubland. **Distribution** Widespread from Namaqualand to Cape Peninsula and Makhanda.

Joubertina, Langkloof 08.10.2011

Oedera imbricata

imbricata = overlapping like tiles; referring to the closely packed leaves

Reseeding, erect, tough, closely leafy shrub to 50 cm. **Leaves** Broadly lanceolate to ovate, hard, 1.5 cm long × 0.5 cm wide, glandular-scabrid on margins. **Flowerheads** Several crowded together in a false head, to 4 cm wide, with yellow rays. Flowering in spring. **Habitat** Transitional Shrubland, and Grassy and Arid Fynbos. **Distribution** Widespread from the Cape to Makhanda.

Engelandkop, Kouga 19.04.2012

Oedera speciosa (= *Relhania speciosa*)

speciosus = splendid, showy; referring to the large, spectacular flowerheads

Resprouting, erect, sparsely branched shrub to 1.5 m. **Leaves** Lanceolate, spreading, 5–9-veined, 1.5 cm long × 0.6 cm wide, with sharp points. **Flowerheads** Solitary at leafy branch tips, large, 4 cm wide, bracts golden brown, rays yellow above and orange below. Flowering in spring and early summer. **Habitat** Found on rocky slopes and ridges in Grassy and Mesic Fynbos. **Distribution** Widespread on mountains between Villiersdorp and Gqeberha (Port Elizabeth).

Sewefontein, Kouga 22.09.2010

Oedera squarrosa Vierkantperdekaroo

squarrosa = rough with stiff leaves; referring to the recurved leaves

Reseeding, erect, twiggy shrub to 1 m. **Leaves** Mostly 4-ranked, obovate, recurved, mucronate, 1.5 cm long × 0.7 cm wide. **Flowerheads** Clustered together in dense terminal umbels 5 cm wide, each to 1 cm wide, with yellow rays. Flowering in spring. **Habitat** Transitional Shrubland and sometimes Grassy Fynbos. **Distribution** Widespread from Namaqualand to Gqeberha (Port Elizabeth).

Stompie se Nek, Baviaans 14.09.2010

OSTEOSPERMUM

Boneseed, bietou

osteon = bone, *sperma* = seed; referring to the hard fruits

Involucral bracts green with a somewhat variable arrangement of 1–4 rows of almost equal length, often with translucent margins. ± 80 species mostly in Africa; 7 species in Baviaanskloof.

fruit

Osteospermum calendulaceum (= *Oligocarpus calendulaceus*) **Boegoebossie**

calendulaceus = similar to the genus *Calendula*

Annual, sprawling, foetid, to 40 cm. **Leaves** Oblanceolate, glandular-hairy, toothed, 2 cm long × 0.4 cm wide. **Flowerheads** Few at branch tips, radiate, 0.6 cm wide, yellow. Achenes warty. Flowering in winter and spring. **Habitat** Dry rocky slopes in Thicket. **Distribution** Widespread from Worcester to Cradock, Eastern Cape. **Other name** Umayibuye (X).

Kariega 17.05.2019

Osteospermum junceum **Iyeza lomoya (X)**

junceus = rush-like; derivation unclear as the plant is not really rush-like. Linnaeus, who named it, may have only had a specimen of the uppermost flowering section of the plant, which is rush-like.

Reseeding, sturdy, upright, with long, woolly flower stems, to 3m tall. **Leaves** Leathery, 16 cm long × 5 cm wide, white-woolly when young; large lower down, becoming smaller, more linear, ascending the long flowering stems. **Flowerheads** Large, to 6 cm wide, with yellow rays and woolly bracts. Fruit dark, shiny and fleshy. Flowering from winter to summer. **Habitat** Mesic and Arid Fynbos on upper slopes. **Distribution** Widespread from Stellenbosch to Makhanda.

Enkeldoring, Baviaans 22.09.20

Osteospermum moniliferum

(= *Chrysanthemoides monilifera*) **Bietou, bosluisbessie, skilpadkos**

moniliferum = bearing a necklace; referring to the ring of glossy black berries that look like a necklace

Reseeding, much-branched, rounded, leafy shrub to 1.5 m. **Leaves** Leathery, 4 cm long × 2 cm wide, white-woolly when young, cobwebby when older. **Flowerheads** Clustered in terminal corymbs to 12 cm wide, radiate, each head to 4 cm wide. Fruit green, turning black and glossy, bird-dispersed. Flowering all year round but mainly autumn. **Habitat** Occurring mainly in Fynbos and Transitional Shrubland. **Distribution** Common and widespread from Namaqualand to tropical Africa. **Notes** Variable, with several subspecies recognised. Useful in restoration and gardens as a fast-growing, short-lived pioneer. Berries are edible, sweet and contain vitamin C. Naturalised in Australia where it is a weed in dune fields.

flowerhead showing 2 rays

Zandvlakte bergpad, Kouga 12.10.2010

Osteospermum polygaloides

polygaloides = like genus *Polygala*; the glabrous leaves and
their neat arrangement on the stem are reminiscent of *Polygala*.

Reseeding, lanky, upright, leathery-leaved shrub to 2 m. **Leaves** Closely
packed, ascending, oblong to ovate, recurved at tips, smooth, 8 cm long
× 3 cm wide at base of plant, 1.5 cm long × 1 cm wide near top of plant,
but size and arrangement highly variable, grey-green. **Flowerheads**
Single on short, rough-hairy peduncles, radiate, 3 cm wide, yellow.
Achenes ribbed and pitted. Flowering mainly in spring and early
summer. **Habitat** Common in first few years after fire in Transitional
Shrubland and Fynbos, most often Grassy Fynbos. **Distribution**
Widespread from Citrusdal and the Cape Peninsula to KwaZulu-Natal.
Notes A variable species with a stunted form near the coast.

Bosrug, Baviaans 27.07.2011

Osteospermum scabrum (= *Gibbaria scabra*)

scabrum = scabby or scaly; referring to the rough-hairy leaves

Resprouting, many-branched from the base, rigid shrub
to 30 cm. **Leaves** Filiform, ascending to spreading,
sometimes slightly curving, rough-hairy, 2 cm long
× 0.1 cm wide. **Flowerheads** Mostly solitary on short
peduncles, radiate, 1.5 cm wide, yellow. Flowering in
spring. **Habitat** Grassy and Mesic Fynbos. **Distribution**
Narrow distribution from Knysna to Kariega (Uitenhage).

Bosrug, Baviaans 27.07.2011

OTHONNA   Bobbejaankool

othonne = linen or cloth; referring to the soft leaves of some species, or the long, cottony-soft,
pappus bristles

This genus has smooth bracts in a single row, fused at the base and without
calycles. The pappus has long, fine, barbed bristles. ± 140 species from Namibia to
KwaZulu-Natal; 7 species in Baviaanskloof.

fruit

Othonna gymnodiscus

gymno = naked, flat, *discus* = disc; referring to the smooth flat disc that
supports the fruit. *Gymnodiscus* is also a genus of daisy, but it does not
resemble this species.

Erect, fleshy, leathery herb to 30 cm, with a tuberous root. **Leaves**
Clasping the stem, held at an upward angle, dull green, 5 cm long
× 2 cm wide. **Flowerheads** Solitary, on a long peduncle, without
rays, 1 cm wide, yellow, with green calyx, 1 cm long. Flowering from
autumn to spring. **Habitat** Loamy soils on bedrock of south-facing
slopes in Grassy Fynbos and Transitional Shrubland. **Distribution**
Very widespread from Namaqualand to Gqeberha (Port Elizabeth).

Damsedrif, Baviaans 19.05.2021

Othonna parviflora Bobbejaankool

parviflora = small flowers; referring to the relatively
small flowerheads

Robust, fast-growing, short-lived, erect,
reseeding shrub to 2 m. **Leaves** Leathery,
smooth, sometimes with toothed margins,
to 12 cm long × 6 cm wide, grey-green.
Flowerheads In large clusters sometimes
over 20 cm wide, with stunted rays, 1 cm
wide, yellow. Flowering mainly from winter
to early summer. **Habitat** Occurring mostly
in Grassy Fynbos in the first few years after
fire. **Distribution** Widespread from Calvinia to
Baviaanskloof. **Other name** Bokveldharpuis (A).

side view of flowerhead
showing some rays

bracts
fused

Mistkraal, Baviaans 23.08.2010

Othonna retrofracta

retrofracta = twisted or turned backwards;
referring to the leaves that tend to be slightly
twisted or curving; the branches also tend to bend.

Deciduous, fleshy-leaved shrublet to 30 cm
tall, but mostly 10 cm. Resprouting from
a fleshy caudex that is usually hidden
underground in fire-prone vegetation. **Leaves**
In loose tufts at branch tips, oblanceolate, pinnatifid, 3 cm long
× 1.5 cm wide, with young leaves, midrib and margin tinged
purple. **Flowerheads** Single on forked peduncles, without rays,
1 cm wide, yellow with red stigmas. Flowering from autumn to
spring. **Habitat** Found in Baviaanskloof in Transitional Shrubland
and Grassy Fynbos. **Distribution** Widespread from Worcester
to Baviaanskloof. **Notes** Popular in pots because of the swollen
caudex; good for bonsai.

Langkloof 07.06.2014

Othonna triplinervia

triplus = triple, *nervus* = nerve; referring to the
three veins coming from the middle of the leaf

Erect, sparsely branched, smooth-stemmed shrub to
1.5 m. **Leaves** Obovate, leathery and fleshy, slightly
lobed, large, 8 cm long × 5 cm wide, blue-grey.
Flowerheads With several heads in a terminal corymb to
15 cm wide, on a long peduncle, each flowerhead with
only 5 long, yellow rays, flowerhead 3 cm wide. **Habitat**
Cliffs and rocky ledges of the ravines in Fynbos Woodland.
Distribution From Baviaanskloof to Makhanda. **Notes**
Much potential in the garden and in containers.

Bosrugkloof, Baviaans 12.07.2016

PEGOLETTIA

Thought to be named after Francesco Balducci Pegolotti, 1310–1347, a prominent Florentine merchant and politician

Often confused with *Pteronia*, but *Pegolettia* bracts are green and the anthers have tails. Rich in essential oils. 9 species from Africa to India; 2 species in Baviaanskloof.

fruit

Pegolettia baccaridifolia Draaibos

Baccharis = plant genus, *folia* = leaf; with leaves like those of *Baccharis*, a genus named after Bacchus, the Roman god of wine

Resprouting, twiggy, rounded shrub, to 70 cm. **Leaves** Ovate, petiolate, 1.5 cm long × 0.5 cm wide, with toothed margins and a strong lemon scent. **Flowerheads** Solitary, no ray florets, 1 cm wide, pale yellow, with green bracts. Flowering after rain, mainly in spring. **Habitat** Dry rocky slopes in Transitional Shrubland, Thicket and Succulent Karoo. **Distribution** Widespread from the Little Karoo to Graaff-Reinet and to Makhanda. **Other name** Braambos (A).

solitary with no ray florets

Krakeel, Langkloof 23.10.2011

Pegolettia retrofracta Geelbergdraaibos

retrofracta = twisted or turned backwards; referring to the leaves that are twisted

Tough, resprouting, twiggy, shrublet to 50 cm. **Leaves** Entire, obovate, slightly twisted, aromatic, 1 cm long × 0.5 cm wide, lime-green, with wavy margins. **Flowerheads** Solitary, without rays, 0.5 cm wide, yellow. Flowering all year round, mainly after rain. **Habitat** Uncommonly encountered in Baviaanskloof, in sandy and loamy soils in most dry and open habitats, but not in Fynbos. **Distribution** Widespread and common throughout the drier parts of South Africa and Namibia.

Engelandkop, Kouga 30.10.2011

PENTZIA Skaapkaroo

Named after Carolus Johannes Pentz, 1738–1803, a Swedish botanist and student of Thunberg; he published *Dissertatio de diosma* in 1797.

No ray florets in this genus. Pappus comprises scales or lacking. ± 23 species in southern Africa; 2 species in Baviaanskloof.

fruit

Pentzia dentata Grootskaapkaroo

dentata = toothed; pertaining to the leaf margins that are palmately toothed at the upper edge

Upright, silver-felted, aromatic shrub to 60 cm. **Leaves** Petiolate, deltoid, toothed above, 1 cm long × 0.5 cm wide, silver-grey, margins golden brown. **Flowerheads** Several in a dense corymbose synflorescence 2.5 cm wide, without rays, each flowerhead 0.5 cm wide, yellow. Flowering mainly in spring and summer. **Habitat** Stony slopes in Transitional Shrubland. **Distribution** Widespread from Worcester to Baviaanskloof. **Notes** The leaves look very similar to those of *Helichrysim excisum*, but that species has recurved bracts while *P. dentata* has straight bracts that are silver-felted.

Rus en Vrede 4X4, Kouga 20.01.2022

Pentzia incana Ankerkaroo

incanus = hoary, white; referring to the fine, cobwebby hairs covering the stems and young leaves

Twiggy, aromatic shrub, to 50 cm, with branches curving down, rooting and forming new plants. **Leaves** Divided with linear lobes, to 1 cm long × 0.5 cm wide, grey-green. **Flowerheads** Single on long peduncles, rounded, without rays, 1 cm wide, yellow. Flowering mainly in summer, or after rain. **Habitat** Common in disturbed, loose, clay-rich soil in most dry habitats, but not in Fynbos. **Distribution** Widespread from Namaqualand through the Little and Great Karoo to Komani (Queenstown). **Notes** Useful for restoring overgrazed and barren ground in dry areas, and often colonising such areas naturally. Browsed by stock and game. Naturalised in Australia. **Other name** Mohantsoana (S).

no ray florets

Rooihoek, Baviaans 29.12.2017

PHYMASPERMUM

Bankrotbos

phyma = swelling, *sperma* = seed; referring to the papillated achenes

Sometimes confused with *Cymbopappus*, but the fruits of *Phymaspermum* are without scales. 17 species in southern Africa; 1 species in Baviaanskloof.

Phymaspermum oppositifolium

oppositifolium = with opposite leaves

Reseeding, upright, much-branched shrub to 1.5 m. **Leaves** Small, scale-like, lanceolate, appressed, 0.3 cm long × 0.1 cm wide. **Flowerheads** Solitary at branch tips, 1 cm wide, yellow centres with white or magenta ray florets. The ray 'petals' fold down at night, away from the disc florets. Flowering in summer. **Habitat** Rare on rock ledges in Grassy Fynbos and Fynbos Woodland. **Distribution** Endemic, known from only 3 collections near Riverside on the Kouga River. **Notes** Differs from *P. appressum* by the opposite leaves and lanceolate involucral bracts. **Status** Rare.

rays fold down at night

Kouga Kliphuis, Kouga River 30.09.2018

PRINTZIA

Named after Jacob Printz, 1740–1779, a Swedish botanist who worked with Linnaeus

Leaves alternate, woolly or felted below, pappus feathery. 6 species in South Africa; 1 species in Baviaanskloof.

Printzia polifolia

Polium = genus name, *folia* = leaf; with leaves like *Polium*, which is now *Teucrium*, alluding to the grey-green leaves that are white-woolly below

Resprouting shrub with rigid branches to 1 m. **Leaves** Obovate, petiole winged, 1 cm long × 0.5 cm wide, green-hairy above, white-woolly below, margins wavy. **Flowerheads** 4 cm wide, centre yellow, rays mauve-blue, bracts woolly. Flowering late winter and spring. **Habitat** Infrequent on shale bands in Grassy Fynbos and Transitional Shrubland. **Distribution** Widespread from Nieuwoudtville to the Cape Peninsula and to Gqeberha (Port Elizabeth).

Krakeel, Langkloof 10.08.2009

PTERONIA

Gombos

pteron = wing; alluding to the wind-dispersed seeds that fly without wings

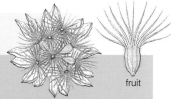

Leaves contain fragrant oils and are scented when crushed, often used as a culinary herb. The genus never has ray florets. ± 80 species in southern Africa; ± 35 species in the Cape flora; 12 species in Baviaanskloof.

fruit

fruiting head shown from above, with involucral bracts dried and open

Pteronia adenocarpa

adeno = gland, *carpa* = fruit; referring to the sticky
surface of the bracts and fruits (cypselas)

Resprouting, much-branched, twiggy shrub to 50 cm. **Leaves** Opposite,
flat (not fleshy), ovate, 1 cm long × 0.5 cm wide, scabrid on margins only,
with recurved tip. **Flowerheads** Oblong, sticky, fragrant, 2.5 cm long ×
1 cm wide, pinkish; obtuse bracts greenish with shades of pink and purple,
bract tips membranous, sometimes jagged. Cypselas sticky, pappus straw-
coloured. Flowering in spring. **Habitat** Occasional on exposed rocky slopes
in Succulent Karoo and Thicket. **Distribution** Widespread from Roggeveld
through the Great Karoo to Baviaanskloof.

Kamassiekloof, Kouga 18.01.2022

Pteronia camphorata Sandgombos

camphorata = smelling of camphor; leaves are camphor-scented.

Reseeding, woody, erect, densely leafy shrub to 2 m. **Leaves** Linear to
filiform, tightly packed, ascending, 2 cm long × 0.2 cm wide, green to lime-
green. Bracts yellowish. **Flowerheads** 1 to a few at branch tips, without rays,
1.5 cm wide, yellow. Flowering in spring. **Habitat** Seasonally damp areas
in Mesic Fynbos. **Distribution** Widespread from Anysberg to the Kouga
Mountains. **Notes** Used traditionally to treat colds, influenza, tuberculosis
and other ailments. Research has found that the plant contains substances
with antimicrobial, anti-inflammatory and antipyretic properties.

East of Kougakop 16.04.2012

Pteronia glauca Boegoekaroo

glaucus = blue-grey; pertaining to the grey-woolly leaves

Reseeding, much-branched shrub to 60 cm tall with branches that arch over and
root where they touch the ground. **Leaves** Woolly, 0.5 cm long × 0.3 cm wide,
blue-grey to greyish. **Flowerheads** Relatively narrow, 0.5 cm wide, yellow. Bracts
non-sticky, with a narrow translucent line down the middle, slightly hairy. Flowering
in spring. **Habitat** Clay soils in Succulent Karoo. **Distribution** Widespread from
Namaqualand to Eastern Cape and Free State, mostly in the arid interior.

translucent
line

Damsedrif, Kouga 11.10.2016

Pteronia glomerata Gombossie

glomerata = collected closely together into a head;
pertaining to the tightly packed leaves or the involucral bracts

Resprouting, low, twiggy shrub to 30 cm. **Leaves** Opposite, in tight clusters,
hard, fleshy, small, 0.3 cm long, dark green. **Flowerheads** Discoid, 0.5 cm
wide, yellow, with slightly sticky involucral bracts, rounded at apex and
minutely ciliolate. Flowering in spring. **Habitat** Common on clay soils in
Succulent Karoo. **Distribution** Widespread from Namaqualand through the
Little and Great Karoo to western end of Baviaanskloof.

Above Nuwekloof Pass,
Baviaans 08.10.2016

Pteronia incana Asbossie, kraakbos

incanus = hoary, white; referring to the leaves and stems felted silver-grey

Reseeding, much-branched, grey-felted shrub to 1 m, flowering prolifically, turning the bush bright yellow in spring. **Leaves** Small, 0.5 cm long × 0.3 cm wide, white- to grey-woolly. **Flowerheads** 1 cm wide, yellow-green, with yellow florets and glabrous bracts. **Habitat** Common and abundant on clay-rich, red soils, mostly in Transitional Shrubland, but also Thicket and Succulent Karoo. **Distribution** Widespread from Namaqualand through Little Karoo to Peddie. **Notes** Important nurse plant for a wide range of succulents and other plants. Asbosveld was named after this plant. It contains essential oils and has been used in traditional medicine for treating flu, backache, fever and kidney problems. Useful as part of restoration in dry areas where grasses have been compromised. **Other name** Ibhubhusi (X).

Zandvlakte bergpad, Kouga 19.09.2010

Pteronia membranacea

membranacea = membranous; referring to the margins of the involucral bracts

Reseeding shrub to 50 cm. **Leaves** Opposite, oblong, slightly keeled, scabrid-hairy, 0.7 cm long × 0.4 cm wide, blue-grey. **Flowerheads** Large, 2.5 cm long × 1.5 cm wide, bracts greenish, tinged red and with opaque, membranous margins, florets yellow. Flowering in summer. **Habitat** Occurring sporadically on loamy soils in Transitional Shrubland. **Distribution** From Montagu to Makhanda and Middelburg, Eastern Cape.

Zandvlakte, Kouga 30.10.2011

Pteronia paniculata Gombossie

paniculata = with panicles; referring to the paniculate inflorescence

Reseeding, much-branched shrub to 1 m. **Leaves** Subterete, sticky, 2 cm long × 0.3 cm wide, shiny green. **Flowerheads** Bunched together in several panicles to 3 cm wide, non-sticky, each flowerhead 0.5 cm wide, florets yellow and bracts yellow-green. Flowering mostly in summer. **Habitat** Shale in Succulent Karoo and Thicket. **Distribution** Widespread from Namibia and the Little Karoo to Eastern Cape. **Notes** Important nurse plants for small succulents.

Damsedrif, Kouga 11.10.2016

Pteronia staehelinoides

Staehelin = a plant genus, *oides* = resembling; *Staehelin* is a daisy genus from the Mediterranean that has similar-looking flowers. The genus was formed by Linnaeus in 1737 and named after Benedikt Staehelin, 1695–1750, a Swiss physician and naturalist, and professor of physics at the University of Basel.

Rigid, tough, much-branched, scabrid shrub to 1 m. **Leaves** Oblong-lanceolate, keeled, 1 cm long × 0.3 cm wide, dark green, rough-hairy on margins. **Flowerheads** Discoid, solitary at branch tips, 3 cm long × 1.2 cm wide, yellow, with acuminate reddish-brown bracts with finely fringed margins. Flowering in spring. **Habitat** Infrequent on dry sandstone and shale slopes in Transitional Shrubland and Succulent Karoo. **Distribution** Narrow distribution from Swartberg to Baviaanskloof.

Doringkloof, Kouga
29.07.2011

bracts with finely fringed margins

Pteronia teretifolia

teres = rounded, *folium* = leaf; referring to the leaves that are almost round in cross section

Resprouting, much branched from the base, upright shrub, to 50 cm. **Leaves** Keeled, subterete, 1 cm long × 0.1 cm wide, lime-green. **Flowerheads** Bunched together at the branch tips with inflorescence 2 cm wide, each flower to 0.4 cm wide, cream-white. Flowering in summer and autumn. **Habitat** Fairly common in Fynbos, mostly Grassy Fynbos. **Distribution** Widespread from Potberg and Outeniquas to Gqeberha (Port Elizabeth).

Geelhoutbos Ruggens, Kouga 18.04

PULICARIA
Fleabane

pulex = flea, *aria* = suffix to indicate an agent of use; referring to the use of some species to repel fleas and other insects

A genus of herbs with sessile leaves that are sometimes eared. The achenes are 4-ribbed and the pappus in 2 rows. ± 80 species, mostly in northern parts of Africa, Mediterranean and Asia; 1 species in South Africa.

Pulicaria scabra Fleabane

scabra = scabrid, rough; referring to the roughly textured and hairy leaves and stems

Reseeding, single-stemmed, slender, upright, herbaceous shrub to 60 cm. **Leaves** Lanceolate, auriculate at the stem, scabrid above, foetid-smelling, hairy, 5 cm long × 1.5 cm wide, dull or pale green. **Flowerheads** Single, 1 cm wide, with many short rays, yellow, on hairy stalks 10 cm long. Flowering from summer into autumn. **Habitat** Stream banks and in marshes. **Distribution** A very widespread species from the Cape to tropical Africa. **Notes** Used by traditional healers. **Other names** Aambeibos (A), ubuhlungu bomlambo (X), isithaphuka (Z).

Zandvlaktekloof, Baviaans 18.02.201

Senecio angulatus   Cape ivy

angulatus = angled; referring to the tendency for the flowering stem to hang down while the cluster of flowerheads faces upwards

Climbing creeper, fleshy-leaved, glossy, to 20 m. **Leaves** Ovate to lanceolate, petiolate, green, coarsely toothed, the upper leaves less so. **Flowerheads** In branched, roughly flat-topped corymbs to 8 cm wide, each flowerhead 1.5 cm wide, with rays, yellow, bracts calycled. Flowering in autumn, winter and spring. **Habitat** Thicket and on Forest margins. **Distribution** Widespread along the coast

Elandsberg road, Patensie 25.10.2018

from George to the Wild Coast. **Notes** Popular in gardens, becoming naturalised on the Cape Peninsula. A problematic weed in parts of Australia and New Zealand where it smothers the indigenous vegetation. **Other names** Indindilili, ichongwane, iphungwana (X).

Senecio chrysocoma   Umthithimbili wentaba (Z)

Chrysocoma = plant genus; the discoid flowerheads are reminiscent of the daisy genus *Chrysocoma*.

Reseeding, erect shrub to 80 cm. **Leaves** Filiform, 7 cm long × 0.3 cm wide, green. **Flowerheads** In loose, branched corymbs, discoid, 1 cm wide, yellow, with calycled involucral bracts. Flowering in spring and summer. **Habitat** Occurring sporadically in young or disturbed Grassy Fynbos. **Distribution** Widespread from Swartberg and Langeberg to KwaZulu-Natal. **Notes** This plant is poisonous.

Combrink's Pass, Baviaans 26.10.2018

Senecio coronatus   Sybossie

coronatus = crowned, adorned with wreaths; referring to the way the leaves grow up around the woolly rootstock; could also refer to the wool on the rootstock as the crown

Resprouting from a woolly rootstock, stemless, cobwebby perennial to 40 cm. **Leaves** Basal, erect, petiolate, obovate, leathery, 15 cm long × 8 cm wide, green, margins crenulate and undulating. **Flowerheads** On sparsely leafy stalks to 30 cm tall, in loose corymbs, large, rays long and yellow, with calycled involucre. Flowering in spring, only in the first few years after fire. **Habitat** Shale bands in Grassy Fynbos and Sour Grassland. **Distribution** Widespread from Sedgefield to tropical Africa. **Other names** Woolly grassland senecio (E), ikhubalo lesikhova, indlebe yebokwe (X), lehlomane (S), izonkozonko (Z).

Blue Hill Escape, Kouga 05.04.2020

Senecio crassiusculus

crassiusculus = moderately thick; referring to the fleshy, slightly succulent leaves

Resprouting, loosely branched, softly woody, glabrous shrublet to 30 cm. **Leaves** Very variable in shape and size, first leaves emerging after fire spathulate, then becoming linear to narrowly lanceolate, sparsely pinnate-lobed and toothed or irregularly lobed, the lobes and teeth acute, fleshy, leaves up to 8 cm long × 1 cm wide, dull green. **Flowerheads** On sparsely leafy, glabrous stalks 20 cm long, few in a loose corymb, with rays, each flowerhead 1.5 cm wide, yellow, with minutely calycled involucre. Flowering from summer to autumn, usually only in first few years after fire. **Habitat** Rocky slopes and ridges in Grassy Fynbos. **Distribution** Widespread from Langeberg to Makhanda.

Zandvlakte, Baviaans 29.04.2018

Senecio crenatus

crenatus = crenate, having rounded teeth, scalloped; referring to the crenate margins of the leaves

Resprouting, with few erect, rod-like, ribbed stems from a woody base, blue-grey, thinly-woolly shrub to 70 cm. **Leaves** Folded, elliptic, shortly petiolate, conspicuously net-veined, 4.5 cm long × 1.5 cm wide, margins closely toothed, recurved at tips. **Flowerheads** In branched corymbs to 10 cm wide, with rays, each flowerhead 1.5 cm wide, yellow, with calycled involucre. Flowering mostly in summer and autumn. **Habitat** Rocky sandstone slopes in Grassy Fynbos. **Distribution** Widespread from Swellendam to Gqeberha (Port Elizabeth).

Zandvlakte, Baviaans 29.04.2018

Senecio deltoideus Canary creeper

deltoideus = shaped like a triangle; referring to the shape of the leaves

Climbing herb with zigzag branching habit, flowering prolifically. **Leaves** Triangular, eared at the base, petiolate, soft, 5 cm long × 3.5 cm wide, fresh green, unequally toothed. **Flowerheads** In corymbose, divaricately branched panicles to 4 cm wide, arising from the axils, without rays, flowerhead 1 cm long × 0.3 cm wide, yellow, involucre with only 5 bracts. Flowering in autumn and winter. **Habitat** Forest, Fynbos Woodland and Thicket. **Distribution** Very widespread from Swellendam to tropical Africa. **Other names** Undenze, ikhubalo lesikova (X).

Zandvlaktekloof, Baviaans 17.04.2012

Senecio erubescens (= *S. lanifer*)   Uvelemonti (X)

erubescens = reddening, blushing; referring to the leaves that are often tinged reddish; *lanifer* = wool-bearing; alluding to the fine glandular hairs covering the plant

Resprouting, glandular-hairy, leaves basal in a rosette, to 60 cm. **Leaves** Oblanceolate, slightly lobed, petiolate, 10 cm long × 3 cm wide, toothed and sometimes lacerated. **Flowerheads** Discoid, in open, flat-topped terminal corymbs, flowerhead 1.2 cm wide, magenta. Flowering in spring and autumn. **Habitat** Frequent in most Fynbos habitats, especially on south-facing slopes. **Distribution** Very widespread from Cederberg to Durbanville and to southern tropical Africa.

Sipoonkop, Kouga 30.04.2018

Senecio glastifolius   Waterdissel

glastum = woad (a herb), *folius* = leaf; referring to the superficial similarity of the leaves to those of dyer's woad or glastum (*Isatis tinctoria*)

Reseeding, upright, rigid, glabrous, softly woody, short-lived shrub to 1 m. **Leaves** Lance-shaped, coarsely toothed, clasping at the stem, large, to 15 cm long × 4 cm wide, glossy green. **Flowerheads** Showy, in branched corymbs, each flowerhead 4 cm wide, with mauve rays and yellow centres, and a calycled involucre. Flowering in spring, mostly in the first few years after fire. **Habitat** Occurring in Grassy and Mesic Fynbos, in the Baviaanskloof, restricted to the south-facing slopes of the Kouga Mountains. **Distribution** Fairly narrow distribution from George to Humansdorp. **Notes** An attractive plant that is popular in gardens. It has become naturalised and weedy in Australia and New Zealand, and local gardeners should be cautioned.

Koud Nek's Ruggens, Kouga 07.11.2011

Senecio hollandii

Named after William Jacob Holland, 1848–1932, a zoologist and palaeontologist

Resprouting, woody, densely white-felted shrub to 1 m. **Leaves** Oblanceolate to obovate, 3 cm long × 1.5 cm wide, discolorous, shiny green above, white-felted below, sharply toothed, with revolute margins. **Flowerheads** In branched corymbs to 10 cm wide, each flowerhead radiate, 1 cm wide, yellow. Flowering in spring and summer. **Habitat** Frequent in Grassy and Arid Fynbos. **Distribution** Widespread from Matjiesfontein to Kariega (Uitenhage).

Rus en Vrede 4X4, Kouga 25.04.2018

Senecio junceus   Sjambokbos

junceus = rush-like; referring to the
almost leafless, succulent stems that
make the plant look rush-like

Resprouting from a woody
rootstock, seemingly leafless
shrub, with many erect, rod-like,
green succulent stems to 1 m.
Leaves Small and scale-like,
drying and falling off early. **Flowerheads** In terminal clusters,
with rays, 1 cm wide, yellow. Flowering in autumn. **Habitat** Found
growing up rocky ridges in Grassy and Mesic Fynbos, but usually
found in dry, rocky non-Fynbos habitat in other areas. **Distribution**
Widespread from Namaqualand to Makhanda.

Rus en Vrede 4X4, Kouga 25.02.2011

Senecio lineatus

lineatus = marked by fine parallel lines;
pertaining to the parallel veins on the leaves

Resprouting from a woody rootstock, erect,
densely leafy, grey-felted shrub to 60 cm.
Leaves Ascending to spreading, 3-veined
from the base, grey-felted, becoming nude
above with age, 6 cm long × 2 cm wide, with
a few small narrow lobes at the base, and
slightly toothed near the apex. **Flowerheads**
With 12–30 heads in a branched corymb
at the tips of the stems, flowerheads with

Ridge west of Elandsvlakte, Baviaans
16.02.2016

yellow rays, 1.5 cm wide. Flowering in summer and autumn. **Habitat** Uncommon on the higher rocky
slopes in Grassy and Mesic Fynbos. **Distribution** Widespread from Cape Peninsula to Makhanda.

Senecio linifolius

linifolius = flax-leaved; referring to the long filiform leaves

Reseeding, with several loose branches from
the base, softly woody, glabrous, using other
shrubs for support, to 2 m. **Leaves** Filiform,
fleshy, soft and flimsy, grooved above, 6 cm
long × 0.2 cm wide, green. **Flowerheads** In
branched corymbs on naked peduncles, with
rays, each flowerhead 1.2 cm wide, yellow,
involucre with tiny calycles. Flowering mainly
from autumn to end of spring, or after rain.
Habitat Common in Thicket, or on the edge
of bush clumps in Transitional Shrubland.
Distribution From Baviaanskloof to Mthatha.

Geelhoutbos bergpad, Kouga 31.10.2011

Senecio othonniflorus   Idwara (X)

othonniflorus = with flowers like those in genus *Othonna*;
the flowers, especially the involucral bracts, look
remarkably like those of *Othonna* spp.

Resprouting from a woody base, woolly at base
of stem, erect, to 60 cm. **Leaves** Mostly basal,
linear-lanceolate, glabrous, with some ascending
the flowering stalk, 8 cm long × 0.6 cm wide,
green. **Flowerheads** Several in a loose corymb on
few-leaved stems, without rays, 1.5 cm wide, yellow,
with broad bracts. Flowering in summer and autumn, mostly after fire.
Habitat Grassy Fynbos and Sour Grassland, sometimes Transitional
Shrubland in grassy habitat. **Distribution** Widespread from Knysna to
Mpumalanga. **Other name** Lehlomane-le-lenyenyane (S).

Bergplaas, Baviaans 22.01.2022

Senecio pauciflosculosus   (= *S. oligantha*)

pauci = small group, *flosculus* = floret; referring to the flowerheads,
which, in this species, have few florets; *oligantha* = few-flowered

Resprouting, with several erect stems
from the base, densely leafy, grey-
felted shrublet to 35 cm. **Leaves**
Closely packed, 2 cm long × 1 cm
wide, white-grey-woolly. **Flowerheads**
Clustered into tightly packed, terminal
synflorescence 2.5 cm wide, discoid,
flowerheads 0.4 cm wide, yellow. Flowering in autumn. **Habitat**
Occurring sporadically in Grassy Fynbos. **Distribution** Widespread
from Du Toitskloof to Makhanda.

Quaggasvlakte, Kouga 12.02.2016

Senecio rhomboideus   Lekoto-la-litsoene (S)

rhomboideus = rhombic-shaped; the leaves have a rhombic shape.

Resprouting, fleshy rooted, hairless, tufted to 40 cm, with basal
leaves and a bare flowering stalk. **Leaves** Rhomboid or ovate, fleshy,
9 cm long × 3.5 cm wide, grey-green. **Flowerheads** Few on 40 cm-
long, naked flower stalks, discoid,
1 cm wide, yellow, sparsely
calycled. Flowering in summer
and autumn. **Habitat** Occasional
in Grassy and Mesic Fynbos.
Distribution Widespread from
the Baviaanskloof through the
escarpment to Limpopo. **Notes** In
Lesotho, the plant is used to treat
sterility in women.

calycles

Zandvlakte, Baviaans 29.04.2018

Senecio robertiifolius Greater groundsel

Robertium = plant genus, *folius* = leaf; the plant resembles
some species of *Robertium*, a European genus.

flowerhead
showing
outer rays

calycles

Resprouting from a horizontal rhizome, tufted,
slightly sticky herbaceous perennial to 40 cm. **Leaves**
Oblanceolate, pinnatifid, lacerate, 12 cm long × 3.5 cm
wide, green. **Flowerheads** Few on glabrous — and
sometimes slightly sticky — scapes, with rays, 2.5 cm
wide, yellow, scarcely calycled. Flowering in spring.
Habitat Sporadic in Grassy Fynbos, on rocky south-facing
slopes. **Distribution** Widespread from Namaqualand
through the Little Karoo extending to Humansdorp.

Rus en Vrede 4x4, Kouga,
25.04.2018

Senecio serratuloides Two-day cure

serratul = serrated, *oides* = possessing; pertaining to the regularly
toothed leaf margins; the common name refers to the quick-acting
healing properties of this plant.

Erect, leafy shrub to 1.5 m.
Leaves Leaf folded lengthwise,
base of leaf with pairs of narrow
leaflets, to 10 cm long × 2 cm
wide, green, paler below, margins
neatly toothed. **Flowerheads**
Many in a dense terminal
corymb, each flowerhead 0.7 cm

Riverside, Kouga River 23.05.2014

wide, yellow. Flowering in autumn. **Habitat** Sandy boulder beds at the edge of perennial streams.
Distribution Widespread from Baviaanskloof to Zimbabwe. **Notes** Used traditionally to treat skin sores.
Other names Ichazampukane, insukumbili (Z).

Senecio speciosus Idambiso (X)

speciosus = showy; the flowers are spectacular.

Resprouting, tufted, glandular-hairy herb to
40 cm. **Leaves** Oblanceolate, in a rosette,
lobed to lacerate, large, 15 cm long × 4 cm
wide. **Flowerheads** 3 cm wide, several
in a loose corymb, rays mauve, involucre
hairy. Flowering in summer after rain.
Habitat Occurring on loam-rich soils in
Grassy Fynbos and Transitional Shrubland.
Distribution Widespread from Cape
Peninsula to Mozambique. **Notes** Sought
after in the garden. Great potential for
medicinal use in treating various ailments.
Other name Ibohlololo (Z).

Cockscomb, Groot Winterhoek 17.02.2016

SERIPHIUM

seriph = stroke, line of a letter; referring to the thin curved leaves and the linear, spike-like inflorescence

Only a single tiny disc flower in each flowerhead; flowerheads small, linear, with golden brown bracts. Petal lobes of disc flowers are tiny, inconspicuous, and dark-coloured. 6 species in southern Africa; 2 species in Baviaanskloof.

Seriphium burchellii

Named after William John Burchell, 1781–1863, an English explorer who collected plants in southern Africa and Brazil. He amassed over 50,000 plant specimens in southern Africa and donated them to Kew Gardens, each record with meticulous, detailed notes on habit and habitat. He described his journey in *Travels in the Interior of Southern Africa*, a 2-volume work, published in 1822 and 1824.

Reseeding, much-branched, erect shrub to 2 m. **Leaves** In fascicles, granular, glabrous, to 0.1 cm long, dark green. **Flowerheads** Acute, in congested clumps in the axils of upper branches, forming spike-like inflorescences, without rays, bracts golden brown. Flowering in autumn. **Habitat** Disturbed areas in Transitional Shrubland and Grassy Fynbos. **Distribution** Nearly endemic from Little Karoo to Baviaanskloof. **Notes** There has been little support for including this species in a wider concept of *S. plumosum*.

Kouga Wildernis 16.05.2021

Seriphium plumosum (= *Stoebe plumosa*) Slangbos

plumosum = plumose, feathery; pertaining to the fluffy hairs on the leaves

Reseeding, much branched, tangled, mostly erect in Baviaanskloof, elsewhere more decumbent, to 1.5 m tall. **Leaves** Closely packed in clusters on the wiry stems, tiny, 0.1 cm long × 0.1 cm wide, silvery. **Flowerheads** Tiny, linear in congested clumps in the axils of upper branches, forming spike-like inflorescences, without rays, bracts golden brown. Flowering mostly in autumn. **Habitat** Most Fynbos habitats, usually near to seepages or streams, or areas that become damp after rain. **Distribution** Widespread throughout southern Africa.

Scholtzberg Peak, Baviaans 24.07.2011

STOEBE
Hartebeeskaroo

stoibe = stuffing; referring to slangbos, a species that has been used for making mats or mattresses to sleep on; slangbos has subsequently been moved to *Seriphium*.

Only a single disc floret per flowerhead. Flowerheads small, arranged into larger, rounded compound inflorescences. Usually brightly coloured. 16 species in South Africa; 2 species in Baviaanskloof.

fruit

Stoebe aethiopica   Knoppiesslangbos

aethiopica = from sub-Saharan Africa; in classical times this region had not yet been explored by Europeans.

petal lobe

involucral bract

capitulum is a cluster of many small florets

Tough, rigid, densely leafy shrub to 1.5 m, with grey-woolly stems. **Leaves** Needle-like, twisted and recurved, pungent, 0.5 cm long × 0.2 cm wide, glabrous above, white-woolly below. **Flowerheads** Discoid, many crowded in a dense terminal synflorescence, 1.5 cm wide, capitulum 0.15 cm wide, with white florets and brown bracts. Flowering in spring. **Habitat** Rocky slopes in Grassy and Arid Fynbos. **Distribution** Widespread from Bokkeveld (Nieuwoudtville) to Baviaanskloof.

Kammanassie Mountains
30.11.2019

Stoebe microphylla

micro = tiny, *phyllon* = leaf; referring to the minute leaves

Reseeding, much-branched, rounded shrub to 50 cm. **Leaves** Adpressed, 0.4 cm long × 0.1 cm wide, woolly below. **Flowerheads** Many bunched together in a small rounded synflorescence to 1 cm wide, at the branch tips, individual capitulum 0.3 cm wide, with pink-purple florets and white bracts. Flowering in summer. **Habitat** Occasional on loamy soils in Grassy Fynbos and Transitional Shrubland. **Distribution** Widespread from Mossel Bay to Kariega (Uitenhage).

Bokloof 4x4, Kouga 31.01.2020

SYNCARPHA
Everlasting, sewejaartjie

syn = together, *carpha* = scale; pertaining to the joined pappus bristles; the Afrikaans name, meaning 7 years, refers to the lifespan of the dormant seed of these short-lived, post-fire ephemerals.

Shrublets with grey-felted leaves, often with rust-coloured hairs on the margins. Heads non-radiate with large, showy involucres. 21 species in southern Africa; 4 species in Baviaanskloof.

fruit

Syncarpha canescens Pienksewejaartjie

cansecens = covered with dense, fine, greyish-white hairs;
referring to the leaves

Reseeding, erect, sparsely
branched, grey-felted shrublet
to 50 cm. **Leaves** Pressed to the
stems, 0.5 cm long × 0.3 cm
wide, silvery-grey. **Flowerheads**
Showy, single at branch tips,
papery, 2.5 cm wide, pink to red, fading to white,
with acuminate bracts. Flowering mostly in summer.
Habitat Grassy Fynbos. **Distribution** Widespread from
Kamiesberg to Humansdorp.

Misgund, Langkloof 21.05.2012

Syncarpha eximia Strawberry everlasting

eximius = striking, most beautiful, distinguished,
excellent; referring to the brilliant flowerheads

Reseeding, single-stemmed, robust, densely leafy,
silvery-felted shrub to 1.5 m. **Leaves** Ascending,
overlapping, ovate, 5 cm long × 3 cm wide.
Flowerheads In dense corymbs nested in the
leaves, obtuse, hemispherical, 2.5 cm wide, with
bright red papery bracts. Flowering in summer.
Habitat Occasional on cool, shaded upper
slopes and ridges in Mesic Fynbos. **Distribution**
Widespread from Riviersonderend Mountains to
Kariega (Uitenhage). **Notes** The plant has been
used in traditional medicine to treat biliousness,
jaundice, liver disease, croup and diptheria.

Peaks above Coleskeplaas, Baviaans 06.03.2012

Syncarpha ferruginea

ferruginea = ferruginous or rusty; referring to
the rusty involucral bracts as well as the
bracteate peduncles

Reseeding, densely leafy, grey-felted shrub
to 50 cm. **Leaves** 1 cm long × 0.5 cm
wide, grey-woolly, with dry brown bristles.
Flowerheads Single, acuminate, on short,
bracteate peduncles, 3 cm wide, coppery-
brown when fresh, fading yellow, with papery
bracts. Flowering in spring and summer.
Habitat Dry rocky slopes in Arid and Grassy
Fynbos. **Distribution** Widespread from
Matjiesfontein to Gqeberha (Port Elizabeth).

Beacosnek, Kouga 22.09.2010

Syncarpha milleflora   Knoppiessewejaartjie

milleflora = a thousand flowers; referring to the large
numbers of flowerheads in the synflorescence

Reseeding, single-stemmed, sparsely branched,
robust, densely leafy shrub to 2 m. **Leaves**
Lanceolate, 7 cm long × 3 cm wide, woolly, silver-
grey. **Flowerheads** Many, obtuse, crowded in dense
terminal corymbs to 7 cm wide, each flowerhead to 0.8 cm
wide, white to pink, with papery bracts. Flowering from winter
to summer. **Habitat** Dry north-facing slopes in Grassy and Arid
Fynbos. **Distribution** Widespread from Ladismith to Makhanda.

Bergplaas, Combrink's Pass 26.10.2018

TARCHONANTHUS   Camphor tree, kanferbos

tarcho = funeral rites, *anthos* = flower or plant; the leaves smell of camphor, which has been used in the
preparation of corpses for burial.

Leaves camphor-scented when crushed, fruit densely hairy. 6 species in Africa; 2 species in
Baviaanskloof.

Tarchonanthus littoralis   Coastal camphor bush

littoralis = pertaining to the seashore; referring to the coastal distribution
of this species, although it does extend quite far inland

Resprouting, dioecious, bushy tree to 6 m. **Leavesü ü** strongly
of camphor, 10 cm long × 3 cm wide, discolorous, grey-felted
below, dull green above. **Flowerheads** In large panicles 15 cm
long × 10 cm wide, at the branch tips, each flowerhead 0.5 cm
wide, creamy white. The achenes are woolly, used by some
birds to line their nests. Flowering in late summer and autumn.
Habitat Associated with bush clumps in Transitional Shrubland,
Fynbos Woodland and Grassy Fynbos, also in Thicket and Forest.
Distribution Very widespread, mostly coastal, from the Cape to
KwaZulu-Natal. **Notes** The leaves were once smoked like tobacco and have also been used to treat
lung problems, and for pain relief. **Other names** Kuskanferbos (A), isiduli selinde, umathola, igqange
(X).

Bergplaas, Combrink's Pass 20.03.2010

URSINIA   Bergmagriet

Named after Johannes Heinrich Ursinus, 1608–1667, a
German theologian and author of *Arboretum Biblicum*,
which detailed botanical references in the Bible.

Bracts are arranged in many overlapping rows,
and are generally thin and dry. 38 species, mainly
in southern Africa; 5 species in Baviaanskloof.

dry, papery
pappus of
scales

achene

fruiting head   basal hair tuft

Ursinia anethoides

anethoides = like *Anethum*; *A. graveolens*, commonly known as dill, also has divided leaves that vaguely resemble those of this species.

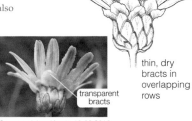

Reseeding, short-lived, erect shrublet to 60 cm.
Leaves Lobes linear to filiform, 2 cm long, fresh green.
Flowerheads Single, on long peduncles, 2 cm wide, with yellow rays. Flowering in summer and autumn. **Habitat** South-facing slopes in Grassy and Mesic Fynbos.
Distribution Widespread from Caledon to Makhanda.

transparent bracts

thin, dry bracts in overlapping rows

Quaggasvlakte, Kouga 13.02.2016

Ursinia anthemoides

anthem = the tribe, *oides* = resembling, similar to some other members of the daisy family

Annual, erect shrublet to 40 cm. **Leaves** Dissected, 5 cm long × 1 cm wide, green fading reddish after flowering. **Flowerheads** Single on long 15 cm-long stalks, usually nodding in bud, each flowerhead 2 cm wide, yellow or orange. Flowering in spring. **Habitat** Fairly common in the lowlands in disturbed areas in Succulent Karoo and Thicket. **Distribution** Widespread from Namibia to Gqeberha (Port Elizabeth).

Langkloof 08.09.2011

Ursinia heterodonta

heterodonta = having different kinds of teeth; referring to the lobes of the leaves; some are simple, others are forked.

Reseeding, erect, scarcely branched shrublet to 50 cm. **Leaves** Ascending, pinnatisect, lower lobes simple, upper lobes variously forked at the tips, 2 cm long × 0.7 cm wide, slightly grey-felted. **Flowerheads** Single on 10 cm-long wiry peduncles, each flowerhead 2 cm wide, with yellow rays. Flowering in spring. **Habitat** Upper slopes in Mesic Fynbos. **Distribution** Widespread from the Hex River Mountains to Kariega (Uitenhage).

membranous bracts

Ridge west of Elandsvlakte, Baviaans 16.02.2016

Ursinia scariosa

scariosus = of thin, dry membranous texture and not green; referring to the golden brown involucral bracts

Reseeding, upright, to 1.7 m, with reddish young branches. **Leaves** Pinnatisect, with linear lobes, 2.5 cm long × 1 cm wide, green fading maroon. **Flowerheads** Single on fairly short peduncles, to 5 cm wide, with yellow rays, bracts papery, rounded and golden brown. Flowering in spring and summer. **Habitat** South-facing slopes in Grassy and Mesic Fynbos.
Distribution Widespread from Paarl to Gqeberha (Port Elizabeth).

involucral bracts

Koud Nek's Ruggens, Kouga 26.09.2011

BIGNONIACEAE

Bignonias

± 85 genera and 860 species. This cosmopolitan family has a centre of diversity in the Amazon, especially of lianas. A well-known member of this family from Brazil, the jacaranda (*Jacaranda mimosifolia*) is popular in gardens and along many South African avenues — especially in Pretoria — because of its showy flowers. Yellow bells (*Tecoma stans*) is an invasive garden escape in South Africa.

RHIGOZUM

Wildegranaat

rhigios = stiff, *ozos* = a branch; alluding to the rigid branches of this plant

7 species in Africa and Madagascar; 1 species in Baviaanskloof.

Rhigozum obovatum Karoo gold

obovatum = egg-shaped; referring to the egg-shaped leaflets that are wider near the tip

Rus en Vrede 4X4, Kouga 06.09.2006

Resprouting, rigid, much-branched shrub or small tree to 4 m. **Leaves** In clusters along spine-tipped branches, obovate, 1 cm long × 0.5 cm wide, dull or grey-green, falling easily. **Flowers** Funnel-shaped, 2 cm wide, golden yellow. Seed capsule flattened, 5 cm long, with papery-winged seeds 1 cm wide. Flowering prolifically after rain in spring and summer. **Habitat** Common on dry, exposed slopes on shale in Thicket and Succulent Karoo. **Distribution** Widespread from Worcester to Zimbabwe. **Notes** Browsed by game and stock. A useful plant for restoring barren, overgrazed slopes. **Other name** Granaatbos (A).

BORAGINACEAE

Borage, forget-me-nots

± 146 genera and 2,000 species worldwide. Trees, shrubs and herbs with coarsely hairy leaves and 5-lobed flowers. 11 genera and 44 species in South Africa; 5 genera and 10 species in Baviaanskloof.

ANCHUSA

anchousa = paint; referring to the use of the root of *A. tinctoria* to make a dye used in cosmetics

About 35 species in Europe, North Africa and West Asia; 2 species in southern Africa; 1 species in Baviaanskloof.

Anchusa capensis Cape forget-me-not, vergeetmynietjie

capensis = from the Cape; referring to the geographical origin of the species

Vigorous annual or short-lived herb to 1 m. **Leaves** In a basal tuft, up to 20 cm long × 2 cm wide, bright green, covered with stiff hairs, rough to the touch. **Flowers** Many in elongated cymes, 0.5 cm wide, blue or dark blue, with white feathery scales at the centre. Flowering in spring and summer. **Habitat** Occurring in sandy soils in disturbed places such as roadsides or dry riverbeds. **Distribution** Very widespread from Namibia, through drier parts of South Africa to Mpumalanga. **Other name** Ossetong (A).

Nuwekloof Pass, Baviaans 16.01.2022

CORDIA

Named after Valerius Cordus, 1515–1544, German physician and botanist. He was a prolific collector and writer, and discovered several scientifically unrecorded species and described them as new; he also discovered the way to synthesise ether. He died of malaria at the age of 29.

7 species in South Africa; 1 species in Baviaanskloof.

Cordia caffra Septee-tree

caffra = from the southeastern part of South Africa, where it was first recorded

Multi-stemmed tree to 10 m tall with smooth, light brown, flaking bark, which resembles bark of the guava tree. **Leaves** Thin, ovate, 10 cm long × 4 cm wide, fresh green above, paler below, long-flimsy-petioled, margins serrated. **Flowers** In dense, terminal heads, sweetly scented, each flower 1 cm long, white, on thin stalks. Fruit roundish and fleshy, with a cup-shaped calyx 1.2 cm wide, yellow to orange. Flowering in spring. **Habitat** Dry stream banks in the narrow kloofs in Forest. **Distribution** Widespread from Baviaanskloof to Mozambique. **Other names** Ouhout (A), umhlovu-hlovu, umnofunofu (X).

Damsedrif, Baviaans 01.05.2018

EHRETIA

Named after George Dionysius Ehret, 1708–1770, a German botanist and entomologist, best known for his botanical illustrations

± 50 species in the world; 3 species in South Africa; 1 species in Baviaanskloof.

Ehretia rigida Puzzle bush

rigida = stiff; referring to the stiff lateral branchlets, or the tough leaves. The common name alludes to the tangled, arching or drooping branches, giving the tree an untidy look. Often tangled up with other trees and shrubs growing in thickets.

Multi-stemmed, tangled, stiff-branched shrub or small tree, to 6 m, with pretty, fragrant flowers. **Leaves** Obovate, small, 2.5 cm long × 1 cm wide, green above, paler below with pockets of hairs in axils below, margins entire. **Flowers** In dense heads a few cm wide, on short lateral branchlets, sweetly scented, each flower 0.7 cm wide, pale bluish to white. Fruit roundish, less than 1 cm wide, bright orange-red, edible. Flowering in spring and summer. **Habitat** Thicket clumps in Thicket and Transitional Shrubland. **Distribution** Widespread from Baviaanskloof to tropical Africa. **Notes** Browsed by game and used medicinally. **Other names** Deurmekaarbos (A), ungobogobana, umbotshane, umhleli (X), mokhalo (S), umkele (Z).

Bergplaas, Combrink's Pass 13.09.2006

Lithospermum papillosum

papilla = nipple, *osum* = abundance; the leaves are covered with papillae.

Resprouting, multi-stemmed, erect, leafy shrublet to 25 cm. **Leaves** Ascending, lanceolate, 2 cm long × 0.4 cm wide, becoming smaller higher up the stem, villous above, only midrib is hairy below. **Flowers** In leafy cymes, with tube as long as the calyx, to 1 cm wide, white. Flowering in summer. **Habitat** Shale or loam-rich soils in Grassy and Mesic Fynbos. **Distribution** Widespread from the Little Karoo to Mpumalanga.

Kouenek, Kouga 26.09.2011

Lithospermum sylvestre

silva = wood, *estris* = place of growth; referring to the plants' forest habitat

Semi-erect, perennial herb to 40 cm. **Leaves** Sessile, elliptic to lanceolate, strigose, 10 cm long × 2 cm wide, fresh green above, paler below, midrib raised below. **Flowers** In terminal or axillary cymes, less than 1 cm long, corolla creamy white, with pale greenish-yellow throat and tube. Nutlets 0.3 cm long, white to silvery grey, glossy. Flowering in spring and summer. **Habitat** Rocky stream beds in Forest and bush in the narrow kloofs. **Notes** A recently recognised species (2016), currently known only from the western part of the Baviaanskloof Mountains. Easy to see near the entrance to Uitspankloof.

seeds

Uitspan, Baviaans 12.03.2022

Lobostemon marlothii

Named after Hermann Wilhelm Rudolf Marloth, 1855–1931, a German-born South African pharmacist, chemist and botanist. He produced the 6 volumes of *Flora of South Africa* from 1913 to 1932, and found many species new to science. Not only are several plants named after him, but also a park and a nature reserve.

Resprouting, sometimes multi-stemmed shrub, to 80 cm. **Leaves** Oblong-lanceolate, hairy, 6 cm long × 1 cm wide, fresh green. **Flowers** In cymes, hairy outside and on style, 2 cm long, blue, sometimes with a pink tube and throat. Flowering in spring. **Habitat** Grassy and Arid Fynbos. **Distribution** Widespread from Matjiesfontein to Humansdorp. **Notes** Easily confused with *L. fruticosus*, which has a western distribution and does not occur in Baviaanskloof. Another species east of the study area, *L. trigonus*, is common near Gqeberha (Port Elizabeth).

Enkeldoring, Baviaans 22.09.2011

BRASSICACEAE (= CRUCIFERAE)   Mustard, cruciferous plants, cabbages

A medium-sized family with 372 genera and 4,060 species, including many economically important plants. Some of the better-known vegetables include *Brassica oleracea* (with common cultivars such as kale, broccoli, cabbage and cauliflower), *B. rapa* (turnip and Chinese cabbage), *Raphanus sativus* (radish) and *Armoracia rusticana* (horseradish). Canola oil is produced from the seeds of *B. napus*. Cruciferae means 'cross-bearing' and refers to the 4-sepalled, 4-petalled shape of the flowers in this family. 29 genera and 74 species in southern Africa, mostly in *Heliophila*; also several weedy species in South Africa.

HELIOPHILA   Sporrie

helio = sun, *phila* = loving; referring to the sun tracking habit of the flowers

± 82 species in South Africa; 6 species recorded in Baviaanskloof.

Heliophila cornuta var. squamata

cornuta = horned; referring to the horned sepal of var. *cornuta*; this variety (*squamata*) lacks horns.

Shrublet to 1.5 m. **Leaves** Filiform, fleshy and subterete, 5 cm long × 0.3 cm wide. **Flowers** 1 cm wide, white, mauve or blue. Fruit to 10 cm long. Flowering in winter and spring. **Habitat** Rocky slopes in Grassy Fynbos. **Distribution** Widespread from southern Namibia to western Kouga Mountains.

Blue Hill Escape, Kouga 04.10.2020

Heliophila elongata

elongata = elongated; referring to the relatively long styles, or the inflorescence that is elongated when in fruit

Thin, upright shrublet to 80 cm tall with grooved stems. **Leaves** Stipule-like, firm, sometimes lacking, ascending, 3 cm long × 0.3 cm wide, green tinged purple. **Flowers** On thin, flimsy peduncles, 0.8 cm long, white or yellowish. Fruit 6 cm long, linear, hanging. Flowering in autumn or summer. **Habitat** Occasional in Grassy Fynbos. **Distribution** Widespread from Riversdale and Swartberg to KwaZulu-Natal.

Zandvlakte, Baviaans 29.04.2018

Heliophila suavissima Ruikpeperbossie

suavissima = sweetest, very sweet; alluding to the sweetly scented flowers, especially fragrant in the evenings

Straggling shrub to 60 cm tall, with annual stems from woody skeleton. **Leaves** Linear, 5 cm long × less than 0.3 cm wide, fleshy. **Flowers** Fragrant, 1.5 cm wide, blue or purple. Fruit to 6 cm long, weakly indented. Flowering in autumn and winter. **Habitat** Lower-lying areas in Succulent Karoo, Thicket and Transitional Shrubland. **Distribution** Widespread from Montagu to KwaZulu-Natal.

Zandvlakte bergpad, Kouga 28.05.2010

Heliophila subulata Ungcilikinde (X)

subulata = resembling an awl; referring to the linear fruit or the shape of the anthers

Resprouting, slender, upright shrublet to 50 cm. **Leaves** Fleshy, ascending, filiform to lanceolate. **Flowers** In loose racemes, 1.5 cm wide, yellow, pink or blue. Fruit linear, to 6 cm long. Flowering in spring, summer and autumn. **Habitat** Sporadic on rocky slopes in Grassy Fynbos. **Distribution** Widespread from Kouebokkeveld Mountains to Kei Mouth.

Noagashoogte, Groot Winterhoek 13.08.2011

BRUNIACEAE

12 genera and 75 species, nearly endemic to the winter-rainfall region of South Africa, with only 1 species from KwaZulu-Natal, which has rainfall in summer. Ericoid shrubs with small, narrow, scale-like leaves that have tiny black tips. Several species are popular as cut flowers. Only 3 species in Baviaanskloof.

BERZELIA   Kol-kol

Named after Jöns Jacob Berzelius, 1779–1848, a Swedish chemist acclaimed as one of the founders of modern chemistry

15 species in Western and Eastern Cape; 2 species in Baviaanskloof.

Berzelia intermedia

intermedia = intermediate; referring to the similarity to 2 other related species

Erect shrub to 1.5 m, resprouting from a woody caudex. **Leaves** Tiny, ascending to spreading, curved in halfway along the leaf, 0.4 cm long × 0.1 cm wide, green. **Flowers** Massed into compact, round heads, grouped in corymbs at the branch tips, each flower 1 cm wide, white. Flower stalks swollen, red and fleshy. Flowering in spring. **Habitat** Infrequent on upper, south-facing slopes in Mesic and Subalpine Fynbos. **Distribution** Widespread from Cederberg to Kariega (Uitenhage). **Notes** The other species found in Baviaanskloof, *B. commutata*, has spreading leaves and smaller flowerheads.

Bergplaas, Combrink's Pass 09.03.2010

BRUNIA

Named after Dr Cornelius Brun, c. 1652–c. 1726, a contemporary of Linnaeus; he was an apothecary and plant collector who travelled in Russia and the Levant in the 17th century.

36 species; 1 species in Baviaanskloof.

Brunia noduliflora Fonteinbos, stompie

nodulus = small knot, *flora* = flower; referring to the flowers that are borne in nodules

Resprouting, rounded shrub to 1.5 m tall, with minutely-hairy branches. **Leaves** Adpressed to the stem, tiny, 0.3 cm long, dull green or grey-green, sometimes tinged yellow-orange. **Inflorescence** Corymbose. **Flowers** Many in globose heads, 2 cm wide, white. Flowering in autumn. **Habitat** Grassy and Mesic Fynbos in deeper soil. **Distribution** Widespread from Clanwilliam to Gqeberha (Port Elizabeth).

Koud Nek's Ruggens, Kouga 18.04.2012

CACTACEAE*   Cacti

± 125 genera and 2,000 species, mostly in North America. There are no indigenous cactuses in South Africa, however there are several problematic weedy cactus species in Baviaanskloof.

ECHINOPSIS*   Hedgehog cactus, sea urchin cactus

echinos = hedgehog, *opsis* = appearance; referring to the spiny covering and the globose growth form of some species

128 species native to South America; 1 garden escape species in Baviaanskloof.

*Echinopsis oxygona**

oxygona = sharp-angled; referring to the many spine-tipped, angled ridges or ribs on the stems

Clumped cactus. **Stems** Globose to elongate, 30 cm tall × 25 cm wide, with 12–15 ribs that have sharp spines. **Flowers** Large, showy, funnel-shaped, faintly scented, 20 cm long × 13 cm wide, pink. Flowering in summer. **Habitat** Garden escape in Thicket. **Distribution** Native to South America, with potential to be invasive in several areas from Oudtshoorn to KwaZulu-Natal.

Cambria, Grootrivierpoort 26.10.2018

OPUNTIA*

Prickly pear

Named after the ancient Greek city of Opus; an edible cactus is said to have grown there.

Native to North America, with over 100 species; 2 invasive weed species in Baviaanskloof.

*Opuntia ficus-indica** Sweet prickly pear

ficus = fig, *indica* = Indian; referring to the indigenous peoples of Mexico, who ate the fruit of this species

Robust, succulent tree to 5 m tall with a thick, woody trunk. **Stems** Divided into flattened, elliptical leaf paddles, grey to grey-green, covered in spines. **Flowers** Showy, bright yellow or orange. Fruit edible, egg-shaped, covered in spines, green becoming yellow, orange, red or purple. Flowering in early summer. **Habitat** Native to Mexico. **Distribution** Introduced around the world for the edible fruit, but it can become an invasive weed covering vast areas. **Notes** Land stewards have introduced several biological controls such as a host-specific cochineal species, *Dactylopius opuntiae*, and the cactus moth, *Cactoblastis cactorum*. In addition, a stem-boring weevil,

Elandsberg road, Patensie 25.10.2018

Metamasius spinolae, has reduced dense stands to a few scattered individuals. Most of the lower barren slopes in Baviaanskloof were covered with prickly pear plants in the 1920s. **Other names** Boereturksvy (A), umthelekisi (Z).

TRICHOCEREUS*

tricho = hair, *cereus* = cactus; pertaining to the hairiness of some members of the genus

± 36 species, mostly in South America; 2 species introduced in South Africa; 1 species in Baviaanskloof.

*Trichocereus spachianus** (= *Echinopsis spachiana*) Torch cactus

Named after Edouard Spach, 1801–1879, a French botanist

Multi-stemmed, succulent shrub to 2 m. **Stems** Columnar with 10—15 ribs, covered with many amber-coloured spines in groups of 8—10, with the central spine longest. **Flowers** Opening at night (thus the common name), 20 cm long, white, tube covered with long hairs. Flowering in summer and autumn. **Habitat** Native to Argentina. **Distribution** An invasive weed in arid parts of South Africa.

Verlorenrivier, Baviaans 19.05.2021

CAMPANULACEAE

Bellflowers

55 genera and over 900 species in this cosmopolitan family, which is more prevalent in temperate and subtropical regions. A family of herbs, shrubs and very rarely trees. Several species are popular in gardens (such as members of *Campanula* and *Platycodon*) and a few are eaten as vegetables. Lobeliaceae is a closely related family or subfamily of Campanulaceae, but it is treated separately in this book. 10 genera and 250 species in South Africa with a high degree of endemism, especially in the Western Cape; 2 genera and 10 species in Baviaanskloof.

WAHLENBERGIA

African bluebell, blouklokkie

Named after George Wahlenberg, 1781–1851, a Swedish naturalist from Uppsala University. He studied the flora of Sweden and published *Flora lapponica* in 1812. This genus now includes *Lightfootia* and *Theilera*.

± 250 species, mostly in southern temperate regions; ± 170 species in South Africa; 9 species in Baviaanskloof.

Wahlenbergia cinerea

cinerea = grey-haired; referring to the woolly hairs on the stems and leaves

Non-resprouting, slender, erect shrublet to 50 cm tall, with white-woolly stems and leaves. **Leaves** Ascending, crowded, linear-lanceolate, 1 cm long × 0.2 cm wide, margins revolute. **Flowers** In the upper axils, 0.5 cm wide, white with tinged reverse. Flowering in summer and autumn. **Habitat** Rocky slopes in Grassy Fynbos. **Distribution** Widespread from Potberg to Port Alfred.

Kouenek, Kouga 18.04.2012

Wahlenbergia guthriei (= *Theilera guthriei*)

Named after Francis Guthrie, 1831–1899, a botanist and professor of mathematics at the University of Cape Town. He was friends with Harry Bolus and collected extensively on the Cape Peninsula.

Slender shrublet to 30 cm. **Leaves** Bunched, ascending, 0.5 cm long × 0.2 cm wide. **Flowers** Tube thin, 1 cm long, flower 0.7 cm wide at its widest point, blue to purple, sometimes with a white ring at the throat. Flowering in summer, autumn and winter. **Habitat** Occasional on rocky, south-facing ridges in Mesic Fynbos. **Distribution** Widespread from the mountains of the Little Karoo to the Kouga Mountains.

Quaggasvlakte, Kouga 12.02.2016

Wahlenbergia neorigida (= *Lightfootia rigida*)

neo = new, *rigida* = rigid; alluding to the rigid stems and leaves; the prefix indicates that the species has been renamed.

Rigid, erect shrublet to 60 cm. **Leaves** Ovate-lanceolate, concave, recurved, 0.4 cm long × 0.3 cm wide. **Flowers** At the tips of branches, single or in small groups, 0.6 cm wide, white, fading yellow. Flowering in summer and autumn. **Habitat** Frequently encountered on rocky sandstone in Arid and Grassy Fynbos, and Transitional Shrubland. **Distribution** Widespread from Kouebokkeveld to Kariega (Uitenhage).

Gannalandkloof, Zandvlakte 17.10.2011

Wahlenbergia rubens

rubens = reddish; the leaves and calyx are sometimes tinged reddish.

Non-resprouting, erect shrub to 50 cm tall, becoming untidy and tangled with age. **Leaves** In clusters, linear-lanceolate, minutely and sparsely toothed, 0.3 cm long × 0.1 cm wide. **Flowers** In small groups in upper axils, 0.7 cm wide, white with darker reverse, fading yellow, reverse sparsely hairy, stigma pale mauve, calyx often fading reddish. Flowering in spring, summer and autumn. **Habitat** North-facing, rocky sandstone slopes in Arid and Grassy Fynbos. **Distribution** Widespread from Bredasdorp to Gqeberha (Port Elizabeth).

Quaggasvlakte, Kouga 13.02.2016

Wahlenbergia thunbergii (= *W. uitenhagensis; Lightfootia divaricata*)

Named after Carl Pehr Thunberg, 1743–1828, a Swedish naturalist who studied under Linnaeus before he collected at the Cape. He was the author of *Flora Capensis* and a pioneer of botanical research in South Africa.

Sprawling, tangled, straggling shrublet to 50 cm. **Leaves** Elliptic to lanceolate, recurved, sparsely toothed, 0.4 cm long × 0.2 cm wide. **Flowers** In upper axils, white or blue, 0.6 cm wide. Flowering in spring, summer and autumn. **Habitat** Lower south-facing slopes in Grassy Fynbos. **Distribution** Widespread from De Hoop to KwaZulu-Natal. In Baviaanskloof, recorded only on southern side of Kouga Mountains.

Nooitgedacht, Kouga 28.09.2011

CANNABACEAE
Hemp

11 genera and 170 species in this small family of trees and herbaceous plants. *Celtis* is the biggest genus with ± 100 species. Economically important genera include *Cannabis* (hemp, marijuana, dagga), and *Humulus* (hops), used in making beer.

CELTIS

Latin name used by Pliny. *Celtis* was previously placed in Ulmaceae, the elm family.

± 100 species; 3 species in southern Africa; 1 species in Baviaanskloof.

Celtis africana White stinkwood

africana = from Africa; referring to the continent where the species originated

Handsome, deciduous tree to 30 m tall, with smooth pale grey bark. **Leaves** Oblique, toothed, 3-veined from the base, 5 cm long × 3 cm wide, dark green above, paler below. **Flowers** Small, inconspicuous, greenish, with male flowers in clusters and female flowers solitary. Fruit small, 0.6 cm wide, yellow to brown, slender stalks. Flowering in spring. **Habitat** Thicket and Forest. **Distribution** Very widespread from the Cape Peninsula to tropical Africa. **Other names** Witstinkhout (A), umvumvu (X), lesika (S), indwandwazane (Z).

Ysrivier, Cambria 23.01.2022

CAPPARACEAE

Capers

33 genera and ± 700 species. Closely related to the Brassicaceae and Cleomaceae. Species have sulphur-containing compounds known as mustard oils that act as a defence mechanism against diseases and predators. Plants are usually woody. Flowers have many anthers, usually exserted, and the fruits are fleshy. The 4 largest genera are each represented by 1 species in Baviaanskloof.

BOSCIA

Named after Louis Augustin Guillaume Bosc, 1759–1828, a French botanist, invertebrate zoologist, entomologist and professor of agriculture

Trees and shrubs. 37 species, mostly in Africa; 8 species in southern Africa; 1 species in Baviaanskloof.

Boscia oleoides Bastard shepherd's tree

Olea = plant genus, *oides* = resembling; leaves vaguely resemble those of the olive tree.

Small, sturdy, gnarled, sometimes multi-stemmed tree, to 7 m tall, with pale bark, usually neatly 'pruned' below by browsers. **Leaves** Oblanceolate to elliptic, leathery, rigid, venation barely visible, 4 cm long × 1.5 cm wide, grey-green. **Flowers** In dense clusters on short side shoots, sweetly scented, 0.6 cm wide, with 4 white petals. Fruit roundish, 1 cm wide, green becoming yellowish. Flowering in spring. **Habitat** Deep shale soils in Thicket and Succulent Karoo. **Distribution** Restricted to the Eastern Cape from Graaff-Reinet and Baviaanskloof to East London. **Notes** It closely resembles *B. albitrunca*, which differs by having small white petals and being restricted to the southeastern part of South Africa. **Other names** Karoo shepherd's tree (E), basterwitgat, karoowitgat (A), umgqamagqama, umgqomo-gqomo, umphunzisa (X).

Zandvlakte 14.09.2022

CADABA

Derived from the Arabic name Kadhab, for *C. rotundifolia*

± 30 species in Africa and Australasia; 4 species in South Africa; 1 species in Baviaanskloof.

Cadaba aphylla Bobbejaanarm, swartstormbos

cadaba = without, *phyllon* = leaf; referring to the leafless habit

Resprouting, tough, multi-stemmed tangled shrub to 2 m. **Stems** Older stems sometimes blackened, young stems dark green or purplish. **Flowers** Long, clustered on side shoots, stamens and stigma to 4 cm, each flower 1 cm long, red. Flowering in spring, summer and autumn. **Habitat** Clay-rich soils in Thicket and Succulent Karoo. **Distribution** Very widespread throughout arid parts of subtropical Africa. **Other name** Usitorhom (X).

Zandvlakte, Baviaans 18.01.2022

CAPPARIS

Capers

kapparis = caper fruit, or *Kypros* = place name; derived from the Greek name for a caper or the island of Cyprus where capers grow in abundance

The well-known caper is a pickled flower bud of *C. spinosa*, a species native to the Mediterranean. ± 65 species, mostly in tropical and subtropical regions; 4 species in South Africa; 1 species in Baviaanskloof.

Capparis sepiaria var. *citrifolia*   Caper

sepiarius = growing in hedges; referring to the scrambling habit, growing in or through dense clumps of thicket. The other variety, var. *subglabra*, grows only in Limpopo.

Thorny scrambler to 5 m. Thorns (modified stipules) paired and hooked, 0.3 cm long. **Leaves** Stems and leaves finely felted, with leaves arising between the thorns, elliptic, margins revolute, midrib raised below, 5 cm long × 2 cm wide, dark green above and paler below. **Flowers** In sprays at branch tips, 1.5 cm wide, white with pink pollen. Flowering in winter and spring. **Habitat** Thicket and Forest. **Distribution** Very widespread from Knysna to tropical Africa, India and Malaysia. **Notes** The flower buds can be pickled and eaten as capers. Used in traditional medicine to treat coughs, colds, impotence and infertility. **Other names** Kaapse-kappertjie (A), imfishlo, intshihlo, upasimani, uqapula (X), usondeza (Z).

Grasnek, Kouga 02.02.2020

MAERUA

Bush cherry

Maerua = derivation unresolved; probably the Arabic name for one of the species

100 species in Africa and Asia; 11 species in South Africa; 1 species in Baviaanskloof.

Maerua cafra   Common bush cherry

cafra = from the southeastern part of South Africa where the plants occur

Shrub or tree to 9 m tall, with pale bark. **Leaves** Digitately 3–5-foliate, petiole to 6 cm long, dark green, leathery. **Flowers** In sprays at branch tips, fragrant, mostly a tuft of spreading anthers, 3 cm long, greenish. Fruit egg-shaped, 4 cm long, green, edible. Flowering in early spring. **Habitat** Thicket and Forest. **Distribution** Widespread from Baviaanskloof to Zimbabwe and Mozambique. **Notes** Roots are ground up; also used as a leaf vegetable, flavouring and beverage enhancer. **Other names** Wildeboshout (A), umphunzisa, umqomoqomo (X), untswantswane (Z).

Grootrivierpoort 23.01.2022

CARYOPHYLLACEAE

Pinks, carnations

A large cosmopolitan family of 81 genera and 2,625 species, represented mostly by herbaceous plants in temperate regions. Many species are grown as ornamental plants and there are also many widespread weeds. Fairly easy to recognise by the regular flowers with notched or fringed tips, opposite leaves and swollen nodes. 22 genera and 87 species in southern Africa; 7 species in Baviaanskloof.

DIANTHUS

Pinks, angelier

dios = divine, *anthos* = flower; referring to the pleasant scent and pretty flowers

± 300 species in the Old World; 15 species in South Africa; 3 species recorded in Baviaanskloof.

Dianthus caespitosus

caespitosus = tufted, forming dense patches; referring to the tufted habit

Resprouting, tufted perennial to 40 cm, with grass-like leaves. **Leaves** Linear, 7 cm long × 0.4 cm wide, grey-green. **Flowers** Several in axillary scapes, petals always toothed, calyx up to 7 cm long, flowers usually white with purple throat. Flowering in spring and summer. **Habitat** Occasional in Transitional Shrubland and Thicket. **Distribution** Widespread from Worcester to Kariega (Uitenhage).

Zandvlakte, Baviaanskloof Valley 04.12.2015

Dianthus thunbergii Ungcana (X)

Named after Carl Pelu Thunberg, 1743–1828, a Swedish naturalist who studied under Linnaeus before he collected at the Cape. He was the author of *Flora Capensis* and a pioneer of botanical research in South Africa.

Resprouting, loosely tufted perennial to 40 cm. **Leaves** Linear, 5 cm long × 0.3 cm wide, blue-grey. **Flowers** Petal tips toothed, calyx to 3 cm long, flower 1.5 cm wide, pink. Flowering in spring and summer. **Habitat** Common in Grassy Fynbos and Transitional Shrubland. **Distribution** Widespread from Caledon to Komani (Queenstown). **Other names** Ubulawu obumhlophe, inkomoyentaba, indlela zimhlope (X).

Doodsklip, Kouga Dam 29.12.2017

POLLICHIA

Waxberry, aarbossie

Named after Johan Adam Pollich, 1740–1780, a German doctor and author of a book on the history of plants in Palatinate, a historical region of Germany

1 species in Baviaanskloof.

Pollichia campestris

campestris = pertaining to flat areas or plains; referring to the lowland flats habitat

Perennial herb to 50 cm tall, sometimes prostrate, with a deep taproot. **Leaves** Oblanceolate with pointy tips, silky-hairy, 0.5 cm long × 0.2 cm wide, blue-grey. **Flowers** In dense clusters, minute, becoming covered with waxy white bracts in fruit. Flowering all year round. **Habitat** Exposed sandy flats in the lowlands. **Distribution** Very widespread from Namaqualand to the Arabian Peninsula. **Notes** The fruit is edible. **Other names** Utywala, utywala behlungulu, amangabangaba (X), ukudla kwamabhayi, umhlungulu (Z).

Suuranysberg, Kareedouw 13.05.2021

SILENE
Campion, catchfly

sialon = saliva; referring to the sticky substance on the stems, which traps small insects – thus the common name catchfly

Largest genus in the family with ± 500 species; 2 species recorded in Baviaanskloof.

Silene burchellii Gunpowder plant

Named after William John Burchell, 1781–1863, an English explorer who collected plants in southern Africa and Brazil. He amassed over 50,000 plant specimens in southern Africa and donated them to Kew Gardens, each record with meticulous, detailed notes about the habit and habitat of the species. He described his journey in *Travels in the Interior of Southern Africa*, a 2-volume work published in 1822 and 1824.

Tuberous-rooted perennial to 70 cm. **Leaves** Mostly in a basal tuft, oblanceolate, 2 cm long × 0.5 cm wide, hairy. **Flowers** Several in a raceme-like cyme, 1 cm wide, white to purple, petals deeply notched, calyx tube with maroon ribs and up to 2 cm long. Flowering in spring and summer. **Habitat** Occurring on loamy soils in all habitats, but most prevalent in Transitional Shrubland. **Distribution** Widespread throughout Africa. **Notes** 4 subspecies are recognised; this is subsp. *pilosellifolia*. **Other names** Oggendblommetjie (A), iyeza lehashe (X).

Baviaanskloof 24.08.2010

Silene undulata Large-flowered campion

undulata = undulating; referring to the petals that are usually wavy

Upright perennial to 60 cm, with thick roots. **Leaves** Opposite, oblanceolate, mostly tufted near the base, glandular-hairy, 15 cm long, thinner and half as long higher up the stem. **Flowers** Petals notched at the tips, calyx glandular-sticky and 3 cm long, white or pink. Flowering in spring, summer and autumn. **Habitat** Occasional in dry riverbeds in Thicket. **Distribution** Very widespread throughout southern Africa to tropical Africa. **Notes** A sacred plant – the root is used to generate dreams said to connect one with the ancestors. The other species to be found in Baviaanskloof is the oggendblommetjiie (*S. burchellii* subsp. *pilosellifolia*). **Other names** Wildetabak (A), ubulawu obumhlope, unozitholana, iinkomo yentaba (X), ugwayana, ugwayelaso, ugwayintombe, umjuje (Z).

Zandvlaktekloof, Baviaans 12.10.2010

CELASTRACEAE

Staff vines, bittersweet

Herbs, vines, shrubs and small trees with 96 genera and ± 1,350 species. All in Baviaanskloof are resprouting woody bushes or trees, with leaves that show elastic threads when broken, and with small flowers with a nectar-secreting disc, and a fruit capsule that splits to reveal seeds with an aril. 24 genera and ± 100 species in southern Africa; 9 genera and 16 species in Baviaanskloof.

GYMNOSPORIA

Common spikethorn, pendoring, stinkdoring

gymno = naked, *sporia* = seed; seeds, partly covered with a yellow aril, appear naked when the capsule opens.

Over 100 species in Africa, Madagascar, Asia and Australia; ± 28 species in South Africa; 4 species in Baviaanskloof.

Gymnosporia buxifolia (= *Maytenus heterophylla*)

Buxus = plant genus, *folia* = leaf; leaves resemble those of *B. sempervirens*.

Resprouting thorny bush or small tree to 7 m. **Leaves** In bunches, obovate, scarcely toothed towards the tip, variable in size, 4 cm long × 2 cm wide, usually pale grey-green. **Flowers** Numerous in axillary cymes, foetid-smelling, attracting flies. Capsules ball-like, warty, 0.7 cm wide, brown. Flowering in spring, summer and autumn. **Habitat** Various lower-lying habitats, especially along stream banks in Thicket. **Distribution** From the Cape to tropical Africa. Browsed by kudu and other game. **Other names** Umhlongwe, umqaqoba, umagcengenene (X), ingqwangane (Z).

Bruintjieskraal, Grootrivierpoort 30.12.2017

three-lobed capsule
orange seed
yellow aril

Gymnosporia linearis subsp. *linearis*

linearis = linear; referring to the narrow and linear leaves

Resprouting, small tree to 4 m. **Leaves** Single, linear, with obscure teeth on the margins, 6 cm long × 0.5 cm wide, dull green or grey-green. **Flowers** Small, few in short, axillary cymes, petals cream, centre yellow. Fruit oval, 0.7 cm long. Flowering in summer. **Habitat** Infrequent along dry stream beds in Thicket. **Distribution** Widespread from southern Namibia to Kariega (Uitenhage). **Notes** 2 subspecies: subsp. *linearis* has been recorded in Baviaanskloof and subsp. *lanceolata* is restricted to the Northern Cape.

Bruintjieskraal, Grootrivierpoort 04.02.2020

MAYTENUS

Derived from the Mapuche Native American name for a similar plant, the mayten tree (*M. boaria*), which is found along waterways in South America

± 150 species, mainly in the tropics and subtropics of both hemispheres; ± 11 species in southern Africa; 3 species in Baviaanskloof.

Maytenus acuminata Silky bark, sybas

acuminatus = acuminate; referring to the shape of the leaf tips

Non-resprouting tree to 10 m tall, with relatively smooth bark. **Leaves** Producing silky threads when broken and pulled apart, ovate to lanceolate, shallowly toothed, 7 cm long × 2 cm wide, glossy green or olive-green above, paler below, new growth and young branches sometimes tinged red. **Flowers** In the axils, 0.5 cm wide. Capsules yellow. Flowering from winter to summer. **Habitat** Forest. **Distribution** Widespread from the Cape to tropical Africa. **Other names** Inqayi, umzungulwa (X), inama elimhlophe, isinama (Z).

Ysrivier, Cambria 22.01.2022

Maytenus oleoides Klipkershout

Olea = plant genus, *oides* = resembling; referring to the superficial resemblance to a particular species of *Olea*

Resprouting tree to 4 m tall, bark becoming rough and corky with age. **Leaves** Obovate to lanceolate, leathery, lateral veins invisible, 3 cm long × 1 cm wide, margins rolled under. **Flowers** Small, in short axillary cymes, whitish. Capsules roundish, yellow to brown or orange, 1 cm wide. Flowering from autumn to spring. **Habitat** Occasional on rocky outcrops in Grassy Fynbos or at Forest margins. **Distribution** Widespread from Richtersveld to Cape Peninsula and to Kariega (Uitenhage).

Kouga Wildernis 14.05.2021

Maytenus undata Kokoboom

undatus = wavy; referring to the leaf blade which is slightly undulating

Resprouting, multi-stemmed tree to 10 m. **Leaves** Ovate, leathery, 6 cm long × 3 cm wide, glossy, dark green above, white-waxy bloom below, net veins distinct below, toothed. **Flowers** Small, in dense clusters in the leaf axils, yellowish. Fruit a 3-lobed capsule. Flowering from spring through summer to autumn. **Habitat** Forest and Thicket. **Distribution** Very widespread from the Gamka Mountains to tropical Africa, Madagascar and the Comores. **Other names** Koko tree (E), inqayi-elibomvu, umkokuza (X), idohame (Z).

open empty seed capsule

Rooikloof, Kouga 28.04.2018

PTEROCELASTRUS

Cherrywood, kershout

pteron = a wing, *Celastrus* = plant genus; referring to the horns (wings) on the fruits; *C. paniculatus* is a tree in this family native to India, also with 3-chambered capsules, but these are without 'horns'.

3 species in South Africa; 1 species in Baviaanskloof.

Pterocelastrus tricuspidatus   Rooikershout

tricuspidatus = three-cusped; referring to the three-chambered capsules that each have 3 cusps or horns

Resprouting, often multi-stemmed tree to 7 m. **Leaves** Obovate, leathery, shiny, with a short pinkish petiole, net veins invisible, leaf 8 cm long × 4 cm wide, dark green above, paler below, breaking of leaf is audible. **Flowers** Several clustered in axillary cymes, sweetly scented, 0.4 cm wide. Fruit 1 cm wide, yellow-orange, 3-lobed, each lobe with 2 or 3 horns. Flowering in winter and spring. **Habitat** Common on rocky slopes and outcrops in Transitional Shrubland, Fynbos Woodland and Grassy Fynbos. **Distribution** Widespread from the West Coast to Mpumalanga. **Notes** 1 of the most common trees in Baviaanskloof. **Other names** Itywina, ibholo, umgobandlovu (X), udwina (Z).

Zandvlaktekloof, Baviaans 19.04.2012

PUTTERLICKIA

Basterpendoring

Named after Aloys Putterlick, 1810–1845, an Austrian botanist and physician who was head of the Natural History Museum in Vienna; he specialised in bryology.

4 species in southern Africa; 1 species in Baviaanskloof.

Putterlickia pyracantha   False spike-thorn

pyracantha = fire-thorn; refers to the pain when pricked or to the red stems and thorns

Resprouting, multi-stemmed, spiny shrub or small tree to 2 m. Young stems reddish with raised dots, spines straight, very sharp, to 5 cm long. **Leaves** Usually bunched, leathery, smallish, 3 cm long × 1 cm wide, dark green. **Flowers** Several in branched heads, small, to 1 cm wide, white. Capsules reddish, 3-lobed; seeds covered by an orange aril. Flowering in summer. **Habitat** Common in patches of bush in Thicket, Succulent Karoo and Transitional Shrubland. **Distribution** Widespread from Velddrif and the Cape Peninsula to Peddie, Eastern Cape. **Other names** Intlangwana, umhlangwe, umqaqoba (X).

Sundays River Mouth 04.12.2021

CONVOLVULACEAE

Bindweed, morning glory

± 60 genera and 1,650 species in the world. Typical of the family is the funnel-shaped, symmetrical flowers with 5 fused petals. Mostly vines but also several other growth forms. Sweet potato (*Ipomoea batatas*) is a well-known member. 17 genera and 130 species in South Africa, also several invasive species and many cultivated.

CONVOLVULUS

Bindweed

convolvere = to wind; referring to the twining growth habit

± 250 species around the world; 13 species in southern Africa; 2 species in Baviaanskloof.

Convolvulus farinosus

farinosus = mealy; referring to the obscurely hairy surfaces of the plant

Profusely twining climber to 3 m. **Leaves** Cordate-deltoid, pubescent, 7 cm long, green. **Flowers** 1.5 cm long, white to mauve. Flowering in summer and autumn. **Habitat** Various lowland habitats, but seldom seen in Fynbos. **Distribution** Very widespread from the Cape Peninsula throughout Africa and to the Mediterranean. **Notes** The other species found in Baviaanskloof, *C. capensis*, has bigger flowers, appearing in spring. **Other names** Klimop (A), uboqo, usinga lamaxhegwazana, inabulele, uboqo wabadlezana|nabulele (X), umkhokha wehlathi (Z).

Skuinspadkloof, Baviaans 01.02.2020

CRASSULACEAE

Stonecrops

A cosmopolitan family of ± 33 genera and more than 1,500 species, largely in semi-arid regions. Mostly leaf and stem succulents ranging from annuals and geophytes to dwarf, thick-set trees. Some have showy flowers, attracting butterflies or birds; many are cultivated as waterwise plants. Some, such as the ornamental *Cotyledon*, are popular in gardens worldwide. Cotyledons are used medicinally. 5 genera and ± 450 species in South Africa; 6 genera in Baviaanskloof.

ADROMISCHUS

adros = thick, *mischus* = stalk; referring to the thickened stalks of the inflorescences

Common garden plants, easily grown from cuttings. Endemic in southern Africa; 5 species in Baviaanskloof.

Adromischus cristatus var. *cristatus*

cristatus = crested; referring to the leaves with undulating tips or crests

Small succulent with exposed fine red roots, and a flowering stalk to 40 cm. **Leaves** Obovoid-cuneate, tip undulating, glandular-hairy, 5 cm long × 2.5 cm wide. **Flowers** Inflorescence a spicate 15 cm-long cyme, each flower 1 cm long, with pale green or maroon calyx, petals white and cylindric, anthers included. Flowering in late summer. **Habitat** Rocky ridges and cliffs in the eastern part of the region. **Distribution** Widespread from Willowmore to the Eastern Cape. **Notes** 5 varieties recognised: var. *cristatus* and var. *schonlandii*, the rare endemic, occur in Baviaanskloof.

Joubertina, Kouga River 09.05.2006

Adromischus sphenophyllus

sphen = wedge, *phyllon* = a leaf; alluding to the wedge-shaped leaves

Relatively robust succulent to 35 cm tall, forming clumps.
Leaves Obovate, 6 cm long × 3 cm wide, grey-waxy,
with hardened leaf margin. **Flowers** In a tall spike, each
flower 1 cm long, petals pink with reddish calyx tube.
Flowering in summer. **Habitat** Common and frequently
encountered on rocky slopes in Transitional Shrubland,
Arid Fynbos and Thicket. **Distribution** Widespread from
Baviaanskloof to East London.

Die Krokodil, Speekhout 24.12.2017

Adromischus subdistichus Rondeblaar-brosplakkie

sub = half, *distichus* = distichous; pertaining to the leaves, which are almost
opposite and in 2 rows

Dwarf, creeping leaf succulent up to 20 cm. **Leaves**
Orbicular or obovate, almost distichous in young plants,
2.5 cm long × 2 cm wide, greyish green. **Flowers**
With 1- or 2-flowered cymes, corolla red-purple on
green-yellow background, lobes pink. Flowering in late
summer. **Habitat** On cliff faces in the western parts of
Baviaanskloof. **Distribution** Fairly narrow from Prince
Albert to Baviaanskloof. **Notes** Easily seen at Raaskrans
in Nuwekloof Pass.

Speekhout, Baviaans 24.12.2017

BRYOPHYLLUM*

bryo = to sprout, *phyllon* = a leaf; referring to the new plantlets sprouting from the notches in the leaves

± 40 species native to Africa, Madagascar and Asia; 1 exotic species in Baviaanskloof.

*Bryophyllum delagoense** Mother of millions, chandelier plant

Named after Delagoa Bay at Maputo in Mozambique; it was erroneously named after
this place, because the plant actually originates in Madagascar.

Erect, greyish-mottled succulent to 50 cm. **Leaves** Terete, with
apical teeth where the tiny plantlets develop; each leaf 10 cm
long × 0.5 cm wide, pinkish grey, mottled dark green. **Flowers**
Drooping, bunched together at the top of the stems, bell-
shaped, 3 cm long, bright red. Flowering in winter. **Habitat** Along
dry stream beds in Thicket. **Distribution** Native to Madagascar,
introduced as an ornamental, but a garden escape that has
spread through many parts of South Africa. Also invasive in
many other parts of the world. **Notes** The plant is toxic and
can cause fatal poisoning in cattle and humans. **Other names**
Kandelaarplant (A), intelezi yobushwa (X).

Kleinplaat, Baviaans 11.07.2017

kotyledon = cup-shaped hollow; referring to the shape of the flowers

Popular garden plants. All are toxic, but the juice of some leaves has been used medicinally. 12 species in Africa and the Arabian Peninsula; 7 species in Baviaanskloof.

Cotyledon campanulata

campanulata = bell-shaped; alluding to the shape of the flowers

Low succulent to 30 cm. **Leaves** Mostly basal, ascending, glandular-hairy, 10 cm long × 2 cm wide, red-tipped. **Flowers** In terminal pendulous clusters, sticky, 2.5 cm long, yellow, with petals curved back. Flowering in early summer. **Habitat** Sporadic on rocky slopes in Transitional Shrubland. **Distribution** Eastern distribution from Baviaanskloof to East London.

Zandvlakte, Kouga 30.10.2011

Cotyledon gloeophylla   Sticky-leaved cotyledon

gloia = glue, *phyllon* = leaf; referring to the sticky leaves

Much-branched succulent shrub to 60 cm. **Leaves** Broadly obovate, 5 cm long × 2 cm wide, covered with sticky glandular hairs. **Flowers** Solitary at the branch tips, pendulous, 3 cm long, orange-red. Flowering in winter and early spring. **Habitat** Thicket. Known only from the steep slopes around Kouga Dam. **Notes** Recorded for the first time in 1990 and described in 2015 by Ernst van Jaarsveld. Closely related to *C. woodii*. **Status** Rare.

Kouga Dam 19.09.2014

Cotyledon orbiculata   Pig's ears

orbiculus = round, *ata* = possessing; pertaining to the roundish leaves in the more common varieties

Branched shrub to 1.5 m. **Leaves** Obovate to slender ovoid, variable, with a grey bloom, margins sometimes reddish. **Flowers** In a fairly loose nodding cyme, 1.5 cm long, orange to red. Flowering in spring and early summer. **Habitat** Mostly in Succulent Karoo and Thicket. **Distribution** Very widespread throughout southern Africa. **Notes** Used medicinally, popular in gardens. 5 varieties are recognised in South Africa, 2 of which occur in Baviaanskloof: var. s*puria* and var. *orbiculata*. **Other names** Plakkie (A), ipewula (X), serelile (S), intelezi (Z).

Grasnek, Kouga 29.12.2017

Cotyledon tomentosa

tomentum = stuffing material of a pillow, *osa* = abundance; referring to the
leaves and flowers that are evenly and densely covered with short hairs

Stunted succulent shrublet to 20 cm. **Leaves**
Ascending, cuneate to ovoid, toothed above, 5 cm
long × 1 cm wide, silvery-hairy. **Flowers** Several
in a cyme, not hanging, 1.5 cm long, orange-red.
Flowering in winter. **Habitat** Rock ledges and cliffs
in the narrow ravines. **Distribution** Uncommon,
with small isolated populations from Barrydale to
Baviaanskloof. **Notes** 2 subspecies are recognised:
subsp. *ladismithensis*, which occurs around Ladismith,
and subsp. *tomentosus*, which has been recorded in
Baviaanskloof. **Status** Vulnerable.

Baviaanskloof 01.03.2019

Cotyledon velutina

velutina = velvety; alluding to the flowers,
which are covered with short hairs, and the
velvety, grey-waxy covering on the leaves

Robust, succulent shrub to 2 m. **Leaves**
Obovate, lower leaves sometimes
auriculate (eared at the stem), large,
10 cm long × 4 cm wide, greyish green,
with margins sometimes tinged reddish.

Geelhoutbos, Kouga 28.12.2017

Flowers Several in a flat-topped, pedunculate cyme, hanging, 1.7 cm long, yellowish with reddish
pink. Flowering in spring and early summer. **Habitat** Exposed stony slopes in Thicket, often using other
shrubs for support. **Distribution** Widespread from Willowmore to southern KwaZulu-Natal.

Cotyledon woodii   Wood's cotyledon

Named after John Medley Wood, 1827–1915, British botanist who collected the unrecorded species near East London

Much-branched, relatively
densely leaved succulent
shrub, to 60 cm. **Leaves**
Obovate, 2.5 cm long ×
1 cm wide, grey-green,
sometimes with a reddish
margin. **Flowers** 1 or 2 at
the branch tips, pendulous,
1.5 cm long, reddish.
Flowering in summer and
autumn. **Habitat** Rocky
ledges mostly in Thicket.
Distribution Widespread
from Anysberg to Peddie.

Grasnek, Kouga 06.05.2006

CRASSULA
Stonecrop

crassus = thick; pertaining to the fleshy, succulent leaves

± 200 species, mostly in southern Africa; 35 species in Baviaanskloof.

Crassula atropurpurea

atropurpurea = dark purple; referring to the leaves, which tend to be reddish or dark purple

Dwarf succulent, up to 60 cm, but usually stemless in Baviaanskloof, and only 20 cm when flowering. **Leaves** Obovate-lanceolate, 3 cm long × 1 cm wide, often reddish (especially in drought), new growth green. **Flowers** In roundish clusters on 10–20 cm-long spikes, each flower 0.4 cm long, creamy yellow. Flowering in spring and early summer. **Habitat** Shallow, stony sandstone ledges in Fynbos and Transitional Shrubland. **Distribution** Very widespread from Namibia to Gqeberha (Port Elizabeth).

Voetpadskloof, Baviaans 27.04.2018

Crassula biplanata

biplanata = on two planes; referring to the opposite leaves

Small, erect succulent with papillate branches. **Leaves** Ascending, opposite, round in cross section, 1 cm long × 0.2 cm wide, sometimes with a silvery wax coating. **Flowers** In small clusters, white, petals 0.2 cm long. Flowering in autumn. **Habitat** Rock ledges in Fynbos. **Distribution** Widespread from Stellenbosch to Kariega (Uitenhage).

Coleskeplaas, Baviaans 06.03.2012

Crassula capitella subsp. *thyrsiflora*   Red pagoda, campfire

caput = head, *ella* = diminutive; alluding to the small flowerheads

Very variable, attractive, smallish, symmetrically leaved succulent to 40 cm. **Leaves** Linear-lanceolate, becoming smaller up the stem, 5 cm long × 1.5 cm wide, with hairy margins. **Flowers** In clusters up the spike, tubular, each 0.5 cm long, white to pink, attracting insects. Flowering in summer and autumn. **Habitat** Dry rocky slopes in various habitats, mostly Transitional Shrubland. **Distribution** Widespread throughout southern Africa. **Notes** 5 subspecies recognised in South Africa, one of which is described here. **Other name** Aanteelrosie (A).

Grasnek, Kouga 17.04.2012

Crassula cotyledonis   Bergplakkie

Cotyledon = plant genus, *onis* = resembling; leaves
look like those of some species of *Cotyledon.*

Low succulent with basal leaves to 30 cm tall when in flower.
Leaves Narrowly to broadly ovate, 6 cm long × 2.5 cm
wide, grey-green. **Flowers** In rounded heads on an
elongated stalk, tiny, to 0.5 cm long, reddish. Flowering
in spring and summer. **Habitat** Stony soil on rocky ridges
in Fynbos and Transitional Shrubland. **Distribution**
Widespread from Namibia to Worcester, Makhanda and
Cradock, Eastern Cape. **Notes** Popular in pots.

Smitskraal, Kouga 09.03.2012

Crassula cremnophila

kremnos = cliff, *philos* = friend; pertaining to the
habitat on shaded cliffs or rock ledges

Dwarf, stemless succulent to 10 cm. **Leaves** Flat on
the ground, opposite, symmetrically stacked, 3 cm
long × 3 cm wide, with hairy leaf margins. **Flowers**
Many in a showy, rounded head, 0.7 cm long, pink. Flowering
in summer. **Habitat** Uncommon on shaded cliffs and ledges in
Fynbos Woodland and Mesic Fynbos. **Distribution** Endemic to the
Baviaanskloof Mountains. **Status** Rare.

Bakenkop, Baviaans 23.12.2011

Crassula cultrata

cultrata = knife-shaped; referring to the lanceolate leaves

Sparsely branched succulent shrublet to 50 cm tall, with slightly woody
stems. **Leaves** Oblanceolate, 5 cm long × 2 cm wide, yellowish green
with red margins. **Flowers** In dense clusters on a branched cyme, tiny,
to 0.5 cm long, corolla white, calyx yellowish green. Flowering in summer.
Habitat Exposed rocky slopes in Thicket and Transitional Shrubland.
Distribution Widespread from Swellendam to KwaZulu-Natal.

Joubertina 26.12.2012

Crassula ericoides

Erica = plant genus, *oides* = resembling; the fine leaves
and tubular flowers are reminiscent of *Erica* spp.

Small, delicate, thin-stemmed succulent to 20 cm.
Leaves Ericoid, tightly packed around the stem, small,
0.5 cm long × 0.2 cm wide. **Flowers** 1—several at branch
tips, partly hidden in the leaves, each 0.3 cm long, white. Flowering in
summer. **Habitat** Sandy soil in Grassy and Mesic Fynbos. **Distribution**
Widespread from Bredasdorp to KwaZulu-Natal.

Rus en Vrede 4X4, Kouga 20.01.2022

Crassula expansa subsp. expansa

expansa = expanding, spreading; referring to the spreading habit

Delicate, sprawling, untidy succulent herb to 20 cm. **Leaves** Small, 1 cm long × 0.3 cm wide, light green. **Flowers** On flimsy stalks, small, 0.3 cm long, white. Flowering in spring and summer. **Habitat** Open ground in shaded situations in Thicket and Forest. **Distribution** Widespread from Namaqualand to KwaZulu-Natal. **Notes** 4 subspecies recognised: but only subsp. *expansa* has been recorded in Baviaanskloof.

Zandvlakte 19.01.2022

Crassula hemisphaerica

hemi = half, *sphaerica* = sphere; referring to the hemispherical shape of the heaped rosette of leaves in cross section

Dotty-leaved succulent, to 15 cm, flat when not flowering. **Leaves** Opposite, in a rosette, 2 cm long × 2.5 cm wide, margin clear-membranous, recurved. **Flowers** Partly hidden by leafy bracts on a spike, tiny, 0.3 cm long, white. Flowering in spring. **Habitat** Gravel flats in Succulent Karoo and Thicket. **Distribution** Widespread from Knersvlakte to Baviaanskloof.

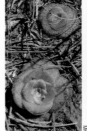

Modderfontein, Zandvlakte 02.01.2003

Crassula lactea   Taylor's parches

lactea = milky; referring to the milky white flowers, the waxy coating on the leaves and the white dots along the perimeter of the leaves

Procumbent, sparsely branched succulent shrub to 40 cm. **Leaves** Thick, opposite, 5 cm long × 2.5 cm wide, with a row of small white dots along the margins. **Flowers** Several in a cyme, star-shaped, fragrant, to 1 cm wide, white, anthers purple. Flowering in winter. **Habitat** Shaded slopes in Grassy Fynbos, Transitional Shrubland and Thicket. **Distribution** Widespread from Mossel Bay to East London.

Zandvlakte, Baviaans 15.07.2011

Crassula mollis

mollis = softly hairy; alluding to the leaves, which are covered in very short hairs, giving them a velvety grey-green coat

Much-branched succulent shrublet to 50 cm. **Leaves** Opposite, linear-elliptic, 2 cm long × 0.3 cm wide, velvety grey-green. **Flowers** In globular clusters 1.5 cm wide, spaced along a stalk, small, each flower 4 cm long, cream-coloured. Flowering in summer. **Habitat** Uncommon on stony ground in Succulent Karoo, Thicket and Transitional Shrubland. **Distribution** Widespread from Riversdale to Addo.

Guerna, Kouga 16.02.2016

Crassula muscosa Lizard's tail

muscus = moss, *osa* = abundance; referring to the small scale-like
leaves covering the thin stems, and the superficially moss-like habit

Thin-stemmed, brittle, closely leafy, suberect to decumbent,
irregularly branched shrublet, to 30 cm. **Leaves** Tiny,
scale-like, 4-ranked, closely packed. **Flowers** Minute, partly
hidden in the axils, pale yellow. Flowering in summer. **Habitat**
Common in rocky areas in various habitats. **Distribution** Very
widespread from Namibia to Cathcart, Eastern Cape. **Notes**
4 varieties recognised: var. *muscosa* has been recorded in
Baviaanskloof. **Other name** Akkedisstert (A).

Modderfontein, Zandvlakte 02.01.2003

Crassula nemorosa

nemus = wood, *osa* = abundance; referring to the habitat, growing
in the shade of woody vegetation or forest

Tuberous-rooted, soft, delicate succulent to 10 cm. **Leaves**
Suborbicular-reniform, opposite, 3 or 4 pairs, to 1 cm wide,
grey-green. **Flowers** Few, cup-shaped, nodding, 0.5 cm
wide, yellowish, petal tips recurved. Flowering in winter
and early spring. **Habitat** Found on the shaded cliffs of the
narrow ravines in Forest. **Distribution** Widespread from the
Richtersveld to East London.

Waterkloof, Bo Kloof 9.09.2015

Crassula nudicaulis var. *platyphylla* Bohobe-ba-balisana (S), sekopo (S)

nudus = bare, *caulis* = stem; referring to the flowering stems that are almost leafless

Low, clump-forming succulent shrublet, to 10 cm tall when not flowering or 30 cm tall in flower. **Leaves**
Broadly oblong to orbicular, 6 cm long × 3.5 cm wide, blue-grey, with marginal hairs. **Flowers** In dense
globular clusters on a spike, tiny, to 0.5 cm long, yellowish, bracts reddish. Flowering in spring and
summer. **Habitat** Grassy Fynbos and Transitional Shrubland. **Distribution** Widespread on the inland
mountains bordering the southern Karoo from Namaqualand to Eastern Cape. **Notes** 3 varieties are
recognised: var. *platyphylla* has been recorded in Baviaanskloof.

Drinkwaterskloof, Kouga 09.11.2011

Crassula orbicularis   Klipblom

orbiculus = round, *aris* = pertaining to; referring to the roundish basal rosette of leaves

Flat on the ground, social succulent to 15 cm tall when flowering. **Leaves** Narrowly elliptic, oblanceolate, 8 cm long × 2 cm wide, in a basal rosette, green to red and sometimes with purple or red undersides, margins densely hairy. **Flowers** In clusters on an elongated stalk, tiny, each 0.4 cm long, white or yellow. Flowering in winter and spring. **Habitat** Rocky ledges on steep south-facing slopes and cliffs in Fynbos Woodland and Thicket. **Distribution** Widespread from Montagu to KwaZulu-Natal. **Other name** Umadinsane (Z).

Endkrale, Kouenek, Kouga 12.10.2010

Crassula ovata   Narrow-leaved crassula

ovata = oval; referring to the shape of the leaves

Thick-stemmed, rounded, succulent shrub or tree to 3 m tall. **Leaves** Oval, 6 cm long × 3 cm wide, dark green, speckled with minute white dots on the upper surface. **Flowers** Many in dense, branched heads 8 cm wide, each flower 1 cm wide, white to light pink. Flowering in winter and early spring. **Habitat** Very common on clay-rich soils in Thicket and Transitional Shrubland. **Distribution** Widespread from De Rust to KwaZulu-Natal. **Notes** Easily confused with *C. arborescens*, the only other *Crassula* species that grows into a tree, and which has a western distribution, not occurring east of Prince Albert. Further, *C. ovata* has narrower leaves, which do not produce a waxy bloom. **Notes** Very popular, grown in pots around the world. **Other names** Kurkei, kerkeibos, boereblitz (A).

Rooikloof, Kouga 12.07.2016

Crassula pellucida   Inyamayamakhwenkwe (X)

pellucida = translucent but not hyaline; referring to the transparent and membranous leaf margins

Short-lived, soft, sprawling succulent, often hanging off cliffs or other plants. **Leaves** Opposite, well-spaced along the trailing stems, clasping the stem, to 1 cm wide, margins often tinged red. **Flowers** In dense terminal clusters, to 1 cm wide, white, sometimes tinged pink. Flowering in spring and summer. **Habitat** Shaded situations, often on rocks or cliffs in Thicket, Fynbos Woodland and Forest margins. **Distribution** Very widespread from Nieuwoudtville to Ethiopia.

Modderfontein, Baviaans 06.10.2010

Crassula perfoliata

per = through, *folium* = leaf, *ata* = possessing;
referring to the leaf blades that cover the stems

Tough, sparsely branched, erect succulent shrub
to 1.5 m. **Leaves** Lanceolate-triangular, 6 cm
long × 2 cm wide, greyish, sometimes with purple
blotches. **Flowers** Numerous in flat-topped clusters 6 cm
wide, showy, each flower 0.6 cm long, white, pink or red.
Flowering in summer. **Habitat** Transitional Shrubland and
Thicket. **Distribution** Widespread from Baviaanskloof to
Limpopo. **Notes** 4 varieties recognised: var. *minor* has been
recorded in Baviaanskloof.

Combrink's Pass, Baviaans 22.01.2022

Crassula perforata   Concertina plant, sosaties

perforo = bore into or through; referring to the sessile, opposite
leaves that come together around the stem

Sprawling succulent shrub to 60 cm. **Leaves** Fused around
the stem, opposite, 1.5 cm long × 1 cm wide. **Flowers** In
relatively loose, rounded clusters at the branch tips, each
flower 0.3 cm long, cream or yellowish, sometimes tinged
pink. Flowering in summer and autumn. **Habitat** Fairly
common on rocky slopes in Thicket. **Distribution** Widespread
from Villiersdorp to KwaZulu-Natal. **Notes** 2 subspecies.
subsp. *perforata* has been recorded in Baviaanskloof; subsp.
kougaensis is a rare endemic from cliffs around Kouga
Dam. It is a smaller and more compact plant with a shorter
inflorescence and tiny 2 mm spaces between the leaves.

Bakenkop, Baviaans 23.12.2011

Crassula pubescens subsp. *radicans*

pubesco = become hairy; referring to the finely hairy leaves of
subsp. *pubescens*; subsp. *radicans* (described here) is hairless.

Sprawling, clumped, succulent shrub with lateral
branches that root at the nodes, to 30 cm.
Leaves hairless, 2 cm long × 1 cm wide, often
red. **Flowers** Small clusters on a short, red stalk,
petal tips with round appendages 0.3 cm long,
white. Flowering in spring. **Habitat** Open, rocky
areas in Thicket. **Distribution** Widespread from
Nieuwoudtville to Addo, Eastern Cape. **Notes**
3 subspecies: subsp. *radicans* has been recorded
in Baviaanskloof. The other subspecies that can
be found in Baviaanskloof, subsp. *pubescens*, is
an erect plant and does not spread laterally like
subsp. *radicans*.

Geelhoutbos bergpad, Kouga 31.10.2011

Crassula pyramidalis   Rygbossie

pyramis = pyramid, *alis* = pertaining to; referring to the pyramid-like habit of this plant

Compact, dwarf succulent to 20 cm. **Leaves** Opposite, triangular, closely packed, 1 cm long, forming 4-angled columns, brownish. **Flowers** Crowded at the tips, 1 cm long, white. Flowering in spring. **Habitat** Occasional on rocky outcrops with shallow clay- or loam-based soils in Succulent Karoo and Transitional Shrubland. **Distribution** Widespread from Kouebokkeveld to Baviaanskloof.

Verlorenrivier, Baviaans
15.02.2016

Crassula rogersii

Named after Frederick Arundel Rogers, 1876–1944, an English cleric and botanist who collected widely in Africa and Iran; he lived in South Africa, and in 1909 published *Provisional List of Flowering Plants and Ferns of Albany and Bathurst.*

Erect, succulent shrublet to 30 cm. **Leaves** Opposite, oblanceolate to club-shaped, densely hairy, 2 cm long × 1 cm wide. **Flowers** In dense clusters on an elongated stalk, small, 0.4 cm long, yellowish. Flowering in summer and autumn. **Habitat** Lower-lying areas in Succulent Karoo and Thicket. **Distribution** Widespread from Matjiesfontein through the Great Karoo to Graaff-Reinet and to Baviaanskloof.

Kleinpoort, Kouga 01.12.2015

Crassula rubricaulis   Red-stem crassula, rooistingelplakkie

rubri = red, *caulis* = stem; referring to the striking red stems

Compact, densely leafy, clump-forming succulent to 30 cm. **Leaves** 2.5 cm long × 1 cm wide, fresh green, often tinged red near the tips, margins finely hairy. **Flowers** Several in a dense head, on red, leafy stalks, each flower 0.5 cm long, white, calyx red. Flowering in autumn and winter. **Habitat** Localised on rocky outcrops in Grassy and Mesic Fynbos. **Distribution** Widespread from Swellendam to Gqeberha (Port Elizabeth).

Louterwater, Langkloof
20.01.2013

Crassula rupestris   Concertina plant, sosaties

rupes = rock, *estris* = place of growth; referring to the rocky outcrop habitat where the plants prefer to grow

Much-branched, rounded, succulent shrub to 60 cm tall, with woody stems. **Leaves** Thick, fused at the base, 1.5 cm long × 1 cm wide, grey-green, margins tinged red. **Flowers** Pretty, in dense, rounded, terminal heads 4 cm wide on a short stalk, each flower 0.4 cm wide, pinkish. Flowering in winter and spring. **Habitat** Common on rocky outcrops on hot dry slopes in Succulent Karoo and Thicket. **Distribution** Very widespread from Namibia to the Cape Peninsula and to Graaff-Reinet and Makhanda. **Notes** 3 subspecies recognised: subsp. *rupestris* has been recorded in Baviaanskloof; subsp. *marnierana* is rare and only found in certain parts of the Little Karoo; subsp. *commutata* only occurs near the Orange River mouth.

Zandvlakte bergpad, Kouga 2

Crassula saxifraga

Saxifraga = plant genus; the leaves look similar
to those of a certain species of *Saxifraga*.

Resprouting, tuberous geophyte to 20 cm. **Leaves** In 1 or 2 pairs,
lying flat, suborbicular and thin, 3 cm wide, with crenate margins. **Flowers**
Neatly arranged in a loose umbel at the end of a slender stalk, 0.8 cm long,
white to pink. Flowering in autumn and winter. **Habitat** Damp south-facing
slopes on rock ledges in Mesic and Grassy Fynbos and Fynbos Woodland.
Distribution Widespread from the Richtersveld to Gqeberha (Port Elizabeth).

Nooitgedacht, Kouga 12.05.2021

Crassula tecta

tego = cover; pertaining to the leaves
that are covered with hard papillae

Stunted, tough, low, clump-forming succulent to 10 cm.
Leaves Tufted, in a rosette, oblanceolate, covered with hard,
white papillae, 3 cm long × 1 cm wide. **Flowers** Packed
into a tight, rounded head 1.5 cm wide, each flower tiny,
0.4 cm long, creamy white. Flowering in autumn and winter.
Habitat Gravel slopes in Succulent Karoo. **Distribution** Nearly
endemic from the Little Karoo to Steytlerville.

Nuwekloof Pass, Baviaans 08.05.2006

Crassula tetragona subsp. *acutifolia*   Karkai

tetra = four, *gony* = knee; referring to the symmetrical arrangement of the
leaves, giving the stems a square look

Erect, branching, succulent shrub to 60 cm. **Leaves** Opposite,
lanceolate, 5 cm long × 0.5 cm wide. **Flowers** In compact heads
to 2 cm wide, at the tips of a long stalk, each flower small, 0.3 cm
long, white. Flowering in summer and autumn. **Habitat** Fairly frequent

Rus en Vrede 4X4, Kouga 09.03.2010

in Transitional Shrubland, Grassy Fynbos and Thicket. **Distribution**
Widespread from Namaqualand to Makhanda and Komani (Queenstown), Eastern Cape.
Notes 6 subspecies recognised, but only subsp. *acutifolia* has been recorded in Baviaanskloof.

Crassula umbella

umbella = parasol, umbrella; referring to
the umbrella-shaped flowers

Tuberous-rooted, small succulent to 15 cm. **Leaves** Held
horizontal, opposite, kidney-shaped. **Flowers** Few on thin stalks,
star-shaped, 0.5 cm wide. Flowering in early spring. **Habitat**
Shade on carbon-rich and rocky soils under thicket or other
woody vegetation in Succulent Karoo and Thicket. **Distribution**
Widespread from Richtersveld to Humansdorp.

07.09.2020

KALANCHOE

Kalanchoe = a Chinese name for one of the species

± 200 species mostly in Old World tropics; 13 species in South Africa; 1 species in Baviaanskloof.

Kalanchoe rotundifolia Nentabos

rotundi = round, *folia* = leaf; referring to the rounded leaves

Erect, succulent shrublet to 1 m. **Leaves** Opposite, elliptic to ovate, narrowing towards the base, 4 cm long × 1.5 cm wide, usually blue-green. **Flowers** Numerous in a terminal, rounded cyme 5 cm wide, each flower 1 cm long × 0.4 cm wide, orange to red. Flowering in autumn. **Habitat** Partial shade under trees in Thicket. **Distribution** Widespread from Baviaanskloof to North Africa and the Arabian Peninsula. **Other names** Nenta kalanchoe (E), plakkie (A), umfayisele yasehlatini, imphewula (X), idambisa (Z).

Ysrivier, Cambria 23.01.2022

CUCURBITACEAE

Gourds, cucurbits

± 965 species in 95 genera. This family has the most species that are used by humans for food of all the plant families. Well-known and economically important genera include *Cucurbita* (squash, pumpkin, zucchini and some gourds), *Lageneria* (calabash), *Citrullus* (watermelon) and *Cucumis* (cucumber). Plants occur mostly in the tropics and subtropics, since all species are sensitive to frost. 18 genera and 75 species in southern Africa; 3 genera and 4 species in Baviaanskloof.

COCCINIA

coccinus = scarlet; referring to the red fruits

± 30 species in Africa and Asia, mostly tropical; 1 species in Baviaanskloof.

Coccinia quinqueloba Bobbejaankomkommer

quinque = five, *loba* = lobes; referring to the 5-lobed leaves

Dioecious tuberous climber to 10 m. **Leaves** Palmate, with sparsely toothed lobes, hairless, 10 cm long × 5 cm wide. **Flowers** Single in the axils, 5 cm wide, yellowish. Fruit ellipsoid, green with white streaks, becoming scarlet. Flowering in summer and autumn. **Habitat** Occasional in Thicket and Forest. **Distribution** Widespread from Patensie to KwaZulu-Natal. **Notes** Traditionally used as a remedy for diarrhoea; the fruit is administered to ailing infants. The tendrils are tied around the ankles of breast-feeding mothers. **Other names** Ivy gourd (E), ithangazana (X), ufokwe (X).

Patensie 04.02.2020

KEDROSTIS

Kedrostis = the Greek name for *Bryonia dioica*, a similar-looking cucurbit from central and southern Europe

± 23 species, mostly in African tropics; 2 species in Baviaanskloof.

Kedrostis capensis

capensis = from the Cape; referring to the geographical origin of the species

Monoecious tuberous perennial climber to 50 cm. **Leaves** Palmatisect, usually with slender, linear lobes, sometimes scabrid-hairy, 7 cm long. **Flowers** 0.5 cm long, greenish yellow, males in bunches, female solitary. Fruit berry-like, oblong, glabrous, 2 cm long, red. Flowering in summer and autumn. **Habitat** Patches of bush in Transitional Shrubland and Thicket. **Distribution** Widespread from southern Namibia to Baviaanskloof.

Rus en Vrede 4X4, Kouga
28.07.2011

Kedrostis nana Ystervarkpatat

nanus = dwarf, referring to the smallish stature of this creeper

Dioecious tuberous perennial climber to 50 cm tall, with a strong smell of rotten eggs. **Leaves** Palmatisect, sometimes only slightly lobed, 4 cm long × 3 cm wide, bright green. **Flowers** To 1 cm wide, pale yellow or greenish, male flowers in a branched inflorescence, female flowers solitary. Fruit 3 cm long, bright yellow. Flowering in autumn. **Habitat** Valleys in Thicket. **Distribution** Widespread from Saldanha to KwaZulu-Natal. **Notes** 3 varieties: var. *schlechteri* is described here. The tuber is used as an emetic. **Other name** Uthuvishe (X).

Zandvlakte, Baviaans 26.03.2012

ZEHNERIA

Named after Joseph Zehner, a 19th-century Austrian botanical illustrator

± 35 species in Africa, Asia and Australia; 1 species in Baviaanskloof.

Zehneria scabra Utangazana (X)

scabra = rough, scabrid; referring to the roughly hairy top surface of the leaves

Dioecious, perennial climber to 6 m. **Leaves** Heart-shaped, slightly lobed and toothed, soft-textured, rough-haired above and softly hairy below, 6 cm long, green. **Flowers** Up to 8 in umbels in the axils, each 0.5 cm wide, cream-coloured, becoming yellow. Fruit roundish, 1 cm wide, green becoming red. Flowering in spring, summer and autumn. **Habitat** Occasional on Forest margins in the narrow ravines. **Distribution** Widespread from Cape Peninsula to tropical Africa. **Notes** Used in traditional medicine for treating skin disorders and other ailments. **Other names** Itanga, usimbene, ukalimela, ithuvana (X).

Zandvlaktekloof, Baviaan 19.01.2022

CUNONIACEAE

27 genera and ± 300 species of trees and shrubs, mostly in tropical and temperate regions of the southern hemisphere. Only 2 species in South Africa; 1 species in Baviaanskloof.

CUNONIA   Rooiels
Named after John Christian Cuno, 1708–1783, a German botanist, poet and merchant

± 17 species in New Caledonia; 1 species in southern Africa.

Cunonia capensis Butterspoon tree
capensis = from the Cape; referring to the geographical origin of the species

Large tree to 30 m, with rough, dark fissured bark. **Leaves** Compound, opposite, with leaflets sharply toothed, glossy, large, 20 cm long, leathery, dark green, leaflets 8 cm long × 3 cm wide, leaf buds enclosed in a large spoon-shaped stipule. **Flowers** Small, crowded on a dense spike to 20 cm long, white, sweetly scented. Flowering in autumn. **Habitat** Locally common along perennial streams in the narrow ravines in Forest, especially at higher altitude on south-facing slopes, near the water sources. **Distribution** Widespread from Tulbagh to Mpumalanga. **Other names** Rooiels (A), igqwakra, umqwashube (X), umaphethu, umlulamomkhulu (Z).

Zandvlaktekloof, Baviaans 09.02.2016

CURTISIACEAE

A family with just 1 species, endemic to southern Africa.

CURTISIA
Named after William Curtis, 1746–1799, an English botanist and entomologist, and founder of *The Botanical Magazine*, the world's longest-running, continuously published botanical journal

This genus was previously placed in the dogwood family (Cornaceae). The single species is recorded in Baviaanskloof.

Curtisia dentata Assegai tree
dentata = toothed; referring to the serrated leaf margins; the common name refers to the old practice of making arrows from the shoots, although this fact is sometimes disputed.

Tree to 12 m tall, with dark grey bark and often with straight slender shoots from near the base. **Leaves** Large, opposite, ovate, toothed, 10 cm long × 7 cm wide, glossy above, paler below, young parts covered with tiny reddish hairs. **Flowers** In terminal panicles, inconspicuous

Boskloof, Kouga Wildernes 15.05.2021

and cream-coloured. Fruit a round fleshy berry to 1 cm wide, white to pink, sometimes red. Flowering in late summer. **Habitat** Forest. **Distribution** Widespread from the Cape Peninsula to Zimbabwe. **Notes** The timber was regarded as optimal for various parts of wagons, especially the wheels. The bark is used in traditional medicine to treat stomach ailments and diarrhoea, and as a blood purifier and aphrodisiac. A protected tree in South Africa. **Other names** Assegaaiboom (A), umgxina, uzintlwa, umdlebe, umlahleni, umguna (X), injundumlahleni, umagunda (Z).

DIDIEREACEAE

6 genera and 20 species in Africa and Madagascar. Succulent plants growing in arid or semi-arid areas. Most members are endemic to Madagascar, where the plants are spiny and form prickly thickets.

PORTULACARIA

Portulaca = plant genus, *aria* = similar to; pertaining to the superficial resemblance of spekboom to some members of the purslanes

Spekboom was formerly placed in the purslane family (Portulacaeae), however DNA research by several botanists from 2000 to 2010 has clearly shown that *Portulacaria* belongs to the family Didiereaceae. *Portulacaria* is endemic to southern Africa and consists of 7 species, of which the following are confined to the Namib region: *P. armiana, P. carrissoana, P. fruticulosa, P. longipedunculata, P. namaquensis* and *P. pygmaea. P. afra* is the most widespread species and the only one to grow in the eastern parts of South Africa, including the Baviaanskloof. The other species are slow-growing and do not occur in dense stands.

Portulacaria afra Spekboom

afra = from Africa; endemic to southern Africa

Resprouting, succulent shrub or small tree to 3 m tall, young stems red. **Leaves** Opposite, obovate, fleshy, 4 cm long × 2 cm wide, edible, sometimes bitter. **Flowers** Many in a branched inflorescence, 0.5 cm wide, pink. Flowering in summer and after rain. **Habitat** The dominant shrub in Spekboomveld, an arid kind of Thicket. Also found in dry, rocky situations in Transitional Shrubland, and Grassy and Arid Fynbos. **Distribution** Widespread from Ladismith to Mozambique. **Other names** Olifantskos (A), igwanishe, umfayisele wehlathi, igwanitsha (X), indibili (Z).

winged seeds

Speekhout, Baviaans 14.02.2016

Spekboom – the unassuming supersucculent

Spekboom (*Portulacaria afra*), also known as pork bush or elephant's food, is a supersucculent of the Baviaanskloof. Typically regarded as part of the Thicket biome, it has an ascending growth with a thick smooth trunk, and it can reach the size of a small tree (3 m). However, it usually takes the form of a many-branched shrub, with ascending branches and a lower skirt of leaves spreading on the ground or rocks below.

Spekboom grows relatively rapidly compared to other species in the genus, and it may be seen as a pioneer species. Historically, populations in the region were dramatically reduced due to incorrect farming practices, but many farmers are rehabilitating the landscape and restoring the Thicket to its prior state of dominance. Spekboom is being planted en masse in areas where it used to be common.

Spekboom plants are dioecious, with separate male and female plants. The species usually flowers during the hot months, from January to March, but it also flowers after rain. The seeds are small, winged, and dispersed by wind. Seedlings are common after a good rainy season, but they soon succumb in warm, dry weather.

The superpower of spekboom

Like so many other succulents, spekboom has an astonishing ability to save water. All plants need to breathe and absorb carbon dioxide for the photosynthesis process. Their 'lungs' take the form of minute breathing pores (stomata) on the photosynthetic parts of the plant, but if the spekboom plant were to keep its stomata open during the day, it would lose too much water by evaporation.

Spekboom has overcome this problem by creating a type of 'battery' that accumulates acids (similar to apple acids) during the night, which it uses for photosynthesis in the daytime, enabling the plant to close its stomata and conserve water. Those who have tasted spekboom leaves in the morning will know that the leaves taste sour, but by the afternoon they are far less acidic, because the plant's battery has 'run down', ready to cycle again through the stages the following day. Refreshing the 'battery' at night and breaking the acids down during the day provides the plant with the carbon dioxide it needs to manufacture food, together with sunlight, chlorophyll and water.

There is a popular misconception that spekboom is superior at trapping carbon. Although the species does this successfully, it is no more effective at this than other succulents (such as the kerkei, prickly pair and agave) in southern Africa or abroad.

Survival tactics

Spekboom often grows in dense dominant stands. The Thicket where it naturally grows usually has a high degree of spinescent and toxic plants. Unlike these plants, spekboom lacks defensive structures. It is unarmed, with soft, palatable leaves and stems, making it a favourite for browsing by larger herbivores including elephant, buffalo and kudu.

To protect itself from being decimated by grazing and trampling game, the spekboom employs the power of passive resistance and vegetative reproduction: it allows itself to be grazed by herbivores but puts the subsequent damage to good use. As the detached branches fall on the soil, they take root to establish new plants. The succulent nature of the leaves and branches enables the plant to continue its functions, while it roots and grows. This ability is not confined to spekboom alone – it is seen in most other succulents such as crassulas, *Adromischus* spp. and succulent senecios.

When herbivores migrate to adjacent regions, the browsed thicket is given a chance to recover. The succulent leaves, stems and roots make the spekboom drought tolerant, giving it an advantage over non-succulent trees and shrubs such as the jacket plum (*Pappea capensis*) and guarri (*Euclea undulata*) (the latter 2 have deep root systems and slow growth).

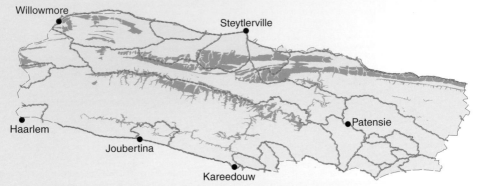

Willowmore

Steytlerville

Haarlem

Patensie

Joubertina

Kareedouw

Areas in green on the map delineate the original distribution of Spekboomveld, mostly confined to well-drained, rocky slopes and cliffs in the valleys and foothills of the mountains.

The spekboom's versatile growth behaviour depends on its habitat. On cliffs it often droops and hangs, with the branches surrendering to gravity. Stems touching ledges will also take root. When grown in a shady place, such as below a tree, the branches may become semi-climbing, leaning on the host plant. When grown in crevices of bedrock, the plants are often dwarfed, forming natural, thick-stemmed bonsai only a few centimetres tall.

Spekboom has ways of infiltrating neighbouring biomes. Although it is typically part of the Thicket biome, it is widely dispersed by wind, and when conditions are favourable it establishes in Grassy Fynbos and Transitional Shrubland, outcompeting and displacing the more flammable grasses and shrubs. Young spekboom plants are generally slow growing, and the saplings are often scorched and killed by veld fires. However, in certain situations, spekboom can grow faster than normal, quickly developing a dense skirt of low branches around the main trunk, which form an insulating barrier. Consequently, if fire occurs, the main trunk is relatively protected. Part of the plant may be scorched, but if the main trunk survives, that individual will have a leading edge on the post-fire recovery. Spekboom therefore acts as a catalyst for the development of patches of thicket in fire-prone vegetation. Spekboom and other tree succulents, such as the kerkei (*Crassula ovata*) and krantz aloe (*Aloe arborescens*), are excellent species for planting as fire barriers.

Top: Spekboom shedding scorched bark, exposing 'fresh skin'. Above: Spekboom surviving a fire in Grassy Fynbos in the Kouga Mountains.

Spekboom also has unusual bark, with skin-like layers on its stem. If the outer covering is scorched, the damaged layers can be shed, and the sheet below becomes the new bark.

Spekboom not only protects the soil from erosion, it also adds carbon and nutrients to the soil when the thick bed of organic matter created by its fallen leaves decomposes. This is an essential part of thicket recovery on degraded and eroding slopes, and bolsters the plant's reputation as a supersucculent.

DIPSACACEAE

Teasel, scabious

Herbs and shrubs in 11 genera and 350 species in temperate climates of Europe, Asia and Africa. The dense flowering heads make them easy to confuse with the daisy family, but the tips of the calyx have five bristles, a feature absent in daisies. 2 genera and 3 species in Baviaanskloof.

SCABIOSA

Pincushion flowers

scabiosa = scabies; the plants have been used medicinally to treat scabies, and the bristles on the calyx are itchy. The dense flowerheads look like pincushions.

9 species in South Africa; 2 species in Baviaanskloof.

Scabiosa albanensis Isilawu (X)

alban – Albany, *ensis* = from; referring to the district in the Eastern Cape where the plants were first recorded

Non-resprouting, slender, erect, herbaceous shrublet to 30 cm. **Leaves** Opposite, ascending, pinnatisect to bipinnatisect, 6 cm long, light green. **Flowers** In pedunculate heads, 2 cm wide, mauve. Flowering in summer. **Habitat** Uncommon, mostly after fire, in Grassy Fynbos and Sour Grassland. **Distribution** Endemic to the Eastern Cape, from Cockscomb to Makhanda.

Cockscomb, Groot Winterhoek 17.02.2016

Scabiosa columbaria Jongmansknoop

columba = pigeon, *aria* = place for; referring to the flowerheads with many flowers that resemble pigeonholes

Resprouting, tufted perennial herb, to 80 cm tall when flowering. **Leaves** Crowded at the base, pinnatisect, hairy, 12 cm long × 6 cm wide, toothed or incised. **Flowers** In dense heads at the end of long stalks, each flowerhead 3 cm wide, white to mauve. Flowering in spring and summer. **Habitat** Common after fire in Grassy Fynbos and Transitional Shrubland. **Distribution** Widespread through Africa, Europe and Asia. **Other names** Wild scabiosa (E), iyeza lamehlo (X).

Joubertina 19.10.2019

DROSERACEAE

Sundews

Carnivorous plants with 3 genera and 180 species. The Venus flytrap (*Dionaea muscipula*), native to North America, is a famous member. Sundews secrete a sticky substance from the glandular hairs on the leaves. Resembling dewdrops, the secretion is also an efficient insect trap. The plants derive nutrients from their catch, allowing them to thrive in nutrient-poor sites such as sandstone seepages.

DROSERA

Sundew

droseros = dewy; referring to the dewy leaf glands

20 species in South Africa; 2 species in Baviaanskloof.

Drosera aliciae

Named after Alice Rasse, who encouraged French botanist Raymond Hamet, 1890–1972, to study sundews

Tufted perennial to 20 cm when in flower. **Leaves** Basal in a rosette, spathulate, 2 cm long × 0.7 cm wide, often red, covered with long, sticky leaf glands. **Flowers** Several on a wiry scape, each 1 cm wide, usually pink, sometimes white. Flowering in summer. **Habitat** Uncommon on peaty sandstone seepages and stream banks in Grassy Fynbos. **Distribution** Widespread from Cape Peninsula to Van Staden's Mountains, Eastern Cape.

Nooitgedacht, Kouga 28.09.2011

Drosera capensis

capensis = from the Cape; referring to the geographical origin of the species

Rhizomatous perennial herb to 50 cm. **Leaves** With a petiole, linear to elliptic, spaced along the stems, 5 cm long × 1 cm wide. **Flowers** Several on a long stalk, 1 cm wide, pink. Flowering in summer. **Habitat** Marshes and seepages, and on the edge of perennial streams in narrow ravines near Forest. **Distribution** Widespread from Cederberg to Kareedouw.

Kleinrivier, Kouga Wildernis 18.12.2011

EBENACEAE

Ebony

Trees and shrubs with 2 genera and ± 770 species across tropical and warm, temperate parts of the world. Many are valued for their wood, notably ebony (*Diospyros ebeneum*). 2 genera in South Africa.

DIOSPYROS

Persimmon, tolbos

dios = divine, *Pyrus* = plant genus; *Pyrus* is the pear genus in the rose family, referring to the edible fruits of some species.

A large genus of roughly 500 species found throughout the world in tropics and subtropics. Usually dioecious woody trees or shrubs, with fruit a fleshy berry. Several species are used in the timber industry and some for their edible fruits. 18 species in South Africa; 6 species in Baviaanskloof.

Diospyros austro-africana Fire-sticks, kritikom

austro = southern, *africana* = from Africa; referring to the origin in southern Africa.
Historically, the wood was used to make fire by friction, thus the common name.

Resprouting, much-branched, evergreen, dioecious, bushy shrub
to 3 m. **Leaves** Oblanceolate, velvety, leathery, 2 cm long × 1 cm
wide, felted below, veins sunken above and raised below, often an
unusual brownish-grey-green colour. **Flowers** Hanging down on
slender 2 cm-long stalks, tube 0.6 cm long, pink or cream flushed
with pink. Fruit round, 1 cm wide, softly and shortly hairy. Flowering
in spring. **Habitat** A drier, cooler habitat in the northwestern part
of Baviaanskloof, in Arid Fynbos and Transitional Shrubland.
Distribution Widespread from Nieuwoudtville to the Cape
Peninsula and Mpumalanga. Easily seen in Nuwekloof Pass, the
key to Baviaanskloof. **Other name** Umbongisa (X).

Nuwekloof Pass, Baviaans 08.10.2016

Diospyros dichrophylla Poison star-apple

dichrophylla = discolorous leaves

Shrub or small tree to 4 m. **Leaves** Oblanceolate, petiolate, entire, hairier
below, 5 cm long × 1 cm wide, margins rolled under, dark green above, paler
below, with raised midrib below. **Flowers** Hanging down on 2.5 cm-long
stalks, each flower 1 cm long, creamy white, with reflexed petals and hairy
calyx. Fruit large, roundish, velvety-haired, 3 cm wide, pale orange when
ripe, calyx large and curling up. Flowering in summer and autumn. **Habitat**
Forest and Thicket in sheltered situations. **Distribution** Very widespread from
Western Cape to Limpopo. **Notes** The fruit is said to be poisonous. **Other
names** Poison peach (E), bloubos (A), umbongisa (X).

Kamassiekloof, Zandvlakte
21.04.2012

Diospyros lycioides Bloubos, Karoo bluebush

lycioides = resembling genus *Lycium*; the similarities between this species and *Lycium* are
not obvious, may be referring to the fruit or the leaves crowded at the branch tips.

Small tree to 7 m tall, often multi-stemmed, with smooth, dark grey bark. **Leaves**
Crowded at the branch tips, variable in size, mostly 5 cm long × 1.5 cm wide,
with midrib raised below and secondary veins almost invisible
below. **Flowers** Hanging down on 3 cm-long stalks, each flower
0.7 cm long, creamy white or yellow, with reflexed petals. Fruit
round to oval, 2 cm long, yellow to red when ripe, calyx bent
upwards. Flowering in spring and summer. **Habitat** Common
on the valley floor in or at the edge of Forest and in Thicket.
Distribution Very widespread from the Little Karoo to tropical
Africa. **Notes** Some authors say that the fruit is edible, while
others maintain it is inedible. Twigs and roots are used as chewing
sticks to clean teeth. 4 subspecies are recognised, but only subsp.
lycioides occurs in Baviaanskloof. **Other names** Star apple (E),
swartbas (A), umbongisa (X).

Kamassiekloof, Zandvlakte 16.10.2011

Diospyros whyteana Bladdernut

Named after Alexander Whyte, 1834–1908, a Scottish plant collector and horticulturist who worked in the West Indies, Sri Lanka, Malawi and Uganda

Reseeding, single-stemmed shrub or tree to 7 m tall, bark dark grey to almost black. **Leaves** 5 cm long × 2.5 cm wide, shiny green above, paler below, margin with a fringe of hairs and sometimes wavy. **Flowers** Few on 2 cm-long stalks, each flower 1 cm long, creamy white, sweetly scented. Fruit oval, to 2 cm long, red, entirely enclosed by large green calyx (appearing like a swollen bladder – thus the common name) that turns brown and splits open when the fruit is mature. Flowering in spring. **Habitat** Occurring as an understorey tree in Forest and Thicket. **Distribution** Widespread from Tulbagh to the Cape Peninsula to Mpumalanga. **Other names** Bostolbos (A), umkhaza, umbongisa, umgugunga, umtenatene, intsanzimane (X).

Waterkloof, Kouga Wildernis 14.05.2021

EUCLEA

eukleia = glory or 'of good report'; referring to the sought-after ebony-type wood of some species

Dioecious trees, shrubs or subshrubs. Leaves are alternate and entire, with fine-mesh veining giving them a translucent appearance when held up to the sun. Fruit a berry with large seeds. 20 species in southern Africa; 6 species in Baviaanskloof.

Euclea crispa subsp. *crispa* Bloughwarrie

crispa = crisped, irregularly waved and twisted, kinky, curled; referring to the leaf margins

Resprouting, dioecious shrub or tree to 4 m tall, rusty-granular on young parts. **Leaves** Obovate to lanceolate, papery, 5 cm long × 1 cm wide, covered in rusty granules (especially the new growth), margins wavy and finely and irregularly scalloped (crisped). **Flowers** In axillary racemes, hairy, cream-white, ovary with bristles, pleasantly scented. Fruit 0.5 cm wide, brown to black. Flowering in spring and summer. **Habitat** Uncommon and sporadic on, or at the base of, sandstone outcrops or in kloofs, in Grassy Fynbos and Fynbos Woodland. **Distribution** Widespread from Riversdale to Zimbabwe. **Notes** There are many different forms of this species that differ mainly in leaf size and shape. 2 subspecies are recognised but only subsp. *crispa* occurs in Baviaanskloof.

Quaggasvlakte, Kouga 13.02.2016

The plant is used in traditional medicine, mainly as a purgative, with the bark of the root being the part most used. Some authors indicate the berries to be edible, while others suggest they may be poisonous. The plant is also used to make a black dye. **Other names** Iyeza lokuxaxuzisa, umgwali, umnquma (X).

Euclea daphnoides Witstam, white-stem guarri

Daphne = old genus name, *oides* = resembling; *Daphne* is an old genus name for the bay laurel (or bay tree), but the resemblance between the 2 is not strong and bay leaves are more similar to *E. natalensis*. Previously regarded as a subspecies of *E. schimperi*, with which it has been confused in the past.

Resprouting, shrub or small tree to 5 m tall, but mostly to 2 m tall, with pale grey bark. **Leaves** Linear to narrowly oblanceolate, sometimes whorled at branch tips, hairless, hard and leathery, 7 cm long × 0.8 cm wide, bright green above, paler below, margins not undulating. **Flowers** In axillary racemes, deeply cleft, glabrous, cream-white, ovary glabrous. Fruit roundish 0.8 cm wide, bright red to purplish. Flowering from summer to autumn. **Habitat** Rocky koppies in Thicket and Transitional Shrubland. **Distribution** Widespread from Baviaanskloof to Mozambique. **Notes** The wood splits and crackles when burned. African folklore cautions that one should never use this species as firewood because it will bring about conflicts and unhappiness in the home. **Other name** Umkaza (X).

Duiwekloof, Baviaans 14.02.2016

Euclea natalensis subsp. *natalensis* Hairy guarri, swartbasboom

natalensis = from Natal; sometimes plants are named after a region where the first specimen was recorded, but later, as in this case, it is established that the species has a wider distribution.

Resprouting shrub or small tree to 6 m tall, but in Baviaanskloof often only to 1 m. **Leaves** Elliptic, stiffly leathery, hairy, petiole to 1 cm long, leaves relatively large, 10 cm long × 4 cm wide, margins wavy. **Flowers** In axillary racemes, cream-white, ovary hairy. Fruit 1 cm wide, red to black. Flowering from autumn to summer. **Habitat** Transitional Shrubland and Grassy Fynbos. **Distribution** Widespread from the Western Cape to tropical Africa. **Notes** 6 subspecies are recognised in South Africa, but only subsp. *natalensis* is found in Baviaanskloof. The bark of the root is used in traditional medicine. A black dye can be made from the roots. The twigs are used as toothbrushes. **Other names** Bergghwarrie (A), umtshekesane, umkaza, umgwali, intlakotshane enkulu (X), inkunzemnyama (Z).

Guerna, Kouga 16.02.2016

Euclea polyandra Baviaanskers, kersbos

poly = many, *andra* = anther or male flower; referring to the numerous anthers (up to 30) on the male flower

Resprouting, erect, tough, bushy, dioecious shrub to 2 m. **Leaves** Elliptic to ovate, 6 cm long × 1.5 cm wide, grey-hairy. **Flowers** In axillary racemes, rusty-haired, petals shallowly lobed, white, scented. Flowering in late spring. Fruit roundish, 1 cm wide, covered with rusty hairs. **Habitat** Rocky outcrops in the mountains in Fynbos, often on rocky ridges and crests. **Distribution** Widespread from Kogelberg to Makhanda.

Engelandkop, Kouga
18.07.2011

Euclea schimperi Glossy guarri

Named after Andreas Franz Wilhelm Schimper, 1856–1901, a German botanist and phytogeographer. Some people believe this species to be a variety of *E. racemosa*. A careful revision of the genus is recommended.

Resprouting shrub or small tree to 5 m tall, mostly to 2 m tall, with pale grey bark. **Leaves** Sometimes whorled at branch tips, obovate to broadly oblanceolate, hairless, thinly leathery, 10 cm long × 3 cm wide (but in exposed situations considerably smaller), bright green above, paler below, margins sometimes undulating. **Flowers** In axillary racemes, glabrous, cream-white, deeply cleft, ovary glabrous. Fruit roundish, 0.8 cm wide, bright red to purplish. Flowering from summer to autumn. **Habitat** Uncommon in Baviaanskloof, occurring in Forest and mesic Thicket, but also on rocky slopes in Grassy Fynbos. **Distribution** Widespread from Baviaanskloof to tropical Africa and to Yemen. **Other name** Umkaza (X).

Nooitgedacht, Kouga 12.05.2021

Euclea undulata Guarri

undulata = undulating; pertaining to the undulating margins of the leaves

Resprouting tree to 7m tall, but mostly to 3 m, with an enormous tap root. **Leaves** Oblanceolate, 4 cm long × 1.5 cm wide, always very wavy or undulating. **Flowers** In axillary racemes, petals deeply lobed, cream-white. Fruit reddish purple, turning black, edible. Flowering in summer and autumn. **Habitat** Common on rocky, shale soils in Thicket and Succulent Karoo. **Distribution** Widespread throughout drier parts of southern Africa. **Notes** Leaves and fruit browsed by game

Speekhout, Baviaans 14.02.2016

and stock. The wood is hard and heavy and makes good coals. Medicines made from the roots and leaves are used to treat a variety of ailments. Indicator species for Guarriveld in Succulent Karoo. **Other names** Umgwali (X), mokwere kwere (S), gwanxe, umshekisane (Z).

ERICACEAE   Heath

A large family found in most parts of the world, with ± 4,250 species in ± 124 genera. Well-known members include cranberry, blueberry, huckleberry and rhododendron. Most species are able to grow in infertile or acidic soils because the roots grow in association with mycorrhizal fungi that allow them to extract nutrients from the soil. Only 2 genera in southern Africa, *Erica* and *Vaccinium*, the latter represented by a single species in the Drakensberg, *V. exul*. The superdiverse genus *Erica* has over 860 species in the world, and over 90 per cent of this diversity is native to southern Africa.

Erica is by far the biggest genus in Fynbos, with ± 800 taxa. The Baviaanskloof area hosts ± 61 *Erica* species, making up around 9 per cent of the genus. Ericas are not only diverse in species, but some of them are also extremely abundant in the landscape, often dominating the vegetation cover, especially on the cooler upper mountain slopes. Erica bushes in flower are spectacular. Even those with little to no interest in flowers or nature will be impressed when seeing a splash of pink ericas in bloom on a distant mountain slope.

Recognising an erica is relatively easy, but accurately identifying the species can be difficult. Ericas have tubular flowers and the leaves are very small and fine, with recurved margins. To positively identify an *Erica* species you will need a 10× hand lens. The key features to look at include: the arrangement of the leaves on the stem (opposite, or in whorls of 3 or 4 leaves), the habit, the habitat, the shape, colour and size of the flower, the hairiness of all the parts, the number of stamens, whether the stamens have appendages, and whether the stamens and stigma are enclosed in the flower.

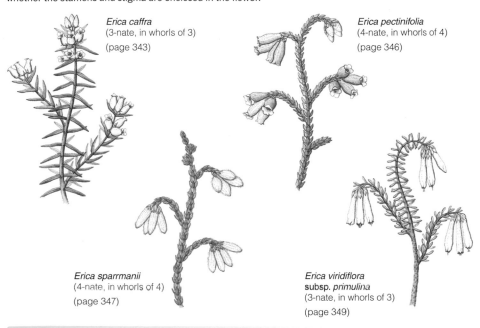

Erica caffra
(3-nate, in whorls of 3)
(page 343)

Erica pectinifolia
(4-nate, in whorls of 4)
(page 346)

Erica sparrmanii
(4-nate, in whorls of 4)
(page 347)

Erica viridiflora
subsp. *primulina*
(3-nate, in whorls of 3)
(page 349)

ERICA

ereiko = to break; the name used for a heath by Pliny and Theophrastus, referring to the brittle stems that are easy to break

Erica affinis

affinis = near to or neighbouring; pertaining to the closely packed leaves that clasp the stem, although it could also refer to similarities shared with closely related species *E. baccans* and *E. excavata*

A tough, robust, erect shrub to 90 cm. **Leaves** 4-nate, tightly packed around the stem, shiny-smooth, hairless, 0.6 cm long. **Flowers** In groups of 4 at branch tips, hanging down from hairy pedicels, urn-shaped, hairless, 0.5 cm long, white or pink. Sepals same size and colour as corolla. Each flower encloses stigma and 8 anthers with broad and jagged edged crests. Flowering in early summer. **Habitat** Uncommon and sporadic on rocky outcrops and ridges on mid- to upper slopes in Grassy and Mesic Fynbos. **Distribution** Near endemic to this region, from Uniondale to Kariega (Uitenhage). **Notes** Slow to mature. Killed by fire.

Elandsvlakte, Baviaans 03.11.2011

Erica andreaei

Named after Dr Hans Karl Christian Andreae, 1884–1966, chemist and naturalist who discovered this species for science. He assisted Dr Rudolf Marloth with plant collections, but his main interest was beetles.

Reseeding, erect bush to 60 cm. **Leaves** 3-nate, rigid, glossy, margins slightly short-haired. **Flowers** In groups of 3 at branch tips, bell-shaped, hanging down on hairy peduncles; each flower 0.3 cm long, white. Calyx white and hairy. Anthers crested and slightly exserted, stigma exserted, ovary hairy. Flowering in spring. **Habitat** Uncommon on higher peaks of mountains that receive less rainfall, in Mesic Fynbos. **Distribution** Narrow distribution from Swartberg to Baviaanskloof.

West of Bosrug, Baviaans 12.10.2016

Erica angulosa (= *Acrostemon fourcadei*)

angulosa = with prominent angles; referring to the various angles at which flowers are arranged on the stem, and the various angles at which the stigma and stamens stick out of the flowers

Thin, erect, reseeding shrublet to 50 cm tall. **Leaves** 3-nate, clasping the hairy stems, finely hairy, very small, to 0.2 cm long. **Flowers** Very tiny, tubular, cup-shaped, cream to pale pink. Calyx grooved, as long as the corolla, covered with long hairs giving flowers a fuzzy appearance. 4 anthers without appendages, or with very short awns, well exserted. Stigma broadened, cup- to plate-shaped, and extending out beyond the anthers. Extremely floriferous in autumn. **Habitat** Forming dense stands on open, exposed slopes of mountains in Grassy and Mesic Fynbos. **Distribution** Endemic to the Kouga and Baviaanskloof mountains.

Kouenek, Kouga 16.07.2011

Erica articularis

articularis = with joints; pertaining to the much-branched habit and very short lateral branches at the tips of the main stems, where flowers form a spike-like inflorescence

A small resprouting plant with many erect stems arising from a woody rootstock, forming a dense, rounded shrub, up to 40 cm, typically with brown flowers amongst the fresh ones. **Leaves** 3-nate, neatly arranged clasping the stem, non-hairy, to 0.5 cm long. **Flowers** Small, 0.4 cm long, white to pale pink and constricted at the throat. Calyx white, covering most of the flower. Crested anthers included. Flower turns brown soon after opening. Flowering in summer and autumn. **Habitat** Growing across a range of habitats. In Baviaanskloof, restricted to the upper south-facing mountain slopes in Mesic Fynbos. **Distribution** Widespread throughout the Cape Fold Mountains from the Cederberg and Cape Peninsula to Humansdorp.

Sipoonkop, Kouga 30.04.2018

Erica bolusanthus (= *Thoracosperma nanum*)

Named after Harry Bolus, 1834–1911, a plant collector, author and founder of the Bolus Herbarium at the University of Cape Town

Small, erect, much-branched, reseeding shrublet. **Leaves** 3-nate, clasping the white-felted stems, tips hair fringed, edges dotted with round glands, to 0.2 cm long. **Flowers** Small, cup-shaped, pale pink to pink. Calyx shorter than corolla and hairy like the leaves. 4 anthers, well exserted, without awn; stigma slightly swollen at its tip, extending beyond the anthers. Flowering in summer. **Habitat** Forming dense stands on the drier rocky, lower and middle slopes in the mountains in Grassy and Mesic Fynbos. **Distribution** Endemic to Baviaanskloof and Kouga mountains.

Scholtzberg, Baviaans 17.11.2011

Erica caffra Water heath

caffra = from the southeastern part of South Africa, pertaining to wide range and area of origin; the common name refers to the stream banks habitat.

Tree up to 4 m. **Leaves** 3-nate, almost perpendicular to the stem ,very finely hairy. **Flowers** Conical, finely hairy, 0.6 cm long, cream-white. Tailed anthers and stigma included. Flowering prolifically from winter to spring. **Habitat** Found on stream banks and usually associated with Forest. However, near Makhanda and further north it can be found in grasslands. **Distribution** Very widespread from Gifberg to Drakensberg. **Other name** Waterheide (A).

Kleinrivier, Hankey 21.09.2019

Erica chamissonis Makhanda heath

Named after Adelbert von Chamisso, 1781–1838, a German poet and botanist. He collected this plant at the Cape in 1818 while on an expedition around the world; he took it to JF Klotzsch, 1805–1860, who described it.

A magnificent sight when in full bloom, this erect or somewhat gnarled shrub produces many small side branches off several main stems. **Leaves** 3-nate, semi-open-backed, 0.3 cm long. **Flowers** 3 hanging at branchlet tips, bell-shaped, with hairless corolla to 0.5 cm long. Calyx small, to 0.25 cm long, fused for more than half its length, densely villous. Anthers included, without awns, surface rough; stigmas pale, exserted slightly. Flowering all year round, but mainly in spring. **Habitat** Found on damp, upper south-facing mountain slopes in Mesic Fynbos. **Distribution** Eastern distribution from Kouga Mountains to Makhanda. **Notes** A spectacular pseudo-raceme of pink, highly favoured in gardens.

Kouenek, Kouga 16.04.2012

Erica curviflora Waterbos, waterheide

curviflora = curved flower; pertaining to the curved floral tube

Handsome, erect, much-branched, resprouting shrub to 1.8 m.
Leaves 4-nate, bunched together around the stems, sparsely
hairy, to 1 cm long. **Flowers** Curved, tubular, glabrous or hairy,
to 3 cm long, varying in colour from yellow to orange or red.
Calyx small and leaf-like. Brown stamens, sometimes with short
awns, almost exserted; stigma clearly visible. **Habitat** Damp
or wet areas along streams or seepages in Grassy and Mesic
Fynbos. **Distribution** Widespread throughout the Cape Fold
Mountains from Gifberg and the Cape Peninsula to Fish River,
Eastern Cape.

Quaggasvlakte, Kouga 13.02.2016

Erica demissa

demissa = low, humble, drooping; pertaining to a tendency of the flowers to hang down. However, the word also
means 'of low altitude', and this species is most common on lower, more arid slopes; it is often the first erica
encountered on a walk up the mountain.

An upright, reseeding bush to 1.5 m tall. **Leaves**
3-nate, clasping the stem, shiny, 0.3 cm long.
Flowers In umbels hanging down at the tips of
the branchlets, small, 0.3 cm long, white. Anthers
narrow, without awns, protruding from the corolla,
stigma white with a pink tip. Flowering mostly in
late summer and autumn. **Habitat** Dominates
the cover on lower south-facing slopes; probably
the most common and abundant *Erica* species in
Baviaanskloof. **Distribution** This species is most
prolific on the mountains between Uniondale and
Makhanda, although there are a few records on the
mountains as far west as the Hex River.

Zandvlakte, Baviaans 31.07.2011

Erica diaphana

diaphanus = colourless or nearly transparent; pertaining
to the corolla tubes that are slightly transparent

A spectacular, robust, sturdy, resprouting shrub.
Leaves 3-nate, rather fleshy, symmetrically arranged
at a 45-degree angle to the stem. **Flowers** Very sticky,
tubular corolla, hanging straight down, up to 2.5 cm
long, bright red. Awnless anthers included, sometimes
with tiny awns, stigma extending just beyond corolla
lobes. Flowering in summer. **Habitat** Associated with
rocky outcrops and exposed ridges in all kinds of
Fynbos. **Distribution** Widely distributed in the Cape Fold
Mountains between Swellendam and Humansdorp.

Quaggasvlakte, Kouga 13.02.2016

Erica discolor subsp. *speciosa*

discolor = variegated; pertaining to the 2 colours on a single flower, seldom present in Baviaanskloof; *speciosa* = showy; pertaining to the bright flowers

A sturdy, much-branched shrub up to 1.5 m, does not sprout after fire. **Leaves** 3-nate, usually glabrous. **Flowers** Corolla tube to 3 cm long, can be sticky. In this region, flowers tend to be pink through to red, not bicoloured. Anthers included, with small awns. Flowering mainly in spring and summer. **Habitat** Damp slopes in Grassy and Mesic Fynbos. **Distribution** Widely distributed between Caledon and Humansdorp.

Gamtoos Valley, Andrieskraal 14.05.2006

Erica flocciflora

floccus = a tuft of wool, *flora* = flower; pertaining to the woolly calyx

Upright, woody shrublet to 1 m tall. **Leaves** 4-nate, ascending, closely packed. **Flowers** Arranged in umbels of 3–8 at the branch tips, urn-shaped, 0.6 cm long, creamy yellow, finely hairy and slightly sticky, calyx and peduncle densely covered in white-woolly hairs. Anthers included, with pink crests. **Habitat** On cliffs or at the base of rocky outcrops in Arid and Mesic Fynbos. **Distribution** Endemic to the Kouga Mountains. **Notes** Listed as Near Threatened.

Below Saptoukop, Kouga 12.10.2014

Erica kougabergensis

kougabergensis = from the Kouga Mountains, where it was first scientifically recorded

Scruffy, wiry stemmed, reseeding shrublet to 0.8 m tall, with slightly sticky stems, leaves and sepals. **Leaves** 3-nate, ovate, open-backed, usually slightly recurved, 0.2 cm long, with gland-tipped hairs dotted along the margins. **Flowers** Small, a few mm long, varying from white to pale pink. Anthers protruding with tiny awns, stigma with a swollen tip with 5 globular lobes, resembling a miniature wine gum sweet. Flowering in spring. **Habitat** Found on mid- to upper slopes in relatively arid mountains. **Distribution** Endemic to the Kouga and Baviaanskloof mountains.

Quaggasvlakte, Kouga 13.02.2016

Erica nabea

Named after William McNab, 1780–1848, Scottish horticulturist at the Royal Botanic Garden at Edinburgh and its expert on the propagation of ericas

An erect, sparsely branched, reseeding shrub up to 1.5 m. **Leaves** 3-nate, linear, 1 cm long, with a few hairs. **Flowers** Very small corolla, 0.3 cm long, hidden under 1.6 cm-long light green calyx, together pointing upwards along the stem, forming a spike-like inflorescence. Anthers elongated, without awns, also hidden inside calyx; stigma sometimes visible, bending over and pointing down into the flower. Flowering in winter. **Habitat** Uncommon on rocky, south-facing ridges in Mesic Fynbos. **Distribution** In the mountains from George to Kariega (Uitenhage). **Notes** The flattish, wind-dispersed seeds are the largest in the genus.

Scholtzberg, Baviaans 24.07.2011

Erica newdigateae

Named after Caroline Newdigate, 1857–1937, a Plettenberg Bay
resident who was the first person to collect the species for science

A showy, erect, reseeding, much-branched shrub up to
1 m. **Leaves** 3-nate, loosely arranged, 0.2 cm long, margins
with fine hairs. **Flowers** Cup-shaped, white to pink. Calyx
only partially fused, non-hairy, distinguishing it from similar-
looking *E. chamissonis*. Anthers included, without awns,
style exserted. Flowering in spring. **Habitat** Forms dense
stands on mid- to upper south-facing slopes in Grassy and
Mesic Fynbos. **Distribution** Widespread from Knysna to
Makhanda. A white-flowered form seems to be the most
prevalent at the top of the Baviaanskloof Mountains.

Ridge north of Enkeldoring, Baviaans 24.09.2011

Erica passerinae

Passerina = plant genus; named after this
genus since the spreading petals might
vaguely resemble those of some of the species

Spectacular in bloom, a much-branched
shrub sprouting from a woody rootstock; stem, leaves
and sepals covered in white woolly hairs. **Leaves** 3-nate,
0.2 cm long, semi-open-backed. **Flowers** Cup-shaped, to
0.6 cm long. Calyx white, constrasting starkly with pink
corolla. Anthers (without awns) and stigma included.
Flowering in winter and spring. **Habitat** Finer-textured
soils of shale bands on the damper south-facing slopes
of the mountains in Grassy Fynbos. **Distribution** Near
endemic, also found in Kammanassie Mountains.

Rus en Vrede 4X4, Kouga 28.07.2011

Erica pectinifolia

pectinis = comb-like; pertaining to the comb-like leaves and sepals

Resprouting, upright, slender, woody
shrub, to 1.5 m tall, sparsely branched.
Leaves 4-nate, 0.3 cm long × 0.2 cm wide,
tightly packed around the stem, with stiff,
translucent hairs on the margins. **Flowers**
1.5 cm long, white to pink, covered with
bristly hairs. Sepals with unusual hairy
margins. Flowering all year round. **Habitat**
Common in Arid Fynbos, extending high
into the mountains on exposed ridges in
Grassy and Mesic Fynbos. **Distribution**
Widespread from Swartberg Mountains to
Gqeberha (Port Elizabeth).

Bosrug, Baviaans 22.07.2011

Erica rosacea subsp. *rosacea* (= *Thoracosperma rosaceum*)

rosacea = a human skin disease that causes reddening of parts of the face; referring to the deep pink flowers, and the tendency for the leaves to turn red when drought-stressed

A small, erect, reseeding shrublet to 0.6 m. **Leaves** 3-nate, clasping the stem, smooth, tiny, to 0.2 cm long. **Flowers** Very small, 0.3 cm long. Anthers awned and exserted; style simple, exserted. Flowering in autumn, winter and spring. **Habitat** Common on rocky and exposed slopes, tolerating dry conditions on the north-facing slopes, in Grassy and Arid Fynbos. **Distribution** From the Langeberg and Little Karoo mountains to Hankey.

Scholtzberg, Baviaans 24.07.2011

Erica sp. nov.

sp. nov. = new species, undescribed; abbreviation commonly used to indicate taxa that have not yet been formally described and/or published

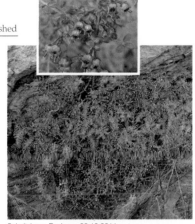

Low, reseeding, spreading shrublet to 30 cm. **Leaves** 3-nate, ovate, held perpendicular to the stem and well-spaced, sparsely covered with tiny, glandular hairs. **Flowers** In groups of 3–5, usually hanging on dark pink, glandular-hairy peduncles (to 1 cm long), sticky, 0.3 cm long, pale pink; anthers included, with tiny appendages, stigma exserted, often bent near the tip. Flowering in spring. **Habitat** Uncommon and sporadic in the shade of rocky overhangs that become seasonally damp, in Mesic and Subalpine Fynbos. **Distribution** Endemic to the Baviaanskloof Mountains on the south side of Scholtzberg and other nearby peaks. **Notes** First recorded in 2011, this species is yet to be published with a formal description.

Scholtzberg, Baviaans 29.10.2011

Erica sparrmanii

Named after Anders Sparrman, 1748–1820, a Swedish botanist and physician, and a pupil of Linnaeus. He visited the Cape from 1772–1776 and was the first to publish an extended account of his travels into the interior.

A sturdy, erect, sparsely branched, lanky, reseeding shrub to almost 2 m. **Leaves** 4-nate, covered in tough, white hairs. **Flowers** 1.8 cm long, yellow to greenish, covered in bristly white hairs. Anthers and stigma included. Flowering all year round. **Habitat** Sporadic on open, exposed slopes in Grassy Fynbos. **Distribution** Between Plettenberg Bay and Humansdorp and inland to the Baviaanskloof Mountains. **Notes** Although this species is quite frequently encountered, it never seems to be abundant.

Bosrug, Baviaans 12.07.2011

Erica thamnoides (= *Thamnus multiflorus*)

thamnoides = bush-like; referring to the bushy habit

Upright, soft, reseeding shrub to 1.2 m. All parts (except corolla and ovary) very finely hairy, with small mole-like nodules (only visible with a 10× lens). **Leaves** 3-nate, linear, very thin, to 0.3 cm long. **Flowers** Pink, with calyx almost half as long. 4 anthers protruding with small awns, stigma simple. **Habitat** Dominating the lower to middle south-facing slopes and ridges in Grassy and Mesic Fynbos. **Distribution** Near endemic, from Sedgefield to Kariega (Uitenhage).

Above Coleskeplaas, Baviaans 06.03.2012

Erica uberiflora (= *Simocheilus* spp.)

uber = superior, *flora* = flower; referring to the ovary situated above the point where the other flower parts emerge

Reseeding, erect shrub to 1 m. **Leaves** 3-nate, clasping the stem, tiny, to 0.2 cm long. **Flowers** Many clustered at branch tips, tiny, 0.4 cm long, white to pink. 4 anthers, exserted, with thin appendages. Flowering in summer and autumn. Fruit indehiscent. **Habitat** Occurring in Grassy and Mesic Fynbos, often on rocky outcrops. **Distribution** Eastern distribution from George to Humansdorp.

Kouga Kliphuis, Kouga River 12.05.2021

Erica umbelliflora

umbellatus = furnished with an umbel; referring to the flowers arranged in umbels

A small, dense, sprawling shrublet to 0.4 m tall, resprouting from a woody rootstock after fire. **Leaves** 3-nate, oval, smooth and shiny, usually hairless, small, 0.2 cm long. **Flowers** In clusters of 3, hanging down from side branches, hairless, 0.3 cm long, pink. Anthers included, sometimes with tiny awns, stigma simple and exserted. Flowering from winter to spring. **Habitat** Quite dry conditions at the arid end of Fynbos, fairly common on rocky and exposed slopes and ridges in Arid and Grassy Fynbos. **Distribution** Nearly endemic, this species extends from the Swartberg to Humansdorp.

Rus en Vrede 4X4, Kouga 28.07.2011

Erica viridiflora subsp. *primulina*

viridis = green, referring to the green flowers;
primulinus = greenish primrose-yellow, referring to the flowers

A tough, gnarled, woody shrublet. **Leaves** 3-nate, spreading, with fine hairs on the margins. **Flowers** Hanging down in 2s or 3s from the tip of branches, tubular, to 2 cm long, yellow-green. Anthers included or slightly protruding, sometimes with small residual awns, stigma simple, extending beyond the floral tube. Flowering from autumn to winter. **Habitat** Grows on cliffs at the western end of the Baviaanskloof Mountains. **Distribution** Found mainly on cliffs in the Swartberg. The population in Uitspankloof (Baviaanskloof) is probably near the eastern limit of its range. **Notes** May be seen at Raaskrans in Nuwekloof Pass.

Nuwekloof Pass, Baviaans 05.05.2006

EUPHORBIACEAE

Spurges, euphorbias

The fifth-largest plant family in the world, with 300 genera and over 7,500 species. Trees, shrubs or herbs, usually with a milky latex that can be toxic. *Euphorbia* is the largest genus, with over 1,500 species all around the world. White mangrove (*Excoecaria agallocha*) produces a latex that can cause temporary blindness if it comes into contact with the eyes. Natural rubber is made from latex of the rubber tree (*Hevea brasiliensis*), native to the Amazon rainforest and now grown in plantations around the world, especially in Asia. ± 49 genera in southern Africa; ± 527 native species.

ADENOCLINE

aden = gland, *cline* = couch; pertaining to the glandular disc in the flower

Endemic to southern Africa; 4 species in southern Africa; 2 species in Baviaanskloof.

Adenocline acuta

acuta = pointed; referring to the shape of the leaves

Soft, hairless, dioecious, squashable scrambler to 1m tall. **Leaves** Deltoid, 5 cm long, light green, finely toothed, 5–7-veined from the base, with long petioles. **Flowers** In drooping racemes, whitish, hardly noticeable. Flowering in spring and summer. **Habitat** Uncommon in Baviaanskloof and only under moist Forest. **Distribution** Widespread from Grabouw to Mpumalanga.

Ysrivier, Cambria 22.01.2022

Adenocline pauciflora

pauciflora = few-flowered; pertaining to the few (1 or 2) flowers per inflorescence

Resprouting from woody rootstock, dioecious, weak-stemmed, procumbent, perennial herb to 20 cm. **Leaves** Linear-lanceolate, though sometimes petioled-ovate below, 1.5 cm long, toothed. **Flowers** Tiny, greenish, males in few-flowered racemes in leaf axils, females solitary. Capsule 3-lobed. Flowering in spring and early summer. **Habitat** Loamy slopes in Grassy and Mesic Fynbos. **Distribution** Widespread from Cape Peninsula to KwaZulu-Natal.

Enkeldoring, Baviaans 21.09.2011

CLUTIA

Bliksembos

Named after Outgers Cluyt (Clutius), 1577–1636, a Dutch botanist, horticulturist, beekeeper and pharmacist. He curated the Leiden Botanic Garden.

All members are dioecious. ± 70 species in Africa and the Arabian Peninsula; 8 species in Baviaanskloof.

Clutia affinis   Oumeisieknie

affinis = allied to; closely related to another species in the genus

Erect, densely hairy shrub to 2.5 m. **Leaves** Lanceolate to elliptic, relatively large, mealy-pubescent, paler below, 8 cm long × 2 cm wide. **Flowers** Several, in the axils, 0.3 cm wide, cream-coloured. Flowering in spring. **Habitat** Uncommon in Forest and forest margins in the narrow ravines. **Distribution** Widespread from Villiersdorp to Mpumalanga. Easily seen on the roadside at Poortjies.

Poortjies, Cambria 26.10.2018

Clutia alaternoides

Alaternus = plant genus, *oides* = resembling; resembling a member of genus *Rhamnus* (= *Alaternus*), commonly known as buckthorn; possibly the similarity lies in the alternate leaves.

Dioecious shrub to 50 cm, resprouting from a woody base. **Leaves** Obovate, 2.5 cm long × 1 cm wide, hairless, paler below, with revolute margins. **Flowers** Male flowers in axillary clusters; female flowers single, 0.5 cm wide, on short flower stalks. Flowering mainly in spring. **Habitat** Frequent but not common in Grassy and Mesic Fynbos. **Distribution** Widespread from Namaqualand to Gqeberha (Port Elizabeth).

Doringkloof, Kouga 29.07.2011

Clutia daphnoides Vaalbliksembos

Daphne = plant genus, *oides* = resembling; referring to the superficial
similarity to some species of *Daphne*, native to Eurasia

Non-resprouting, erect, woody shrub to 2 m. **Leaves** Oblanceolate,
3 cm long, slightly paler below, white-mealy on young parts. **Flowers**
Axillary, cream-coloured. Flowering in winter and spring. **Habitat**
Transitional Shrubland and Thicket. **Distribution** Widespread from
Saldanha to Kenton-on-Sea, Eastern Cape, mostly along the coast on
limestone in dune thickets.

Witruggens, Kouga 03.11.2011

Clutia ericoides

Erica = plant genus, *oides* = resembling; alluding to the narrow leaves that are
rolled under and superficially resemble those of some species of *Erica*

Resprouting from a woody base, an erect shrub with noticeable black
buds in the axils, to 60 cm. **Leaves** Narrow-lanceolate, concave,
hairless, 1 cm long × 0.2 cm wide. **Flowers** Axillary, small, cream-
coloured, males in pairs, females single. Flowering in autumn and
winter. **Habitat** Fairly frequent on mountain slopes in Grassy and Mesic
Fynbos. **Distribution** Widespread from Gifberg to Makhanda.

Quaggasvlakte, Kouga 13.02.2016

Clutia polifolia

polio = polish, *folia* = leaves; referring to the hairless leaves that
are slightly glossy, looking polished

Non-resprouting, erect, densely leafy shrub to 1 m. **Leaves**
Linear, 0.5 cm long × 0.2 cm wide, margins rolled under, glossy
grey-green above, paler below. **Flowers** Axillary, cream-coloured,
males 2 or 3, females solitary. Flowering in winter and spring.
Habitat Found on sandy and loamy soils in Mesic and Grassy
Fynbos, and Transitional Shrubland. **Distribution** Widespread
from Namaqualand to Gqeberha (Port Elizabeth).

Blue Hill Escape, Kouga 29.09.2020

Clutia pulchella Lightning bush

pulchellus = beautiful and little; derivation unclear, referring to
the attractive leaves that fade orange-brown

Reseeding shrub or small tree to 3 m. **Leaves** Petiolate, thin, 3.5 cm
long × 2 cm wide, pale green, older leaves fading to orange-brown.
Flowers Axillary, cream-white. Capsules warty, 0.5 cm wide. Flowering in
spring. **Habitat** Common in the understorey of Forest and moist Thicket.
Distribution Widespread from Cederberg to Cape Peninsula to Limpopo.
Other names Umsiphane, umthungwa, uqadi, umkhondo, umfiyo,
umbumbu, umkhwinti (X), lenkosi, umembesa, ungwaleni (Z).

Skuinspadkloof, Baviaans 01.02.2020

DALECHAMPIA

Named after Jacques Daléchamps, 1513–1588, a French doctor and botanist. He translated botanical texts into Latin and published this information.

The genus has extraordinary flowers with large leafy bracts. Many species are pollinated by resin-collecting mason or leafcutter bees; they use the resin in nest construction. ± 121 species in tropical lowland areas of the Americas, Africa, Madagascar and Asia; 3 species in South Africa; 1 species in Baviaanskloof.

Dalechampia capensis   Inzula (Z)

capensis = from the Cape; referring to the geographical origin of the species

Climbing/creeping, hairy perennial to 3 m. **Leaves** Palmate, 5-lobed, toothed, 4 cm long, lime-green, paler below. **Flowers** Enfolded by 2 large, 2 cm-wide, yellowish leafy bracts. **Flowers** Crowded, greeny-yellowish, bisexual. Flowering in early summer. **Habitat** Rocky slopes in Thicket. **Distribution** Very widespread from Baviaanskloof to Tanzania.

Grootrivierpoort 23.01.2022

EUPHORBIA

Spurge, melkbos

Named after Euphorbus, a hero of Linnaeus, who lived in the 1st century and was a personal physician to King Juba of Mauritania. Euphorbus used plants of the genus *Euphorbia* in medicinal treatments.

Flowers are unisexual, small and cup-like, in a false flower (cyathium) surrounded by 5 glands, each with a petal-like appearance. All members exude a toxic white sap when damaged or broken. Several species have been used for medicinal purposes. ± 1,500 species all around the world; 164 species in South Africa; 16 species recorded in Baviaanskloof.

Euphorbia caerulescens   Noors

caerulescens = bluish; alluding to the blue-grey stems

Resprouting from rhizomes, erect succulent, with many stems to 1.5 m. **Stems** 4–6-angled, spined on the ridges, 5 cm wide, blue-grey, becoming grey. **Leaves** Minute, falling off early. **Flowers** In groups of 3, bunched together along the ridges near the branch tips, 0.5 cm wide, bright yellow. Flowering in early summer. **Habitat** Dry north-facing slopes in Thicket. **Distribution** Widespread from Calitzdorp to Gqeberha (Port Elizabeth) and to Peddie, Eastern Cape. Uncommon in Baviaanskloof; known from a single population near Kouga Dam. The dominant plant in Noorsveld between Kariega (Uitenhage) and Jansenville. **Notes** The plant can be used as fodder for stock if cut into sections and allowed to dry.

Steenbokvlakte, Kariega 17.05.2006

Euphorbia esculenta Vingerpol

esculentus = edible; derivation unresolved as it is unclear what is edible on this plant; when dried it can possibly be used as fodder for stock.

Low, thick stemmed, with a rosette of upward-curving, tuberculated, cylindrical, succulent branches to 75 cm. **Leaves** Minute. **Flowers** Numerous at branch tips, 0.5 cm wide. Fruit red to purple, embedded in wooliness. Flowering in spring. **Habitat** Low hills and plains in Succulent Karoo. **Distribution** Fairly narrow distribution in Eastern Cape from Graaff-Reinet to Jansenville and to Baviaanskloof.

Rus en Vrede, Kouga 01.03.2011

Euphorbia grandidens Tree euphorbia

grandidens = with big teeth; pertaining to the rosette of branches at the top of the stems, reminding Adrian Haworth – in 1825 – of big teeth, but this analogy is not very convincing

Tall tree to 15 m, with several, thick, grey-white, spiny stems and a rosette of dark green, spiny, usually 3-angled, ascending branches, each 3 cm wide, at the tips of the stems. **Leaves** Minute, triangular. **Flowers** 0.5 cm wide, yellow-green. Flowering in winter and early spring. **Habitat** Common in dense Thicket on well-drained, steep rocky slopes. **Distribution** Widespread from Baviaanskloof to Eswatini and Mpumalanga. **Other names** Naboom (A), umhlontlo (X), isiphapha, umhlonhlo (Z).

Zandvlakte bergpad, Kouga 13.07.2016

Euphorbia heptagona (= *E. atrispina*) Klipnoors

hepta = seven, *gona* = angled; referring to the 7-angled or 7-ridged branches

Erect, much-branched, dense succulent to 1 m. **Stems** To 3 cm wide, very spiny, with young spines bright red turning grey. **Leaves** Tiny, seldom seen. **Flowers** 0.5 cm wide, green to red. Flowering from winter to early summer. **Habitat** Common on rocky, stony, hot, north-facing slopes on sandstone in Transitional Shrubland and Thicket. **Distribution** Widespread from Montagu to Jansenville.

Nuwekloof Pass, Baviaans 08.10.2016

Euphorbia mammillaris Duikernoors

mamilla = nipple or small projection, *aris* = having; referring to the knobbly ridges on the stems

Similar to *E. heptagona* but smaller in all regards, to 30 cm. **Stems** 4 cm wide, with up to 15 ridges, and spreading, irregular, branches pale grey-green becoming grey. Spines 1 cm long, pinkish when young, becoming white. **Leaves** Seldom present, 0.5 cm long. **Flowers** 0.4 cm wide, yellowish. Flowering in autumn and winter. **Habitat** Uncommon on clay-rich, stony soils in Transitional Shrubland. **Distribution** Widespread from Albertinia and the Little Karoo to Makhanda, though absent from most of the area. Present on the Andrieskraal Road and near Patensie.

Andrieskraal Road, Patensie 17.05.2021

Euphorbia polygona

poly = many, *gona* = angled; referring to the stout stems that have up to 20 ridges

Thickset, gnarly, spiny, succulent shrub to 2 m. **Stems** With 7–20 prominent ridges, grey or blue-grey. **Flowers** 0.7 cm wide, dark purple. Flowering in autumn and winter. **Habitat** Rocky outcrops in exposed situations in various habitats, especially rocky outcrops in Transitional Shrubland and Spekboomveld. **Distribution** Widespread from the Little Karoo to Makhanda. **Notes** Various forms and sizes exist; 1 form from Kouga Wildernes is known as snowflake and is well regarded by succulent collectors. A small form, *E. horrida*, has been included within this species.

Engelandkop, Kouga 09.07.2011

Euphorbia procumbens Melkbol

procumbens = procumbent; the plant grows flat on the ground and the branches are horizontal.

Dwarf, resprouting, thick-rooted, ground succulent, to 5 cm. **Stems** Tapering, 3 cm long × 0.6 cm wide, with conical tubercles. **Leaves** Linear-lanceolate, 0.5 cm long. **Flowers** 0.6 cm wide, green to reddish. Flowering in spring and summer. **Habitat** Uncommon on loamy soils in Grassy Fynbos and Transitional Shrubland. **Distribution** Widespread from Riversdale to Makhanda, Eastern Cape. **Other names** Intsema, inkamamasane, isihlele (X).

Bosrug, Baviaans 02.07.2011

Euphorbia silenifolia   Melkbol

Silene = plant genus, *folia* = leaf; the leaves look like those of some species of *Silene*; the common name refers to the bulb that exudes a milky sap.

Resprouting, dioecious, tuberous-rooted perennial to 5 cm. **Leaves** Basal, in a rosette, narrowly oblanceolate, folded upwards, 12 cm long × 1 cm wide, grey-green. **Flowers** Few on 5 cm-long peduncles, each flower structure 0.6 cm wide; with tiny flowers clustered in the centre; the 5 yellow false petals are glands. Fruit a hairy, 3-chambered capsule. Flowering in winter and spring. **Habitat** Found sporadically in Grassy Fynbos and Transitional Shrubland. **Distribution** Very widespread from Namaqualand to Port Alfred.

Heights, Langkloof 09.07.2019

Euphorbia spartaria   Bekruipmelkbos

spartos = broom, *aria* = having; referring either to the many branches that become thinner higher up, or to the superficial resemblance to Spanish broom (*Spartium junceum*)

Slender-stemmed, much-branched, succulent shrub to 1 m. **Leaves** Narrow, spathulate, seldom present. **Flowers** Greenish yellow. Flowering in spring, summer and autumn. **Habitat** Rocky slopes in Succulent Karoo and Thicket. **Distribution** Very widespread from Namibia to the Eastern Cape and KwaZulu-Natal.

Verlorenrivier, Baviaans 15.02.2016

Euphorbia triangularis   Tree euphorbia, riviernaboom

triangularis = triangular; referring to the branches that are usually 3-angled

Tall, spiny tree to 15 m tall, with a cylindrical trunk. **Stems** Ascending, 3–5-angled, to 1.5 m long × 10 cm wide, green. **Leaves** Small, ovate, falling early. **Flowers** 0.5 cm wide, yellow. Flowering in winter. **Habitat** South-facing slopes in Thicket. **Distribution** Widespread from Patensie to Eswatini. **Other names** Umhlonthlo (X), umhlonhlwane (Z).

Patensie 25.10.2018

JATROPHA

iatros = physician, *trophe* = food; pertaining to the medicinal properties; used to treat tuberculosis and as a remedy for ringworm

175 species, mostly subtropical; 12 species in South Africa; 1 species in Baviaanskloof.

Jatropha capensis Traanbos

capensis = from the Cape; referring to the geographical origin of the species

Resprouting, erect shrublet to 2 m tall, with pale bark and sticky young parts. **Leaves** Arrow-shaped, thinly leathery, 5 cm long × 1 cm wide, fresh green. **Flowers** Several in a branched head 2.5 cm wide, unisexual, each flower 0.5 cm wide, greenish. Capsule 3-chambered, 1.5 cm wide, green. Flowering in summer. **Habitat** Occasional on rocky slopes in Thicket and Transitional Shrubland. **Distribution** Fairly narrow from Baviaanskloof to Peddie.

Zandvlakte bergpad, Kouga 01.01.2018

RICINUS*

Ricinus = Latin name for the castor oil plant; it is also the Latin name for a tick, which the seeds resemble.

1 species worldwide.

*Ricinus communis** Castor oil plant, kasterolieboom

communis = common; referring to the abundance of this plant around the world

Softly woody shrub or small tree to 4 m tall, without milky sap. **Leaves** Palmate, 5–9-lobed, 30 cm wide, with serrated margins. **Flowers** On long stalks, reddish. Fruit a 3-lobed capsule covered with soft bristles. Flowering all year round. **Habitat** Occurring along watercourses and dry stream beds. **Distribution** Originally from northeast Africa and the Mediterranean basin, but now a problematic weed worldwide. **Notes** The whole plant is toxic, especially the seeds, which are lethal if consumed. Castor oil is produced from the seed and has a wide variety of uses. **Other names** Umhlavuthwa, umkakuva (X), mohlafotha (S), umhlakuva (X, Z).

Baviaanskloof River, Beacosnek 26.04.2018

TRAGIA   Noseburns, spurge

Named after Hieronymus Bock (Latinised to Tragus), 1498–1554, a German botanist who wrote *Kreutterbuch* (literally 'plant book') in 1539, which included 700 German herbal plants. It contained the first known attempt at a classification system for plants. The common name pertains to the stinging hairs on the plant.

± 150 species worldwide; 1 species in South Africa, including Baviaanskloof.

Tragia capensis (= *Ctenomeria capensis*)

capensis = from the Cape; referring to the geographical origin of the species

Twining perennial to 2.5 m tall, with stinging hairs. **Leaves** Cordate, petiolate, slightly toothed, soft and flimsy, 6 cm long × 2.5 cm wide, paler below. **Flowers** In thin, axillary spikes, greenish. Flowering in summer and autumn. **Habitat** Forest. **Distribution** Widespread from George to Mpumalanga.

Ysrivier, Cambria 23.01.2022

FABACEAE

Peas, legumes, beans

The third-largest plant family in the world, with 770 genera and ± 19,600 species. Legumes may well be the most diverse plant family in terms of growth forms, from trees to shrubs, and creepers to annuals. Peas are found almost everywhere in a myriad forms. Some species may dominate vegetation communities and they can become invasive weeds. Peas in the Faboideae (a subfamily of the Fabaceae) are easily recognised by their distinctive flower structure, made up of a standard petal or 'sail', 2 wing petals and a keel petal. Almost all pea plants have stipules (small parts at the base of the leaves), which vary from being leaf-like to thorny or inconspicuous. These are a useful aid to identification. The pea family includes economically important agricultural crops such as soya beans, peas, chickpeas, peanuts, carob and liquorice.

In the Cape flora, the Fabaceae are the second-largest family, with ± 43 genera and 770 species; 25 genera and 86 species in Baviaanskloof, plus 6 exotic species.

Pea flower structure

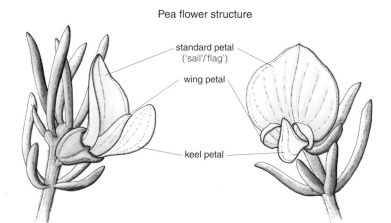

standard petal
('sail'/'flag')

wing petal

keel petal

ACACIA*

akakia = thorn; an appropriate name for the genus until it was reclassified in 2011, not without controversy

The genus now comprises over 900 thornless species mostly confined to Australia. The few indigenous species in Africa have been assigned to *Vachellia* and *Senegalia*. According to the rules of taxonomy, the evidence suggested that *Acacia* should have been retained in Africa, while the Australian acacias should have been renamed *Racosperma*, but the Australian community was reluctant to accept this. *Acacia* flowers are made up of many tiny flowers clustered into globular or cylindrical heads, giving the inflorescence a fuzzy appearance. A study in 2018 estimated that there were 141 introduced species in South Africa, more than double the number of species recorded 40 years prior to this; 5 species in Baviaanskloof.

*Acacia cyclops** Rooikrans, redeye

Cyclops = one-eyed giant of Greek mythology; referring to the seed surrounded by an orange-red aril, resembling the eye of Cyclops

Reseeding, rounded shrub or small tree to 4 m. **Leaves** Known as phyllodes, 6 cm long × 1 cm wide, bright green, with 3–7 longitudinal veins. **Flowers** Few in a cluster, each 1 cm wide, bright yellow and globular; pods flat, undulate or twisted, splitting to reveal several brown to black seeds surrounded by a fleshy bright red aril. Flowering all year round. **Habitat** Native to Australia, invading Grassy Fynbos. **Distribution** Widespread from Namaqualand to Eastern Cape. **Notes** A gall-forming midge, *Dasineura dielsi*, is a biological control agent used to reduce the amount of seeds a tree makes, thereby slowing the spread of this alien invader.

Bergplaas, Combrink's Pass 18.05.2021

*Acacia mearnsii** Black wattle

Named after Edgar Alexander Mearns, 1856–1916, an army surgeon and field naturalist in the USA

Tall, erect tree, growth tips golden-hairy, to 10 m. **Leaves** Bipinnate short leaflets, 0.3 cm long, dark olive-green; leaf stem (rachis) has pairs of raised glands in the junctions of the leaflet pairs on the underside. **Flowers** In large fragrant sprays, each flower 0.5 cm wide, pale yellow or cream, globular. Pods dark brown, finely hairy, constricted between the seeds. Flowering in spring. **Habitat** A wide range of habitats, mostly in drainage lines and on riverbanks. Originally from Australia, the species is a serious alien invader in South Africa. **Distribution** Invading fast in the Kouga Mountains and along the Kouga River. **Notes** A seed-feeding weevil, *Melanterius maculatus*, and gall midge, *Dasineura rubiformis*, are 2 biological control agents in use to slow down seed production.

Kouga Kliphuis, Kareedouw 12.05.2021

AMPHITHALEA

amphi = around, *thallos* = a green stalk; referring to the flowers clustered around the leafy stems

Leaves are simple, usually without stalks or stipules. Flowers are relatively small, pods are 1- or 2-seeded and have a fleshy appendage. 42 species in South Africa; 5 species in Baviaanskloof.

Amphithalea phylicoides

Phylica = plant genus, *oides* = resembling; silvery-white hairs on the leaves and stems are reminiscent of some *Phylica* species.

Tall, resprouting, diffusely much-branched (untidy) silvery shrub to 1.5 m. **Leaves** Oval to lance-shaped, very silky-hairy above, densely villous beneath, 0.5 cm long, silver, margins recurved. **Flowers** Small, 0.6 cm long, white with magenta. Calyx teeth unequal, shorter than the floral tube. Pods golden hairy, 0.8 cm long. Flowering from summer to winter. **Habitat** Occasional on upper, south-facing, rocky ridges in Mesic Fynbos. **Distribution** Widespread from George to Makhanda.

Kouenek, Kouga 16.07.2011

Amphithalea williamsonii

Named after William Crawford Williamson, 1816–1895, an English naturalist and palaeobotanist

Resprouting, erect shrub to 2 m. **Leaves** Ovate, flattish to concave, silky, 1 cm long. **Flowers** 0.7 cm long, pale mauve and purple. Flowering in winter. **Habitat** Uncommon in Mesic and Grassy Fynbos. **Distribution** Narrow from Plettenberg Bay to Kariega (Uitenhage).

Bosrug 12.07.2011

ARGYROLOBIUM

argyros = silver, *lobos* = legume pod; referring to the silver pods of some *Argyrolobium* species

Easily recognised by the trifoliate leaves and paired stipules at the base of the leaf stalk. 3 of the 5 lobed calyces are fused higher up than the other 2, and the pods are held erect. ± 70 species in Africa, also from the Mediterranean across to India; 6 species recorded in Baviaanskloof.

Argyrolobium collinum

collinum = of the hills; referring to the habitat where the plant grows, on open slopes and on ridges in the mountains

Resprouting, silvery, dwarf shrublet with few erect stems to 40 cm. **Leaves** Trifoliate, 1.5 cm long, leaflets narrowly to broadly obovate, densely covered with fine silky hair, margins and new leaves sometimes golden-felted. **Flowers** Solitary, yellow, fading to reddish orange. Pod 3.5 cm long. Flowering in spring and summer. **Habitat** Frequent but never abundant in Grassy Fynbos, Sour Grassland and Transitional Shrubland. **Distribution** Fairly widespread from the Baviaanskloof to Makhanda.

Engelandkop, Kouga 19.04.2012

Argyrolobium incanum

incanus = grey or hoary; pertaining to the
silvery-grey leaves

Resprouting, silvery shrub with few rod-like
stems, sparsely branched, to 2 m. **Leaves**
Trifoliate, 1 cm long, leaflets obcordate, apex
round, densely silver-felted on both surfaces.
Flowers In a raceme with up to 8 flowers, each
1 cm wide, yellow fading to russet. Pod silky-hairy,
to 6 cm long. Flowering from winter to summer.
Habitat Quite common in Grassy Fynbos and
Sour Grassland; especially noticeable in first few
years after fire. **Distribution** Widespread from the
Baviaanskloof to Makhanda.

Zandvlakte, Kouga 07.07.2011

Argyrolobium parviflorum

parvus = small, flora = *flower*

Resprouting, much-branched, dense, rounded
shrub to 1 m. **Leaves** Trifoliate, 0.8 cm long,
distinctive olive-green colour, leaflets broadly
obovate, with veins raised beneath, netted vein
grooves visible on upper surface, rusty-felted on
margins and young shoots. **Flowers** Relatively
small, pale yellow. Flowering mainly in summer
and autumn. **Habitat** Flat plateaus with red soils
in Grassy Fynbos and Grassland. **Distribution**
Endemic to the Baviaanskloof Mountains. The
main population is on Elandsvlakte, with a few
others on Enkeldoorn and other flattish plateaus
in the vicinity. **Status** Rare.

Elandsvlakte, Baviaans 16.02.2016

Argyrolobium pumilum

pumilum = dwarf; pertaining to the low-growing habit

Resprouting, dwarf shrublet with few slender
stems, to 20 cm. **Leaves** Trifoliate, 1 cm long,
discolorous, leaflets ovate with a sharp pointed
tip, margins and underside silver-felted. **Flowers**
Solitary, yellow, fading to russet. Pods straight,
slender, 4 cm long. Flowering mainly in late
summer and autumn, or after rain. **Habitat**
Rocky slopes, also often found on hot north-
facing slopes in Grassy Fynbos. **Distribution**
Widespread from the Baviaanskloof to
Somerset East.

Engelandkop, Kouga 19.04.2012

ASPALATHUS

Cape gorse

aspalathos = a scented bush that grew in Greece; now in *Astragalus*, a large pea genus in the northern hemisphere, the name of which is derived from an ancient Greek word meaning 'ankle bone'. The ankle bones of hoofed animals were precursors of game dice and *Astragalus* seeds rattling inside their pods sounded like rolling dice; this is not the case for *Aspalathus*.

Aspalathus linearis is the source of rooibos tea. Leaves are 1–3-foliate, often linear and spine-tipped, forming clusters on the stems. Calyx lobes are equal, pods with 1–8 seeds. Species are notoriously tricky to identify. The genus has been under-collected in these mountains, with 5 species recently described by Charles Stirton and Muthama Muasya. Some species are short-lived, fast-growing, post-fire ephemerals. For identification, key features are the calyx teeth and hairiness of all parts of the flower, as well as habit and habitat. ± 285 species in South Africa; 23 species recorded in Baviaanskloof.

Aspalathus aciphylla

akis = pointed object, *phyllon* = leaf; referring to the sharp-pointed leaves that are extremely prickly

Reseeding, rigid, erect, stiff-branched, untouchable shrub to 2 m. **Leaves** Obscurely trifoliate, leaflets awl-shaped, with extremely hard and sharp tips. Calyx glabrous, with spine-tipped lobes. **Flowers** Single, nested in the leaves of upper stems, bright yellow, non-hairy. Flowering in spring. **Habitat** Middle to upper rocky slopes in Grassy and Mesic Fynbos. **Distribution** Widespread from Agulhas to Baviaanskloof.

Bosrug, Baviaans 12.10.2016

Aspalathus chortophila

chortus = green herbage, *phila* = loving; derivation unresolved

Resprouting, sprawling, dense shrub to 50 cm, spreading to over 1 m wide. Young branches silky-hairy. **Leaves** Trifoliate, leaflets dark, linear, straight or slightly incurved, spine-tipped and usually glabrous. **Flowers** Solitary, yellow, scattered; keel hairy, back of standard hairy, with a red mucron turning black. Calyx silver-felted, calyx teeth dark red, becoming black, shiny and glabrous. Flowering in spring and early summer. **Habitat** Rocky ridges in Mesic and Grassy Fynbos. **Distribution** Widespread from Baviaanskloof Mountains to KwaZulu-Natal.

Enkeldoring, Baviaans 21.09.2011

Aspalathus collina subsp. *collina*

collina = hill; referring to the plant's habitat on mountain slopes

Reseeding, rigid, tangled, dense, prickly shrub to 1.3 m.
Leaves Trifoliate, leaflets subterete, short, straight, stiff and
pungent. **Flowers** Single or paired at branch tips or on short
shoots, yellow, glabrous, although margin of the standard
with a fringe of hairs. Calyx glabrous, lobes triangular and
spine-tipped. Flowering from midwinter to early summer.
Habitat Common and sometimes abundant on lower to
middle slopes in Grassy Fynbos and Transitional Shrubland.
Distribution Narrow from Swartberg to Baviaanskloof.
Notes The subsp. *lucenta* has a woolly calyx.

Quaggasvlakte, Kouga 13.02.2016

Aspalathus fourcadei

Named in honour of Henri Georges Fourcade, 1866–1948, French-
born South African forester, surveyor and plant collector. He collected
extensively in the southern Cape and discovered and described many
new species; 22 plant species have been named after him.

Reseeding, robust, erect, tall, lanky shrub to 3 m.
Leaves Trifoliate, leaflets relatively long and straight,
dark green, needle-like, terete, spine-tipped, slightly
hairy becoming hairless. **Flowers** Single on short shoots,
scattered, yellow, keel silky and beaked. Calyx woolly,
lobes hairy and awl-shaped. Flowering from midwinter
to early spring. **Habitat** Uncommon and localised on very steep
south-facing dip slopes with rock slabs, in Grassy and Arid Fynbos.
Distribution Near endemic from Tsitsikamma Mountains to Kouga
and Baviaanskloof mountains. Seen only near base of Scholtzberg
Peak in Baviaanskloof Mountains. **Status** Rare.

Scholtzberg, Baviaans 24.07.2011

Aspalathus glabrescens

glabresco = becoming glabrous, smooth, free from hair; referring to the
tendency for parts of this plant to become less hairy as they age

Reseeding, robust, erect, tall, much-branched
shrub to 2 m. **Leaves** Trifoliate, grey-green,
leaflets needle-like, fairly rigid, silvery-silky,
usually non-hairy towards the tip. **Flowers** 1 or
2 on short shoots, scattered, pale yellow, tinged
orange. Calyx finely silky, lobes awl-shaped.
Flowering in spring. **Habitat** Uncommon on rocky,
arid, north-facing slopes in Transitional Shrubland.
Distribution Fairly narrow from Outeniqua
Mountains to Baviaanskloof, representing an
eastward extension of its range. **Status** Endangered.

Smitskraal, Kouga 09.03.2012

Aspalathus hirta subsp. *stellaris*

hirta = hairy, *stellaris* = stellate, starry; referring to the hairs on some parts of the plant; the yellow, pointy calyx lobes are reminiscent of a star.

Reseeding, untidily and sparsely branched, erect shrub to 2 m. **Leaves** Trifoliate, leaflets relatively long, terete, needle-like, non-hairy, lime-green with translucent spined tips. **Flowers** Solitary on short shoots, scattered, bright yellow, non-hairy but back of standard sparsely hairy. Calyx shortly hairy, lobes awl-shaped and spine-tipped. Flowering in spring and summer. **Habitat** Arid and Grassy Fynbos on rocky, north-facing slopes. **Distribution** Widespread from Outeniqua to Kouga Mountains.

Quaggasvlakte, Kouga 13.02.2016

Aspalathus hystrix

hystrix = hedgehog; the clusters of very spiny leaves are reminiscent of a hedgehog.

Reseeding, erect, robust, densely branched shrub to 2 m. **Leaves** Trifoliate, needle-like, often black-spine-tipped, silvery grey-hairy. **Flowers** Solitary on short shoots, scattered, bright yellow, non-hairy but back of standard sparsely hairy. Calyx and lobes white-hairy. Flowering in spring. **Habitat** Transitional Shrubland, and Grassy and Arid Fynbos. **Distribution** Widespread from Witteberg to western end of Baviaanskloof.

Beacosnek, Kouga 22.09.2010

Aspalathus kougaensis

Kouga = the mountain range, *ensis* = denoting origin; this species is common in the Kouga Mountains. It also occurs on other mountain ranges in the area.

Reseeding or resprouting, sprawling or erect, silvery grey-leaved shrub to 1 m. **Leaves** Not spine-tipped, grey-hairy, relatively short. **Flowers** Pale lemon-yellow, keel hairy, back of standard hairy. Calyx densely white-hairy, calyx teeth short, blunt and hairy. Flowering in spring, summer and autumn. **Habitat** Mostly in Grassy Fynbos, but also in all other Fynbos habitats. **Distribution** Fairly narrow from Swartberg, throughout the Baviaanskloof, to Humansdorp.

Engelandkop, Kouga 30.10.2011

Aspalathus modesta

modestus = modest, unassuming; referring to its superficial similarity
to *A. karooensis*, but a modest or smaller version thereof

Reseeding, erect, silvery shrub to 60 cm, with a vase-shaped habit. **Leaves** Linear,
0.6 cm long, densely silvery-woolly. **Flowers** Small, pale yellow, very hairy on all
parts. Flowering in late summer. **Habitat** Rare on shale bands in Grassy Fynbos.
Distribution Known from a single population in Baviaanskloof Mountains, on a
saddle to the west of Bosrug. **Notes** Recognised as unusual by Douglas Euston-
Brown in 2012; described by Stirton and Muasya in 2016. **Status** Vulnerable.

Bosrug, Baviaans
12.10.2016

Aspalathus nivea

niveus = snowy, snow-white; leaves and young branches are
covered with white hairs, giving the plant a silvery grey hue.

Resprouting, lanky, often multi-stemmed shrub, to 3 m. **Leaves**
1.5 cm long × 0.2 cm wide, terete, covered with silvery hairs.
Flowers Single to a few on slender peduncles, pale yellow,
wings and keel hairy; calyx silvery-silky, lobes awl-shaped.
Flowering from spring to autumn. **Habitat** Uncommon on
rocky, exposed slopes in Transitional Shrubland. **Distribution**
Nearly endemic from Uniondale to Kariega (Uitenhage).

Doodsklip, Baviaans 29.12.2017

Aspalathus rubens

rubens = reddish; referring to the tendency for the flowers to turn red after maturing

Resprouting, sprawling shrublet to 25 cm tall, with spreading horizontal
branches over 1 m wide. **Leaves** Short, subterete, straight or slightly incurved,
glabrous or silvery-silky. **Flowers** Solitary and scattered, 0.7 cm long, yellow
turning orange and/or red; calyx shortly-hairy. Flowering all year round.
Habitat Common on upper, rocky slopes in Mesic and Subalpine Fynbos.
Distribution Widespread from Karoo Mountains to Gqeberha (Port Elizabeth).

Bakenkop, Baviaans 23.12.2011

Aspalathus setacea

setacea = bristle-like or bristly; referring to the spine-tipped leaves

Reseeding, erect, small tree to 2.5 m. **Leaves** 1.3 cm
long × 0.3 cm wide, subterete, needle-like, spine-tipped,
glabrous or silver-hairy. **Flowers** Single, usually clustered
near branch tips, yellow fading brownish, keel and wings
silky-hairy; calyx shortly-hairy, lobes narrow-triangular,
spine-tipped. Flowering in winter, spring and summer.
Habitat Steep and rocky lower to middle slopes in Grassy
Fynbos and Transitional Shrubland. **Distribution** Fairly
widespread from Knysna to Makhanda.

Zandvlakte, Baviaans 29.04.2018

Aspalathus sp. nov. (*chlorina* ms.)

chlorinus = yellowish green; referring to the distinctive lime or pale yellow-green leaves

Reseeding, erect, slender shrub to 1 m. **Leaves** Trifoliate, 2 cm long, bright lemon-green, thread-like leaflets, soft, hairy. Calyx silky-hairy and yellowish, with long thin lobes. **Flowers** Partly hidden in the leaves, 1 cm long, pale yellow, keel slightly hairy. Pods silky-hairy. Flowering in summer. **Habitat** Localised on middle slopes in Grassy Fynbos. **Distribution** Known from a single population near Guerna at the eastern end of the Kouga Mountains. Recognised and recorded as new to the scientific world in 2012. This species has been formally described, but is yet to be published. **Status** Endangered.

Guerna, Kouga 09.02.2012

Aspalathus sp. nov. (*dupreezii* ms.)

Named after Brian Du Preez, a South African botanist

Tall tree to 4 m, with corky bark, allowing it to survive and resprout after fire. **Leaves** Trifoliate, leaflets needle-like, slightly curved and silvery grey. **Flowers** Single, scattered, yellow, keel and standard hairy, especially below; calyx densely white-woolly, lobes green, slightly hairy and awl-shaped. Flowering in spring and summer. **Habitat** Grassy Fynbos and Sour Grassland. **Distribution** Narrow from Baviaanskloof to Patensie. Easily seen near Combrink's Pass. **Notes** A recently described species, soon to be published.

Combrink's Pass, Baviaans 26.10.2018

Aspalathus sp. nov. (*vulpicaudata* ms.)

Vulpes = genus name for foxes, *caudatus* = ending with a tail-like appendage; referring to the distinctive 'bushy' packing and overall hairiness of the leaves, so that the branch tips look like fox tails

Reseeding, short-lived, erect shrub to 2 m. **Leaves** Trifoliate, densely packed and clustered, stems hidden by the leaves. Leaflets 1.5 cm long × 0.4 cm wide, soft, linear, mostly out-curved, covered in white-villous hairs. **Flowers** 1 cm long, pale to greenish yellow, almost hidden in the foliage; calyx silky, soft, lobes narrowly triangular to almost linear, pink towards the base. Flowering in late summer. **Habitat** Localised on south-facing slopes in Grassy and Mesic Fynbos. **Distribution** Known from only 2 locations in the southeastern part of the Kouga Mountains, near Takkieskraal and Guernakop. **Notes** First collected for scientific purposes in 2012; recently described and soon to be published. **Status** Endangered.

Nooitgedacht, Kouga 13.02.2012

Aspalathus subtingens

sub = somewhat, a little, *tingens* = having the ability to tinge; referring to the tendency for the flowers and leaves to turn orange or reddish as they age

Resprouting, erect, neatly branched shrub to 1.2 m. **Leaves** 0.4 cm long × to 0.1 cm wide, dark green turning reddish brown in dry conditions, terete or sausage-shaped, usually slightly incurved, tip rounded, mostly hairless. **Flowers** Usually single, sometimes paired, yellow fading orange-red, hairless; calyx pale green, hairless. Flowering in spring, summer and autumn. **Habitat** Hot, exposed, rocky lower slopes in Transitional Shrubland, and Arid and Grassy Fynbos. **Distribution** Widespread from Witteberg to Somerset East.

Kouga Mountains 27.08.2010 ü ü

Aspalathus usnoides

Usnea = lichen genus, *oides* = resembling; *Usnea* is a grey lichen, commonly known as old man's beard, that hangs from branches of old trees; referring to the drooping habit and grey colour of this species

Reseeding, slender, small willowy tree to 5 m, with pendulous branches. **Leaves** 1 cm long, silver-green-hairy, linear, incurved, soft. **Flowers** Solitary, scattered, hidden in the leaves, silky-hairy except for the glabrous, recurved wing petals; calyx white-villous, lobes narrowly triangular to linear, green. Flowering from winter to summer. **Habitat** Common but localised in Grassy Fynbos on lower, steep, south-facing slopes between 300 and 400 m above sea level. **Distribution** Narrow endemic known from a few populations above the Kouga River, north of Kareedouw. **Notes** Described by Stirton and Muasya in 2016. **Status** Endangered.

Nooitgedacht, Kouga 28.09.2011

CALOBOTA

kalos = beautiful, *bota* = wineskin bag; referring to the pods of some species, which look like a wineskin

16 species in South Africa; Namibia and North Africa; 1 species in Baviaanskloof.

Calobota pungens

pungens = pungent, terminating in a hard, sharp point; referring to the spine-tipped branches

Tough, spiny, erect, grey-branched shrub to 2 m. **Leaves** 1–3-foliate, 0.5 cm long, flat and dropping early so the plant may appear leafless. **Flowers** 1.2 cm long, bright yellow, petioles short, grouped in spike-like racemes, wing petals shorter than the keel. Flowering in spring and early summer. **Habitat** Guarriveld in Succulent Karoo. **Distribution** Fairly widespread from the Witteberg Mountains southwest of Lesotho, to the western end of Baviaanskloof in the foothills of Antoniesberg.

Elandspoort, Antoniesberg 08.10.2016

CALPURNIA

Slender trees or small shrubs with pinnate leaves and yellow flowers in racemes or panicles

9 species mostly in southern Africa; 1 species in Baviaanskloof.

Calpurnia aurea subsp. *aurea*   Natal laburnum

aurea = yellow; referring to the golden yellow flowers

Shrub or slender tree to 5 m. **Leaves** Pinnate, 10 cm long. Leaflets 8–30, obovate, 2 cm long × 1.5 cm wide. **Flowers** Inflorescence pendulous, 8–40-flowered, petals yellow, 1 cm long. Calyx green, 1 cm long, sparsely hairy to glabrous, lobes straight, 0.4 cm long. Flowering in spring and summer. **Habitat** Thicket and Forest margins. **Distribution** Very widespread from Baviaanskloof to Ethiopia. The population at the Grootrivierpoort represents its western limit. **Notes** 2 subspecies, subsp. *aurea* and subsp. *indica*, the latter only occurring only in southern India. **Other names** Geelkeur (A), umdloli, umbethe (X).

Komdomo, Grootrivierpoort 03.02.2020

CROTALARIA   Geelkeurtjie, rattle pod

krotalos = rattle, *arius* = provided with; pertaining to the dry seed pods that rattle when shaken

± 600 species worldwide; 56 species in South Africa; 1 species in Baviaanskloof.

Crotalaria capensis   Eared rattle pod

capensis = of the Cape; referring to the geographical origin of the species

Erect shrub to 3 m. **Leaves** Trifoliate, leaflets obovate, 5 cm long × 2 cm wide, sometimes hairy, petiole usually shorter than the leaves, with 2 leaf-like stipules at the base of the petiole. **Flowers** In hanging racemes, pea-like, 2 cm wide, bright yellow. Pod 5 cm long. Flowering all year round except late summer. **Habitat** Sheltered areas at the margins of Forest and Thicket. **Distribution** Very widespread from Knysna to tropical Africa. **Other names** Ihlolo getyane (X), ubukheshezane (Z).

Grootrivierpoort 17.05.2021

CYCLOPIA

Honeybush

kyklops = one-eyed monster; derivation unresolved; the common name refers to its use as a palatable tea.

23 species in Western and southern Cape; 2 species in Baviaanskloof.

Cyclopia intermedia Kougabergtee, bergtee

intermedia = intermediate; referring to features that are in-between those of other species

Resprouting, robust, erect shrub, with numerous yellowish stems from the base to 2 m. **Leaves** Trifoliate, oblanceolate, leaflets to 3.5 cm long × 0.7 cm wide, yellowy green, margins slightly recurved. **Flowers** 1.5 cm long, bright yellow, calyx leathery and flat-bottomed. Flowering in spring. **Habitat** Middle to upper south-facing slopes on deeper, loamy soils in Grassy and Mesic Fynbos. **Distribution** Widespread from Witteberg to Van Staden's Mountains. **Notes** Harvested from natural populations on the Kouga Mountains; used to make honeybush tea, which is said to have medicinal and therapeutic properties.

Bosrug, Baviaans 12.10.2016

DIPOGON

di = two, *pogon* = beard; referring to the style that has 2 'beards'

1 species in Baviaanskloof.

Dipogon lignosus Cape sweet pea, bosklimop

lignosus = woody; referring to the plant base that becomes gnarled and woody over time

Resprouting, herbaceous climber to 5 m tall. **Leaves** Relatively large, soft, flat and rhombic, greyish below, green on top. **Flowers** With peduncles on racemes, 1.5 cm long, magenta or pink. Flowering all year round. **Habitat** Forest or forest margins, usually along streams. **Distribution** Very widespread from Saldanha to the Wild Coast, Eastern Cape. **Notes** Attractive, easy to grow from seed, and rewarding to cultivate in the garden or in pots.

Oubos, Krakeel 06.10.2011

INDIGOFERA

indicus = blue dye, *fero* = bearing; these plants sometimes have purplish flowers.

Genus being revised by Brian Schrire and Brian Du Preez at time of publication. ± 730 species worldwide; 12 species in Baviaanskloof.

Indigofera capillaris

capillaris = hair- or thread-like; referring to the thin calyx lobes

Reseeding, sprawling or decumbent shrublet to 15 cm. **Leaves** Pinnate, 5–13-foliate. Peduncles longer than the leaves. **Flowers** In racemes, pink, magenta or copper, with back of standard hairy at the tip. Pods hairless, deflexed. Flowering in spring and summer. **Habitat** Occurring mainly in Arid Fynbos, but also in Grassy Fynbos. **Distribution** Widespread from Nieuwoudtville to the western end of Baviaanskloof.

Vaalwater, Antoniesberg 14.09.2010

Indigofera declinata

declinatus = bent, curved; probably referring to the tips of the calyx lobes that curve inwards, or to the reflexed stipules

Reseeding, prostrate to decumbent shrublet to 20 cm. **Leaves** Pinnate, thinly-hairy, 5–13-foliate. Peduncles longer than the leaves. **Flowers** In racemes, bright pink, back of standard hairy. Pods hairless, spreading to suberect. Flowering in spring and summer. **Habitat** Occasional on upper and steep south-facing slopes in Mesic and Subalpine Fynbos. **Distribution** Widespread from the Cederberg to Baviaanskloof.

Scholtzberg, Baviaans 29.10.2011

Indigofera denudata

denudatus = denuded, stripped, having the leaves or hairs worn off; probably referring to the hairless nature of the plant, or possibly to the sparsely leafy, twiggy habit

Resprouting, much-branched, twiggy shrub to 1.5 m. **Leaves** 1–3-foliate, leaflets linear-oblong to obovate, hairless, variable in size, usually to 1 cm long, dull greyish green. **Flowers** In racemes on a very short stalk, mauve. Pods hairy when young. Flowering all year round. **Habitat** Common and frequently encountered on lower to middle slopes, mainly in Transitional Shrubland and Grassy Fynbos. **Distribution** Narrow distribution from Baviaanskloof to Makhanda. **Notes** Closely related and similar to *I. nigromontana*, but that species does not occur in Baviaanskloof.

Geelhoutboskloof, Baviaans 20.04.2012

Indigofera disticha

distichus = arranged in two opposite rows; probably referring to the pinnate leaves

Decumbent or sprawling shrub to 80 cm. **Leaves** Pinnate, 5–7-foliate, leaflets elliptic-oblong, 1.5 cm long, dull green. **Flowers** In racemes on peduncles, as long as leaves, brick-red, back of standard hairy. Pods pale, sparsely hairy. Flowering from spring to autumn. **Habitat** Occasional in Transitional Shrubland and Thicket. **Distribution** Widespread from eastern end of Baviaanskloof to Cradock and Port Alfred.

Grasnek, Kouga 08.03.2012

Indigofera flabellata

flabellatus = flabellate, fan-shaped; referring to the
subdigitate, 5-foliate leaves that are arranged like a fan

Reseeding, erect, floriferous shrub to 1.5 m. **Leaves**
3–7-foliate, subdigitate, leaflets linear-oblanceolate,
midrib prominent below, margins revolute. **Flowers**
Few in racemes on relatively short peduncles, pink or
purple. Calyx very hairy. Pods spreading, grey-hairy.
Flowering from late summer to spring. **Habitat** Middle to
upper slopes in Grassy and Mesic Fynbos. **Distribution**
Widespread from the Langeberg to Tsitsikamma and
Kouga mountains.

Engelandkop, Kouga 18.07.2011

Indigofera sp. nov. (*vlokii* ms.)

Named after the Oudtshoorn-based botanist Jan Vlok, who first recognised the species as unrecorded

Reseeding, sprawling shrublet to 30 cm. **Leaves** 3–5-foliate, hairy, petioles short and pinkish red, leaflet
tips curved down and with a black mucron. **Flowers** Bright pink with hairy pod and calyx. Flowering in
summer and autumn. **Habitat** Mesic Fynbos. **Distribution** Localised endemic known only from upper peaks
of the Kouga Mountains. **Notes** A recently described species, soon to be published as *Indigofera vlokii*.

Quaggasvlakte, Kouga 13.02.2016

Indigofera sulcata

sulcatus = furrowed or grooved;
referring to the furrowed branches

Reseeding, erect, rigid shrub to 80 cm. **Leaves**
3–5-foliate, subdigitate, leaflets linear, midrib prominent
beneath, margins revolute. **Flowers** In racemes on short
peduncles, mauve, hairy along midrib of back of standard
petal. Pods glabrous. Flowering mainly in winter. **Habitat**
All Fynbos, but mainly Mesic Fynbos. **Distribution**
Widespread from Swartberg to Gqeberha (Port Elizabeth).

Peaks above Coleskeplaas, Baviaans 06.03.2012

LESSERTIA
Balloon pea, blaas-ertjie

Named after Jules Paul Benjamin de Lessert, 1773–1847, a French banker and an ardent amateur botanist. He collected plants for research purposes; those collections were later to form part of a herbarium.

± 55 species in Africa; 1 species in Baviaanskloof.

Lessertia frutescens Kankerbos, belletjie, gansies, eendjies

frutescens = becoming shrubby

Reseeding, erect shrub to 1 m. **Leaves** Pinnate with 6–10 pairs of leaflets, oblong to heart-shaped, sometimes thinly hairy above. **Flowers** In racemes, each to 3.5 cm long, red, wing petals very small and hidden. Pods large, swollen and papery. Flowering from midwinter to midsummer. **Habitat** Open and disturbed areas in all habitats, though seldom seen in Fynbos; often on roadsides. **Distribution** Widespread throughout southern and central South Africa. **Notes** The leaves are used to treat flu, stomach problems and cancer. **Other name** Umnwele (X, Z).

Bosrug, Baviaans 13.09.2006

LOTONONIS

Lotus = plant genus, *Ononis* = plant genus; presumably *Lotononis* shares features of both Fabaceae genera.

Lotononis almost always has a solitary stipule. For identification, another feature to look at is the calyx – the 4 upper lobes are fused and the lower 1 is narrower. ± 150 species in Africa, Eurasia and Pakistan; ± 144 species in South Africa; 4 species in Baviaanskloof.

Lotononis azurea

azurea = pure deep blue; referring to the flower colour

Resprouting, prostrate shrublet to 10 cm. **Leaves** Trifoliate, paired, with stipules unequal. **Flowers** Single on slender peduncles, blue with yellow centre. Flowering in spring. **Habitat** Uncommon in Grassy Fynbos and grassland. **Distribution** Fairly narrow from Uniondale to Gqeberha (Port Elizabeth).

Bergplaas, Baviaans 28.08.2010

Lotononis pungens

pungens = pungent, terminating in a hard and sharp point; the derivation is unclear as the plant does not appear to have spines or sharp points.

Resprouting, prostrate or spreading, dwarf shrublet to 25 cm. **Leaves** Trifoliate, linear to obovate, 2 cm long including petiole, grey-green-hairy. **Flowers** Few, usually at branch tips, single on short peduncles, 1 cm long, yellow. Pods 1 cm long, hairy, green turning pale brown. Flowering from spring to autumn. **Habitat** Hot exposed areas in a wide variety of vegetation types. **Distribution** Widespread from Worcester to Gqeberha (Port Elizabeth) and Tarkastad.

Grasnek, Kouga 08.03.2012

MELOLOBIUM

melos = a joint, *lobos* = lobe or pod; pods are constricted (joint-like) between the seeds.

± 15 species in southern Africa; 3 species in Baviaanskloof.

Melolobium candicans Stroopbos

candicans = becoming pure white; referring to the stems
that are covered with short, silvery white hairs

Reseeding, rigid, thorny shrublet to 50 cm. Thorns yellow-tipped.
Leaves Trifoliate, to 0.5 cm long, leaflets obovate, appearing
hairless but covered with sessile glands. **Flowers** Few near the
tips of the thorns, 0.8 cm long, yellow fading orange, calyx lobes

Andrieskraal Road, Patensie 14.05.2006

acute. Pods straight and hairy. Flowering all year round except late
summer. **Habitat** Uncommon in disturbed areas on clay soils in Transitional Shrubland and Succulent
Karoo. **Distribution** Very widespread from Namaqualand, through the arid interior, to Cradock and
Gqeberha (Port Elizabeth). **Notes** Another species, *M. adenodes*, is also found in Baviaanskloof, but it is
sticky and covered with stalked glands. **Other names** Umakope (X), sehlabane (S).

PODALYRIA

poda = foot, *Lyria* = gastropod genus; the surface of *Lyria* shells are covered in nodules or bumps and the
roots of this legume are covered with nitrogen-fixing nodules. Additionally, Podaleirios was the son of the
Greek god of healing, Asklepios. Derivation unresolved.

Silvery silky-haired shrubs with simple leaves and a concave calyx base. 19 species in South
Africa; 3 species in Baviaanskloof.

Podalyria burchellii

Named after William John Burchell, 1781–1863,
English explorer who collected thousands of
specimens, firstly in South Africa, later in Brazil.
He collected over 50,000 plant specimens in South
Africa, which he donated to Kew Gardens, each with
meticulous and detailed notes on habit and habitat.
He described his journey in *Travels in the Interior of
Southern Africa*, a 2-volume work appearing in
1822 and 1824.

Resprouting, erect, willowy shrub with few
branches to 2 m. **Leaves** Elliptic-suborbicular,
thinly hairy above to shaggy-hairy below, 2 cm
long. **Flowers** To 2 cm long, pink or magenta
with a white centre. Calyx long-lobed and woolly-
hairy. Bracts ovate, mucronate. Flowering
almost all year round. **Habitat** Occurring
sporadically in rocky areas in Grassy and
Mesic Fynbos. **Distribution** Widespread from
Swellendam to Makhanda.

Kouenek, Kouga 18.04.2012

Podalyria myrtillifolia

Myrtus = plant genus, *folia* = leaf; the leaves are
reminiscent of those of some species of myrtle.

Resprouting shrub, with several stems from the
base, usually only to 1 m. **Leaves** Obovate, 1.5 cm
long, silky-hairy above and below. **Flowers** 1.5 cm
long, pink and white. Bracts oblanceolate to
obovate. Flowering in autumn, winter and spring.
Habitat Sporadic in loamy soils in Grassy and Mesic
Fynbos. **Distribution** Widespread from Tulbagh and
the Cape Peninsula to Makhanda.

Nooitgedacht, Kouga 21.05.2010

PSORALEA (= *Otholobium*)   Dotty pea, fonteinbos, bloukeur

psoraleos = warty; pertaining to the sunken or warty glands that cover leaves, stems and calyces

Recent molecular studies (2022) revealed that *Psoralea* and *Otholobium* should be combined into
a single genus. *Psoralea*, as the older name, takes precedence and all the *Otholobium* species were
transferred into *Psoralea*. Species of the enlarged genus can be recognised by their single-seeded
fruits and gland-dotted surfaces. ± 125 species in southern Africa; 16 species in Baviaanskloof.

Psoralea acuminata (= *Otholobium acuminatum*)

acuminata = tapering to a point; referring to the glandular calyx in which the
lower sepal is longer than the flower and curved, tapering to a point

Reseeding, much-branched shrub to 2 m. **Leaves** 3-foliate, digitate,
glands prominent, leaflets to 1 cm long, cuneate-emarginate. **Flowers**
Numerous in dense, rounded heads, 2.5 cm wide at the branch tips,
each flower 0.8 cm long, mauve, purple. Calyx glandular, margins
ciliate. Flowering from spring to autumn. **Habitat** South-facing slopes
in Mesic and Grassy Fynbos, especially after fire. **Distribution** Nearly
endemic from Kammanassie to Van Staden's Mountains.

lower sepal

Koud Nek's Ruggens, Kouga 07.11.2011

Psoralea axillaris

axillaris = axillary; referring to the flowers
that are borne in the axils

Reseeding shrub to 3 m. **Leaves** Trifoliate,
flattish and warty, 2.5 cm long, grey-
green. **Flowers** 1.5 cm long, deep purple
with violet, wing petals white. Flowering in
summer. **Habitat** Localised, sometimes
forming dense stands on stream sides
and seepages, in Forest and Mesic
Fynbos. **Distribution** Widespread from
Barrydale to Van Staden's Mountains.

De Hoek, Joubertina 04.01.2010

Psoralea candicans (= *Otholobium candicans*)

candicans = becoming pure white; referring to the
white flowers of some specimens

Resprouting, erect shrub to 1.5 m. **Leaves** Trifoliate,
digitate, leaflets oblanceolate, to 1 cm long, subglabrous,
stipules with hairy margins. **Flowers** 0.7 cm wide, white,
mauve or pink. Calyx glabrous to densely white-silky.
Flowering in spring. **Habitat** Occasional on shale bands
in Transitional Shrubland or Grassy Fynbos. **Distribution**
Widespread from Tulbagh to Makhanda.

Duiwekloof, Baviaans 28.10.2018

Psoralea imminens

imminens = projecting, overhanging; referring to the tall growth habit with overhanging branches

Reseeding, tall, willowy
tree to 5 m. **Leaves**
7–9-foliate, 6 cm long,
with raised glands.
Flowers Many on
seasonal shoots, axillary
with 1- 5 per axil,
1.5 cm long, white to
pale mauve. Cupulum
3-lobed, two-thirds up
pedicel. Flowering in
spring and early summer. **Habitat** Seepages and
stream sides in Mesic Fynbos. **Distribution** Fairly
widespread from Calitzdorp to Baviaanskloof.

Nooitgedacht, Kouga 05.11.2011

Psoralea kougaensis

kouga = Kouga Mountains, *ensis* = denoting
origin; referring to where this recently
described species was first recorded in 2008

Reseeding, erect, neat shrub to 1.5 m.
Leaves 5–7-foliate, closely packed,
warty. Stipules persistent becoming
woody. **Flowers** 1 cm long, mauve, with
distinct purple nectar guide. Flowering
in spring and early summer. **Habitat**
Uncommon and localised on rocky
south-facing side of high peaks in Mesic
Fynbos. **Distribution** Endemic to the
Kouga and Baviaanskloof mountains;
known from Potjiesberg (near Uniondale),
Sipoonkop and Bosrug. **Status** Rare.

Peak west of Bosrug, Baviaans 12.10.2016

Psoralea oligophylla

oligo = few, *phyllon* = leaf; referring to the sparsely leaved habit

Reseeding, willowy shrub or small tree to 4 m. **Leaves**
Simple or trifoliate, relatively wide, 3 cm long, with
inconspicuous gland dots. **Flowers** 1.5 cm long, off-white,
standard with violet nectar guides. Flowering in spring,
summer and autumn. **Habitat** Uncommon in damp areas
on stream banks or seepages. **Distribution** Widespread
from George to Barkly East.

Zuurberg 14.02.2021

Psoralea picta (= *Otholobium pictum*)

picta = coloured, painted; some flowers have shades of blue or
turquoise in the centre of the cream-white petals.

Resprouting, multi-stemmed shrub to 2 m. **Leaves** Trifoliate, leaflets linear-
oblanceolate with tips mucronate and recurved, 1.5 cm long. **Flowers**
Numerous, in elongated heads at the branch tips, each flower 0.7 cm
long, white with purplish centre. Calyx glandular and black-hairy, with
lowest sepal larger. Flowering in spring. **Habitat** Rare with small, localised
populations on south-facing slopes in loamy soils in Grassy Fynbos.
Distribution Narrow endemic in Kouga and Baviaanskloof mountains.

Nooitgedacht, Kouga 28.09.2011

Psoralea polyphylla (= *Otholobium polyphyllum*)

poly = many, *phyllon* = leaf; alluding to the densely leafy habit

Resprouting, much-branched, erect, closely leafy shrub to 1 m. **Leaves**
Digitately trifoliate, 1 cm long, margins ciliate, leaflets oblong, stipules
falling early. **Flowers** 0.6 cm long, white. Calyx black-hairy, growing after
flowering, lowest sepal larger. Flowering in winter and spring. **Habitat**
Infrequent and localised on steep south-facing slopes in Grassy and
Mesic Fynbos. **Distribution** Narrow endemic on Baviaanskloof and
Cockscomb mountains.

Scholtzberg, Baviaans 29.10.2011

Psoralea prodiens (= *Otholobium prodiens*)

prodiens = coming forth, appearing; alluding to its relatively tall height,
sticking out above the surrounding vegetation; perhaps also referring
to its habit of sprouting after fire

Resprouting, semi-erect, few-stemmed shrub to 2.5 m. **Leaves** Trifoliate,
leaflets 1.5 cm long, obovate and tip hooked, with net veins distinct, and
stipules falling early. **Flowers** To 1 cm long, white with purple nectar guides,
calyx thinly black-hairy. Flowering in winter and spring. **Habitat** Loamy soils
in Grassy Fynbos and Transitional Shrubland. **Distribution** Widespread from
Langeberg to Kouga mountains.

Prince Alfred's Pass, Avontuur
19.06.2019

Psoralea sericea (= *Otholobium sericeum*)

sericea = silky-hairy; referring to the silvery-hairy calyx, stipules and undersides of the leaves

Sprawling shrub to 2 m. **Leaves** Trifoliate, leaflets 4 cm long, elliptic, hairy below and glabrous above, margins rolled under. **Flowers** Crowded in heads 2 cm wide, long peduncles, each flower to 1 cm long, deep blue-purple, calyx black-silky or silvery. Flowering in spring and summer. **Habitat** Uncommon in Transitional Shrubland and Grassy Fynbos. **Distribution** Widespread from Albertinia to Kareedouw.

Kouga Kliphuis, Kouga Valley 02.05.2015

Psoralea stachyera (= *Otholobium stachyerum*)

stachyera = spike-like; referring to the spike-like inflorescence

Reseeding, erect, tree-like shrub to 3 m. **Leaves** Trifoliate, 1 cm long, leaflets obovate-cuneate, midribs and stipules hairy. **Flowers** Numerous in compact, terminal, elongated racemes, each to 1 cm long, violet to mauve. Calyx white- or black-hairy, glandular. Flowering from spring to the end of summer. **Habitat** Localised in seepages, stream sides and forest margins in Grassy Fynbos. **Distribution** Widespread from Caledon to the Amathole Mountains and the Wild Coast.

Nooitgedacht, Kouga 19.12.2011

Psoralea trullata

trullata = trowel; referring to the trowel-shaped curve of the lower sepal as well as the trullate standard petal

Reseeding, tangled, cryptic, creeping, flimsy shrublet to 30 cm tall × more than 2 m wide. **Leaves** Trifoliate, linear-oblong, 4 cm long. Stipules fused for most of length, appressed. **Flowers** Axillary, single, 1 cm long, deep purple with a white nectar guide. Flowering in summer and autumn. **Habitat** Uncommon on upper slopes in Mesic Fynbos. **Distribution** Widespread from Langeberg to Kariega (Uitenhage). **Status** Rare.

East of Kougakop 16.04.2012

RAFNIA

Named after Karl Gottlob Rafn, 1769–1808, a Danish botanist and civil servant who compiled a flora of Denmark

22 species in southern Africa, mostly in Western Cape; 1 species in Baviaanskloof.

Rafnia elliptica

elliptica = elliptical; referring to the shape of the leaves

Resprouting, erect, many-stemmed, angular-branched shrub to 75 cm. **Leaves** Broadly obovate, of variable shape and size, to 5 cm long, becoming narrower and tapering higher up the stems. **Flowers** Single in the axils, to 2 cm long, yellow, with a pair of leafy bracts at the base. Flowering from winter to summer. **Habitat** Grassy Fynbos, Transitional Shrubland and grassland. **Distribution** Widespread from Swartberg to Port Edward.

Pisgoed Vlakte, Combrink's Mountain 23.09.2011

RHYNCHOSIA

rhynchos = beak, snout, horn; pertaining to the keeled petal that ends with a distinct 'nose'

± 200 species in the world, mostly tropical; ± 70 species in South Africa; 2 species in Baviaanskloof.

Rhynchosia caribaea

caribaea = from the Caribbean; a misleading name as this is not where the plant is originally from. It was grown from seed and scientifically described for the first time from plants that had been cultivated in the Caribbean.

Creeper from a woody base, with stems and peduncles hairy, 4 m. **Leaves** Trifoliate, heart-shaped, central leaflet to 6 cm. **Flowers** Several in an axillary raceme, flowers 2 cm long, yellow with dark stripes. Pods glandular-hairy. Flowering all year round. **Habitat** Occurring mainly in Forest. **Distribution** Widespread throughout southern Africa, extending to tropical Africa. **Other name** Monya-mali (S).

Zandvlaktekloof, Baviaans 17.04.2012

SCHOTIA

Boerboon

Named by botanist and chemist Nikolaus Joseph von Jacquin after Richard van der Schot, 1733–1790, who was head gardener at the Imperial Garden in Vienna, Austria

A genus of small trees recognised by their bright red or pink flowers, fused calyx and large woody pods with a persistent rim. 18 species in southern and tropical Africa; 2 species in Baviaanskloof.

Schotia afra Karooboerboon

afra = from Africa

Small tree to 7 m. **Leaves** Pinnate, 5 cm long, with 6–18 pairs of elliptic leaflets, each leaflet 1 cm long × 0.5 cm wide. **Flowers** In dense, many-flowered racemes, each flower 2.5 cm long, red or pink. Flowering from late winter to spring. **Habitat** Usually on dry stream banks or seasonally damp sites in Thicket or Succulent Karoo. **Distribution** Widespread from Anysberg to East London. **Notes** The seeds of var. *afra* are edible when green, roasted or ground into meal. The other variety, var. *angustifolia*, is found only in Namaquland and Namibia. **Other names** Small boerbean (E), intiza, umgxam, umqongci (X).

Andrieskraal Road, Patensie 21.01.2022

Schotia latifolia Bosboerboon

latus = broad, *folium* = leaf; the leaves are broad compared with leaves of other species in the genus.

Small tree, resprouting, to 5 m. **Leaves** Pinnate, 8 cm long, with 3–5 pairs of leaflets, asymmetric, leaflets 6 cm long × 3 cm wide. **Flowers** In racemes at branch tips, pale pink. Pods large, seeds light brown with yellow arils. Flowering in late spring and summer. **Habitat** Thicket and Forest. **Distribution** Widespread from Vleesbaai to the Wild Coast. **Notes** The seeds can be used for food and are edible when green, roasted or ground into meal. Good subject for bonsai. **Other names** Umaphipha, umgxam (X).

Zandvlakte, Baviaans 19.01.2022

TEPHROSIA

tephros = ashen; referring to the grey-haired leaves of many species

The leaflets have parallel veins. ± 400 species, pantropical, mostly in Africa; 2 species in Baviaanskloof.

Tephrosia capensis

capensis = from the Cape; referring to the geographical origin of the species

Resprouting from a woody rootstock, scrambling, spreading shrub, to 30 cm. **Leaves** Pinnate, 5–8 pairs of leaflets, each leaflet 1.5 cm long, with dark parallel veins visible below. Stipules small and triangular. **Flowers** Scattered on racemes, 1 cm long, bright pink or purple. Flowering all year round. **Habitat** Frequent in Transitional Shrubland and Grassy Fynbos, often on rocky north-facing slopes. **Distribution** Widespread from the Cape Peninsula to Makhanda. **Notes** The roots are used in traditional medicine.

Doringkloot, Baviaans 13.09.2006

Tephrosia grandiflora   Rooi-ertjie

grandi = great, *flora* = flower; referring to the large showy flowers

Resprouting, sparsely branched shrub to 1.5 m, with large deciduous bracts. **Leaves** Pinnate, 8 cm long, with large, many-veined stipules. **Flowers** Bunched in racemes, to 2 cm long, bright pink or magenta. Flowering in autumn, winter and spring. **Habitat** Sour Grassland and grassy Transitional Shrubland. **Distribution** Widespread from Kareedouw to KwaZulu-Natal. **Other names** Ihlozane, iqwense (Z).

Suuranysberg, Kareedouw 11.05.2021

VACHELLIA

Named after George Harvey Vachell, 1789–1839, who worked for the British East India Company in Macau and collected plants in China

± 160 species in the tropics and subtropics; 1 species in Baviaanskloof.

Vachellia karroo Doringboom, soetdoring, sweet thorn

karroo = old spelling of karoo, a Khoisan word describing the arid interior of South Africa, where this plant is common

Tree or shrub, resprouting, to 10 m. The long, straight white thorns are modified stipules and grow in pairs. **Leaves** Bipinnate, 5 cm long × 2.5 cm wide, leaflets 0.5 cm long. **Flowers** In globose heads, 1 cm wide, sweetly scented. Pods sickle-shaped, constricted between the seeds. Flowering in early summer. **Habitat** A wide variety of habitats, but most abundant in Doringveld. **Distribution** Widespread throughout southern Africa. **Notes** Particularly useful fodder for game and stock. Bark was used for tanning. Gum has been used in confectionery and as an adhesive. Other uses are as a source of nectar for honey, timber for construction and fencing, sewing needles, and for pinning insects collected by early naturalists. Popular as braai wood in Baviaanskloof. **Other names** Umngampunzi, intlaka (X), leoka, mooka (S), umunga (X, Z).

Baviaanskloof River, Speekhout 25.12.2017

VIRGILIA

Named after Roman poet Virgil, 70–19 BC

2 species of southern African trees, in Western and Eastern Cape.

Virgilia divaricata Keurboom

divarico = spread out, diverging; referring to the widely spreading branches. The Afrikaans name keurboom ('choice tree') marks the species as a good choice for cultivation. This tree has been introduced in gardens around the world, where it is known as the Cape lilac or tree-in-a-hurry because of its rapid growth.

Reseeding, widely branched tree to 15 m. **Leaves** Pinnate, 10 cm long, with 5–10 pairs of leaflets, elliptic, leaflets 2.5 cm long × 0.5 cm wide. **Flowers** In racemes, 2 cm long, pale pink to pink but sometimes almost white, with keel tip dark purple. Pods leathery and felted, 5 cm long × 1 cm wide, splitting to reveal several brownish seeds. Flowering in spring. **Habitat** Fynbos and Forest, usually associated with streams or damp habitats on forest margins; forms dense stands in favourable situations. **Distribution** Widespread from George to Kariega (Uitenhage). **Notes** The pods can be heard splitting open on warm, dry days in winter. **Other name** Umzitsikama (X).

Poortjies, Cambria 26.10.2018

GENTIANACEAE

Gentian

Cosmopolitan plants found mostly in temperate zones, with 87 genera and ± 1,600 species. Some are cultivated as ornamentals and there are others with medicinal uses. In southern Africa there are 9 genera and ± 84 species. Baviaanskloof has 2 well-known genera, *Chironia* and *Sebaea*, represented by 5 species.

CHIRONIA

Centaury, bitterwortel

Named after Chiron, the great mythological centaur, in reference to the medicinal properties of the genus, as Chiron studied the healing properties of plants. The arrangement of the stigma and anthers is a specialised adaptation to ensure cross-pollination by carpenter bees.

16 species in South Africa; 3 species in Baviaanskloof.

Chironia baccifera Bitterbossie, aambeibessie, Christmas berry

baccifer = berry-bearing; the Afrikaans name aambeibessie refers to the plant's use for treating haemorrhoids.

Much-branched, dense, rounded shrub to 60 cm. **Leaves** Linear, opposite and spreading, 1.5 cm long × 0.5 cm wide. **Flowers** 1.5 cm wide, pink. Fruit 0.5 cm wide, berry-like, becoming bright red. Flowering in summer. **Habitat** Occasional in sandy soil in Grassy Fynbos and Transitional Shrubland. **Distribution** Widespread from Namaqualand to KwaZulu-Natal. **Notes** Used medicinally as a blood purifier and for the treatment of piles. Toxic to sheep.

Bergplaas, Combrink's Pass 26.09.2006

Chironia linoides

Linum = plant genus, *oides* = resembling; the leaves bear a resemblance to those of some species of *Linum* (flax).

Reseeding, upright, variable, floriferous, dense shrublet to 50 cm. **Leaves** Linear, opposite, 1.5 cm long × 0.3 cm wide. **Flowers** 2 cm wide, pink, anthers yellow, stigma white. Flowering in late spring and summer. **Habitat** Mountain slopes in Grassy and Mesic Fynbos, especially after fire. **Distribution** Widespread from Namaqualand to Baviaanskloof.

Damsedrif, Baviaans 12.10.2016

Chironia melampyrifolia

melas = black, *pyri* = fire, *folia* = leaf; derivation unresolved

Reseeding, spreading, diffuse, slightly sticky shrublet, to 30 cm, with angled stems. **Leaves** Opposite, well-spaced, spreading, ovate-lanceolate, 2 cm long × 0.8 cm wide, tips hooked. **Flowers** 3 cm wide, corolla tube 1 cm long, pink, sticky. Flowering in spring and summer. **Habitat** Occasional in damp, shaded areas on stream banks in Forest and Thicket. **Distribution** Widespread from Grabouw to Port Alfred.

Poortjies, Cambria 26.10.2018

GERANIACEAE
Pelargoniums

There are 11 genera and 800 species in subtropical and temperate regions across the world. These plants grow easily and several species have medicinal properties. Leaves are often fragrant when compressed. Several *Pelargonium* species are cultivated commercially for oil production. 5 genera and ± 315 species in southern Africa; 1 *Geranium* species, 1 *Monsonia* species and 30 *Pelargonium* species in Baviaanskloof.

GERANIUM
Geranium, cranesbill

geranos = crane (bird); referring to the fruiting body that looks like a crane's bill in some species

Geranium is distinguished from the genus *Pelargonium* by having a symmetrical flower comprising 5 similarly sized petals. In *Pelargonium* 2 of the upper petals are larger, making an asymmetrical flower. Originally Linnaeus treated the 2 genera as 1, and to this day pelargoniums are sometimes commonly referred to as geraniums, causing some confusion. Geraniums have palmately-lobed leaves, usually circular in form. The flowers are colourful and spectacular with prominent veins on the petals. The plants are easy to grow and popular in gardens and nurseries around the world. 422 species worldwide, mostly in the eastern part of the Mediterranean; 28 species in South Africa; 1 species in Baviaanskloof.

Geranium canescens   Ubukhubele (X)

canescens – covered with fine, dense hairs; leaves and stems are finely felted with greyish-white hairs.

Reseeding, diffuse perennial, soft and delicate, scrambling subshrub, to 30 cm. **Leaves** 3–5-lobed on long petioles, finely hairy, greyish green. **Flowers** 1 or 2 on long, hairy, axillary peduncles, to 2 cm wide, white to pale pink, sometimes with pink veins. Flowering in spring. **Habitat** Uncommon on shale in Grassy Fynbos. **Distribution** Widespread from the Cape to Baviaanskloof. **Other name** Utywala beentaka (X).

Nooitgedacht, Kouga 28.09.2011

MONSONIA
Parasol

Named by Linnaeus after Lady Anne Monson, 1726–1776, great-granddaughter of Charles II. She was an English botanist and plant and insect collector who corresponded with Linnaeus. She visited Cape Town briefly en route to India, and collected around Cape Town with Thunberg, from where she sent specimens of *Monsonia* to Kew Gardens in 1774.

The dried stems of some species can be used as candles and firelighters.

Monsonia emarginata   Dysentery herb, sambreeltjie

emarginata = shallowly notched; referring to the curved leaf margins

Resprouting, tuberous-rooted, small shrublet to 20 cm. **Leaves** Entire, margins wavy and shallowly notched, ovate to obovate, 4 cm long × 2 cm wide, long-hairy. **Flowers** Single on thin peduncles, more than 2 cm wide, white, cream or pink, cup-shaped. Flowering from spring to autumn. **Habitat** Frequent but not abundant in openings in Grassy Fynbos, Transitional Shrubland and Thicket. **Distribution** Widespread from the Cape Peninsula to East London. **Other names** Geitabossie (A), igqitha, unoboyana (X).

Guerna, Kouga 16.02.2016

PELARGONIUM

Storksbill, pelargonium, malva

pelargos = stork; fruiting bodies resemble a stork's bill.

± 275 species in Africa, Madagascar, the Middle East and Australia, with ± 255 in South Africa. The genus is remarkable for its diversity of growth forms and the variety of habitats it manages to occupy. Many species have spectacular flowers and some have wonderfully scented leaves. Easy to grow from cuttings and very popular in the garden. ± 150 species in the Cape flora; 32 species recorded in Baviaanskloof.

Pelargonium alchemilloides   Pink trailing pelargonium, rankmalva

Alchemilla = plant genus, *oides* = resembling; the palmately-lobed and toothed margins are reminiscent of some species of *Alchemilla*, a genus in the rose family from Europe.

Sprawling perennial, resprouting from a woody rootstock, to 20 cm. **Leaves** Roundish, soft, sometimes lobed or toothed, 7 cm wide, with sparse white hairs, sometimes with circular reddish bands. **Flowers** Inflorescence umbel-like, flower 1.5 cm wide, white, yellow or pink; calyx tube up to 3.5 cm long (longer than the short pedicel). Flowering mostly in spring. **Habitat** Infrequently encountered in Transitional Shrubland. **Distribution** Widespread from the Cape to North Africa. **Notes** Roots and leaves are used in traditional medicine. It has become a problematic weed in the southern parts of Western Australia. **Other names** Umtetebu (X), bolila-ba-litsoene (S), inkubele (X, Z), umangqengqe (Z).

Nooitgedacht, Kouga 28.09.2011

Pelargonium carneum

carneus = flesh-coloured; referring to the pale orange flowers

Resprouting from a tuberous rootstock with papery bark, leaves in a basal tuft, to 40 cm tall in flower. **Leaves** Pinnate or bipinnate, on long-hairy petioles, entire leaf to 20 cm long, the linear segments 5 cm long × 3 cm wide, dry at flowering. **Flowers** Inflorescence umbel-like, on branching peduncles, flower 4 cm wide, white to pinkish with wine-red feather markings on 2 upper petals; calyx tube to 7 cm long, pink. Flowering mostly in late summer and autumn. **Habitat** Occasional in Grassy Fynbos and Transitional Shrubland on rocky slopes with fine-textured soil. **Distribution** Widespread from Worcester to the Kouga Mountains.

Guerna, Kouga 16.02.2016

Pelargonium carnosum   Vetplantmalva

carnosum = fleshy, succulent; referring to the succulent stems

Shrub with thick succulent stems to 1 m. **Leaves** Fleshy, 20 cm long, grey-green, much divided into thin, channelled lobes. **Flowers** 1.5 cm wide, white to pink, 2 upper petals with faint markings. Flowering in spring, summer and autumn. **Habitat** Clay-rich soils in Thicket and Transitional Shrubland. **Distribution** Widespread from Namibia to Gqeberha (Port Elizabeth).

Cambria 04.02.2020

Pelargonium cordifolium   Heart-leaved pelargonium, hartblaarmalva

cordatus = cordate, heart-shaped, *folium* = leaf; referring to the heart-shaped leaves

Reseeding, erect shrub with considerable variation in leaf shape and hairiness, to 1 m. **Leaves** Cordate, 5 cm wide, aromatic, dark green above, white-hairy below, margins toothed, often reddish and sometimes wavy. **Flowers** Inflorescence umbel-like, on short axillary peduncles, flower 4 cm wide, magenta, upper petals much wider and marked with dark purple lines; calyx tube 1 cm long, shorter than the pedicel. Flowering from early spring to summer. **Habitat** Damp, sheltered areas on south-facing slopes in Mesic Fynbos. **Distribution** Widespread from Bredasdorp to Qonce (King William's Town). **Notes** Popular in gardens and nurseries.

Scholtzberg, Baviaans 29.10.2011

Pelargonium exhibens

exhibens = to exhibit, naked in public; referring to the anthers and stigma that are prominently exserted, a feature that distinguishes it from *P. trifidum* with reproductive parts more hidden within the flower

Reseeding, scrambling, sparsely branched shrub with smooth stems, to 30 cm. **Leaves** 3- or 4-lobed, the lobes pinnatisect, 5 cm wide, bright green, appearing glabrous, but sparsely silky-hairy under a microscope. **Flowers** Inflorescence umbel-like with few flowers, flower 3 cm wide, white with wine-red lines on the much larger upper petals; calyx tube 1 cm long, longer than pedicel and both with short glandular hairs. Flowering in summer. **Habitat** Infrequent in arid Thicket. **Distribution** Known in only a few locations from Graaff-Reinet to Baviaanskloof and Makhanda. **Notes** A recently described species (1986) that is not well known. Discovered near Zandvlakte by Magriet Kruger, a find which extended this species' known range by several hundred kilometres. **Status** Near Threatened.

Vlakkie, Zandvlakte 21.12.2011

Pelargonium inquinans   Scarlet geranium

inquinatus = fouled, stained; referring to a tendency for the leaves to turn brown when touched

Semi-succulent shrub to 2 m. **Leaves** Round to heart-shaped, slightly lobed, velvety-hairy, 8 cm wide. **Flowers** Clustered into heads of 5–30 flowers, 2.5 cm wide; tube 4 cm long, bright red, sometimes with white at the centre. Flowering all year round. **Habitat** Clay soils in Thicket and Transitional Shrubland. **Distribution** Widespread from Cambria to Mthatha. **Notes** Traditionally used as a deodorant. Also made into a powder to treat headaches and colds. **Other name** Ibhosisi (X).

Ysrivier, Cambria 03.02.2020

Pelargonium laevigatum

laevigatus = smooth and polished; referring to the smooth, hairless leaves of this plant

Resprouting, sprawling, usually rounded shrublet, to 50 cm. **Leaves** Trifoliate, segments terete, smooth, fleshy, 3 cm long, grey-green, not scented. **Flowers** Usually solitary, 2.5 cm wide, pale pink with bright purple markings on upper petals; calyx tube much longer than the pedicel. Flowering all year round. **Habitat** Occurring mostly in Mesic and Grassy Fynbos. **Distribution** Widespread from the Little Karoo to Somerset East.

Kouenek, Kouga 02.12.2009

Pelargonium laxum

laxus = loose, flaccid; referring to the softish leaves that tend to droop

Reseeding, succulent-stemmed shrublet, with smooth whitish stems, to 30 cm. **Leaves** Pinnatisect, somewhat leathery, to 12 cm long × 6 cm wide, margins pale, tending to droop. **Flowers** Inflorescence many-flowered on branched peduncles, flower small, white or pale pink with anthers and stigma prominent, exserted beyond the flower; calyx lobes as long as the petals. Flowering from winter through summer to autumn. **Habitat** The valley floor in open Thicket and Succulent Karoo. **Distribution** Widespread from the Little and Great Karoo to Makhanda. **Notes** Popular as a container plant and good for bonsai.

Zandvlakte, Baviaans 26.08.2010

Pelargonium myrrhifolium Wildemalva

myrio = very many, countless, *folium* = leaf; referring to the finely divided leaves, made up of numerous segments

Resprouting, sprawling shrublet to 30 cm. **Leaves** Bipinnatisect, slightly hairy, petiole hairy and up to 10 cm long, leaf blade to 5 cm long × 3 cm wide. **Flowers** Inflorescence umbel-like, up to 5 flowers; flower 2.5 cm wide, 4-petalled, upper petals larger with feather-like markings, white, pink or magenta; calyx tube to 1 cm long, often reddish. Flowering in spring and summer. **Habitat** Fairly frequent in open, sandy areas amongst rocks in Fynbos and Transitional Shrubland. **Distribution** Widespread from Kamiesberg to Kariega (Uitenhage).

Rus en Vrede 4X4, Kouga 28.07.2011

Pelargonium odoratissimum Scented geranium

odoratissimus = most fragrant; referring to the pleasant smell released by the leaves at the slightest touch

Reseeding, aromatic, straggling, leafy perennial with slightly tuberous roots, to 30 cm. **Leaves** Heart-shaped to roundish, velvety-hairy, to 12 cm wide, strongly apple-mint-scented, margins crenulate. **Flowers** Inflorescence umbel-like, with up to 10 flowers; flower small, 1.5 cm wide, white; calyx tube to 1 cm long, as long as the pedicel. Flowering almost all year round. **Habitat** Occasional in well-shaded areas at the base of rocks or under Forest or Thicket. **Distribution** Widespread from Ceres to KwaZulu-Natal. **Notes** Essential oils from this species have antibacterial properties and have been used to treat a variety of ailments, and in aromatherapy. Popular in gardens because of the wonderful scent. Several cultivars and hybrids each have a distinct fragrance.

Gert Smitskloof, Baviaans 17.01.2022

Pelargonium ovale subsp. *ovale*

ovalis = oval, elliptic; pertaining to the oval leaves

Resprouting, tufted, rhizomatous shrublet to 30 cm. **Leaves** 4 cm long × 3 cm wide, elliptic, toothed and densely hairy, petiole 2 times as long as leaf blade. **Flowers** Inflorescence umbel-like, with up to 7 flowers, peduncles stout and branched; flower to 4 cm wide, various shades of pink, 2 upper petals with markings, and larger than 3 lower petals; calyx tube to 1.3 cm long, shorter than pedicel. Flowering from spring to autumn. **Habitat** Common in Grassy and Mesic Fynbos, especially in the first few years after fire. **Distribution** Widespread from Tulbagh to Kariega (Uitenhage).

Koud Nek's Ruggens, Kouga 07.11.2011

Pelargonium ovale subsp. veronicifolium

veronicinus = veronica-violet, *folium* = leaf; derivation unclear as it is the flowers that have violet-coloured markings

Similar to subsp. *ovale*, but different in having larger upper petals, with more striking violet markings. Nearly endemic from Kammanassie to Baviaanskloof.

Rus en Vrede 4X4 08.09.2011

Pelargonium panduriforme

panduriforme = fiddle-shaped; referring to the shape of the leaves

Reseeding, erect, leafy shrub to 1.2 m. **Leaves** Cordate, 3–5-lobed, soft and very hairy, more so below, 6 cm long × 5 cm wide, balm-scented. **Flowers** Inflorescence umbel-like, up to 20 flowers, fairly compact, on short, stout, axillary peduncles; flower 4 cm wide, pink, upper petals with dark red markings; calyx tube 1.3 cm long, peduncle shorter. Flowering in spring and summer. **Habitat** Fairly frequent on rocky slopes in Grassy Fynbos and Transitional Shrubland. **Distribution** Widespread from Baviaanskloof to Port Alfred.

Bosrugkloof, Baviaans 24.10.2011

Pelargonium peltatum   Rankmalva, ivy-leaved pelargonium

peltatum = having the petiole coming from the centre of the leaf rather than the edge

Reseeding, climbing, semi-succulent creeper to 4 m. **Leaves** 5–7-lobed, rounded, leathery, with reddish zonal markings, to 7 cm wide. **Flowers** Inflorescence umbel-like, with up to 9 flowers; flower to 5 cm wide, pale pink to magenta with purple markings on the upper petals; calyx tube relatively long, to 5 cm. Flowering in spring and early summer. **Habitat** Common in clay-rich soils in Thicket, or in bush clumps in Transitional Shrubland. **Distribution** Widespread from Worcester to East London. **Notes** Popular in gardens, with many different cultivars. **Other names** Umnewana, ityholo (X).

Stompie se Nek, Kasey 17.10.2011

Pelargonium pulverulentum

pulverulentus = powdered, dusty, covered with powder; alluding to the leaves that lie flat on the ground and are often rather sandy

Resprouting, geophytic, woody-tubered, stemless herb, to 20 cm tall in flower. **Leaves** Cordate, often deeply incised, sparsely hairy, flat on the ground, to 10 cm long. **Flowers** Inflorescence umbel-like, with up to 14 flowers on an erect flower stalk; flower to 2 cm wide, petals black or maroon with pale yellow margins; calyx tube to 5 cm long. Flowering in spring and summer. **Habitat** Loamy or clay soils in grassy Transitional Shrubland. **Distribution** Widespread from Baviaanskloof to KwaZulu-Natal. **Other names** Ikubalo likamlanjeni, indlebe yebokwe (X).

Engelandkop, Kouga 30.10.2011

Pelargonium quercifolium Muishondbossie

Quercus = oak genus, *folium* = leaf; leaves have a shape reminiscent of some oak leaves.

Reseeding, erect shrub to 1.5 m. **Leaves** Deeply lobed, roughly hairy, sometimes sticky, margins toothed, leaves to 8 cm long, and wavy, aromatic. **Flowers** Inflorescence umbel-like, with up to 6 flowers, on short, stout axillary peduncles; flower 3 cm wide, magenta, with darker markings on the 2 upper petals; calyx tube to 1 cm long. Flowering in spring and summer. **Habitat** Dry rocky slopes and in dry riverbeds at the base of ravines. **Distribution** Fairly narrow from Oudtshoorn to Baviaanskloof.

Witrivier, Cambria 18.05.2021

Pelargonium radens

radula = rasp or file; referring to the sandpapery feel of the stiff hairs covering the leaves

Reseeding, erect shrub to 1 m. **Leaves** Bipinnatisect, finely divided, 6 cm wide, rose-scented. **Flowers** Inflorescence umbel-like, with up to 8 flowers on axillary peduncles; flower 2 cm wide, magenta; calyx tube 0.8 cm long, about as long as pedicel. Flowering in spring and summer. **Habitat** Damp slopes in Mesic Fynbos. **Distribution** Widespread from Tulbagh to Makhanda.

Cockscomb, Groot Winterhook 17.02.2016

Pelargonium reniforme Rabassam, rooirabas

reniformis = kidney-shaped; referring to the leaf shape

Resprouting, tuberous-rooted shrublet, to 40 cm. **Leaves** On long thin petioles, kidney-shaped, margins crenulate, shortly hairy, 2–7 cm wide, grey-green, paler below. **Flowers** Inflorescence umbel-like, up to 15 flowers on branched peduncles; flower 2 cm wide, bright pink; calyx tube to 4 cm long, longer than the pedicel. Flowering all year round. **Habitat** Fairly frequent in open areas in Transitional Shrubland and Grassy Fynbos. **Distribution** Widespread from Outeniqua Mountains to Mthatha. **Notes** Harvested from wild populations and used in traditional medicine to treat a number of ailments. A form from Kleinrivier (Kouga Wildernes Farm) with small leaves and a much longer hypanthium needs to be investigated. **Other names** Iyeza lesikhali, ikubalo, umsongelo, umkumiso (X). **Status** Near Threatened.

Cockscomb, Groot Winterhoek
17.02.2016

Pelargonium ribifolium

Ribes = currant genus, *folium* = leaf; leaves are reminiscent of those of some species of currant.

Reseeding, erect shrub to 1.2 m. **Leaves** Palmate, roughly hairy, 5 cm wide, lime-green. **Flowers** Inflorescence umbel-like, to 12-flowered, on short axillary peduncles; flower small, 2 cm wide, white; calyx tube to 0.8 cm long, about as long as pedicel, with a knee-like bump where the tube meets the pedicel. Flowering in spring. **Habitat** Dry stony stream banks in the side ravines or in Transitional Shrubland. **Distribution** Widespread from Oudtshoorn to Hogsback. **Notes** Easily confused with hoenderbos (*P. scabrum*), which is also found in Baviaanskloof — the difference being that the latter does not have the kink where the calyx tube meets the pedicel.

Suuranysberg, Kareedouw 12.05.2021

Pelargonium schizopetalum Muishondbossie

schizo – to split, *petalum* = petal; referring to the petals, which are divided into numerous thin lobes

Resprouting from a large woody tuber, tufted, to 30 cm. **Leaves** Pinnate, hairy and soft, 15 cm long × 7 cm wide. **Flowers** Inflorescence umbel-like, with up to 20 flowers; flower petals finely dissected, to 4 cm wide, pale yellow with some parts tinged purplish, calyx tube to 7 cm long, longer than the pedicel. Flowering in summer. **Habitat** Infrequent on shale bands in Grassy Fynbos. **Distribution** Eastern distribution from George to Hogsback. **Other name** Indlebe yebokwe (X).

Koud Nek's Ruggens, Kouga 20.12.2011

Pelargonium sidoides Kalwerbossie

Sida = plant genus, *oides* = resembling; leaves resemble those of *S. rhombifolia*.

Resprouting from swollen roots, tufted perennial, to 30 cm. **Leaves** Crowded, cordate, 3 cm wide, grey-silvery-velvety, margins slightly lobed and crenate, on long petioles. **Flowers** Inflorescence umbel-like, on branched peduncles, with up to 14 flowers; flower 1.5 cm wide, dark maroon to almost black; calyx tube to 3.5 cm long, longer than the pedicel. Flowering in late spring and summer. **Habitat** Transitional Shrubland and Grassy Fynbos where the soil is fine-textured and grass cover is well developed. **Distribution** Widespread from Baviaanskloof to Gauteng. **Notes** Used in traditional medicine for treating a sore throat and congestion, and in the pharmaceutical industry for treating bronchitis. **Other names** Umsangela, umsongelo (X).

Geelhoutbos bergpad, Kouga 14.07.2011

Pelargonium tetragonum Square-stemmed pelargonium

tetra = four, *gonia* = angled; referring to the 4-angled, square stems

Reseeding, soft-fleshy stemmed shrub to 1.5 m; it uses other bushes for support. Stems are green and 4-angled, 0.7 cm wide. **Leaves** 4 cm wide, deeply lobed, sometimes with red zonal markings, falling early. **Flowers** Inflorescence 2-flowered; flower showy, 4 cm wide, with large, white or pale pink upper petals with reddish lines. Flowering in spring and summer. **Habitat** Bush clumps where there is clay in the soil, most frequent in Thicket. **Distribution** Widespread from Worcester to Makhanda.

Geelhoutbos, Kouga 31.10.2011

Pelargonium zonale Horseshoe geranium

zona = band or belt; referring to the prominent and characteristic reddish, circular band around the leaf

Reseeding, robust, erect, leafy shrub to 2 m. **Leaves** Relatively large, 12 cm wide, scented, roundish, sparsely hairy, with red zonal markings, margins shallowly lobed. **Flowers** Inflorescence umbel-like, in crowded clusters of up to 70 flowers; flower 3.5 cm wide, pink; calyx tube to 4.5 cm long, pedicel very short. Flowering mainly in spring and summer. **Habitat** Frequent and sometimes abundant in sandy stream beds. **Distribution** Widespread from Piketberg to the Eastern Cape and Underberg, KwaZulu-Natal. **Notes** Fast growing, popular in gardens. **Other names** Iyeza lendlebe, umsongelo (X).

Rus en Vrede 4X4, Kouga 07 11 2011

GESNERIACEAE

A family with colourful and showy flowers. ± 152 genera and ± 3,450 species, mostly in tropical and subtropical areas around the world. Well-known members of the family are African violet (*Saintpaulia*) and gloxinia (*Sinningia speciosa*), both popular container plants for shade. There are 8 genera in Africa, but in South Africa only *Streptocarpus* is present.

STREPTOCARPUS   Wild gloxinia

streptos = twisted, *carpos* = fruit; pertaining to the much-twisted capsules

± 125 species of herbs, mostly tropical and subtropical; 56 species in South Africa; 2 species in Baviaanskloof.

Streptocarpus meyeri

Named after either Ernst Heinrich Friedrich Meyer, 1791–1858,
or Carl Anton Meyer, 1795–1855, both German botanists

Low, resprouting herb, to 20 cm when in flower. **Leaves** In a basal rosette,
elliptic, velvety, variable in size, 8 cm long × 4 cm wide, margins toothed.
Flowers Few, 3 cm long, on glandular-hairy peduncles, tube cylindrical,
violet with white lobes. Flowering in late summer and autumn. **Habitat**
Fairly frequent, but localised on shaded moss banks, on south-facing rocky
banks and ledges in Mesic and Grassy Fynbos, and Transitional Shrubland.
Distribution Widespread from Baviaanskloof to Mpumalanga. **Notes** The first
species in the genus to be described, *S. rexii,* has longer leaves than *S. meyeri*,
and is the only other species that might be found in Forest in Baviaanskloof.
Leaves and roots are used medicinally and also feature in local folklore.

Kouga Kliphuis, Kouga Valley
12.05.2021

GISEKIACEAE

A recently recognised family (2003). 1 genus and 7 species confined to Africa and Asia.

GISEKIA

Named after Paul Dietrich Giseke, 1741–1796, a German botanist, physician, teacher and librarian. He
was a friend and student of Linnaeus and published his notes taken during Linnaeus's lectures. His book
included a map-like drawing of plant families.

4 species; 2 species in Baviaanskloof.

Gisekia pharnaceoides

Pharnaceum = plant genus, *oides* = resembling; referring to the resemblance to some species of *Pharnaceum*

Sprawling, fleshy, annual herb, to
5 cm tall and quite wide. **Leaves**
In tufts, linear to obovate, 2 cm
long, often tinged red. **Flowers**
Several in small umbels in axils and
on branch tips, tiny, 0.2 cm wide,
purplish, with 5 anthers. Flowering
in spring, summer and autumn.
Habitat Disturbed areas on dry,
sandy riverbeds. **Distribution**
Very widespread from drier parts
of southern Africa to tropical
Africa, Madagascar and Asia.
Notes Much used in medicine for
various ailments, most commonly
diarrhoea. The plant is also eaten
as a condiment. *G. africana*, the
other species that may occur in
Baviaanskloof, has up to 20 anthers.

Baviaanskloof River, Zandvlakte 07.04.2012

GUNNERACEAE

Rhubarb

Contains a single genus with 63 known species, growing in most parts of the southern hemisphere. The family is known for the mammoth leaves on *Gunnera manicata* and for the edible rhubarb, which is a hybrid of *Rheum rhabarbarum* and *R. rhaponticum*. Various species have a long history of medicinal use. It is the only family that has developed a symbiotic relationship with a nitrogen-fixing blue-green algae, allowing its members to flourish in nutrient-poor wetland environments. It is also one of the oldest families of angiosperms and was probably browsed by dinosaurs.

GUNNERA

River pumpkin

Named after the Norwegian bishop and botanist Johann Ernst Gunnerus, 1718–1783, who collected specimens and published prolifically on natural history

1 species in South Africa.

Gunnera perpensa   Pampoenblaar

perpenso = weigh exactly, consider carefully; pertaining to its similarity to other species, thus care must be taken in identification

Rhizomatous, tufted perennial to 50 cm. **Leaves** Kidney-shaped, long-petioled, large, 15 cm wide, with toothed margins. **Flowers** On spike-like racemes 20 cm long, each flower tiny, greenish. Flowering in spring and summer. **Habitat** Gently flowing water of streams, marshes and seeps. **Distribution** Widespread from Cape Peninsula to North Africa. Easy to see from the road at Poortjies. **Notes** Used medicinally for various stomach ailments. **Other names** Iphuzi lomlambo, ithangazana (X), qobo (S), uxobo (X, Z), imfeyesele, ugobho, uklenya (Z).

Sewefontein, Studtis 02.04.2012

HYPERICACEAE

St John's wort

± 9 genera and 700 species. Plants often have black, orange or translucent spots, which contain a photosensitive compound that causes blistering of the muzzle in stock. The common St John's wort (*Hypericum perforatum*), is the most toxic species in the family. It originated in Eurasia and is now a problematic weed in various parts of the world. 1 genus, 8 indigenous species and a few naturalised species in southern Africa.

HYPERICUM

St John's wort

hyper = above, *eikon* = picture; referring to the age-old tradition of hanging the plant above religious icons on St John's Day. The common name derives from the time of year when it flowers and is harvested, which coincides with the summer solstice in the northern hemisphere (24 June).

6 species and 2 exotic species in South Africa; 1 species in Baviaanskloof.

Hypericum lalandii

Named after Pierre Antoine Delalande, 1787–1823, French naturalist, explorer and painter from Versailles. He travelled and collected extensively in southern Africa. A crayfish, a crab and a shark have been named after him.

Resprouting from a woody rootstock, stems thin, 4-angled, to 40 cm. **Leaves** Opposite, linear to elliptic, faintly gland-dotted, 1 cm long × 0.3 cm wide. **Flowers** Few at the branch tips, 1.5 cm wide, yellow, with numerous stamens. Flowering in summer and autumn. **Habitat** Shale on damp slopes in Grassy Fynbos, especially after fire. **Distribution** Very widespread from Somerset West to North Africa.

Cockscomb, Groot Winterhoek
17.02.2016

KEWACEAE (= *Hypertelis*)

A recently founded family with 1 genus and 8 species. These species were previously classified in the genera *Pharnaceum* and *Hypertelis* in the Molluginaceae. Mostly growing in southern Africa, but also on St Helena and Madagascar.

KEWA

Named after the Royal Botanic Gardens, Kew in London

1 species in Baviaanskloof.

Kewa salsoloides Braksuring

Salsola = plant genus, *oides* = resembling, the leaves bear a superficial resemblance to those of some species of *Salsola*.

Short-lived, small, tufted, blue-grey shrublet to 15 cm. **Leaves** Terete, fleshy, 2.5 cm long × 0.3 cm wide. **Flowers** In few-flowered umbels at the end of long peduncles, to 1 cm wide, white to pink, calyx reddish gland-dotted. Flowering mostly in spring or after good rain. **Habitat** Common in open, disturbed, arid, slightly saline ground in Succulent Karoo and Thicket. **Distribution** Very widespread from Namibia to the Karoo and Zimbabwe.

Rus en Vrede, Kouga 25.04.2018

LAMIACEAE

Mint, sage

A large family of ± 240 genera and 6,500 species, recognised by the square stems, aromatic leaves and irregular flowers, usually 2-lipped. Many species of medicinal and horticultural value, with mint (*Mentha longifolia* the best known. 35 genera and ± 255 species in South Africa; ± 6 genera and 13 species in Baviaanskloof.

COLEUS (= *Plectranthus*)

koleos = a sheath; referring to the fused stamens in most species

A recently reinstated genus based on molecular studies; previously assigned to *the* genus *Plectranthus*. *Coleus* flowers differ by having the pedicel asymmetrical on the calyx (centred in *Plectranthus*), a shorter upper lip, and an S-shaped corolla tube; rarely with the corolla tube swelling at the base. 294 species in the world; 29 species in South Africa; 2 species in Baviaanskloof.

Coleus madagascariensis (= *Plectranthus madagascariensis*)   Isikholokotho (X)

madagascariensis = from Madagascar; a misnomer as this is not where this species was first scientifically recorded; it occurs naturally in Baviaanskloof.

Sprawling perennial to 1 m tall, rooting at the nodes. **Leaves** Petiolate, 3 cm long, with rounded teeth. **Flowers** 1.5 cm long, white or mauve to purple, calyx hairy. Flowering in spring, summer and autumn. **Habitat** Semi-shade of Forest and Thicket. **Distribution** Widespread from Knysna to Mpumalanga. **Notes** Extremely popular around the world, easily grown in pots; cultivars with white-rimmed leaves are favoured. **Other name** Ibozane (Z).

Gonjah, Patensie 17.05.2021

Coleus subspicatus (= *Plectranthus spicatus*)

sub = somewhat, *spicatus* = with flowers in spikes; referring to flowers in spikes

Sprawling succulent with long flowering spikes, to 60 cm. **Leaves** Fleshy, 2 cm long × 1.5 cm wide, green sometimes tinged reddish, well-spaced on the stems, bluntly toothed. **Flowers** On elongated spikes, 1 cm long, purple. Flowering in autumn. **Habitat** Infrequent on open, rocky sandstone slopes in Transitional Shrubland and Thicket. **Distribution** Widespread from Baviaanskloof to Mpumalanga.

Geelhoutbos bergpad, Kouga 20.04.2012

LEONOTIS

leon = lion, *otis* = ear; alluding to a single flower that resembles a lion's ear

± 10 species in Africa; 3 species in South Africa; 2 species in Baviaanskloof.

Leonotis leonurus Wildedagga, minaret flower

leonurus = lion-coloured; the flower colour of some forms resembles a lion's mane, also referring to the fur-like hairs on the flowers. The common name alludes to the old Khoisa practice of smoking the dried leaves and flowers to produce a mildly euphoric sensation.

Reseeding, upright shrub to 2 m. **Leaves** Lanceolate, 10 cm long × 2 cm wide, margins toothed. **Flowers** Many clustered near the branch tips at the leaf nodes, 4 cm long, velvety orange, attracting sunbirds. Flowering in summer and autumn. **Habitat** Damp areas on riverbanks in Grassy Fynbos and Forest margins. **Distribution** Widespread from Clanwilliam to Gauteng. **Notes** Used medicinally for many ailments. **Other names** Utywala bengcungcu, imvovo, umfincafincane (X), lebake (S).

Langkloof, Kareedouw 13.05.2021

Leonotis ocymifolia Klipdagga

Ocimum = plant genus, *folia* = leaf; referring to the leaves that look similar to those of basil (genus *Ocimum*)

Reseeding, slender, erect shrub to 2 m. **Leaves** Ovate, 6 cm long × 3.5 cm wide, toothed, with a herbal scent if crushed. **Flowers** In clusters around the stem, each flower 4 cm long, velvety orange. Flowering from summer to winter. **Habitat** Sporadic in rocky areas in Thicket, Transitional Shrubland and Grassy Fynbos. **Distribution** Widespread from the Cederberg to tropical Africa. **Notes** Used medicinally for chest complaints and skin problems. **Other names** Umuncwane, umfincafincane, isihlungu sedobo, intlolokotshane, isigagisa samahlati (X).

Kamassiekloof, Zandvlakte 02.12.2015

MENTHA   Mint

The Latin name for mint, one of the oldest-known plant names, after a Greek nymph that was transformed into a mint plant

± 30 species in temperate regions, 2 species in South Africa; 1 species in Baviaanskloof.

Mentha longifolia Wild spearmint, horse mint

longi = long, *folia* = leaf; alluding to the long leaves

Semi-aquatic, rhizomatous, strongly mint-scented, straggling perennial herb with 4-angled stems, to 1 m. **Leaves** Lanceolate, 8 cm long × 1.5 cm wide, toothed margins. **Flowers** Numerous, crowded in a spike, 7 cm long × 1 cm wide, each to 0.4 cm long, white or mauve. Flowering in summer and autumn. **Habitat** Perennially damp areas along stream banks and ponds in the narrow ravines. **Distribution** Extremely widespread from Africa to Europe. **Notes** Age-old medicinal and culinary herb with various cultivars. **Other names** Balderja (A), inxina, inzinziniba (X), kerena (S).

Zandvlaktekloof, Baviaans 19.01.2022

PLECTRANTHUS

Plectranthus, spur-flower

plektron = spur, *anthos* = flower; referring to the spurred flowers in this genus

± 350 species in Africa, Asia and Australasia; 45 species in South Africa; 3 species in Baviaanskloof.

Plectranthus verticillatus

verticillatus = whorled; referring to the flowers on the spiked inflorescence

Sprawling, delicate, semi-succulent herb to 30 cm. **Leaves** Ovate, glossy, hairless, 3 cm wide, green, sometimes purple below, toothed. **Flowers** In whorls, 1.5 cm long, white to mauve, calyx hairless. Flowering in summer and autumn. **Habitat** Uncommon on damp, shaded rocks and cliffs under Forest and Thicket. **Distribution** Widespread from Knysna to Limpopo.

Poortjies, Baviaans 15.12.2019

PSEUDODICTAMNUS (= *Balotta*)

Horehound, kattekruie

pseudo = false, *Dictamnus* = plant genus; superficially resembling some species of *Dictamnus*, a European genus in the family Rutaceae. (Previously included in genus *Balotta*, which was the Greek name used by Dioscorides, possibly derived from *ballein* = to throw or reject, as a result of the bad-smelling leaves.)

± 33 species, mostly in the Mediterranean and Eurasia, but also in Africa; 1 species in Baviaanskloof.

Pseudodictamnus africanus (= *Balotta africana*)   **Catmint**

africanus = from Africa; referring to the origin of the species

Erect, weedy, aromatic, soft, greyish shrub to 1 m. **Leaves** Hairy, soft, 4 cm long × 3 cm wide, coarsely toothed. **Flowers** Bunched around the stem above the leaves, 1 cm wide, pink. Flowering in winter and spring. **Habitat** Rocky, disturbed areas in dry riverbeds in Succulent Karoo. **Distribution** Widespread from Namibia to Graaff-Reinet and Hlanganani. **Notes** Used for colds and flu.

Uitspan, Baviaans 19.02.2016

SALVIA

salvere = to heal; referring to the healing properties of some species

± 900 species worldwide; 22 species in South Africa; 4 species in Baviaanskloof.

Salvia namaensis   **Klipsalie**

namaensis = from Namaqualand; the plant was first recorded in Namaqualand.

Much-branched, leafy, slightly sticky aromatic shrub to 1 m. **Leaves** Rough, deeply lobed, 4 cm long × 2 cm wide, margins toothed. **Flowers** 1 cm long, white, mauve or blue, with white markings on the lower lip. Flowering in spring, summer and autumn. **Habitat** Rocky slopes and dry, sandy stream beds in Succulent Karoo and Thicket. **Distribution** Widespread from Namibia to Baviaanskloof. **Other name** Kamfersalie (A).

Gannalandkloof, Baviaans18.09.201

Salvia stenophylla · Blue mountain sage

stenophylla = narrow-leaved; pertaining to the narrow leaves

Resprouting perennial, from a woody rootstock, with square stems, to 60 cm. **Leaves** Pinnatifid or pinnatisect, rough, aromatic, 5 cm long × 1 cm wide. **Flowers** Small, to 1 cm long, pale blue. Flowering in spring, summer and autumn. **Habitat** Infrequent on sandy, brackish soils in the valley floodplain in Thicket and Succulent Karoo. **Distribution** Very widespread from Namibia and Botswana to Baviaanskloof and Limpopo. **Notes** The plant is widely used in traditional medicine. The essential oils are said to have healing properties.

Rus en Vrede, Kouga
28.08.2010

STACHYS

stachus = a spike; referring to the spike-like inflorescence

± 450 species worldwide, mainly in temperate and subtropical regions; 40 species in South Africa; 1 species in Baviaanskloof.

Stachys aethiopica · African stachys, katbossie

aethiopica = from sub-Saharan Africa; in classical times this region had not yet been explored by Europeans.

Delicate, square-stemmed, glandular-hairy shrub to 50 cm. **Leaves** Opposite, ovate, aromatic, distinctly quilted above, 3 cm long × 1.5 cm wide, toothed. **Flowers** Clustered around the stems above the leaves, 1 cm wide, white, pink or mauve with darker markings (nectar guides). Flowering in spring and summer. **Habitat** Sheltered and damp situations in loamy soils in Transitional Shrubland and Grassy Fynbos. **Distribution** Widespread from Calvinia to Cape Peninsula to Eswatini. **Other names** Wild sage (E), isihawuhawu, itshilisi yabantsundu, igangatshane (X), bokhatha, bolao-ba-litaola (S).

Zandvlaktekloof, Baviaans
18.04.2012

TEUCRIUM

Possibly named after Teucer, the first king of Troy, who may have used the plant for medicinal purposes

± 200 species; 3 species in South Africa; 1 species in Baviaanskloof.

Teucrium africanum · Paddaklou, aambeibossie

africanum = from Africa; referring to the origin of the species. The Afrikaans name 'paddaklou' means frog claw and refers to the shape of the leaves.

Soft, greyish shrublet to 30 cm. **Leaves** Deeply 3-lobed, thinly-hairy, 2.5 cm long, grey-green. **Flowers** Single on short stalks, to 1 cm long, white. Flowering in summer and autumn. **Habitat** Grassy Fynbos and Transitional Shrubland. **Distribution** Widespread from De Hoop to Makhanda. **Notes** The plant has been used in traditional medicine to treat haemorrhoids. **Other names** Bitterbos (A), ubuhlungu benyushu (X).

Vlakkie, Zandvlakte
07.04.2012

LIMEACEAE

Limeum

A small, single-genus family of 26 species. A recently described family, previously grouped with Molluginaceae.

LIMEUM

Lizard's foot, koggelmandervoet

loimos = a plague, pestilence; referring to the poisonous nature of some species, however this is not evident in the South African species

26 species, from the Western Cape to tropical Africa and India; 1 species in Baviaanskloof.

Limeum aethiopicum Koggelmandervoet, aarbossie

aethiopica = from sub-Saharan Africa; in classical times this region had not yet been explored by Europeans.

Low, decumbent, resprouting from a woody base, to 15 cm. **Leaves** Oblong to elliptic, 2 cm long × 0.5 cm wide. **Flowers** Numerous in terminal and axillary clusters, 1.5 cm wide × 0.3 cm wide, green-and-white. Flowering in late summer. **Habitat** Rocky slopes with some clay in Succulent Karoo and Thicket. **Distribution** Very widespread from Namibia to the Little Karoo to subtropical Africa. **Other names** Umula, inceba (X).

Doringkraal, Kouga 09.03.2012

LINACEAE

Flax

14 genera and 250 species. Flax or linseed (*Linum usitatissimum*) is cultivated to produce linseed oil. Seeds and fibres from the plant are woven into linen. The largest genus is *Linum* with ± 200 species.

LINUM

Flax

linon = flax; derived from Linon, the name used by Theophrastus for flax; the name given to the unspun fibres of the flax plant

± 200 species, from subtropical and temperate regions; 14 species in South Africa; 3 species in Baviaanskloof.

Linum africanum African flax

africanum = from Africa; referring to the origin of the species

Resprouting, with several slender stems, to 60 cm. **Leaves** Opposite, linear-lanceolate, 1.5 cm long × 0.3 cm wide, blue-grey, with small round black stipules. Calyx usually with raised glands along the margin. **Flowers** In elongated panicles, 1.5 cm wide, yellow; styles fused halfway up. Flowering in summer. **Habitat** Occasional on rocky sandstone in Grassy Fynbos and Transitional Shrubland. **Distribution** Widespread from West Coast to Baviaanskloof. **Notes** Another similar-looking species that may be found in Baviaanskloof, *L. thunbergii*, has unfused styles.

Roelf se Put, Kouga 04.12.2015

Linum esterhuyseniae

Named after Elsie Elizabeth Esterhuysen, 1912–2006, a humble botanist described as 'the most outstanding collector of plants in South Africa'. She contributed over 36,000 specimens to the Bolus Herbarium. 56 species and 2 genera have been named after her.

Resprouting, with few slender, villous stems, to 50 cm. **Leaves** Ovate, ascending, closely packed, 1.5 cm long, with stipular glands. **Flowers** On short side shoots and in terminal panicles, 2.5 cm wide, yellow, styles free or partly fused. Flowering in summer. **Habitat** Uncommon in sheltered and damp situations at the base of rock slabs in Mesic Fynbos. **Distribution** Fairly narrow from Ladismith to Kouga Mountains.

Quaggasvlakte, Kouga
13.02.2016

LINDERNIACEAE

Relatively small, mostly neotropical family, with 13 genera and 195 species. A recently recognised family (2007) with members that were previously grouped in Scrophulariaceae or Plantaginaceae. 2 genera and 13 species in southern Africa; 1 species in Baviaanskloof.

LINDERNIA

Named after Franz Balthasar von Lindern, 1682–1755, a German botanist, author and doctor, and director of the university botanical garden at Strasbourg

± 80 species, mostly in warm regions of Africa and Europe; 11 species in southern Africa; 1 species in Baviaanskloof.

Lindernia parviflora

parvi = small, *flora* = flowers; referring to the small flowers

Diffusely branched annual herb to 15 cm. **Leaves** Opposite, variable, to 2 cm long, sometimes slightly toothed. **Flowers** Lower lip 3-lobed, 1 cm long, white to pale blue with darker blue markings. Flowering in summer and autumn. **Habitat** Wet sand on marshy stream banks. **Distribution** Widespread from Cape Peninsula to Mpumalanga.

Doodsklip, Kouga Dam 02.02.2020

LOBELIACEAE   Lobelias

30 genera and 1,260 species worldwide. Mostly herbs with irregular flowers. Some diagnostic characteristics of this family include simple, alternate leaves, milky sap, tubular flowers with upper lip mostly 2-lobed, lower lip mostly 3-lobed, and anthers fused into a tube. 6 genera and 187 species in southern Africa.

CYPHIA   Baroe

kyphos = bent; referring to the twining habit of many species

± 65 species in Africa, mostly southern Africa; 4 species in Baviaanskloof.

Cyphia digitata Vlaktebaroe

digitata = digitate; referring to the leaves that have several
lobes, making them appear finger-like

Resprouting, geophytic, twining perennial to
60 cm. **Leaves** 3–7-digitate with linear lobes,
sometimes slightly toothed. **Flowers** In the axils,
1 cm long, white with purple markings. Flowering
in winter and spring. **Habitat** Grassy Fynbos,
Transitional Shrubland and Thicket. **Distribution**
Widespread from Namaqualand to Gqeberha
(Port Elizabeth).

Rus en Vrede 4X4, Kouga 25.04.2018

Cyphia sylvatica Bosbaroe

sylvaticus = of the woods; referring to the habitat
that is usually somewhat woody

Resprouting, geophytic, twining perennial to 50 cm. **Leaves**
Simple, linear to lanceolate, obscurely toothed. **Flowers** 1 cm
long, white to mauve, pedicels and calyx hairless. Flowering
in spring. **Habitat** Found on rocky slopes in Thicket and
Transitional Shrubland. **Distribution** Widespread from Swartberg
to Makhanda.

Zandvlakte, Baviaans 01.05.2018

LOBELIA   Itshilizi (X)

Named after Matthias de L'Obel, 1538–1616, a Flemish botanist and court physician to King James I of
England. He was the first botanist to identify a difference between monocotyledons and dicotyledons.

± 450 species, mostly in tropical and warm temperate regions of the world. Some members are
toxic, while others are used in traditional medicine. Lobelias in southern Africa are generally
small herbs, but in the East African mountains there are giant species, of which *L. deckenii* is an
example. 69 species in South Africa; 7 species in Baviaanskloof.

Lobelia anceps

anceps = two-edged; referring to the winged stems

Sprawling perennial to 5 cm. **Leaves**
Oblanceolate, tapering, sparsely toothed and
hairless. **Flowers** 1.5 cm long, blue, mauve
or white; the 2 upper lobes distinctly narrow
and curved back. Flowering in summer and
autumn. **Habitat** Damp marshy areas, often
on stream banks. **Distribution** Widespread
from Cape Peninsula to KwaZulu-Natal.

Doodsklip, Kouga Dam 28.12.2017

Lobelia erinus Edging lobelia

erinos = an unidentified plant; alluding to the numerous
species and varieties in the genus, making identification tricky

Annual or perennial, erect or spreading, to 10 cm. **Leaves**
Ovate-elliptic, crowded basally, becoming smaller and
more linear up the stem, usually sparsely hairy, shallowly toothed.
Flowers Several on long stalks in a loose inflorescence, white to dark
blue, with white markings. Flowering in spring and early summer.
Habitat Damp situations in various habitats, mostly along stream
banks. **Distribution** Very widespread from Nieuwoudtville to tropical
Africa. **Other names** Yomqala (X), mahlo-a-konyana, tsoinyane (S),
impenjana (Z).

Poortjies, Cambria 26.10.2018

Lobelia linearis

linearis = linear; referring to the small, linear leaves, or to the rigid,
green stems that tend to have ridges or lines along their length

Resprouting, erect shrub to 1 m, stems usually grey-green. **Leaves**
Tiny, linear and obscure. **Flowers** 1.5 cm long, blue to purple with
white markings; calyx straw-coloured with dark ridges. Flowering
in spring, summer and autumn. **Habitat** Dry, stony slopes in Grassy
Fynbos and Transitional Shrubland. **Distribution** Widespread from
Cederberg to Baviaanskloof.

Suuranysberg, Kareedouw 21.01.2022

Lobelia neglecta Wild lobelia

neglecta = neglected; alluding to the diversity of the genus, which
sometimes results in some species being neglected or forgotten

Reseeding, upright, sparsely branched shrub to 50 cm. **Leaves**
Oblong, usually spreading, margins rolled under, hairy below, 1 cm
long × 0.4 cm wide. **Flowers** Few on long stalks, held horizontal,
1.5 cm long, bluish; calyx and tube with transparent hairs. Flowering
all year round. **Habitat** South-facing slopes in Mesic and Subalpine
Fynbos. **Distribution** Widespread from Mossel Bay to Makhanda.

12.10.2021

Lobelia tomentosa

tomentosa = hairy; leaves and stems are covered with fine, spreading hairs.

Reseeding, tufted shrublet, branching from woody base to 30 cm.
Leaves Ascending, narrow, toothed, hairy, 2.5 cm long × 0.3 cm
wide. **Flowers** Few on long stalks, 2.5 cm long, blue, pink or white,
with white markings. Flowering in summer and autumn. **Habitat**
Stony slopes in Grassy Fynbos. **Distribution** Widespread from Cape
Peninsula to southern Mozambique.

28.05.2022

MONOPSIS

mon = one, single, *opsis* = appearance; referring to the regular flowers in some species

13 species in southern and tropical Africa; 2 species in Baviaanskloof.

Monopsis unidentata Wild violet

uni = one, *dentata* = tooth; pertaining to the leaves
that have a single tooth

Sprawling perennial herb to 15 cm. **Leaves** 1 cm long × 0.3 cm
wide, usually with a single pair of teeth. **Flowers** 1 cm wide, purple
to coppery with darker centre. Flowering in spring, summer and
autumn. **Habitat** Shaded, damp situations, usually on stream
banks. **Distribution** Very widespread from Riviersonderend to
KwaZulu-Natal.

Riverside, Kouga River 24.05.2014

WIMMERELLA

Named after Franz Elfried Wimmer, 1881–1961, a Viennese botanist, naturalist and Roman Catholic priest
who studied the Lobeliaceae. A recently described genus (1999) that was formerly placed in *Laurentia*.

10 species in South Africa; 1 species in Baviaanskloof.

Wimmerella hederacea

hederacea = of or pertaining to ivy; alluding to the
leaves that are reminiscent of some species of ivy

Soft, delicate, decumbent, hanging perennial to 5 cm. **Leaves**
and stems densely hairy. **Leaves** Cordate-reniform, 5–9-lobed,
1 cm long. **Flowers** On long-hairy peduncles, 0.4 cm long, white.
Flowering in spring, summer and autumn. **Habitat** Occasional in
cracks on shaded cliffs in the narrow ravines in Thicket or Forest.
Distribution Widespread from Knysna to Makhanda.

Uitspankloof, Baviaans 19.02.2016

LORANTHACEAE   Showy mistletoes

77 genera and ± 950 species worldwide, mostly in tropical regions. Woody plants, most of them
hemiparasites. The largest genus is *Loranthus* with ± 500 species. 13 genera and ± 34 species in southern
Africa; 1 species in Baviaanskloof.

MOQUINIELLA

Named after Christian Horace Benedict Alfred Moquin-Tandon, 1804–1863, a French naturalist and prolific
author. He was the director of the botanic garden of Toulouse and a professor of botany in Paris.

1 species in Baviaanskloof.

Moquiniella rubra Vuurhoutjies

rubra = red; referring to the bright red flowers; the common name refers to the flowers' resemblance to matchsticks.

Stem parasite to 1 m wide. **Leaves** Elliptic, 4 cm long × 2 cm wide, smooth. **Flowers** In axillary clusters, tubular, splitting and lobes then coiling back, 3 cm long, orangey-red, tip green becoming black. Berry red. Flowering in autumn. **Habitat** Thicket and Succulent Karoo, on thorn trees and species of *Diospyros*, *Euclea* or *Searsia*. **Distribution** Widespread from Namaqualand to Komani (Queenstown) and Makhanda, Eastern Cape. **Other names** Lighted candles mistletoe (E), voëlent (A).

Geelhoutbos, Kouga 20.04.2012

MALVACEAE
Mallow, hibiscus, cotton

A large family of ± 244 genera and 4,225 species around the world. The common name is derived from *Malva*, a well-known genus and the origin of the word mauve, probably because many of the flowers have this colour. Large and well-known genera include *Hibiscus*, *Sterculia*, *Dombeya*, *Pavonia* and *Sida*. Plants of commercial importance include cotton, okra, cacao and durian.

ABUTILON
Wildemalva

Abutilon = the Arabic name for a species in this genus

150 species, cosmopolitan; 20 species in South Africa; 1 species in Baviaanskloof.

Abutilon sonneratianum Yendlebe (X)

Named after Pierre Sonnerat, 1745–1814, French naturalist, explorer, artist and author. Many species have been named after him, including the grey junglefowl (*Gallus sonneratii*).

Reseeding, erect, velvety shrub to 2 m. **Leaves** 4 cm long × 2 cm wide, paler below, margins toothed. **Flowers** 1.5 cm wide, usually yellow or orange. The seed pod is symmetrically attractive, with numerous pointed ridges. Flowering in summer. **Habitat** Disturbed areas in Thicket and Transitional Shrubland. **Distribution** Widespread from Bredasdorp to Zimbabwe. **Notes** Apparently used to strengthen cattle bulls in spring. **Other names** Wild hibiscus (E), ibhosisi (X).

Witruggens, Kouga 04.11.2011

ANISODONTEA

African mallow, bergroos

aniso = unequal, *odontus* = toothed; pertaining to the irregularly toothed leaf margins

20 species in southern Africa; 1 species in Baviaanskloof.

Anisodontea scabrosa   Sandroos

scabrosa = distinctly scabrous; pertaining to the leaves that feel rough to the touch

Reseeding, erect, glandular-hairy shrub to 2 m. **Leaves** Variable, usually 3-lobed, toothed. **Flowers** 3 cm wide, pale pink to pink with darker markings. Flowering in spring and summer. **Habitat** Found sporadically in Grassy Fynbos. **Distribution** Widespread from Saldanha to KwaZulu-Natal. **Other names** Wildestokroos, harige malva (A).

Bosrug, Baviaans 14.03.2012

GREWIA

Cross-berry, kruisbessie

Named after Nehemiah Grew, 1641–1712, an English botanist who pioneered the study of plant physiology; he is recognised as one of the founders of the science of plant anatomy.

400 species in Africa, Asia and Australia; 24 species in South Africa; 2 species in Baviaanskloof.

Grewia occidentalis

occidentalis = western; referring to the geographic origin of the species

Resprouting shrub or small tree to 3 m with tendency to spread through other bushes. **Leaves** Petiolate, 3-veined from the base, hairless, toothed, 5 cm long × 3 cm wide. **Flowers** 4 cm wide, pink or purple, rarely white. Fruit 4-lobed, becoming reddish, attracting birds, especially mouse birds and bulbuls. Flowering in spring. **Habitat** Margins of Forest or in Thicket. **Distribution** Widespread from Cape Peninsula to Zimbabwe. **Notes** Fast growing, good for hedges. **Other names** Umvilani, umnqabaza, uhlolo oluncinci, unyenye (X), morretla (S).

Kamassiekloof, Zandvlakte 18.10.2011

Grewia robusta   Karookruisbessie

robusta = robust; pertaining to the relatively stout habit and hardy nature

Resprouting, much-branched, tough shrub to 3 m. **Leaves** Often clustered on short shoots, ovate, 3-veined from the base, 2 cm long × 1 cm wide, grey-velvety below, finely toothed. **Flowers** 1.5 cm wide, bright pink. Fruit 4-lobed, purplish. Flowering in spring and summer. **Habitat** Fairly common in dense bush, mostly in Thicket and Transitional Shrubland. **Distribution** Widespread from Oudtshoorn to Beaufort West and to Peddie. **Other name** Umnqabaza (X).

Geelhoutbos bergpad, Kouga 19.04.2012

HERMANNIA

Doll's rose, poprosie

Named after Paul Hermann, 1646–1695, Dutch botanist, herbalist and professor of botany at Leyden. He explored in Africa, India and Sri Lanka, and made one of the earliest scientific plant collections in the Cape; it is now housed at the Sloane Herbarium, British Museum of Natural History.

The common name pleisterbos refers to the healing properties of some species. ± 200 species, mostly from southern Africa; 18 species in Baviaanskloof.

Hermannia althaeifolia

Althaea = plant genus, *folia* = leaf; the leaves resemble those of some species of *Althaea*, a genus in the same family, native to Europe and Asia.

Reseeding, erect or spreading, softly hairy shrublet to 50 cm. **Leaves** 3 cm long × 1.5 cm wide, grey-green, densely hairy, lobed and toothed. **Flowers** In terminal and axillary clusters, 1 cm wide, calyx inflated, becoming cream-coloured. Flowering in spring. **Habitat** Clay or loam-rich soils in Transitional Shrubland. **Distribution** Narrow range from the Little Karoo to Kareedouw.

28.05.2014

Hermannia althaeoides

Althaea = plant genus, *oides* = resembling; probably referring to the resemblance of the leaves to some species of *Althaea*

Reseeding, sprawling, slightly mealy shrublet to 60 cm. **Leaves** Petiolate, ovate, 2 cm long × 1 cm wide, paler below, finely toothed. **Flowers** In loose terminal or axillary clusters, 0.5 cm wide, yellow fading red. Flowering in spring and early summer. **Habitat** Found on rocky sandstone slopes in Grassy Fynbos and Transitional Shrubland. **Distribution** Eastern distribution from Humansdorp to Komani (Queenstown).

Engelandkop, Kouga 23.07.2011

Hermannia coccocarpa   Pink dollsrose, moederkappie

coccus = round, *carpa* = fruit; pertaining to the round fruit

Low, spreading, twiggy shrublet to 25 cm. **Leaves** Linear-oblanceolate, hairless, 2.5 cm long × 0.8 cm wide, toothed. **Flowers** 1 or 2, nodding on slender peduncles, 0.8 cm wide, white or pink or purple. Capsules oblong. Flowering in spring, summer and autumn. **Habitat** Succulent Karoo. **Distribution** Widespread through drier parts of southern Africa. **Other name** Umbovu (X).

Uitspan, Baviaans 30.01.2020

Hermannia filifolia

fili = thread, *folia* = leaf; referring to the thin, thread-like leaves

Resprouting, twiggy shrublet to 1 m. **Leaves** In tufts, linear, 2 cm long × 0.2 cm wide, margins revolute. **Flowers** Few on slender branches, 0.5 cm wide, orange to red, calyx spreading. Flowering in spring. **Habitat** Infrequent on heavier loams or clays in Transitional Shrubland and Succulent Karoo. **Distribution** Very widespread from Nieuwoudtville to the Free State.

Elandsvlakte, Baviaans 16.02.2016

Hermannia flammea   Poprosie

flammeus = flame-coloured, fiery red; referring to the bright red flowers

Resprouting, semi-erect, sparsely branched shrub to 60 cm. **Leaves** Oblanceolate to cuneate, 2 cm long × 0.3 cm wide, sometimes toothed. **Flowers** 0.5 cm wide, orange to red, calyx spreading. Flowering in spring. **Habitat** Common in Grassy Fynbos. **Distribution** Widespread from Wellington to Makhanda, Eastern Cape. **Other name** Umsolo wetafa (X).

Langkloof 19.01.2019

Hermannia gracilis

gracilis = thin, slender; probably referring to the lanky growth form or to the thin leaves

Reseeding, slender, sparsely branched shrub to 2 m. **Leaves** Oblanceolate, 3 cm long × 0.5 cm wide, grey-velvety. **Flowers** On slender, terminal racemes, 0.5 cm wide, yellow and reddish, calyx shortly-lobed and cup-shaped. Flowering in spring. **Habitat** Rocky slopes in Thicket and Succulent Karoo. **Distribution** Widespread from Calitzdorp to Makhanda.

Zandvlakte, Baviaans 06.07.2011

Hermannia holosericea

holo = entire, *sericea* = silky; referring to the foliage that is completely covered in silky hairs

Reseeding, erect, greyish shrublet to 1.5 m. **Leaves** 2.5 cm long × 0.7 cm wide, silvery-grey-green, usually sparsely toothed near the apex. **Flowers** Several in racemose clusters, relatively small, 0.3 cm wide, yellow. Flowering in spring. **Habitat** Clay or loam soils in Transitional Shrubland, especially after fire. **Distribution** Widespread from Worcester to Jeffreys Bay.

Engelandkop, Kouga 23.07.2011

Hermannia mucronulata

mucron = abruptly terminated by a hard and short
point, *ulata* = having; referring to the mucronate leaf tips

Reseeding, erect, twiggy shrub to 2 m. **Leaves** Oblanceolate, entire,
black-mucronate, lime-green. **Flowers** Urn-shaped, deep yellow,
calyx very inflated. Flowering in spring. **Habitat** Localised on loamy
slopes in Grassy Fynbos and Transitional Shrubland, especially
after fire. **Distribution** Eastern Cape endemic from Baviaanskloof to
Kariega (Uitenhage).

Enkeldoring, Baviaans 22.09.2011

Hermannia saccifera

saccatus = bag-shaped, pouched, saccate, *ifera* = having;
referring to the calyx and the fruiting capsule

Resprouting, multi-stemmed, sprawling shrublet to 40 cm.
Leaves Hairless, slightly resinous, 4 cm long × 1 cm wide,
bright green, toothed. **Flowers** Flaring to 1.5 cm wide, pale
yellow, with cup-shaped bracts. Flowering in spring. **Habitat**
Common on richer soils with rocks in Grassy Fynbos and
Transitional Shrubland, especially after fire. **Distribution**
Widespread from Riviersonderend to Gqeberha (Port Elizabeth).

Rus en Vrede 4X4, Kouga 25.04.2018

Hermannia salviifolia

Salvia = plant genus, *folia* = leaf; referring to the leaves that
look like those of *S. farinosa*

Reseeding, erect, rusty-yellow velvety shrub to 2 m. **Leaves** Obovate,
entire, 1.5 cm long × 0.7 cm wide, sometimes slightly toothed,
and velvety-hairy. **Flowers** In dense terminal clusters, yellow or
orange, 0.5 cm wide, calyx inflated. Flowering in spring. **Habitat**
Locally abundant after fire on loam-rich soils in Grassy Fynbos and
Transitional Shrubland. **Distribution** Widespread from the Cape
Peninsula to Makhanda. Often confused with *H. involucrata*, but that
species has larger leaves (3 cm long) that are often toothed.

Bosrug, Baviaans 27.07.2011

Hermannia stipulacea

stipulaceus = of stipules; referring to the prominent stipules

Reseeding, erect, densely hairy shrub to 50 cm. **Leaves** Ovate to
cuneate, ascending, toothed, with large and leafy stipules. **Flowers**
Few in terminal clusters, yellow, with golden-velvety calyx. Flowering
in spring. **Habitat** Sporadic on rocky sandstone slopes in Mesic and
Grassy Fynbos, especially after fire. **Distribution** Widespread from
Riversdale to Gqeberha (Port Elizabeth).

Kouenek, Kouga 26.09.2011

Hermannia velutina

velutinus = velvety; referring to the leaves and calyx that are densely covered with short soft hairs

Reseeding, erect, grey-mealy shrub to 3 m. **Leaves** Narrowly oblanceolate, 3.5 cm long × 1 cm wide, grey-velvety, with leafy and linear stipules. **Flowers** Few in dense terminal clusters, 0.5 cm wide, yellow, calyx lobes acuminate. Flowering in spring. **Habitat** Found on loamy soils in Grassy Fynbos and Transitional Shrubland. **Distribution** Widespread from Joubertina to KwaZulu-Natal.

Guerna, Kouga 16.02.2016

HIBISCUS

hibiskos = the classical name for marshmallow (*Althaea officinalis*), to which this genus bears a vague resemblance

2 species with large showy flowers, *H. syriacus* and *H. rosa-sinensis*, are popular and familiar garden plants. In India, *H. cannabinus* is grown as a fibre plant. Used in traditional medicine. ± 300 species in tropics and subtropics; ± 50 species in South Africa; 3 species in Baviaanskloof.

Hibiscus aethiopicus   Common dwarf wild hibiscus

aethiopicus = from sub-Saharan Africa; in classical times this was the region that had not yet been explored by Europeans.

Resprouting from woody rootstock, dwarf shrublet to 10 cm. **Leaves** Ovate-elliptic, 3–5-veined from base, golden-haired, 2.5 cm long × 1.5 cm wide, toothed. **Flowers** 3 cm wide, white to yellow, sometimes with a dark centre. Flowering in spring and summer. **Habitat** Common after fire in Grassy Fynbos and Transitional Shrubland. **Distribution** Widespread from Wellington to KwaZulu-Natal. **Other names** Iyeza lamasi (X), lereletsane, se-seholo (S), elimhlophe, uvemvane (Z).

29.03.2014

Hibiscus aridus

aridus = arid; referring to the hot, dry and low-rainfall conditions of the preferred habitat

Slender, weak-stemmed, glandular-hairy shrublet to 30 cm. **Leaves** Oval to cordate, petioled, rough to touch, 4 cm long, irregularly toothed. **Flowers** 4 cm wide, bright yellow, on long, hairy flower stalks, calyx hairy. Flowering in summer and autumn. **Habitat** Infrequent and uncommon on clay soils in arid Thicket and Transitional Shrubland. **Distribution** Narrow from Baviaanskloof to Fish River Valley.

Zandvlakte, Baviaans 19.01.2022

Hibiscus pusillus   Dark-eyed dwarf hibiscus

pusillus = very small, dwarf; referring to the small, low habit of this plant

Resprouting from woody rootstock, dwarf shrublet to 30 cm. **Leaves** Variable, 3–5-palmatisect, smooth, hard-textured, and usually prominently toothed. **Flowers** 3 cm wide, white, yellow or pink, often with a dark centre. Flowering in spring and summer. **Habitat** Sporadic in Transitional Shrubland and Grassy Fynbos, especially after fire. **Distribution** Very widespread from Caledon to tropical Africa. **Other names** Umzongwane (X), uguqukile, uvuma (Z).

Engelandkop, Kouga 19.04.2012

MELHANIA

Named after Mount Melhan on the Arabian Peninsula, where one of the species occurs

± 60 species in Africa and India; 13 species in South Africa; 1 species in Baviaanskloof.

Melhania didyma

didyma = in pairs; pertaining to the flowers, which are borne in pairs

Resprouting from a woody base, rough-hairy shrub to 1 m. **Leaves** Large, 10 cm long × 3 cm wide, slightly hairy above, grey-felted below, and shallowly toothed. **Flowers** In pairs, 5 cm long, yellow or orange, calyx densely hairy and cream-coloured. Flowering in summer and autumn. **Habitat** Infrequent on loamy soils in Grassy Fynbos and Transitional Shrubland. **Distribution** Widespread from Baviaanskloof to Mpumalanga.

Doringkraal, Kouga 17.04.2012

SPARRMANNIA

Named after Anders Sparrman, 1748–1820, a Swedish botanist and physician, and pupil of Linnaeus. He visited the Cape from 1772–1776 and was the first to publish an extended account of his travels into the interior. He collected this plant in Baviaanskloof in 1775.

7 species in Africa and Madagascar; 1 species in Baviaanskloof.

Sparrmannia africana   Cape stock rose

africana = from Africa; referring to the geographical origin

Soft-wooded shrub or tree to 4 m tall, with bristly branches. **Leaves** Heart-shaped with up to 9 veins from the base, soft-textured, covered in bristly hairs, 25 cm long × 20 cm wide, margins toothed. **Flowers** Showy, drooping in umbels, white with numerous yellow and red stamens. Capsules covered in bristles. Flowering in winter and spring. **Habitat** Occurring along streams and at Forest margins. **Distribution** Widespread from Riversdale to Gqeberha (Port Elizabeth). **Notes** The stamens puff out when touched. Bristles can be a skin irritant. Popular in gardens. **Other names** Cape hollyhock (E), stokroos (A).

19.06.2019

STERCULIA

Named after Sterculius, the god of manure in Roman mythology, referring to the bad-smelling flowers of some species, notably *S. foetida*, the first member of the genus to be described by Linnaeus in 1753. In India, karaya gum is extracted from *Sterculia* species and mainly used as a thickener and emulsifier in foods.

91 species worldwide; 3 species in South Africa.

Sterculia alexandri   Cape star chestnut, sterkastaiing

Named after Richard Chandler Prior (born Richard Chandler Alexander), 1809–1902, briefly curator of the Fielding Herbarium, Oxford, and the first to collect and record the plant in 1847

Tree to 8 m tall, spreading by suckering. Bark smooth and silvery. **Leaves** Digitately compound, with 3–7 elliptic leaflets diverging from a stout leaf stalk to 5 cm long, leaflets to 13 cm long × 2 cm wide. **Flowers** In axillary cymes, hanging downwards, 2 cm wide, yellow with reddish marking in the centre. Flowering mainly in autumn and winter. **Habitat** Rarely encountered, restricted to rocky slopes in Thicket. **Distribution** Known from a few populations between Kouga Dam, Kariega (Uitenhage) and Van Staden's Mountains. **Status** Rare.

Elandsberg Mountains, Uitenhage 11.06.2009

MELIACEAE   Mahogany

53 genera and ± 600 species worldwide. Well known for the quality timber from some of the tree species. 6 genera and 10 tree species in South Africa; 1 indigenous species and 1 exotic species in Baviaanskloof.

NYMANIA

Named after Carl Fredrik Nyman, 1820–1893, a Swedish botanist who worked on European plants

1 species in Baviaanskloof.

Nymania capensis   Klapperbos, Chinese lantern bush

capensis = from the Cape; referring to the geographical origin of the species

Tall, slender, rigid shrub to 5 m. **Leaves** Clustered on short shoots, oblanceolate, 2.5 cm long × 1 cm wide. **Flowers** Single, in the axils, 1 cm wide, dull red. Fruit inflated and papery. Flowering in spring or after good rain. **Habitat** Occurring in Thicket and Succulent Karoo. **Distribution** Very widespread from Namibia to Addo, Eastern Cape.

Rus en Vrede 4X4, Kouga 28.07.2011

MELIANTHACEAE

A small family of 3 genera (*Melianthus*, *Bersama*, *Greyia*) and 11 species in tropical and southern Africa.

MELIANTHUS

meli = honey, *anthos* = flower; referring to the flowers that produce copious amounts of black nectar, attracting sunbirds

1 genus and 2 species in Baviaanskloof.

Melianthus comosus   Kruidjie-roer-my-nie

comosus = bearing a tuft of hairs or leaves; pertaining to the densely leafy nature of the plant. The common name refers to the unpleasant smell produced when the leaves are disturbed.

Reseeding, upright shrub to 1.5 m. **Leaves** Dissected, slightly hairy, 30 cm long, margins toothed. **Flowers** 2.5 cm wide, red marked with black. Flowering in spring and early summer. **Habitat** Sporadic on slopes and stream banks in Thicket and Succulent Karoo. **Distribution** Very widespread from Namibia to Bavlaanskloof and northwards to Venterstad. **Other names** Touch-me-not (E), ubutyayi, ubuhlungu bemamba, irhabiya, isidwadwa (X).

Rus en Vrede, Baviaans 05.09.2015

MOLLUGINACEAE   Carpet weed

Herbs and dwarf shrubs with fleshy leaves. The family has recently been re-organised, based on molecular data. Currently includes 8 genera and 72 species in southern Africa, mostly in the Western Cape, especially Namaqualand; 2 species in Baviaanskloof.

PHARNACEUM   Sneeuvygie

Named after Pharnaces II, 63–47 BC, a monarch of Greek and Persian ancestry, son of the King of Pontus in northern Anatolia on the Black Sea

25 species in southern Africa; 1 species in Baviaanskloof.

Pharnaceum dichotomum

dichotomum = having divisions always in pairs; referring to the branching pattern of the inflorescence

Small, reseeding shrublet to 30 cm. **Leaves** Mostly basal, but also in whorls on the stem, linear, 0.4 cm long × 0.1 cm wide. **Flowers** In dichotomously branched inflorescences on long peduncles, each flower 0.3 cm wide, white. Flowering in winter and spring. **Habitat** Dry, rocky slopes in Thicket and Succulent Karoo. **Distribution** Very widespread from Namaqualand and Karoo to Eastern Cape and Free State.

Voetpadskloof, Baviaans 27.04.2018

PSAMMOTROPHA

psamos = sand, *trophos* = feeder; pertaining to the plants' preferred habitat in sandy soil

11 species from Tanzania to southern Africa; 1 species in Baviaanskloof.

Psammotropha quadrangularis Kruisblaartjie

quadrangularis = four-angled; referring to the 4-ranked leaves

Tiny, gnarled, closely leafy shrublet to 20 cm. **Leaves** 4-ranked, ascending, overlapping, 0.4 cm long × 0.1 cm wide, spine-tipped. **Flowers** Tiny, in dense clusters, 0.3 cm wide, greenish cream to pinkish. Flowering in spring. **Habitat** Sporadic in open, stony areas, often on conglomerate outcrops, in Succulent Karoo and Thicket. **Distribution** Widespread from Namaqualand to Baviaanskloof.

Uitspan, Baviaans 05.05.2006

MONTINIACEAE

± 14 genera and 230 species of cosmopolitan plants; 1 species in southern Africa.

MONTINIA

Named after Lars Jonasson Montin, 1723–1785, a Swedish botanist and pupil of Linnaeus

1 species in Baviaanskloof.

Montinia caryophyllacea Bergklapper, wild clove bush

Caryophyllus = plant genus; pertaining to the flowers that superficially resemble those of cloves (*Syzygium aromaticum*) (– *C. aromaticus*)

Resprouting, root-suckering, erect, rigid, sparsely leaved, dioecious shrub to 1.5 m. **Leaves** Oblanceolate, hairless, 3.5 cm long × 1.5 cm wide, grey-green. **Flowers** 0.7 cm wide, white. Fruit a persistent, woody, 2-lobed capsule, 2 cm long, with wind-dispersed seeds. Flowering mainly in winter. **Habitat** Common on rocky slopes in sandy-loam soils in Grassy Fynbos and Transitional Shrubland. **Distribution** Very widespread from Angola to Cape Peninsula to Makhanda.

Zandvlakte, Kouga 18.01.2022

MORACEAE

Mulberry, fig

± 38 genera and 1,100 species, mostly in tropical and subtropical regions. Well-known for edible fruits including the fig, banyan, breadfruit and mulberry. *Ficus* (fig) is the largest genus, with 750 species; 25 species of *Ficus* in South Africa.

FICUS

Fig

ficus = the Latin name for the cultivated fig, *F. carica*

Plants have a milky latex, simple leaves that are usually alternate, and flowers inside the fleshy fruit or fig. Small wasps go inside the fruit to lay their eggs and pollinate the flowers. 2 species in Baviaanskloof.

Ficus burtt-davyi Veld fig, nou-tou

Named after Joseph Burtt Davy, 1870–1940, a British-born botanist who worked at Kew Gardens; 11 plants have been named after him. He founded organisations in South Africa that have since evolved into SANBI, and published field manuals to plants from the former Transvaal province and Eswatini in 1926 and 1932 respectively. The Afrikaans common name alludes to the use of the long, strong roots as tow ropes.

Long-rooted, often cliff dwelling, with spreading branches to 5 m. **Leaves** Elliptic, leathery, 3 cm long × 1.5 cm wide, exuding milky latex. **Flowers** Many tiny flowers are inside the fruit. Fruit 0.7 cm wide, green becoming yellowish. Flowering in summer. **Habitat** Rocky outcrops and cliffs of narrow ravines. **Distribution** Widespread from Oudtshoorn to KwaZulu-Natal. **Other names** Umdendekwana (X), uluzi (X, Z), umthombe (Z).

Kloof to west of Bosrug, Baviaans 14.07.2016

Ficus sur Cape fig

Named after the Sur region in Ethiopia

Majestic, tall, dark grey-barked, massive-rooted tree to 30 m. **Leaves** Elliptic to ovate, relatively large, with broad teeth. **Flowers** Many tiny flowers are inside the fruit. Figs in bunches hanging from trunk and branches, 4 cm wide, becoming red. Flowering in summer. **Habitat** Common along perennial streams, especially in the narrow ravines in Thicket. **Distribution** Widespread from Knysna to North Africa and the Arabian Peninsula. **Other names** Umkhiwane, uluzi (X, Z), mphayi (S).

Zandvlaktekloof, Baviaans 18.02.2016

MYRICACEAE

Wax-myrtle

4 genera and ± 57 species in this small family of shrubs and trees. Leaves and stems are usually aromatic, with resinous gland dots, and there are nitrogen-fixing nodules in the roots. 1 genus and 10 species in South Africa.

MORELLA (= *Myrica*)

Morus = plant genus, *ella* = diminutive; *Morus* is the genus for mulberries, alluding to the fruits, which resemble small mulberries.

± 50 species worldwide; 10 species in South Africa; 1 species in Baviaanskloof.

Morella serrata Waterolier

serrata = serrated; referring to the serrated leaf margins

Resprouting, dioecious, upright, resin-dotted shrub or small tree to 4 m. **Leaves** With net veins clearly visible, usually gland-dotted below, 8 cm long × 2 cm wide, olive-green, serrated. **Flowers** Tiny, in short spikes to 3 cm long in the axils. Fruit 0.4 cm wide. Flowering in spring. **Habitat** Rocky stream banks of perennial rivers. **Distribution** Widespread from Wellington to Mpumalanga. **Other names** Lance-leaved waxberry (E), berg-wasbessie (A), isibhara (X), umakhuthula (X, Z), iyethi (Z).

Poortjies, Cambria 22.01.2022

MYRSINACEAE

Myrsine

Relatively small family with 35 genera and 1,000 species. *Myrsine africana* has been grown as an ornamental in England. 4 genera and 7 species in South Africa.

MYRSINE

Derived from *murtus*, the Greek name for myrtle, because the leaves vaguely resemble those of this plant

2 species in South Africa; 1 species in Baviaanskloof.

Myrsine africana Cape myrtle, African boxwood

africana = from Africa; referring to the geographical origin of the species

Resprouting, root-suckering, erect, leafy shrub or small tree to 3 m. **Leaves** Small, obovate, 2 cm long, glossy, slightly toothed, with margins rolled under. **Flowers** In axillary clusters, small with anthers exserted. Fruit 0.4 cm wide, fleshy, dark becoming pinkish. Flowering all year round. **Habitat** Rocky areas at the margins of Forest and in Grassy Fynbos and Transitional Shrubland. **Distribution** Very widespread from Nieuwoudtville to tropical Africa and the Azores. **Other names** Umbovini, iyeza lenkomo (X), thakxisa (S).

28.11.2021

RAPANEA

Cape beech, Kaapse boekenhout

Probably derived from the Guinean name for a species in this genus

± 200 species in tropical and subtropical parts of the world; 2 species in South Africa; 1 species in Baviaanskloof.

Rapanea melanophloeos

melano = dark, black, *phloeos* = bark; pertaining to the dark bark of older plants, which can be misleading since younger plants have pale grey bark

Dioecious, upright tree to 20 m tall with purplish twigs at the branch tips. **Leaves** Oblong-elliptic, leathery, 10 cm long × 3 cm wide, dark green above, paler below, with purplish petioles 1.5 cm long. **Flowers** In clusters on old twigs, faintly scented, 0.3 cm long, greenish white. Fruit round, 0.5 cm wide, fleshy, purple, eaten and dispersed by baboons, monkeys and birds. Flowering in winter. **Habitat** Uncommon tree in Baviaanskloof in Forest, but also found stunted on rocky outcrops in Fynbos. **Distribution** Widespread from the Cape Peninsula to tropical Africa. **Notes** Fine-grained wood has been used for cabinets and violins. Bark is used in traditional medicine. **Other names** Isiqalati, isiqwane oohlati, udumo olubomvu, umemezi, umaphipha, isiqwandwemshube (X), intlungunyothu, olphluthe, isicalabi, umaphiphakhubalo (Z).

Boskloof, Kouga Wildernes 15.05.2021

MYRTACEAE

Myrtle, eucalyptus, guava

Large family of nearly 6,000 species in 132 genera. Better-known plants include eucalyptus, guava, allspice and clove. 4 genera and 25 native species in southern Africa, but there are many more naturalised or invasive species. The indigenous genera in South Africa are *Metrosideros*, *Heteropyxis*, *Eugenia* and *Syzygium*. No native Myrtaceae members in Baviaanskloof.

EUCALYPTUS*

Gum

eu = good, *kalypto* = cover; pertaining to the bud cover, or operculum, which is a typical character of the genus

Over 700 species, mostly in Australia; several naturalised and invasive species in South Africa.

*Eucalyptus botryoides** Bangalay

botrys = grapes, *oides* = resembling; the dangling clusters of woody fruit vaguely resemble a bunch of grapes.

Massive tree to 40 m tall with rough, reddish-brown and fissured bark. Resprouting if cut or burnt. Young twigs reddish. **Leaves** Broad-lanceolate, 15 cm long, dark green above and paler below. **Flowers** In groups of 6–11, roundish, white, buds club-shaped and pale green. Fruit a woody, ovoid to cylindrical capsule, 1 cm long, with the valves at, or sunken below, the rim. **Habitat** Invading dry riverbeds. **Distribution** Native to Australia. Invasive in South Africa, but far less common than *E. camaldulensis*.

Nuwekloof Pass, Baviaans 19.05.2021

*Eucalyptus camaldulensis** River redgum

Named after the Camaldoli Monastery in Naples, Italy. The type specimen was grown from seed collected in 1817 in a private garden near the monastery.

Massive tree to 40 m tall with smooth, stripping bark. **Leaves** Lance-shaped, 20 cm long, blue-grey. **Flowers** In groups of 7—11, roundish, white, buds creamy yellow. Fruit a woody, round capsule up to 1 cm wide, with valves raised above the rim. **Habitat** Invading dry riverbeds. **Distribution** Native to Australia, it has spread in many parts of the world and has invaded in most parts of South Africa. It is spreading along the Baviaanskloof River and urgently needs to be eradicated because it sucks up the groundwater.

Zandvlakte, Homestead 19.05.2021

OCHNACEAE

32 genera and 550 species of mostly shrubs and trees in this relatively small, pantropical family. The unusual leaves are quite distinct, mostly with closely set parallel veins and toothed margins. 2 genera and 13 species in South Africa; 1 species in Baviaanskloof.

OCHNA

ochne = wild pear; referring to the leaves that superficially resemble those of the pear

85 species mostly in Africa and Madagascar; 12 species in South Africa.

Ochna serrulata Mickey mouse bush

serrulata = serrulate; alluding to the finely toothed, serrulate leaf margins; the colouring and shape of the shiny black fruit and red sepals bear a vague resemblance to the popular Disney Mickey Mouse character.

Resprouting, stout shrub or small tree with smooth brown bark covered with white dots. **Leaves** Elliptic, 3 cm long, toothed, glossy; midrib and parallel lateral veins visible above. **Flowers** 2 cm wide, yellow. Fruit with up to 6 berries, green becoming shiny-black, attached to bright red, persistent sepals. Flowering in summer. **Habitat** Rocky areas in Grassy Fynbos, Transitional Shrubland and Thicket. **Distribution** Widespread from Western Cape to KwaZulu-Natal. **Notes** Roots are used medicinally by the Zulu people. **Other names** Fynblaarrooihout (A), ilitye (X).

Nooitgedacht, Kouga
24.09.2011

OLEACEAE

A subcosmopolitan family of 25 genera and ± 700 species. Better-known members include olive, ash and jasmine. Mostly trees, shrubs and a few lianas, often with fragrant flowers. 5 genera and 25 species native to southern Africa; 3 genera and 4 species in Baviaanskloof.

OLEA

Olea = Latin name for olive

± 40 species worldwide. Trees and shrubs with opposite leaves and drupe fruit. 4 species in South Africa; 2 species in Baviaanskloof.

Olea capensis subsp. *capensis*   Klein ysterhout, small ironwood

capensis = from the Cape; referring to the geographical origin of the species

Resprouting shrub or small tree to 6 m. **Leaves** Opposite, elliptic-ovate, to 10 cm long × 5 cm wide, glossy and dark green above, paler below. **Flowers** Small, in terminal panicles, sweetly scented, white. Fruit to 1 cm wide, becoming purple-black, consumed by birds. Flowering in summer. **Habitat** Rocky outcrops in Grassy Fynbos or in Forest. **Distribution** Widespread mostly near the coast from Clanwilliam and the Cape Peninsula to KwaZulu-Natal. **Other names** Ugqwangxe, umdlebe, umnqumaswili (X).

Boskloof, Kouga Wildernes 15.05.2021

Olea europaea subsp. *africana*   Wild olive, olienhout

europaea = from Europe, *africana* = from Africa; the species has a wide distribution from the Mediterranean region to Africa and the Himalayas.

Resprouting, gnarled, well-crowned tree to 10 m tall with rough grey bark. **Leaves** Opposite, 5 cm long × 1.5 cm wide, grey-green or dark green above, greyish below. **Flowers** Numerous in racemes from the leaf axils, small, white, fragrant. Fruit a drupe, 1 cm long, green becoming black. **Habitats** Occurring in a wide variety of habitats, but most common in Baviaanskloof in the narrow ravines. **Distribution** Very widespread from southern Africa to the Arabian Peninsula and China. **Notes** There are 6 subspecies: subsp. *africana* is the only native subspecies in South Africa. There are hundreds of cultivars, which are used to produce table olives and olive oil. **Other names** Umnquma (X), umhlwathi, umnqumo (Z).

Zandvlakte bergpad, Kouga 03.12.2015

OROBANCHACEAE

Broomrapes

Mostly parasitic plants with ± 90 genera and 2,000 species worldwide. Some species in the genera *Orobanche* and *Striga* are problematic in agriculture, infesting certain crops. Most are root parasites and lack the ability to photosynthesise. In some genera, the flowers are the only parts that appear above ground. 15 genera and 113 species in southern Africa; 3 genera and 5 species in Baviaanskloof.

GRADERIA

Wild penstemon

Graderia = an anagram of Gerardia; named after John Gerard, 1545–1612, an English surgeon, herbalist and author. His book, published in 1597, *The Herball, or generall historie of plantes*, is still popular with those interested in herbal remedies.

5 species in Africa and Socotra; 3 species in South Africa; 1 species in Baviaanskloof.

Graderia scabra Pink ground-bells

scabra = rough; the leaves are covered with short, rigid hairs that feel rough when touched

Resprouting from a woody rootstock, shrublets to 60 cm. **Leaves** Overlapping, 3–5 veined, covered with rough hairs, 3 cm long, sparsely toothed. **Flowers** In the upper axils, funnel-shaped, 3 cm long, pinkish. Flowering in spring. **Habitat** Grassy Fynbos and Sour Grassland. **Distribution** Widespread from Langeberg to Mpumalanga. **Other names** Uvelabahleke (X), impundu, isimonyo, ugweja (Z).

Cockscomb, Groot Winterhoek 16.02.2016

HARVEYA

Ink flower, inkblom

Named after William Henry Harvey, 1811–1866, Irish botanist and author, regarded as a pioneer of South African botany. He came to the Cape between 1834 and 1838 and was the main author of *Flora Capensis*.

A genus of root parasites that lack chlorophyll. The flowers turn black on maturing and have been used for writing ink, and medical remedies. ± 35 species in Africa and the Mascarenes; 12 species in South Africa; 3 species in Baviaanskloof.

Harveya purpurea

purpurea = purple; a misnomer since the flowers are pale yellow or white to pink, but the calyx is purplish

Root parasitic herb to 15 cm. **Leaves** Reduced to scales on the roots. **Flowers** 3 cm long × 4 cm wide, white to pink with a yellow throat, calyx deeply lobed. Flowering in spring and summer. **Habitat** Occasional in Grassy and Mesic Fynbos. **Distribution** Widespread from the Cederberg to Makhanda. 3 subspecies, but only subsp. *purpurea* occurs in Baviaanskloof.

Heights, Langkloof 30.10.2020

HYOBANCHE

Red broomrape, katnaels, wolwekos

hyobanche = derivation unresolved

8 species in southern Africa; 1 species in Baviaanskloof.

Hyobanche sanguinea

sanguineus = blood-red; referring to the colour of the flowers; they can also be pinkish.

Root parasite, living underground except when in flower. **Leaves** Scale-like with obtuse bracteoles. **Flowers** Densely hairy, red or pink, stigma white and protruding slightly, stamens inside the flower tube. Flowering in spring. **Habitat** Uncommon in Grassy Fynbos. **Distribution** Widespread from southern Namibia to Makhanda.

Groendal, Kariega 07.10.2018

OXALIDACEAE

Wood sorrels

6 genera and 775 species worldwide, mostly within *Oxalis*. Herbs, shrubs and trees with divided leaves that can close at night and open in the morning, a trait known in the plant world as sleep movements or, more formally, nyctinasty. All resprout after fire. The only family of dicots that produces corms (tuberous, bulb-like rootstocks), which are eaten and dispersed by many different animals and birds, especially porcupines and francolins. Well known as sour-sucks (or surings in Afrikaans) amongst children who 'chew' the acidic juice from the flowering stems of *Oxalis pes-caprae*, which is also used to flavour stews, especially waterblommetjie stew. 2 genera in southern Africa.

OXALIS

Sorrels, surings, umncwane (X)

oxis = acid; referring to the sour-tasting sap in some species

The shape of the bulb is the best way to tell the species apart. The styles come in 3 different lengths to ensure cross-pollination. ± 700 species worldwide; ± 270 species in southern Africa; 13 species in Baviaanskloof.

Oxalis caprina Goat's foot

caprina = a goat; possibly referring to the bilobed shape of the leaflets, which reminded Linnaeus of the bilobed hoof of a goat; alternatively, the pointed bulb looks somewhat similar to a goat's toe, or the tracks left by goat hooves.

Small, soft geophyte to 15 cm tall, stem absent or very short. **Corm** Small and rounded, 1 cm wide, with a pointed tip. **Leaves** Trifoliate, bilobed to middle. **Flowers** Few per peduncle, pale pink or white with greenish tube. Flowering in autumn. **Habitat** Disturbed areas in Succulent Karoo. **Distribution** Widespread from the Cape Peninsula to Port Alfred. **Other name** Boksuring.

Misgunst, Antoniesberg 23.04.2018

Oxalis ciliaris Vingersuring

cilium = eyelash; referring to the hairs on the leaf margins

Short-stemmed geophyte to 20 cm tall, shortly hairy
on all parts. **Leaves** Trifoliate, leaflets linear, 1.4 cm
long, hairy especially on the margins. **Flowers** Purple
or pink, with a yellow tube. Flowering in autumn.
Habitat Loamy soils in Transitional Shrubland.
Distribution Widespread from Ceres and Agulhas to
Gqeberha (Port Elizabeth).

Suuranysberg, Kareedouw 12.05.2021

Oxalis fergusoniae

Named after Margaret Clay Ferguson, 1863–1951,
an American botanist known for advocating education
in the botanical field

Low, acaulescent geophyte with a small, shallow,
oblong corm with a pointy tip. **Leaves** Small, usually
flat on ground, leaflets rotund, with water cells
visible below. **Flowers** Pinkish or white with a
yellow tube. Flowering in spring. **Habitat** Occasional
in richer soils on shale in Succulent Karoo,
Transitional Shrubland and Grassy Fynbos. Fairly
Distribution Widespread from Robertson to the
Kouga Mountains.

Rus en Vrede 4X4, Kouga 25.04.2018

Oxalis imbricata var. *violacea*

imbricata = overlapping; referring to the flower buds with overlapping petals

Acaulescent geophyte with a shallow, twisted corm. **Leaves** Trifoliate, leaflets obcordate, silky-hairy.
Flowers On articulated peduncle, pinkish with a yellow-green tube. Flowering in autumn. **Habitat**
Occasional in Transitional Shrubland and Grassy Fynbos. **Distribution** Widespread from Robertson to
Riebeek East. **Notes** 3 varieties; the other 2 have not been found in Baviaanskloof.

Grasnek, Kouga 17.04.2012

Oxalis obtusa Geeloogsuring

obtusa = obtuse; referring to the shape of the leaflets

Acaulescent geophyte with deeply pitted corm. **Leaves**
Trifoliate, leaflets cuneate-obcordate, hairy, variable. **Flowers**
On articulated peduncle, in various colours, but brick-red most
common, with darker veins and yellow tube. Flowering in winter
and spring. **Habitat** Richer soils in various veld types. **Distribution**
Widespread from Namaqualand to Gqeberha (Port Elizabeth).

Doringkloof, Kouga 04.09.2006

Oxalis psilopoda

psilo = bare, naked, *poda* = foot; referring to the tiny bulb
that is difficult to find

Acaulescent geophyte to 20 cm. **Leaves** Trifoliate,
leaflets obcordate, 1 cm wide, silky-hairy below.
Flowers Single on a long, articulated peduncle, white or
pink with a yellow tube. Flowering in autumn. **Habitat**
Uncommon under Forest. **Distribution** Widespread
from Riversdale to Gqeberha (Port Elizabeth).

Kouga Wildernis 28.03.2021

Oxalis punctata Dotted sorrel

punctatus = dotted; referring to the pitted corm and maybe the
leaves that have relatively large epidermal cells that make the
leaf surface appear finely dotted.

Acaulescent geophyte with a small,
angled and pitted corm. **Leaves** Trifoliate,
hairless, with large epidermal cells.
Flowers Single on silvery-hairy peduncles,
white or pink with a yellow tube. Flowering
in autumn. **Habitat** Uncommon in Grassy
Fynbos and Transitional Shrubland.
Distribution Widespread from Cape Peninsula to
Gqeberha (Port Elizabeth).

Suuranysberg, Kareedouw 13.05.2021

Oxalis purpurea Bobbejaansuring

purpurea = purple; pertaining to the flower colour, but they are not always purple

Acaulescent geophyte, corm large, tunics gummy, with peeling, dark brown
flakes. **Leaves** Trifoliate, obovate, hairy, leaflets purple below with hairy
margins. **Flowers** Single, close to the ground, purple, yellow or white, tube
yellow. Flowering in winter and spring. **Habitat** Common in various habitats
but mostly on loamy soils in Grassy Fynbos. **Distribution** Widespread from
Steinkopf in Namaqualand to Gqeberha (Port Elizabeth).

Langkloof 04.09.2020

Oxalis smithiana Klawersuring, rooisuring

Named after Sir James Edward Smith, 1759–1828, English botanist and author who purchased some of Linnaeus's specimens and books and founded the Linnean Society of London

Stemless geophyte to 20 cm. **Leaves** Bilobed almost to the base, appearing 6-foliate, lobes ovate-oblong, elongating and slender. **Flowers** Pale lilac or blue, tube greenish. Flowering in summer, autumn and winter. **Habitat** Loam soils on south-facing slopes in Mesic and Grassy Fynbos. **Distribution** Widespread from George to Mbombela. **Other names** Izotho (X), umuncwane, inkolwane (X, Z), bolila (S), incangiyane (Z).

Zandvlakte, Baviaans 29.04.2018

Oxalis stellata

stellata = star-like; referring to the star-shaped flowers

Short-stemmed geophyte, corm blackish, ovate, with smooth scales. **Leaves** Bilobed, clustered in umbel-like heads at stem tips, sleeping easily (nyctinasty). **Flowers** In small umbels on elongated peduncles, white or pink with yellow tube, sepals callous-tipped. Flowering in summer, autumn and winter. **Habitat** Sporadic and infrequent on lower slopes in Transitional Shrubland, Thicket and Forest. **Distribution** Widespread from Cape Town to Addo.

Rus en Vrede 4X4, Kouga 25.04.2018

PENAEACEAE   Penaea

9 genera and 29 species endemic to South Africa. 7 genera endemic to the Fynbos of the Western Cape, some with only 1 species. The family now also includes 2 forest-dwelling genera, *Olinia* and *Rhynchocalyx*.

OLINIA   Hard pear, hardepeer
Named after Johan Henrik Olin, 1769–1824, Swedish botanist and author

Shrubs and trees with opposite leaves and 4-angled branchlets. Previously placed in a family of its own, Oliniaceae. ± 10 species in Africa; 6 species in South Africa; 1 species in Baviaanskloof.

Olinia ventosa

ventosa = fast, swift; referring to the fast-growing habit

Tree to 15 m, with 4-angled branchlets. **Leaves** Simple, opposite, obovate, 5 cm long, smelling of almonds when crushed. **Flowers** Numerous in dense axillary cymes, white, fragrant. Fruit a red berry, to 1 cm wide. Flowering in spring. **Habitat** Infrequent in kloofs in Forest. **Distribution** Widespread from Cape Peninsula to Port Edward. **Notes** The yellow-brown wood is hard, heavy and suitable for furniture. **Other names** Umkunye, ungenalahle, inqudu, umpanzi (X), ingobamakhosi (Z).

Boskloof, Kouga
Wildernes 15.05.2021

PHYLLANTHACEAE

± 60 genera and 2,000 species. *Phyllanthus* is the largest genus, with 1,200 species.

FLUEGGEA

Named after Johannes Fluegge, 1775–1816, a German physician, botanist and lecturer

± 14 species in pantropical and warm, temperate parts of the world; 2 species in South Africa; 1 species in Baviaanskloof.

Flueggea verrucosa (= *Phyllanthus verrucosus*)

verrucosa = warty; referring to the grey bark that is covered with small lenticels (warts)

Resprouting, much-branched shrub to 2.5 m. **Leaves** Broadly obovate, to 2 cm long, paler below. **Flowers** Single in the axils. Fruit on hanging pedicels, 1 cm long. Flowering in summer. **Habitat** On the margins of thicket clumps in Transitional Shrubland and Thicket. **Distribution** Eastern Cape endemic from Baviaanskloof to Hogsback.

Grasnek, Kouga 06.05.2006

PHYLLANTHUS

phyllon = leaf, *anthos* = flower; in certain species the flowers emerge from leaf-like growths called cladodes.

Over 750 species worldwide; ± 20 species in South Africa; 1 species recorded in Baviaanskloof.

Phyllanthus incurvus Dyebossie

incurvus = incurved; referring to the recurved stipules or the female flower pedicels that curve downwards

Monoecious or dioecious, resprouting from a woody base, sparsely branched shrublet to 30 cm. **Leaves** Lanceolate, 3 cm long × 0.5 cm wide, margins and tips often tinged red. Stipules reddish, paired and recurved. **Flowers** Axillary, males few with 3 stamens, females singular with 6 maroon sepals that have white membranous margins. Fruit on short, pendulous peduncles 0.5 cm long. Flowering mainly in spring. **Habitat** Uncommon and sporadic on loamy soils in Grassy Fynbos. **Distribution** Widespread from Riversdale to Gqeberha (Port Elizabeth) and other semi-arid parts of southern Africa.

Drinkwaterskloof, Kouga 09.11.2011

PHYTOLACCACEAE

Pokeweed

A small family of ± 15 genera and 100 species in temperate to tropical parts of the world; 3 genera and 6 species native to summer-rainfall parts of South Africa.

PHYTOLACCA

Pokeweeds

phyton = plant, *lacca* = a red dye; pertaining to the fruits that contain a crimson red juice that can stain

2 native species in South Africa; 1 exotic species in Baviaanskloof.

*Phytolacca dioica** Ombu, belhambra

dioica = dioecious; the males and females are on separate plants.

Massive tree to 20 m tall, with umbrella-like crown to 15 m wide, and a massive, mound-forming trunk. **Leaves** Large, to 20 cm long, long-petioled, midrib prominent. **Flowers** Many, crowded on 15 cm long × 2.5 cm-wide pendulous racemes, small, creamy white; males with about 20 stamens, females with about 10 sterile stamens. Fruit a segmented berry, to 1 cm wide, purplish black. **Habitat** Found along streams and rivers, and around homesteads and kraals in the valley bottom; planted for shade. **Distribution** Native to the Pampa of South America, a declared invasive weed in South Africa. **Notes** Could be mistaken for *Ficus sur* from a distance. Used in bonsai. The leaf sap is poisonous. **Other names** Isidungamsi, impangapanga, isitinqiliti (X), umzimuka (Z).

Studtis, Baviaanskloof River 26.04.2018

PITTOSPORACEAE

9 genera and 240 species of trees, shrubs and lianas; 1 species in South Africa.

PITTOSPORUM

Cheesewoods

pitta = pitch, *spora* = seed; referring to dark sticky resin covering the seeds

Pittosporum undulatum from Australia has become invasive in South Africa and other parts of the world. 200 species worldwide; 1 species in South Africa.

Pittosporum viridiflorum   Cape cheesewood, kasuur

viridi = green, *florum* = flower; alluding to the green flowers. The Afrikaans name is derived
from candle hour and refers to the time at dusk when the flowers emit their sweet fragrance.

Tree to 10 m tall with brown bark with white lenticels. **Leaves** Simple, crowded at the branch tips,
oblanceolate, up to 10 cm long, margins revolute, net-veining prominent below. **Flowers** In terminal
panicles, sweetly scented, creamy green. Fruit a yellow-brown capsule, releasing sticky red seeds.
Flowering in summer. **Habitat** Bush clumps in Grassy Fynbos, Thicket and Forest. **Distribution** Very
widespread from Swellendam to tropical Africa. **Other name** Umkhwenkwe (X), mosetlela (S). **Status**
A protected tree in South Africa.

Rooikloof, Kouga 28.04.2018

paniculate inflorescence

PLUMBAGINACEAE   Plumbago, leadwort

Herbs, lianas and shrubs in 32 genera and 725 species worldwide. Notable for their ability to colonise arid
and salt-rich soils. Water in the plants' tissues dissolves the salts and excretes them via specialised 'chalk
glands' (microscopic pores on the leaf surface).

PLUMBAGO

plumbum = lead, *ago* = resembling; derivation unresolved but most probably pertaining to the chalk glands
that exude salts, covering the leaves and giving them a lead-grey colour

± 10 species worldwide; 2 species in South Africa; 1 species in Baviaanskloof.

Plumbago auriculata   Syselbos, tokkelossiebos

auriculata = ear-shaped; referring to the
ear-like appendage at the base of the leaves

Scrambling shrub to 3 m. **Leaves** Oblong, thin-textured,
2.5 cm long, leaf base eared and clasping the stem. **Flowers**
In terminal spikes, pale blue, calyx glandular-hairy. Calyx
sticks to animal fur, dispersing the seeds. Flowering in late
summer. **Habitat** Thicket and Forest margins. **Distribution**
Widespread from George to Mpumalanga. **Notes** Popular
in gardens around the world. **Other names** Umabophe,
umatshintshine, umuthi wamadoda (X, Z).

Bosrugkloof, Baviaans 12.07.2016

POLYGALACEAE

Milkworts

± 21 genera and 900 species in this cosmopolitan family of herbs, shrubs and trees. Most of the species are in the genus *Polygala*. ± 4 genera and 200 species in southern Africa.

MURALTIA

Purple-gorse, skilpadbos

Named after Johannes von Muralt, 1645–1733, a prolific Swiss natural scientist who published in many different disciplines. His achievements were mainly in the field of surgery.

The seeds are dispersed by ants. ± 118 species, all in South Africa (except for 1 species in Tanzania) with the vast majority occurring in the Cape flora; 6 species in Baviaanskloof.

Muraltia ericifolia

Erica = plant genus, *folia* = leaf; leaves are reminiscent of species in genus *Erica*.

Reseeding shrub to 70 cm. **Leaves** In bunches, 1.2 cm long × 0.3 cm wide, spine-tipped. **Flowers** Axillary, 0.5 cm long, white to pink. Flowering in summer. **Habitat** Fairly common in Transitional Shrubland and Grassy and Arid Fynbos. **Distribution** Widespread from Robertson to Gqeberha (Port Elizabeth).

Nooitgedacht, Kouga 27.08.2010

Muraltia juniperifolia

juniper = the plant, *folia* = leaf; leaves look superficially like those of the juniper.

Resprouting shrub to 50 cm. **Leaves** Needle-like, 0.5 cm long × 0.2 cm wide, spine-tipped. **Flowers** 0.5 cm long, pink or white. Flowering in winter and spring. **Habitat** Occasional in Transitional Shrubland, and Grassy and Arid Fynbos. **Distribution** Fairly narrow from Kammanassie to Kariega (Uitenhage).

Bo Kouga road, Langkloof 03.10.2009

Muraltia spinosa (= *Nylandtia spinosa*)   **Tortoise berry**

spinosa = spined; referring to the spine-tipped branches

Reseeding, rounded, spiny shrub to 1 m. **Leaves** Single, oblong. **Flowers** 0.5 cm long, white to pink to purple. Fruit red and fleshy, edible. Flowering mostly in winter. **Habitat** Common in sandy and loamy soils in Transitional Shrubland, Grassy Fynbos and Arid Fynbos. **Distribution** Widespread from Namaqualand to Gonubie. **Notes** Browsed by game. **Other name** Skilpadbessie (A).

Above Nuwekloof Pass, Baviaans 14.09.2010

Muraltia squarrosa

squarrosa = spreading at right angles; referring to the habit of leaves and flowers to spread out at right angles to the stem

Reseeding, shrub to 50 cm. **Leaves** Fascicled, hairy when young, 1 cm long × 0.5 cm wide. **Flowers** 0.5 cm long, pink. Capsules without horns. Flowering in spring and summer. **Habitat** Common in Grassy and Arid Fynbos. **Distribution** Widespread from George to Hlanganani.

Rus en Vrede 4X4, Kouga 20.01.2022

POLYGALA

poly = much, *gala* = milk; it was believed that animals grazing in fields with *Polygala* produced more milk; however, this notion probably stems from the sometimes-white apical crest on the lower petal, resembling a fountain of milk.

Found in temperate and tropical parts of the world. The 2 lower petals are fused and extend outward to form a fringed crest at the front of the flower. Mostly pollinated by bumble bees. There has been much taxonomic confusion over the number of species; in 2020 there were 660 accepted species. 84 species in southern Africa, with ± 33 species in the Cape flora and 9 species in Baviaanskloof.

Polygala asbestina   Poeierkwassie

asbestina = asbestos; referring to the grey powdery leaf surfaces

Reseeding, rounded, heavily browsed, sprawling shrublet to 20 cm. **Leaves** Leathery, 1 cm long × 0.5 cm wide, grey-green. **Flowers** 0.5 cm wide, blue. Flowering in late summer after rain. **Habitat** Exposed, dry slopes on heavier soils in Succulent Karoo and Thicket. **Distribution** Widespread from Anysberg to Cradock. **Other name** Ungqenqendlela (X).

Cambria 03.02.2020

Polygala ericaefolia

Erica = plant genus, *folia* = leaf; the thin, narrow leaves vaguely resemble those of some species of *Erica*.

Resprouting, densely leafy shrublet to 40 cm. **Leaves** Linear, ascending, 1 cm long × 0.2 cm wide. **Flowers** In short bunches at the branch tips, to 1 cm wide, purplish, side petals obtuse. Flowering mostly in spring. **Habitat** Uncommon in stony soils closer to the coast in Grassy Fynbos and Transitional Shrubland. **Distribution** Widespread from George to Gqeberha (Port Elizabeth).

Hankey 25.10.2018

Polygala fruticosa

fruticosa = shrubby, bushy; referring to the bushy growth form

Reseeding, erect shrub to 2 m. **Leaves** Opposite, cordate at base, lanceolate to ovate, 2.5 cm long × 1.5 cm wide. **Flowers** At the branch tips, 2.5 cm wide, dark pink. Flowering mostly in spring. **Habitat** Damper situations, often near streams, in Transitional Shrubland and Grassy Fynbos. **Distribution** Widespread from Ceres to KwaZulu-Natal. **Other name** Ithethe (Z).

Kaseykloof, Baviaans 28.10.2018

Polygala myrtifolia var. myrtifolia September bush

myrtifolia = myrtle-like leaves; the leaves bear a superficial resemblance to some species of myrtle. The common name alludes to the flowering time.

Reseeding, erect shrub to 2 m tall, sometimes to 4 m. **Leaves** Variable, alternate, closely crowded, usually 3 cm long × 1.5 cm wide, margins slightly revolute. **Flowers** In clusters at the branch tips, 2.5 cm wide, pink to purple, crest white. Flowering in spring. **Habitat** Transitional Shrubland, Succulent Karoo and Thicket. **Distribution** Widespread from Nieuwoudtville and the Cape Peninsula to KwaZulu-Natal. **Notes** A narrow-leaved form that occurs in the kloofs may be of interest for further research. **Other names** Ulopesi, umabala-bala (X).

Bosrug, Baviaans 22.07.2011

Polygala myrtifolia var. pinifolia Skoenlapperbos, augustusbos

pinifolia = pine-leaved; the plant has leaves that look vaguely like those of some species of pines (*Pinus*).

Reseeding shrub to 1.5 m. **Leaves** Narrow, 2 cm long × to 0.5 cm wide, grey-green, margins revolute. **Flowers** 2.5 cm wide, with white, pink and dark purple. Flowering in spring. **Habitat** Found sporadically on shale in Transitional Shrubland. **Distribution** Narrow distribution from Little Karoo to Baviaanskloof.

Vaalwater, Antoniesberg 14.09.2010

Polygala virgata

virgatus = long and slender, rod-like; referring to the long, slender growth form

Reseeding, slender, rod-like shrub to 2 m. **Leaves** Only near the branch tips, lanceolate-elliptic, 3 cm long. **Flowers** Numerous on long racemes at the branch tips, striking, dark purple. Flowering in early summer. **Habitat** Occasional in Grassy Fynbos, Transitional Shrubland and Thicket. **Distribution** Very widespread from Swellendam to tropical Africa. **Other names** Ithethe, ujulwezinyosi, umphehlwana (Z).

Grasnek, Kouga 08.03.2012

PROTEACEAE

Proteas

± 83 genera with ± 1,660 known species, mostly in the southern hemisphere. This ancient family was already in existence before Gondwanaland split up ± 140 million years ago. The greatest concentrations of species occur in Australia and southern Africa. Well-known genera include *Protea*, *Leucadendron*, *Banksia*, *Grevillea*, *Hakea* and *Macadamia*. Many species are highly sought after in the cut-flower industry. The delicious nuts of *M. integrifolia*, a tree native to Queensland, Australia, are grown commercially. The South African cricket team is known as the Proteas, and the New Zealand rugby team, the Waratahs, take their name from the waratah (*Telopea speciosissima*), which is also a member of the Proteaceae.

Members of the protea family play a crucial role in the proper functioning of the Fynbos ecosystem, as many animal species are dependent on them. For example, three of the five bird species endemic to Fynbos (Cape Sugarbird, Orange-breasted Sunbird and Protea Seedeater) depend on the presence of healthy stands of proteas to provide sufficient resources. The family is relatively well known thanks to *The Protea Atlas Project*, a project that greatly enhanced information on the distribution and population dynamics of particular species.

± 14 genera and 360 species in South Africa, with ± 92 per cent confined to the Cape flora, most of which are endemic to the Western Cape. 6 genera and 34 species in Baviaanskloof, with *Leucadendron* and *Protea* being the most diverse. 2 alien invasives from Australia are *Hakea sericea* and *Grevillea robusta*.

LEUCADENDRON

Cone bush, tolbos

leukos = white, *dendron* = tree; the first specimen that Linnaeus saw was the silver tree (*L. argenteum*), which is endemic to the Cape Peninsula. The name is not particularly apt because this is one of the few species in the genus that grow into trees with silver leaves, the majority of the species are green-leaved shrubs.

Dioecious plants. Often with colourful, bright yellow or red leaves surrounding the flowers. Females bear cones with seeds; males have pollen-producing flowers. Forming dense stands on mountain slopes in Fynbos, so that entire slopes appear bright yellow. 83 species in South Africa, mainly in the Western Cape; 12 species in Baviaanskloof.

Leucadendron album Linear-leaf conebush

album = whitish; pertaining to the silver-felted leaves

Reseeding, erect, woody shrub to 2 m. **Leaves** Linear-oblanceolate, males 4 cm long, females 6 cm long, larger below the flowers, silvery with adpressed hairs. **Flowers** Male flowerheads 1.5 cm wide, female flowerheads 2.5 cm wide, slightly scented. Flowering from late spring to early summer. **Habitat** Restricted to upper slopes of mountains in Mesic and Subalpine Fynbos. **Distribution** Widespread from Swartberg and Langeberg to Groot Winterhoek Mountains. **Other name** Aarbei-silwertolbos (A).

West of Bosrug, Baviaans 27.10.2018

Leucadendron eucalyptifolium   Gum-leaf conebush

Eucalyptus = plant genus, folium = leaf; leaves look similar to those of some species of Eucalyptus.

Reseeding, erect, tree-like, to 6 m. **Leaves** Twisted at the base, linear-lanceolate, 10 cm long × 0.8 cm wide, hairless when mature, with a fine, blunt point at the tip. Involucral leaves few, longer and yellow. **Flowers** Male flowerheads 1.6 cm wide, female flowerheads 1.2 cm wide, with a fruity smell. Cones 4.5 cm long × 2 cm wide. Flowering winter to spring. **Habitat** Common and dominant on steep, south-facing slopes in Grassy and Mesic Fynbos, especially where rainfall is higher. **Distribution** Widespread from Touwsrivier and Potberg to Van Staden's Mountains. **Other name** Grootgeelbos (A).

Ridge north of Enkeldoring, Baviaans 24.09.2011

Leucadendron loeriense

Named after Loerie, where the species was first recorded

Reseeding, erect, bushy below, slender above, to 2.5 m. **Leaves** Narrowly oblong, 3.5 cm long × 0.6 cm wide, with velvety hairs. **Leaves** Involucral leaves slightly larger, whitish green, forming a star-like cup. **Flowers** Male flowerheads conical, 1.5 cm long × 1.2 cm wide; female flowerheads ovoid, 1 cm wide. Cones globose, 2 cm wide. Flowering in summer. **Habitat** Dense stands on upper slopes in Mesic Fynbos. **Distribution** Narrow distribution from eastern end of Baviaanskloof Mountains to Elandsberg and Van Staden's Mountains.

Enkeldoring, Baviaans 21.09.2011

Leucadendron nobile   Karoo conebush

nobile = noble; alluding to the grand stature of this plant and its ability to thrive in relatively hot and dry conditions

Reseeding, stout, erect shrub to 4 m. **Leaves** Needle-like, ascending, males 4 cm long × 0.1 cm wide, females 6 cm long × 0.15 cm wide. **Flowers** Male flowerhead 4 cm long × 1 cm wide, with flowers in a spike. Female flowerhead 2.8 cm long × 1.2 cm wide, with a musty odour. Cones large, 9 cm long × 4 cm wide, bracts yellow with red margins on fresh cones, older cones silvery. Flowering mainly in summer. **Habitat** Steep rocky middle slopes in Grassy and Arid Fynbos. **Distribution** Near endemic to these mountains, its range extends onto the Witteberg and Grootrivier mountains north of Willowmore. **Notes** Probably the most arid-tolerant proteaceous shrub in the region. **Other name** Naaldblaar-tolbos (A).

Bosrug, Baviaans 22.07.2011

Leucadendron pubibracteolatum

pubis = hairiness, *bracteolatum* = provided with bracts; the male floral parts are covered with soft hairs.

Reseeding, relatively small, sparsely branched, leafy shrub to 1.3 m. **Leaves** Sessile, elliptic, hairless, pointing slightly downwards at the base and curved upwards at the tip, purplish grey; male 6 cm long × 2 cm wide, female leaves slightly bigger. Involucral leaves larger, more tightly packed, becoming yellow, basal bracts reddish. Cones 5 cm long × 4 cm wide, seeds released when ripe, non-serotinous. Flowering in winter. Occurring sporadically and with small populations in Mesic Fynbos. **Distribution** Fairly narrow distribution from the Outeniqua Mountains to the eastern Swartberg and to Baviaanskloof. **Status** Near Threatened as a result of the small, fragmented populations, with noted declines in population sizes.

Koud Nek's Ruggens, Kouga 25.09.2011

Leucadendron rourkei   Uniondale conebush

Named after Dr John Rourke, b. 1942, a Cape Town-based botanist; he was curator of the Compton Herbarium and specialised in the Proteaceae.

Reseeding, erect, slender shrub to 5 m tall, tends to be bushy near the base. **Leaves** Twisted at the base, hairless, 2 cm long × 0.4 cm wide, larger in females. Involucral leaves inconspicuous. **Flowers** With a yeasty odour. Cones relatively small, 2.5 cm long × 2 cm wide, serotinous. Flowering in summer. **Habitat** Occasional on middle to upper mountain slopes in Mesic Fynbos, often on stony soils or shale bands. **Distribution** Narrow distribution on the eastern Swartberg, Kammanassie and western Kouga mountains. **Other name** Uniondale-tolbos (A).

Quaggasvlakte, Kouga 13.02.2016

Leucadendron rubrum   Spinning top

rubrum = red; referring to the tips of the bracts that turn red on the cones, the stems and leaf tips are also sometimes tinged red. The common name refers to the shape of the cones.

Reseeding, erect, dense shrub to 2.5 m. **Leaves** Lance-shaped, narrowed and twisted at the base, becoming hairless; male leaves 3 cm long, females to 7 cm long. **Flowers** Male flowerheads prolific, small and narrow without involucral leaves; female flowerheads cone-shaped with a tuft of stigmas at the apex. Cones 5 cm long × 3.5 cm wide, with colourful pointed bracts. Flowering in spring. **Habitat** Relatively dry parts of the mountains in Grassy, Arid and Mesic Fynbos. **Distribution** Widespread from Nieuwoudtville to Baviaanskloof. **Other name** Tolletjiesbos (A).

Kouenek, Kouga 25.09.2011

Leucadendron salignum Common sunshine conebush

salignum = relating to willows, genus *Salix*; leaves look
similar to those of some species of *Salix.*

Resprouting from a woody rootstock, with numerous stems,
usually to 1 m, but sometimes to 2 m. **Leaves** Oblanceolate-
linear, males 2.5 cm long, females 3.5 cm long, tip with a
fine, blunt, red point. Involucral leaves in males only slightly
longer, while in females they are broadened and conceal
the flowerhead, turning yellow or red. **Flowers** With a sweet
or yeasty smell. Cones to 2 cm wide, bracts with soft hairs.
Flowering from autumn to spring. **Habitat** Common in a wide
range of Fynbos habitats but mostly Grassy Fynbos and Sour
Grassland on deeper soils, often more prevalent where shale is
present deeper in the soil profile. **Distribution** Widespread from
Nieuwoudtville to Makhanda. **Notes** A red-leaved form in the
Langkloof is sought after in the cut-flower industry. **Other name**
Knoppiesgeelbos (A).

Quaggasvlakte, Kouga 13.02.2016

Leucadendron sorocephalodes Woolly conebush

Sorocephalus = plant genus, *oides* = resembling; the plant resembles
Sorocephalus, a genus in the Proteaceae with small needle-like leaves.

Reseeding, low, spreading shrub to 30 cm. **Leaves** Ericoid, fleshy,
linear, hairless, females 1.8 cm long × 0.2 cm wide, smaller in
males. Involucral bracts numerous, brown, tip pointed. **Flowers**
Male flowerhead cream-white, woolly, strongly sweet-smelling.
Female with woolly-cream flowers at tip. Cones globose, 1.5 cm
wide. Flowering in late winter. **Habitat** Isolated populations on
south-facing slopes of the high, rocky peaks in Mesic and Subalpine
Fynbos. **Distribution** Fairly narrow distribution from Outeniqua to
Kouga and Baviaanskloof mountains. **Status** Near Threatened.

Kouenek, Kouga 25.09.2011

Leucadendron uliginosum subsp. *glabratum* Tsitsikamma conebush

uliginosum = grows in bogs and swamps; the plant grows in
Mesic Fynbos on upper mountain slopes, so this name is not appropriate.

Reseeding, erect shrub, with a fairly bushy base, to 4 m. **Leaves** Oblong,
becoming hairless, 2.6 cm long × 0.4 cm wide in females, slightly shorter
in males. Involucral leaves numerous, ivory-coloured, forming a
star-like cup. **Flowers** Male flowerheads 1 cm wide, smelling faintly
of banana. Cones 2 cm long × 2 cm wide, hairless. Flowering in
summer. **Habitat** Fairly common on rocky upper slopes in Mesic
Fynbos. **Distribution** Near endemic, found from Tsitsikamma to the
Kouga Mountains. **Notes** The other subspecies, subsp. *uliginosum*,
is found between the Langeberg and Outeniqua mountains and
does not occur in Baviaanskloof. **Status** Near Threatened.

Kouenek, Kouga 20.07.2011

LEUCOSPERMUM

Pincushion, speldekussing

leukos = white, *spermum* = seed; pertaining to the white, waxy aril that covers the seeds

The flowers of this genus are clustered into yellow-red heads that resemble pincushions; also recognised by the horny 'teeth' at the tips of the leaves. Seeds are dispersed by ants. 48 species in South Africa and Zimbabwe, mostly in the Western Cape; 2 species in Baviaanskloof.

Leucospermum cuneiforme Warty-stemmed pincushion

cuneatus = wedge-shaped, *forme* = formed; referring to the wedge-shaped leaves

Resprouting from a woody rootstock, or from a corky barked stem, to 2 m. Stems warty near the base. **Leaves** Wedge-shaped, hairless, 10 cm long, with 3–10 apical teeth. **Flowerheads** Oval, 7 cm wide, yellow, fading red. Flowering from late winter to summer. **Habitat** A wide range of Fynbos habitats, but most common in Grassy Fynbos and Sour Grassland. **Distribution** Widespread from Caledon to Makhanda. **Notes** On the Groot Winterhoek Mountains and Elandsberg there are tree-like specimens to 3 m tall; their thick, corky bark enables them to survive the frequent grassy fires. **Other names** Luisiesbos (A), isiqwane (X).

Bosrug, Baviaans 12.10.2016

Leucospermum wittebergense Swartberg pincushion

wittebergense = from the Witteberg; the first scientific specimens were collected in the Witteberg Mountains near Laingsburg, but the plant has since been found in other mountain ranges as far east as Baviaanskloof.

Reseeding, erect, much-branched, rounded shrub to 1.5 m. **Leaves** Elliptic-lanceolate, densely hairy, 2 cm long × 5 cm wide, silvery, with 1–3 glandular teeth. **Flowerheads** Globose, 2 cm wide, pinkish yellow, style almost 2 cm long, club-shaped, pink to red, tip yellow. Flowering from late winter to midsummer. **Habitat** Dry, arid, rocky slopes in Arid or Grassy Fynbos. **Distribution** Widespread from the Witteberg to Baviaanskloof. **Other name** Swartberg-speldekussing (A).

Engelandkop, Kouga 18.07.2011

PARANOMUS

Cornflower sceptres, doll's bush, poppiesbos

para = beyond or contrary to, *nomos* = custom, law; referring to the unusual leaves, with 2 different leaf shapes on the adult plant

19 species in Western Cape; 3 species in Baviaanskloof.

Paranomus esterhuyseniae

Named after Elsie Elizabeth Esterhuysen, 1912–2006, a humble botanist described as 'the most outstanding collector of plants in South Africa'. She contributed over 36,000 specimens to the Bolus Herbarium. 56 species and 2 genera have been named after her.

Reseeding, small, erect shrub to 0.7 m. **Leaves** Dissected, with thin segments, sometimes entire below the flowerheads, glabrous, 5 cm long, pale green. **Flowerheads** Globose, velvety, 2 cm wide, cream-coloured, style glabrous. Flowering in spring. **Habitat** Occasional and uncommon on rocky ridges of higher peaks in Mesic and Subalpine Fynbos. **Distribution** Near endemic from the Outeniquas to the Kouga and Baviaanskloof mountains. **Notes** *P. dregei* is confined to the western end of the Kouga Mountains and is a larger bush with leaves flattened below the flowerheads. *P. reflexus* is found only on the Elandsberg and Van Staden's mountains; it has obovate-rhomboid leaves below the flowerheads. **Status** Near Threatened.

Scholtzberg, Baviaans 29.10.2011

PROTEA

Protea, sugarbush

Proteus = name of an early Greek god of the sea, or *Proteia* = chief rank or first place; this Greek deity was able to take many forms, alluding to the great diversity and adaptability within the genus. Another possibility is that Linnaeus was so impressed by the first *Protea* specimens sent to him that the name might allude to this genus being 'the best' or ranking very highly in the plant world.

Probably the most well-known and charismatic group of plants in Fynbos. Proteas are easily recognised by the striking involucral bracts that surround the flowerheads, and the hard, woody base where flowers and seeds are stored. Proteas are serotinous, an adaptation whereby they hold onto their seeds, only releasing them in response to a trigger such as fire. ± 115 species in southern and tropical Africa, mostly in the Western Cape; 15 species in Baviaanskloof.

Protea cynaroides   King protea

Cynara = genus name, *oides* = resembling; this species resembles the flower bud of the globe artichoke, *Cynara scolymus*, so Linnaeus called it *cynaroides*.

Resprouting strongly from a woody rootstock to 1 m. **Leaves** Round, ovate or elliptic, 20 cm long on a petiole 10 cm long. **Flowerheads** Bowl-shaped, to 20 cm long × 30 cm wide, with involucral bracts overlapping, mostly lanceolate, usually with silky hairs, cream, pink or red; style curved inwards. Flowering mostly in spring and summer. **Habitat** Uncommon and localised on rocky south-facing slopes in Grassy and Mesic Fynbos. **Distribution** Widespread from Gifberg to the Cape Peninsula and to Makhanda. **Notes** South Africa's national flower, cultivated worldwide for the cut-flower industry. **Other names** Giant protea (E), bergsuikerkan (A), isiqwane esincinci (X).

Joubertina 04.10.2019

Protea eximia   Wide-leaf sugarbush

eximia = distinguished, excellent or extraordinary;
the wide leaves and bright red narrowly spoon-shaped
bracts are spectacular and distinctive.

Reseeding, erect, sparsely branched shrub, 2–5 m. **Leaves**
Spreading, cordate at the base, oblong-ovate, to 10 cm
long × 4 cm wide, grey- to purple-green. **Flowerheads**
14 cm long × 12 cm wide with involucral bracts widely
splayed, oblong to narrowly spoon-shaped, pink, orange-
brown to red, margins white-hairy; perianth awns with
purplish velvety hairs sometimes black-tipped. Flowering
mainly in spring. **Habitat** Upper mountain slopes in
Grassy and Mesic Fynbos. **Distribution** Widespread from
Worcester to Gqeberha (Port Elizabeth). **Other name**
Breëblaarsuikerbos (A).

Kammiesbos, Kareedouw 07.09.2009

Protea intonsa   Tufted sugarbush

intonsus = unshaven, bristly, shaggy; referring to
the white-woolly hairs at the tips of the perianth

Resprouting from scaled, underground stems to 30 cm
tall × 60 cm wide. **Leaves** In tufts, needle-like to linear,
hairless, 30 cm long × 0.5 cm wide, sometimes somewhat
sandpapery. **Flowerheads** 4 cm wide; outer involucral
bracts ovate, becoming reddish brown; perianth tips with
white, woolly hairs. Flowering in late spring. **Habitat** Dry,
exposed rocky ridges and slopes in Arid and Mesic Fynbos.
Distribution Nearly endemic from eastern Swartberg and
Kammanassie mountains to Baviaanskloof.

Ridge west of Elandsvlakte, Baviaans 29.09.2011

Protea lorifolia   Strap-leaf sugarbush

lorifolia = strap-shaped leaves; pertaining to the shape of the leaves

Reseeding, robust, much-branched shrub or small tree
to 5 m. **Leaves** Ascending, 20 cm long, greyish green,
margins thickened, red, leathery and becoming hairless.
Flowerheads Large, to 13 cm long; involucral bracts with
silky hairs; bract tips bearded white or
purple-brown. Flowering mainly in autumn.
Habitat Arid, Grassy and Mesic Fynbos.
Distribution Widespread from Nieuwoudtville
to Makhanda. **Notes** Fairly common,
although too frequent fires and climate
change have contributed to the shrinking or
collapse of some populations. **Other name**
Riemblaarsuikerbos (A).

Zandvlakte, Baviaans, 31.07.2011

Protea mundii Forest sugarbush

Named after Johannes Ludwig Leopold Mund, 1791–1831,
a Berlin pharmacist who collected plant specimens in the
Cape for the Prussian government

Reseeding, erect, tall, slender shrub or small tree to 6 m tall,
sometimes taller. **Leaves** Elliptic to elliptic-lanceolate, 10 cm
long × 3 cm wide. **Flowerheads** Oblong-obconic, 8 cm long
× 5 cm wide; involucral bracts greenish cream or pink, silky-
hairy; tips silky-fringed. Flowering mostly in autumn. **Habitat**

Cockscomb, Groot Winterhoek 17.02.2016

Moist slopes in Mesic Fynbos. Often associated with shale-
derived or deeper soils where Forest can develop. **Distribution** Widespread from Kogelberg to Groot
Winterhoek Mountains, but strangely absent from the Langeberg. Rare in Kouga and Baviaanskloof
mountains, more common on Cockscomb. **Other name** Boswitsuikerbos (A).

Protea neriifolia Narrow-leaved sugarbush

Nerium = plant genus, *folia* = leaves; leaves look similar to those of *N. oleander*.

Reseeding, erect, much-branched shrub to 3 m. **Leaves**
Curving upwards, oblong, 15 cm long × 3 cm wide,
green. **Flowerheads** 13 cm long × 7 cm wide; involucral
bracts cream, pink or reddish brown, rounded tips
white- or black-bearded, with silver hairs below the
beard. Flowering all year round except midsummer.
Habitat South-facing slopes in Grassy and Mesic
Fynbos. **Distribution** Widespread from Hottentots
Holland Mountains to Gqeberha (Port Elizabeth). **Other
name** Baardsuikerbos (A).

East of Kougakop 16.04.2012

Protea nitida Wagon tree

nitida = shining; referring to the silvery-grey leaves that shimmer
in the sunlight; the common name refers to the historical use of
the wood on wagons, for wheel rims and break blocks.

Resprouting, robust, thick-stemmed tree 5–10 m, with thick
white-grey bark enabling survival through fire. New shoots
may show up bright red, looking like large red flowers from
a distance. **Leaves** Oblong to elliptic, leathery, 15 cm long
× 5 cm wide, olive to silvery grey or glaucous. **Flowerheads**
Bowl-shaped, to 16 cm wide, bracts relatively short, silver-
grey to cream, with white styles prominently exserted.
Flowering mostly in autumn and winter. **Habitat** Common
on deeper, slightly richer soils, often on old talus or scree,
on the lower to middle slopes in Grassy Fynbos and Fynbos

Rus en Vrede 4X4, Kouga 22.05.2010

Woodland. **Distribution** Widespread throughout most of the Fynbos biome, from Nieuwoudtville to Van
Staden's Mountains. **Notes** In the past, dark liquid was extracted from the leaves for use as ink. Acid
from the bark has been used for tanning leather. **Other name** Waboom (A).

Protea punctata   Water sugarbush

punctata = dotted, marked with dots; referring to the dot-like pores on the leaves (only visible with a 10× hand lens/loupe)

Reseeding, tall, erect shrub to 4 m. **Leaves** Curving upwards, ovate-elliptic, becoming hairless, 6 cm long, bluish grey. **Flowerheads** 6 cm long, involucral bracts horizontal at flowering, white to pink. Flowering mostly in autumn. **Habitat** Locally abundant, forming dense stands on upper, usually south-facing slopes in Mesic and Subalpine Fynbos. **Distribution** Widespread from Cederberg to Baviaanskloof. **Other name** Waterwitsuikerbos (A).

East of Nuwekloof Pass, Baviaans 15.02.2016

Protea repens   Common sugarbush

repens = creeping; a misnomer assigned by Linnaeus and based on an inaccurate drawing, as this species is an erect bush

Reseeding, erect, well-branched shrub to 4.5 m. **Leaves** Ascending, strap-shaped to narrowly spathulate, 15 cm long. **Flowerheads** 12 cm long × 8 cm wide, bracts cream, green or red, hairless and sticky. Flowering all year round. **Habitat** Common and sometimes locally abundant in Fynbos and Transitional Shrubland. **Distribution** The most widespread protea, from Nieuwoudtville to Makhanda. Although still common, this species has disappeared from vast areas as a result of too frequent burning. Unfortunately there is insufficient

Kouonek, Kouga 20.01.2022

evidence of local extinctions and population collapses across its range over the past few hundred years, so it may have been inaccurately red-listed. **Notes** Flowers produce copious amounts of nectar, which has been used to produce sugar syrup. **Other name** Suikerbos (A).

Protea rupicola   Krantz sugarbush

rupes = rock, cliff, *cola* = dweller; referring to the rocky or cliff habitat on upper levels of high peaks

Reseeding, erect or spreading, much-branched shrub to 2 m. **Leaves** Curving upwards, strap- to lance-shaped, leathery, to 6 cm long, blue-grey to green, sometimes with red-tinged margins. **Flowerheads** To 10 cm wide, slightly hairy, styles pink and curving inwards, involucral bracts reddish brown and inconspicuous. Flowering from spring to late summer. **Habitat** Uncommon and isolated amongst rocks and rock ledges on top of the highest peaks in Subalpine and Mesic Fynbos. **Distribution** Widespread from Tulbagh to Cockscomb Peak. **Other name** Kranssuikerbos (A). **Status** Endangered.

Scholtzberg Peak, Baviaans 24.07.2011

Protea tenax Tenacious sugarbush

tenax = holding fast, tough; referring to the tough,
trailing stems and hardy resprouting habit

Resprouting from a woody rootstock, low,
trailing, sparsely branched to 25 cm. **Leaves**
Ascending from a horizontal stem, linear to
broadly elliptic, to 18 cm long, with a prominent
midrib. **Flowerheads** Bowl-shaped, to 6 cm wide,
involucral bracts greenish yellow, sometimes
reddish, odour yeasty to attract rodents for
pollination. Flowering all year round, mainly
winter and spring. **Habitat** Never abundant, but
frequently encountered in Grassy and Mesic
Fynbos, and Sour Grassland. **Distribution** Fairly
widespread from Outeniqua Mountains to
Cockscomb and Elandsberg.

Onder Kouga road, Joubertina 17.06.2010

RANUNCULACEAE   Buttercups, crowfoot

A cosmopolitan family of mostly herbs consisting of 50 genera and ± 1,500 species. *Ranunculus* is the
largest genus, with ± 600 species. 7 genera and 35 species indigenous in southern Africa; 3 genera and
4 species recorded in Baviaanskloof.

ANEMONE

anemos = wind; anemones grow well in windy areas; in Greek mythology, Anemone was a nymph that was
turned into a flower by a jealous goddess.

± 120 species in temperate parts of the world; 15 species in South Africa; 1 species in Baviaanskloof.

Anemone vesicatoria subsp. *humilis* Brandblaar

vesicare = to raise blisters; referring to the
allergic reaction caused by contact with the
leaves and roots of this species

Resprouting from a woody rootstock, tufted
perennial to 1 m. **Leaves** Trifoliate, ovate,
toothed, leathery, each leaflet 4 cm long
× 3 cm wide, on long petioles. **Flowers** In
long-stemmed umbels, white to yellow or
purplish and pinkish, anthers numerous
and yellow. Flowering in spring. Fruit fleshy,
glabrous, and turning black. **Habitat**
South-facing slopes in Grassy or Mesic
Fynbos. **Distribution** Widespread from
Nieuwoudtville and the Cape Peninsula
to Makhanda. **Notes** 3 subspecies of
A. vesicatoria are recognised.

Nooitgedacht, Kouga 05.11.2011

CLEMATIS Traveller's joy

klema = vine branch, twig or tendril; classical Greek name for a climbing plant

± 230 species worldwide, mostly growing in temperate areas; 4 species in South Africa; 1 species in Baviaanskloof.

Clematis brachiata Old man's beard

brachiata = branched at right angles; referring to the right-angled branching habit

Perennial climber or scrambler to 5 m. **Leaves** Opposite, divided, ovate, 4 cm long × 2 cm wide, toothed. **Flowers** Scented, 2 cm wide, with 4 white sepals and a cluster of white anthers. Flowering in summer and autumn. **Habitat** Forest and Thicket, often near streams. **Distribution** Widespread from Montagu eastward, throughout southern Africa to tropical Africa. **Other names** Umvuthuza, ityolo, ihlonzo leziduli (X), inhlabanhlanzi, umdlandlathi, umdlonza (Z).

Kamassiekloof, Zandvlakte 21.04.2012

RHAMNACEAE Buckthorn

Trees, shubs and some vines with 55 genera and 950 species. More common in tropical and subtropical regions. A family member, the Chinese jujube or date (*Ziziphus jujuba*) (= *Z. zizyphus*), is a popular fruit in China. 9 genera and 203 species in southern Africa, with most species in the genus *Phylica*.

NOLTEA

Named after Ernst Ferdinand Nolte, 1791–1875, a German botanist and physician who worked at Kiel (a German port on the Baltic Sea) and studied Danish flora

Only 1 species in this genus, endemic to South Africa.

Noltea africana Seepblinkblaar

africana = of Africa; a direct translation of the Afrikaans common name means 'soap-shiny leaf', referring to the shiny surface of the leaves, twinkling in the sun; the leaves were used to make soap.

Resprouting, much-branched shrub or tree to 4 m. **Leaves** Leathery, 6 cm long × to 2 cm wide, glossy above, paler below, bluntly serrated, with prominent midrib, shortly petioled. **Flowers** In few-flowered panicles, each flower 0.3 cm wide. Flowering in spring. Fruit a 3-lobed capsule, 1 cm wide, splitting to release black seeds. **Habitat** Usually occurring in damp areas of riparian bush in Forest and Thicket. **Distribution** Widespread from the Cape Peninsula to Port Edward. **Notes** Soap made with the plants was traditionally used for washing laundry. **Other names** Iphalode, umglindi, umkhuthuhla, umaluleka (X).

Ysrivier, Cambria 23.01.2022

phylikos = leafy; pertaining to the dense and abundant foliage covering most species

Important features for identification are the arrangement of the flowers in spikes or heads, and the presence or absence of petals. ± 150 species in Africa, Madagascar and the South Atlantic islands; ± 142 species in South Africa; 9 species in Baviaanskloof.

Phylica abietina

Abies = plant genus, *tina* = like; leaves vaguely resemble those of a fir tree (*Abies* spp.).

Resprouting, upright, leafy shrub to 1 m. **Leaves** Linear-lanceolate, with a smooth surface, 0.6 cm long, margins revolute. **Flowers** In dense, rounded heads surrounded by leaves, white or pinkish. Flowering in autumn and winter. **Habitat** Uncommon in Grassy Fynbos. **Distribution** Narrow distribution from George to Kariega (Uitenhage).

Bergplaas, Combrink's Pass 11.07.2016

Phylica alba

alba = white; pertaining to the white flowers

Resprouting, upright, dense and rounded shrub to 1 m. **Leaves** Linear-lanceolate, 0.6 cm long, margins revolute, surface finely granular. **Flowers** In dense, flat-topped heads. Flowering in autumn and winter. **Habitat** Rare on rocky outcrops on mountain peaks and ridges in Mesic and Subalpine Fynbos. **Distribution** Narrow from Langeberg to the Tsitsikamma and Kouga mountains.

Quaggasvlakte, Kouga 13.02.2016

Phylica axillaris var. *axillaris*

axillaris = axillary; the flowers are in the axils.

Reseeding and sometimes resprouting erect shrub to 1.5 m. **Leaves** 1.5 cm long × 0.5 cm wide, slightly hairy, margins revolute. **Flowers** In the axils at the branch tips, creamy white or pinkish. Flowering in summer and autumn. **Habitat** Grassy Fynbos, Sour Grassland and Transitional Shrubland. **Distribution** Widespread from Agulhas to Hlanganani.

Rus en Vrede 4X4, Kouga 25.04.2018

Phylica axillaris var. hirsuta

axillaris = axillary, *hirsuta* = hairy; the flowers are
in the axils of this finely hairy variety.

Resprouting, upright shrub to 60 cm tall, finely silvery-
hairy. **Leaves** Silky-hairy, 1.5 cm long × 0.5 cm wide,
margins revolute. **Flowers** In the axils at the branch
tips, creamy white. Flowering in summer and autumn.
Habitat Grassy Fynbos and Sour Grassland. **Distribution**
Restricted to the Groot Winterhoek Mountains.

Cockscomb, Groot Winterhoek 16.02.2016

Phylica debilis

debilis = weak; referring to the thin,
semi-decumbent branches

Reseeding, diffuse, inconspicuous shrublet to 40 cm.
Leaves Cordate-lanceolate, 0.3 cm long, held at right
angles to the stem. **Flowers** In small flattened heads,
whitish. Flowering in summer. **Habitat** Common on
south-facing slopes in Mesic and Subalpine Fynbos.
Distribution Widespread from Caledon to Baviaanskloof.

Outeniqua Mountains 20.01.2017

Phylica lachneaeoides

Lachnaea = plant genus, *oides* = resembling; the
flowering heads resemble some species of *Lachnaea*.

Reseeding, upright, densely leafy shrub to 1 m.
Leaves 0.8 cm long × 0.4 cm wide, margins
revolute. **Flowers** In rounded heads, pinkish,
tube relatively long, styles elongated. Flowering
in winter and spring. **Habitat** Common in Mesic
and Grassy Fynbos and Sour Grassland. **Distribution**
Nearly endemic from George to Humansdorp. The most
frequently encountered *Phylica* in Baviaanskloof.

Engelandkop, Kouga 13.07.2011

Phylica lanata

lanata = woolly; referring to the furry flowers

Reseeding, erect shrub to 1 m. **Leaves** Ascending,
0.6 cm long × 0.4 cm wide. **Flowers** Clustered into
heads at the branch tips, 0.5 cm wide, reddish, densely
white-woolly outside. Flowering in spring. **Habitat**
Common on rocky, south-facing slopes in Arid Fynbos.
Distribution Narrow distribution from Witteberg and the
Little Karoo to Baviaanskloof.

Top of Nuwekloof Pass 23.04.2018

Phylica paniculata

paniculata = paniculate; flowers are arranged in a diffuse, leafy, branched inflorescence (or panicle).

Reseeding shrub to 3 m. **Leaves** 3 cm long × 1 cm wide, margins revolute. **Flowers** 0.3 cm wide, scattered on much-branched inflorescences at the branch tips, dull yellow. Flowering in autumn and winter. **Habitat** Common and abundant on south-facing slopes in Grassy Fynbos and Fynbos Woodland, often at or near the bottom of steep mountain slopes. **Distribution** Very widespread from Worcester to Zimbabwe.

Nuwekloof Pass, Baviaans 23.04.2018

RHAMNUS   Buckthorns

rhamnos = tuft of branches; referring to the resprouting habit of some species, resulting in many-stemmed bushes

9 species, found throughout the tropics; 1 species in South Africa and Baviaanskloof.

Rhamnus prinoides   Blinkblaar

Prinos = plant genus, *oides* = resembling; referring to a superficial similarity to some species of evergreen oak in genus *Prinos*

Resprouting, much-branched woody bush or tree to 4 m. **Leaves** Alternate, elliptic, 6 cm long × 3 cm wide, glossy above, dull below, finely toothed. **Flowers** In axillary clusters, small, greenish. Fruit berry-like, 0.5 cm wide, shiny red becoming black. Flowering in early summer. **Habitat** Occasional on the margins of Forest or in riverine scrub. **Distribution** Very widespread from Riversdale to tropical Africa. **Notes** Used in traditional medicine and said to bring good luck. Charcoal from the wood has been used as a fuel in gunpowder. **Other names** Dogwood (E), umglindi (X), umnyenye (X, Z).

Nooitgedacht, Kouga 20.01.2022

SCUTIA

scutum = a shield; the flowers of this species have a cup-shaped calyx that superficially resembles a shield.

5 species worldwide; 1 species in South Africa, including Baviaanskloof.

Scutia myrtina   Cat-thorn

Myrtus = plant genus; similar in some way to myrtle plants

Scrambling shrub, creeper or small tree to 8 m; young stems green and usually with a pair of hooked thorns. **Leaves** Opposite or subopposite, ovate to elliptic, rounded at tip and base, leathery and hairless, up to 6 cm long × 4 cm wide, margin rolled under and sometimes wavy, petiole to 1 cm long, dark green above and slightly paler below. **Flowers** In the axils of the leaves on short stalks, 0.4 cm wide, yellow-green. Flowering in spring and summer. Fruit a round berry, 0.8 cm wide, purple to black. **Habitat** Forest margins and Thicket. **Distribution** Widespread from the Cape Peninsula to Uganda, also in India and Madagascar. **Other name** Katdoring (A), umqapuma (X), isiphingo (X, Z), umsondeza, isibinda (Z).

23.06.2020

ROSACEAE   Roses

± 90 genera and 4,800 species in this economically important family, known for its wide diversity of fruits and ornamental plants. Almonds, cherries, peaches, apricots and plums are in *Prunus*. Other large genera include *Rubus*, *Cotoneaster*, *Alchemilla*, *Sorbus* and *Crataegus*. Most of the diversity is in the northern hemisphere. 8 genera in southern Africa and 165 native species. There are also ± 23 naturalised and 180 cultivated species in South Africa. Only 1 native genus in Baviaanskloof with 14 species and 1 naturalised species.

CLIFFORTIA   Climber's friend

Named after George Clifford, 1685–1760, a Dutch merchant and banker, amateur botanist and zoologist. In 1737, Clifford commissioned Linnaeus to write *Hortus Cliffortianus*, a description of many of the plant species in Clifford's magnificent garden at his country estate, De Hartecamp.

The female flower may be identified by the achene and the male by the number of anthers. ± 140 species in southern Africa, a few extending to tropical Africa; ± 15 species in Baviaanskloof.

Cliffortia cervicornu

cervicornu = stag's horn; referring to the resemblance of the divided leaves to the antlers of a male deer

Reseeding, erect, slender shrub to 1 m. **Leaves** Tiny and divided, 0.3 cm long, leaflets 0.05 cm wide. **Flowers** Female achene to 0.3 cm long, dark brown, with 2–4 ribs. Flowering in October. **Habitat** Localised populations on south-facing slopes of peaks and ridges in Mesic and Subalpine Fynbos. **Distribution** Narrowly distributed on the inland mountains of Swartberg, Grootrivierberg and Baviaanskloof. **Notes** No other species in this genus has such finely divided leaves.

Groot Swartberg, Fullarton 16.11.2019

Cliffortia drepanoides

drepa = sickle, *oides* = resembling; referring to the sickle-shaped leaves

Reseeding, upright, rounded shrub to 1.2 m. **Leaves** Trifoliate, curving up towards the stem, slightly falcate, glabrous, to 2 cm long, relatively broad, grey-green. **Flowers** Female achene 6-ribbed; about 25 male anthers. Flowering in autumn. **Habitat** Frequent in Grassy and Mesic Fynbos. **Distribution** Near endemic from Uniondale to Kariega (Uitenhage). Probably the most common *Cliffortia* in the Kouga Mountains.

Witruggens, Kouga 03.11.2011

Cliffortia graminea   Vleirooigras, wilde-ertjie

graminea = grass-like; referring to the grass-like appearance of the dense stands

Resprouting, spreading, sparsely branched shrub to 2 m. **Leaves** 10 cm long × 0.5 cm wide. **Flowers** Female achene oblong, 0.5 cm long, greenish; about 30 male anthers. Flowering in autumn. **Habitat** Localised in perennial seepages and riparian swamps in Grassy and Mesic Fynbos. **Distribution** Widespread from Cape Peninsula to Makhanda.

Cockscomb, Groot Winterhoek 17.02.2016

Cliffortia ilicifolia   Doringtee, jankoensedoring

Ilex = plant genus, *folia* = leaf; referring to the toothed and ovate leaves that are reminiscent of holly (genus *Ilex*)

Resprouting, spreading clonally with creeping roots, erect shrub to 4 m tall, forming dense stands. **Leaves** Simple, ovate, 2.5 cm long, with 8–12 veins from the base, margins toothed. **Flowers** Female achene to 1 cm long, reddish, with about 20 ribs; about 40 male anthers. Flowering in early summer. **Habitat** Occurring mostly on south-facing slopes in Grassy and Mesic Fynbos. **Distribution** Widespread from Cape Peninsula to Makhanda.

Kouenek, Kouga 16.07.2011

Cliffortia linearifolia

lineari = linear, *folia* = leaf; referring to the linear leaves

Resprouting, upright shrub to 1 m. **Leaves** Trifoliate, leaflets linear, 0.4 cm long × 0.1 cm wide, midrib visible. Flow Female achene reddish brown, smooth, shiny and striated; male with 4 sepals and 4 anthers. Flowering in autumn and winter. **Habitat** Loamy soils in Grassy Fynbos and Sour Grassland in the southeastern part of the study area. **Distribution** Widespread from Knysna to tropical Africa. Other n River rice-bush (E).

Suuranysberg, Kareedouw 11.05.2021

Cliffortia neglecta

neglecta = neglected; this species was previously not
regarded as distinct from another closely related taxon.

Resprouting, low, compact, sprawling shrub to 30 cm.
Leaves Simple, needle-shaped, pointed, 1 cm long. **Flowers**
Female achene ovoid, dark brown, ribbed; male with 6
anthers. Flowering in spring. **Habitat** Found at high altitude
on rocky peaks and ridges in Mesic and Subalpine Fynbos.
Distribution Widespread from Cederberg to Stellenbosch
and to the Kouga and Baviaanskloof mountains.

Bosrug, Baviaans 27.10.2018

Cliffortia paucistaminea **var.** *australis*

pauci = few, *staminea* = stamens; referring to
the few stamens on the male flowers

Resprouting, upright, much-branched, densely leafy shrub
to 2 m. **Leaves** Trifoliate, linear, curved, 1.3 cm long, margins
finely toothed. **Flowers** Female achene light brown, with
12–16 ribs, male with 4 anthers. Flowering in late summer.
Habitat South-facing slopes in Grassy Fynbos. **Distribution**
Narrow distribution from George to Kariega (Uitenhage).

East of Kougakop 16.04.2012

Cliffortia polita

polita = polished; pertaining to the shiny
upper surface of the leaves

Resprouting, low, twiggy, sparsely leafy shrub to 60 cm.
Leaves Trifoliate, linear, shiny, 0.5 cm long. **Flowers** Female
achene irregularly 6-ribbed; male with 6 anthers. Flowering
in summer. **Habitat** Occasional on upper rocky slopes in
Mesic Fynbos. **Distribution** Near endemic from Kamanassie
Mountains to Cockscomb Peak.

Bakenkop, Baviaans 23.12.2011

Cliffortia ramosissima

ramosissima = much branched; pertaining
to the much-branched habit

Reseeding, much-branched, upright or spreading shrub to
2 m. **Leaves** Trifoliate, slightly curved, 0.5 cm long × 0.1 cm
wide. **Flowers** Female achene brown, ribbed; male with
6 anthers. Flowering in summer and autumn. **Habitat** Found
on middle to lower south-facing slopes in Grassy Fynbos.
Distribution Very widespread from Agulhas and Roggeveld
to Soutpansberg, Limpopo.

Riverside, Kouga 23.05.2014

Cliffortia stricta

strictus = very straight, very upright; referring to the
upright habit of the leaves and shrub

Erect shrub to 1.5 m tall, with sheaths and stipules prominent.
Leaves Trifoliate, leaflets linear, to 0.8 cm long. **Flowers** Female
with a smooth, 12-veined receptacle; male with 6 stamens.
Flowering most of the year except in winter. **Habitat** Common
on south-facing and upper slopes in Mesic and Grassy Fynbos.
Distribution Widespread from Cape Peninsula to Baviaanskloof.

Engelandkop, Kouga 18.07.2011

Cliffortia strobilifera   Kammiebos, pypsteelbos, vleibos

strobilifera = cone-bearing; alluding to the cone-like galls
that frequently form on this species, which led early
explorers to think it was a cedar species

Resprouting, upright, much-branched shrub to 3 m. **Leaves**
Simple, linear, usually glabrous, 1.5 cm long × 0.5 cm wide.
Flowers Female achene dark brown, ribbed; male with 15–20
anthers. Flowering in summer. **Habitat** Dense stands on the banks
of perennial streams in Fynbos. **Distribution** Very widespread from
Kamiesberg and Cape Peninsula to Soutpansberg, Limpopo. **Other
names** Cone rice-bush (E), umnwele, umgwele (X).

Cockscomb, Groot Winterhoek 17.02.2016

RUBUS   Brambles, raspberries, blackberries

ruber = red; referring to the red fruit of some species

A large genus of 250–700 species, mostly in temperate regions of the northern hemisphere.
Identification is difficult as polyploidy is common and there are many natural and cultivated hybrids.
± 8 indigenous and several naturalised species in southern Africa; 1 species in Baviaanskloof.

Rubus rigidus   White bramble

rigidus = rigid or ridged; referring to the branches that have ridges running along their length

Resprouting, multi-stemmed, scrambling, thorny
shrub to 2 m. Branches ridged, with recurved
spines to 0.5 cm long. **Leaves** 3–5-pinnate,
toothed, midrib with prickles below. **Flowers**
In terminal or axillary panicles, pale lilac with
purple anthers. Flowering in summer. Fruit
globose, 0.8 cm long, orange becoming blackish.
Habitat Dense stands on the banks of perennial
streams and at Forest margins. **Distribution** Very
widespread from Clanwilliam to tropical Africa.
Other names Braambos (A), umgcunube, idinde
(X), monoko-metsi (S), ugagane, ijingijolo (Z).

Baviaanskloof River, Apieskloof 26.10.2018

RUBIACEAE
Coffee, madder, bedstraw

The fourth-largest family of flowering plants with over 600 genera and 13,000 species. Cosmopolitan, but with most diversity in the tropics and subtropics. *Coffea* is the source of coffee, the popular beverage, while quinine comes from the genus *Cinchona* and is used in the treatment of malaria. *Rubia* is the source of some red dyes. An important and life-changing plant family for humans and biodiversity in general. ± 80 genera and 350 species in southern Africa, with quite a few naturalised and cultivated species; ± 8 genera and 13 species in Baviaanskloof.

ANTHOSPERMUM

anthos = flower, *spermum* = seed; referring to the propensity for some 'male' flowers to have ovaries capable of producing an abundance of seeds

± 40 species in Africa and Madagascar; 26 species in southern Africa; 3 species in Baviaanskloof.

Anthospermum aethiopicum

aethiopicum = from sub-Saharan Africa; in classical times this was the region that had not yet been explored by Europeans.

Reseeding, erect, dioecious shrub to 2 m. **Leaves** Whorled, 3-nate, 0.5 cm long × 0.2 cm wide. **Flowers** Inconspicuous and hidden in the leaves. Flowering in spring and summer. **Habitat** Common in Grassy and Mesic Fynbos. **Distribution** Widespread from Clanwilliam and Cape Peninsula to Makhanda. **Other names** Katstert (A), umthi wamaqhakuva, umsantsana (X).

Vanstadensberg 18.07.2021

Anthospermum galioides

Galium = plant genus, *oides* = resembling; resembling some species of *Galium*

Reseeding, low, sprawling subshrub to 30 cm. **Leaves** Usually curving downwards, 1 cm long × 0.3 cm wide, in whorls or bunches along the stems. **Flowers** Clustered in the axils, tiny, yellowish. Flowering all year round. **Habitat** Rocky slopes in Grassy Fynbos and Transitional Shrubland. **Distribution** Widespread from Nieuwoudtville and the Cape Peninsula to Port Alfred.

Quaggasvlakte, Kouga 13.02.2016

Anthospermum spathulatum Skaapbos

spathulatus = spatula-shaped; referring to the shape of the leaves

Reseeding, erect, dioecious shrub to 1.5 m. **Leaves** Opposite, decussate, 1 cm long × 0.3 cm wide. **Flowers** Female with 2 long stigmas aiding wind pollination; males with 4 anthers. Flowering in spring and summer. **Habitat** Common on loam soils in Transitional Shrubland and Grassy Fynbos, usually in more arid situations than those in which *A. aethiopicum* occurs. **Distribution** Very widespread from Namaqualand to Cape Peninsula and Lesotho.

Kamassiekloof, Kouga 18.01.2022

CANTHIUM

Canti = Malabar name for turkey-berry (*Solanum torvum*)

± 50 species; 13 species in southern Africa; 3 species in Baviaanskloof.

Canthium inerme Common turkey-berry

inerme = unarmed; an ill-suited name as some plants develop long spines

Resprouting small tree to 6 m tall, with smooth, pale bark when young, branching divaricately and sometimes with pairs of spines up to 8 cm long. **Leaves** Opposite, elliptic, glossy, soft-textured, 8 cm long × 4 cm wide, light green. Leaf undersurface with hairy domatia (pits) in the axils of the veins. **Flowers** Numerous in dense axillary clusters, each 3 cm wide, cream to greenish yellow. Flowering in summer. Fruit ovoid, 2-lobed, green becoming dark brown and wrinkled. **Habitat** Occurring mainly as a subcanopy tree in Forest or Thicket. **Distribution** Very widespread from Cape Peninsula to Zimbabwe. **Other names** Gewone bokdrol (A), umvuthwamimi, umnyushulube, isiphingo (X), umvuthwemini (Z).

Zandvlaktekloof, Baviaans 19.01.2022

GALIUM

galion = bedstraw, *gala* = milk; in medieval Europe, the dried plants of *G. verum* were used for stuffing mattresses and to curdle milk; to this day it is still used for colouring cheese.

± 400 species, 12 species in South Africa; 1 species in Baviaanskloof.

Galium tomentosum Kleefgras

tomentosum = hairy; referring to the stems, which are densely white-hairy, and the long, fluffy hanging flower stalks

Perennial scrambler, dioecious, to 3 m. **Leaves** Whorled in groups of 6–8, 3 cm long × 1 cm wide, margins with bristly, hooked hairs. **Flowers** Female flowers on hanging bunches of long, white fluffy stalks. Flowering in spring. **Habitat** Occasional in the shade in Thicket or Forest. **Distribution** Very widespread from Namibia to Eastern Cape and Free State. **Notes** The flower stalks are used by some birds for making nests, thus dispersing the seed to favourable sites in dense bush. **Other names** Old man's beard (E), rooivergeet (A).

Waterkloof, Bokloof 26.04.2018

ROTHMANNIA

Named after Joran Johansson Rothman, 1739–1778, a Swedish botanist and doctor who was a student of Linnaeus

± 25 species in sub-Saharan Africa; 1 species in Baviaanskloof.

Rothmannia capensis   Cape gardenia

capensis = from the Cape; referring to the geographical origin of the species

Reseeding, neatly branched tree to 10 m. **Leaves** Opposite, elliptic, with domatia (pits) clearly visible. **Flowers** Single at the branch tips, cup-shaped, to 8 cm long, cream, sweetly scented. Fruit round, hard, 7 cm wide, green turning brown. Flowering in late summer. **Habitat** Occasional in Forest in the narrow kloofs. **Distribution** Widespread from Swellendam to Limpopo. **Other name** Aapsekos (A).

domatia

Ysrivier, Cambria 03.02.2020

RUBIA

ruber = red; pertaining to the red roots that were used to make dye

± 60 species worldwide; 3 species in southern Africa; 1 species in Baviaanskloof.

Rubia petiolaris   Madder

petiolaris = petioled; pertaining to the long petioles on the leaves. The common name refers to *R. tinctorum,* the roots of which have been the source of a red dye for centuries.

Perennial scrambler, bristly, rough to touch, to 3 m tall, stems 4-angled. **Leaves** Whorled, petiolate, ovate, 3-veined from the base. **Flowers** In axillary clusters, greenish. Flowering in summer. **Habitat** Occasional in Thicket and Forest in the damp ravines. **Distribution** Widespread from Riversdale to Free State. **Other name** Kleefgras (A).

Uitspankloof, Baviaans 02.05.2018

RUTACEAE

Citrus, buchu

± 161 genera and over 1,800 species worldwide. Well known for the economically important genus *Citrus*, which includes oranges, lemons, limes and grapefruit. These fruits are said to originate from the foothills of the Himalayas; however, wild populations no longer exist there. The diverse varieties bred by humans are cultivated in many warmer parts of the world. Characteristic of the family are the oil glands, most visible on the underside of the leaves and giving off a strong scent, which is believed to function as a defence mechanism against browsers. Some species of *Boronia*, a large genus from Australia, have very fragrant flowers that are used commercially in oil production. *Agathosma*, commonly known as buchu, is a large genus of ± 150 species in southern Africa. *A. betulina* and *A. crenulata* are grown commercially for their leaves, which are used in teas. Oil is also extracted from the leaves for use in food colourants, soaps and cosmetics. In addition, these 2 species are also used medicinally for treating renal disorders and chest problems. 23 genera and ± 300 species in southern Africa, mostly in the southern parts of the Western Cape; 10 genera and 32 species in Baviaanskloof, plus 1 exotic species, *Ruta graveolens*.

AGATHOSMA

Buchu, boegoe

agathos = good, *osme* = scent; pertaining to the strong citrus smell of the crushed leaves

Leaf shape and size are distinctive and each species has a unique smell. Oil glands carry the scent and are visible on the back and side of the leaves. The shape and size of the capsules and the number of chambers they have are useful for identification. ± 150 species in southern Africa, mostly in the Western Cape; 17 species in Baviaanskloof.

Agathosma blaerioides

Blaeria = plant genus, *oides* = resembling; *Blaeria* is one of the minor genera in the *Erica* family that has been incorporated within the genus *Erica*. The leaves of this species are reminiscent of some species previously assigned to *Blaeria*.

Reseeding, erect shrub to 1 m. **Leaves** Spreading, often saddle-shaped, small, to 0.5 cm long, sometimes with stalked glands on the margins. **Flowers** In small axillary and terminal clusters, white, sometimes with 1 to a few red gland dots on the petals. Flowering from autumn to midsummer. Fruit 1–3-chambered. **Habitat** Upper slopes in Arid and Mesic Fynbos. **Distribution** Widespread from Mossel Bay to Baviaanskloof.

Nuwekloof Pass, Baviaans 13.10.2016

Agathosma capensis Boegoe, steenbokboegoe

capensis = from the Cape; referring to the geographical origin of the species

Resprouting shrublet to 50 cm, with numerous thin branches from the base, giving off a liquorice, aniseed or sweet spice scent. **Leaves** Small and closely packed, linear to oval, 0.5 cm long × 0.2 cm wide. **Flowers** Clustered into dense heads at branch tips, white, pink or purple. Flowering in spring and summer. Fruit 3-chambered. **Habitat** Found mostly in Grassy Fynbos but also in other Fynbos habitats. **Distribution** Widespread from Namaqualand to Gqeberha (Port Elizabeth).

Combrink's Pass, Baviaans 26.10.2018

Agathosma kougaense

kougaense = of the Kouga; referring to the origin
of the plant in the Kouga Mountains. Kouga is
the original Khoisan name for a leopard.

Reseeding, slender shrublet to 25 cm. **Leaves** Linear-
lanceolate, keeled, 1 cm long, oil glands raised,
margins finely-hairy. **Flowers** In pairs in the axils,
white, hairy. Flowering mainly in spring. Fruit 1- or
2-chambered. **Habitat** Uncommon and localised
on the south-facing slopes of high peaks in Mesic
and Subalpine Fynbos. **Distribution** Endemic to the
Kouga Mountains. **Status** Rare.

East of Kougakop 16.04.2012

Agathosma mucronulata

mucron = pointed tip, *ulata* = possessing;
referring to the leaves, which have a
distinct pointy, stiff mucron, making the
shrub prickly to the touch

Reseeding, much-branched, rounded
shrub to 1 m, turpentine-scented.
Leaves Ascending, closely packed, ovate, with a
sharp, sometimes recurved pointy tip (mucronulate),
0.6 cm long. **Flowers** Clustered at the tips of the
branches, white with purple dots. Flowering in spring.
Fruit 3-chambered. **Habitat** Common on middle
north-facing slopes in Grassy Fynbos. **Distribution**
Endemic to the Kouga and Baviaanskloof mountains.
Notes Closely related to and possibly just an arid
form of *A. martiana*, which also grows in the area.

Engelandkop, Kouga 13.07.2011

Agathosma mundtii Jakkalspisbos

Named after JL Leopold Mund, 1791–1831, a Berlin
pharmacist who was sent to the Cape in 1815 to
collect plant specimens

Resprouting or reseeding wiry shrub to 1 m.
Leaves Finely-hairy and smelling extremely
unpleasant, to 1 cm long × 0.5 cm wide but
usually thinner, with margins strongly revolute.
Flowers Clustered into heads at the branch
tips, white. Flowering in winter and spring. Fruit
2-chambered. **Habitat** Frequent on upper slopes
in dry, rocky areas in Grassy Fynbos, but also in all
other Fynbos habitats. **Distribution** Widespread from
Witteberg to Baviaanskloof and Humansdorp.

Uitslag 4X4, Baviaans 09.07.2013

Agathosma ovata Basterboegoe

ovata = ovate; describing the shape of the leaves

Reseeding, or sometimes resprouting, shrub to 2 m, herb-scented. **Leaves** Variable in shape, here slightly elongated-ovate, 1.2 cm long × 0.6 cm wide. **Flowers** Single to a few in the axils near the branch tips, white, pink or purple. Flowering all year round. Fruit 5-chambered. **Habitat** Common on rocky outcrops in all Fynbos habitats. **Distribution** Widespread from the Witteberg Mountains to Lesotho. **Other names** False buchu (E), umahesakomhlope (Z).

Engelandkop, Kouga 18.07.2011

Agathosma pilifera

pilifera = hairy; referring to the distinctive long hair-like tips of the leaves, incurved, whitish and relatively long

Reseeding, erect shrub to 60 cm, faintly herb-scented. **Leaves** Ovate, to 0.6 cm long, bright green, with a distinct, slender incurved mucron, glands prominent on the margins. **Flowers** In terminal clusters, sepals leaf-like, petals white with purple spots. Flowering in spring. Fruit 1–3-chambered. **Habitat** Uncommon on rocky upper slopes in Mesic Fynbos. **Distribution** Endemic from Baviaanskloof to Elandsberg. **Notes** The leaf-like sepals are a notable feature.

Enkeldoring, Baviaans 21.09.2011

Agathosma puberula

puberula = hairy, puberulous; referring to the puberulous branchlets, peduncles, sepals, and sometimes petals and fruit

Reseeding, erect, handsome, much-branched shrub to 3 m tall, acrid-, herb- or sulphur-scented. **Leaves** Linear-lanceolate, to 2 cm long, mucronate. **Flowers** In terminal clusters, peduncles 0.7 cm long and hairy, flowers white with reddish dots. Flowering from autumn to spring. Fruit 3- or 4-chambered. **Habitat** Uncommon and localised at middle altitudes, often off the edge of the flat-topped grassy plateaus, in Transitional Shrubland or Grassy Fynbos, often at or near the boundary with Thicket. **Distribution** Eastern parts of Baviaanskloof, eastward to Makhanda.

Grasnek, Kouga 17.07.2011

Agathosma pungens

pungens = pungent; referring to the leaves that end with a stiff hard point

Reseeding, erect, much-branched, floriferous shrub to 80 cm tall, pleasantly aromatic. **Leaves** Linear-lanceolate, curving upwards, spine-tipped. **Flowers** Single in the axils at the branch tips, pink or purple, with purple spots. Flowering in winter and spring. Fruit 2-chambered. **Habitat** Upper south-facing slopes, often on rock ledges, in Mesic Fynbos. **Distribution** Fairly narrow distribution from Swartberg and Kammanassie to the Kouga and Baviaanskloof mountains.

Bosrug, Baviaans 12.10.2016

Agathosma recurvifolia   Kanferboegoe

recurvi = recurved, *folia* = leaves; referring to the recurved leaves

Reseeding, erect to spreading shrub to 1 m tall, turpentine-scented. **Leaves** Ovate, spreading, recurved, 0.4 cm long, margins hyaline, tip hair like. **Flowers** Clustered in heads at the branch tips, white, anthers red. Flowering in winter and spring. Fruit 2-chambered. **Habitat** Uncommon in Transitional Shrubland and Thicket, often at the boundaries between vegetation types. **Distribution** From the Swartberg to Kariega (Uitenhage). **Notes** Possibly related or similar to *A. spinosa* from near Uniondale.

Krakeel, Langkloof 07.09.2009

Agathosma unicarpellata

uni = one, *carpel* = central chambered female organ in a flower, *ata* = having; referring to the single-chambered fruit

Reseeding, slender, sparsely branched shrub to 45 cm tall, turpentine-scented. **Leaves** Ascending, lanceolate, 1.5 cm long × 0.3 cm wide, hairless. **Flowers** 1–3 in axils, 0.7 cm wide, white. Flowering from autumn to early summer. Fruit 1-chambered. **Habitat** Middle to upper rocky ridges in Grassy, Arid and Mesic Fynbos. **Distribution** Endemic to Kouga and Baviaanskloof mountains.

Scholtzberg, Baviaans 24.07.2011

Agathosma venusta   Goeieboegoe

venusta = beautiful, graceful; referring to the beauty of this pale mauve-flowered species

Resprouting, upright, neatly branched shrub to 1 m, pleasantly liquorice-scented. **Leaves** Ovate or round, 0.8 cm long, oil glands raised and bumpy on the margins. **Flowers** In clusters in the axils near and on the branch tips, thus forming a spike-like inflorescence, pale mauve. Flowering in spring and summer. Fruit 2–5-chambered, often aborting some, so the number of mature chambers per capsule can vary on the same plant. **Habitat** Middle to upper south-facing slopes in Mesic and Grassy Fynbos. **Distribution** Widespread from Swartberg to Kariega (Uitenhage) and to Graaff-Reinet.

Bosrug, Baviaans 12.10.2016

CALODENDRUM
Cape chestnut, wildekastaiing

kalos = beautiful, *dendron* = tree; referring to the magnificent flowers on this tree

1 species in Africa.

Calodendrum capense

capense = of the Cape; referring to the geographical origin of the species; in the past, *capense* referred to the southern part of Africa in general, not only the Cape provinces.

Tree to 8 m. **Leaves** Simple, wide-elliptic, 20 cm long × 10 cm wide, covered in gland dots, wavy and curling. **Flowers** In dense, showy heads, each flower 5 cm wide, white to pink. Flowering in warmer seasons. Fruit a 5-lobed woody capsule, splitting from above to reveal large black seeds. **Habitat** Sporadic in kloofs in Thicket and Forest. **Distribution** Widespread from Swellendam to tropical Africa. **Notes** Used medicinally for many ailments. **Other names** Umsitshana, umemezi (X), umbhaba (X, Z).

Gonjah, Patensie 17.05.2021

CLAUSENA
Horsewood, perdepis

Named after Peder Claussøn, 1545–1614, a Norwegian priest, author and botanist

± 50 species from Africa to Malaysia; 1 species in Baviaanskloof.

Clausena anisata Perdepis

anisata = aniseed-scented; the crushed leaves have a strong and unpleasant smell of aniseed or horse urine – thus the common name for the genus.

Resprouting shrub or small tree to 5 m tall, sometimes to 10 m. **Leaves** Pinnately compound, 12–17 alternate or subopposite leaflets, and a terminal leaflet. Leaflets sometimes with slightly toothed or wavy margins. **Flowers** Inflorescence a branched axillary spray with small flowers 0.4 cm wide, white with orange-yellow anthers. Flowering in spring. Fruit a black berry dispersed by birds. **Habitat** Thicket and at the margins of Forest. **Distribution** Widespread throughout southern Africa to tropical Africa. **Notes** Used in traditional medicine to treat heart ailments, fevers and internal parasites. **Other names** Umnukambile, isifutho (X), umsanga (Z).

10.07.2020

COLEONEMA
Cape may, confetti bush

koleos = a sheath, *nema* = a thread or filament; pertaining to the staminodes that run up the petals in a sheath

8 species in Western and Eastern Cape; 1 species in Baviaanskloof.

Coleonema aspalathoides

Aspalathus = plant genus, *oides* = resembling; possibly referring to the position in which the flowers are nested in the axils of the leaves and scattered along the branches, which is reminiscent of *Aspalathus*

Reseeding, much-branched, dense, rounded shrub to 1 m tall, faintly aromatic. **Leaves** Needle-like, linear-oblong, tip a fine point, lime-green. **Flowers** Single in the axils of the leaves, sessile, numerous scattered along the branches, to 1 cm wide, bright pink. Flowering from winter to early spring. Fruit 5-chambered. **Habitat** Lower to middle slopes in Grassy Fynbos. **Distribution** Widespread from Potberg to Zuurberg.

Doringkloof, Kouga 29.07.2011

DIOSMA
False buchu

dios = divine, *osme* = scent; the crushed leaves of most species have an interesting smell.

Distinguished by the small flowers up to 1 cm wide and with a green disc on top of the ovary. Styles and anthers are shorter than the petals. ± 30 species in winter-rainfall South Africa; 5 or 6 species in Baviaanskloof.

Diosma apetala

a = without, *petala* = petals; referring to the flowers that drop their petals soon after opening

Reseeding, sprawling, wiry shrublet to 20 cm. **Leaves** Adpressed-erect, lanceolate, closely packed, chunky, small, 0.3 cm long. **Flowers** Single at branch tips, 0.3 cm wide, white with greenish centre, turning brown and falling off early. Flowering mostly in spring. Fruit 5-chambered. **Habitat** Upper rocky slopes in Arid and Mesic Fynbos. **Distribution** Nearly endemic from Swartberg to Kouga and Baviaanskloof mountains.

Nuwekloof Pass, Baviaans 08.10.2016

Diosma prama

prama = abbreviation of *pseudoramosissima*; indicating a superficial similarity to *D. ramosissima*, from Namaqualand

Reseeding, erect, non-hairy, much-branched shrub to 1.5 m tall, sweetly scented. **Leaves** Adpressed, glabrous, 0.3 cm long × 0.1 cm wide, with gland dots mostly along midrib. **Flowers** Single or few at branch tips, 0.4 cm wide, white. Flowering in spring and summer. Fruit relatively large, to 1.5 cm wide, with distinct horns. **Habitat** Common and sometimes abundant on rocky north-facing slopes in Arid and Grassy Fynbos. **Distribution** Widespread from Touwsberg to Kouga and Baviaanskloof mountains.

Engelandkop, Kouga 13.07.2011

Diosma rourkei

Named after Dr John Rourke, b. 1942, a Cape Town-based
botanist; he was curator of the Compton Herbarium and
specialised in the Proteaceae.

Reseeding, erect to spreading, diffusely branched shrub
to 1 m tall, aromatic. **Leaves** Ascending, linear-lanceolate,
acute, curving in very slightly near the tip, very finely-hairy,
with short petioles. **Flowers** 1 or 2 at the branch tips, nested
in the leaves, 0.4 cm wide, white, with green disc. Flowering
in winter and early spring. Fruit 5-chambered, reddish,
horns green. **Habitat** Rocky slopes in Grassy and Mesic
Fynbos. **Distribution** Local endemic to Baviaanskloof.

Zandvlakte, Baviaans 29.04.2018

Diosma sp. nov.

sp. nov. = new species, undescribed; abbreviation used to indicate
taxa that have not yet been formally described and/or published

Reseeding, slender, erect, wiry shrublet to 40 cm. **Leaves**
Adpressed-erect, finely pubescent, small, 0.2 cm long ×
0.1 cm wide. **Flowers** 1 or 2 at the branch tips, cup-shaped,
0.3 cm wide, white, calyx finely-hairy. Flowering in spring
and summer. Fruit 5-chambered, tips slightly recurved.
Habitat Infrequent on rocky, south-facing slopes at 600–
1,000 m above sea level. **Distribution** Local endemic known
only from Baviaanskloof Mountains near Scholtzberg Peak.
Notes First recorded by the author, Douglas Euston-Brown,
in 1992. Yet to be described.

Scholtzberg, Baviaans 29.10.2011

EUCHAETIS

eu = well, fine, *chaite* = long hair or mane; pertaining to hairs on inner side of petals, a distinguishing
feature of the genus

23 species in winter-rainfall South Africa; 2 species in Baviaanskloof, both endemic.

Euchaetis cristagalli

crista galli = name for part of the human skull; derivation unclear,
possibly referring to the collar-like disc around the ovary

Reseeding, erect, slender, sparsely branched shrublet to
60 cm. **Leaves** Ascending, linear-lanceolate, glabrous,
1 cm long. **Flowers** Single to a few at the branch tips, petals
acute, white. Flowering in autumn. Fruit 5-chambered.
Habitat South-facing slopes in Mesic Fynbos. **Distribution**
Endemic to Groot Winterhoek Mountains in the vicinity of
Cockscomb Peak. **Status** Rare.

Cockscomb, Groot Winterhoek 16.02.2016

Euchaetis vallis-simiae

vallis = valley, *simiae* = monkey; referring to the origin of the plant in Baviaanskloof

Reseeding, erect, neatly branched shrub to 1.2 m tall, crushed leaves unpleasantly scented. **Leaves** Ascending, elliptic, slightly recurved, finely bristly-hairy, with apex knobbed, 0.5 cm long, olive-green. **Flowers** 1–4 at branch tips, white. Flowering in spring. **Habitat** Localised populations, sometimes forming dense stands on rocky slopes in Grassy Fynbos. **Distribution** Endemic to Kouga and Baviaanskloof mountains.

Bosrug, Baviaans 22.07.2011

PTAEROXYLON

Sneezewood, nieshout

ptairo = to sneeze, *xylon* = wood; pertaining to the smell of the freshly cut wood, which may provoke sneezing

1 species in southern Africa.

Ptaeroxylon obliquum

obliquum = oblique; referring to the oblique leaflets

Resprouting, dioecious, deciduous shrub or tree to 10 m tall or higher. Bark pale grey to whitish, darkening with age. **Leaves** Opposite, oblique leaflets in 3–7 pairs, leaflets 2.5 cm long × 1.3 cm wide, hairy when young. **Flowers** Axillary, in branched heads, sweetly scented, to 5 cm long × 0.7 cm wide, pale yellow. Flowering in spring and early summer. Fruit a capsule that splits open to release 2 winged seeds. **Habitat** Thicket and Forest. **Distribution** Widespread from Baviaanskloof to tropical Africa. **Notes** An attractive wood used in the production of furniture. It is particularly hard, making it durable enough for use in machine bearings. **Other names** Umthathi, umpafa (X), ubhaqa (Z).

Grootrivierpoort 04.02.2020

VEPRIS
White ironwood, witysterhout

vepres = a bramble; derivation unclear, the only possible similarity to a bramble is in the trifoliate leaves.

15 species in Africa and on Mascarene Islands; 3 species in South Africa; 1 species in the Baviaanskloof.

Vepris lanceolata (= *V. undulata*)

lanceolata = lance-shaped; referring to the lance-shaped leaves; *undulata* pertains to the undulating leaf margins.

Reseeding, evergreen shrub or small tree 5 m tall, reaching 20 m in tall forest. Bark relatively smooth, grey to dark grey. **Leaves** Trifoliate, lemon-scented, on a long petiole not winged, leaflets 8 cm long × 2.5 cm wide, margins wavy. **Flowers** In a dense terminal head or panicle, flowers small and yellowish. Flowering in late summer. Fruit fleshy, black when mature, 0.5 cm wide. **Habitat** Dry forest, also in a few of the forested kloofs in Baviaanskloof. **Distribution** Very widespread from Swellendam to tropical Africa. **Notes** Good source of timber, used to make structural beams and implement handles. The flu virus has been treated with powder produced from the roots. Porcupines are known to eat the bark. **Other names** Umdlebe, umngamazele, umzane (X), isutha (Z).

19.05.2020

SALICACEAE
Willows

A cosmopolitan family of 55 genera and 1,010 species; 10 genera and 26 species native to southern Africa, with several others naturalised or cultivated; 2 genera and 2 species in Baviaanskloof.

SALIX

salignus = willowy, willow-like; pertaining to the flexible branches that hang down, as an adaption to the strong-flowing rivers in which they grow

± 400 species, mainly in the northern hemisphere; 1 species in Baviaanskloof.

Salix mucronata Willow

mucronata = mucronate; referring to the small pointed tip (mucron) on the leaves

Dioecious shrub or small tree to 10 m tall, with rough bark and drooping branches. **Leaves** Lanceolate, 4 cm long × 1 cm wide, silvery-hairy, paler below, margins finely toothed. **Flowers** Numerous on spikes in the axils, small, yellow, with woolly seeds. Flowering in spring. **Habitat** Common on the banks of perennial streams and rivers. **Distribution** Very widespread throughout southern Africa. **Notes** This is subsp. *mucronata*; 3 other subspecies do not occur in Baviaanskloof. **Other names** Rivierwilger (A), umngcunube (X).

Baviaanskloof River,
Saaimanshoek 20.09.2010

SCOLOPIA

skolopes = pointed stick, thorn; pertaining to the sharp spines produced by some species

40 species worldwide; 5 species in South Africa; 1 species in Baviaanskloof.

Scolopia zeyheri Thorn pear

Named after Karl Ludwig Philipp Zeyher, 1799–1858, German botanist who collected in South Africa for 22 years. Many of his specimens were lost in disastrous events such as shipwrecks and fires. Best remembered as co-author of *Enumeratio Plantarum Africae Australis Extratropicae,* a catalogue of South African plants produced with Danish botanical collector CF Ecklon. He died of smallpox.

Resprouting, much-branched bush or small tree to 5 m tall, sometimes taller. Branches with masses of tough, sharp, dark brown spines. **Leaves** Elliptic, 5 cm long × 2 cm wide, paler below, margins obscurely toothed. **Flowers** In spike-like racemes in the axils, white or yellow. Flowering in winter and spring. Fruit a red berry to 1 cm wide. **Habitat** Thicket. **Distribution** Very widespread from Knysna to Limpopo and tropical Africa. **Notes** Fruit edible and hard; wood is heavy, used as fuel. **Other names** Doringrooipeer (A), umqokolo, iqumza, umqaqoba, iqumza elinameva (X).

Elandsberg road, Patensie 25.10.2018

SALVADORACEAE

A small family of 3 genera and 11 species. Only 2 genera and 3 species in South Africa; 1 species in Baviaanskloof.

AZIMA   Needle bush, speldedoring

azim = defender, *azima* = mighty; referring to the sharp spines

4 species in tropics and subtropics; 1 species in Baviaanskloof.

Azima tetracantha

tetra = four, *cantha* = spines; referring to the 4 spines arranged at right angles at the nodes

Resprouting, scrambling, dioecious shrub to 5 m tall, forming dense thickets. Branches with 4 spines arranged in whorls at each node. **Leaves** Opposite, elliptic to suborbicular, 4 cm long × 3 cm wide, bright green. **Flowers** Few in loose clusters in the axils, greenish to yellow. Flowering in summer. Fruit round, 1 cm wide, yellow to white. **Habitat** Occurring at the margins of Forest and Thicket, often in areas exposed to heavy browsing, near water. **Distribution** Very widespread from Malgas to tropical Africa, Madagascar and India. **Other names** Icegceya (X), gecaya (Z).

Waterkloof, Bokloof 26.04.2018

SANTALACEAE

Sandalwood

A medium-sized family of hemiparasitic trees and shrubs with ± 43 genera and 1,000 species worldwide. The famous sandalwood tree (*Santalum album*) produces one of the most expensive kinds of timber in the world. Its oil is used in perfumes because the fragrance lasts for decades. The slow-growing trees in this family have been over-harvested for centuries and several species are now endangered. ± 6 genera and 226 species in southern Africa; 3 genera and 19 species in Baviaanskloof.

LACOMUCINAEA (= *Thesium*)

Named after Ladislav Mucina (1956–), a Slovakian-born botanist who co-authored a vegetation map of southern Africa; he is currently based in Australia.

1 species in southern Africa.

Lacomucinaea lineata (= *Thesium lineatum*) Vaalstorm

lineatus = marked with fine parallel lines; referring to the lines on the grooved branches

Dense, rigid shrub to 2 m tall, with spine-tipped, grey-green branches. **Leaves** Inconspicuous, linear, falling early. **Flowers** Small, on short racemes near the branch tips, whitish. Flowering in spring and summer. Fruit fleshy, 1 cm long, white. **Habitat** Infrequent in Transitional Shrubland, Succulent Karoo and Thicket. **Distribution** Widespread from Namibia to Baviaanskloof. **Other name** Witstorm (A).

parallel lines

Vlakkie, Zandvlakte 06.07.2011

OSYRIS (= *Colpoon*)

ozos = branch; referring to the branching habit of the plants; *kolpos* = breast; referring to the shape of the berries; alternatively, *colpus* = groove, which might refer to grooves in the ovary where the stamens pass through

Molecular studies revealed that the genus *Colpoon* should be included in *Osyris*. 8 species worldwide; 3 species in southern Africa; 1 species in Baviaanskloof.

Osyris compressa (= *Colpoon compressum*) Cape sumach

compressum = laterally flattened; referring to the arrangement of the leaves around the stems, with leaves sometimes appearing compressed

Resprouting, hemiparasitic shrub or small tree to 5 m. **Leaves** Opposite, ovate-elliptic, leathery, with obscure venation. **Flowers** In panicles at the branch tips, greenish. Flowering in summer and autumn. Fruit 1.5 cm long, red becoming black. **Habitat** Common in various habitats but mostly in Grassy Fynbos and Transitional Shrubland. **Distribution** Very widespread from Namaqualand to tropical Africa. **Other names** Pruimbas (A), intekaza (X), umbulunyathi (X, Z), ingondotha-mpethe (Z).

Waterkloof, Bokloof 26.04.2018

THESIM

thes = a hired labourer; alluding to the hemiparasitic nature of the plant, obtaining some resources from the host plant but also able to grow independently

300 species worldwide. Prevalent in first 10 years after fire. 178 species in southern Africa; 14 species in Baviaanskloof.

Thesium flexuosum

flexuosum = bent, curved; referring to the upward-curving stems

Straggling shrublet to 30 cm tall with grooved stems that curve upwards. **Leaves** Few, acute, tiny, 0.1 cm long. **Flowers** In spikes, 0.1 cm wide, whitish, subtended by a dark pointy bract. Flowering mainly in autumn and winter. **Habitat** Dry, north-facing slopes in Grassy Fynbos. **Distribution** Widespread from Montagu to Baviaanskloof.

Bergplaas, Combrink's Pass 18.05.2021

Thesium scandens

scandens = climbing, twining; referring to the scrambling and climbing habit

Scrambler with zigzag, trailing and hanging branches. **Leaves** Tiny, terete, fleshy. **Flowers** Whitish. Flowering mainly in spring. **Habitat** Occasional in Thicket. **Distribution** Widespread from Baviaanskloof to Fort Beaufort.

Groenek, Kouga 08.03.2012

Thesium squarrosum

squarrosus = spread at right angles; referring to the spreading leaves on the angled stems

Dense shrublet to 20 cm tall, with angular branchlets, rough to touch. **Leaves** Spreading, linear, fleshy, tips recurved. **Flowers** On a disc at branch tips. Flowering in summer. **Habitat** Infrequent in Grassy Fynbos and Transitional Shrubland. **Distribution** Widespread from Knysna to Eastern Cape.

Koud Nek's Ruggens, Kouga 18.04.2012

Thesium strictum   Teringbos

strictus = very upright, straight up; pertaining to the tall, lanky growth habit

Reseeding, tall, lanky, bare-stemmed shrub to 2.5 m. **Leaves** Falling easily, adpressed, lanceolate, 1 cm long. **Flowers** In loose terminal heads, 0.3 cm wide, yellowish white. Flowering in spring and summer. **Habitat** Grassy Fynbos. **Distribution** Very widespread from Kuboes in Namaqualand to Makhanda.

Rus en Vrede 4X4, Kouga 28.07.2011

VISCUM

Mistletoe

viscum = sticky, viscid; referring to the sticky berries that pass through a bird's digestive tract intact; when eliminated, the sticky mass of seeds adheres to branches of host plants.

From the Old World tropics. ± 100 species; 19 species in South Africa; 4 species in Baviaanskloof.

Viscum capense Voëlent

capense = from the Cape; referring to the geographical origin of the species

Dioecious stem parasite to 50 cm, with rigid and brittle branchlets. **Leaves** Scale-like. **Flowers** Yellow. Flowering in winter and spring. Berries white, smooth. **Habitat** Common on various shrubs and trees. **Distribution** Widespread from Namibia to Eastern Cape. **Notes** Fruit poisonous to humans.

Roelf se Put, Kouga 01.01.2018

Viscum rotundifolium

rotundus = almost circular, *folius* = leaf; referring to the rounded leaves that are slightly longer than wide

Monoecious stem parasite to 50 cm, sometimes drooping. **Leaves** Opposite, broadly ovate, 1 cm wide. **Flowers** Tiny, creamy green. Flowering in autumn. Berry orange-red when ripe. **Habitat** Common on various trees mostly in Thicket. **Distribution** Widespread in southern Africa.

Nooitgedacht, Kouga 12.05.2021

SAPINDACEAE

Soapberry

A large family of trees, shrubs and lianas, with ± 138 genera and 1,900 species worldwide. Lychee, horse chestnut and maple belong to this family, together with several other economically important tropical fruits. ± 20 genera and 35 species are native to southern Africa; 6 genera and 6 species in Baviaanskloof.

ALLOPHYLUS

allo = different, *phylum* = tribe; suggesting that this group of taxa is quite different from others in the family

± 200 species worldwide. Some plants are trifoliate and can be easily confused with *Searsia*, but *Allophylus* always has hairy pits in the axils of the veins on the underside of the leaves. 9 species in southern Africa; 1 species in Baviaanskloof.

Allophylus decipiens Bastertaaibos

decipiens = deceiving; referring to its similarity to a closely related species, with which it is sometimes confused; could also be referring to *Searsia* species

Resprouting, multi-stemmed shrub or small tree to 4 m. **Leaves** Trifoliate, leaflets 5 cm long × 2 cm wide, soft-textured, paler below, not hairy, but with hairs in the axils of the leaves. **Flowers** In axillary spikes, 6 cm long, white. Flowering in summer. Fruit roundish, 0.6 cm wide, becoming red. **Habitat** Occurring in Forest in the narrow ravines. **Distribution** Widespread from Gourits River to Mpumalanga. **Other names** Small-leaved false currant (E), uthathabani (X), umcandathambo (X, Z), umhlohlela (Z).

Gannalandkloof, Baviaans 30.12.2017

ATALAYA

atalai = the Timorese name for a species in this genus

18 species worldwide; 3 species in South Africa; 1 species in Baviaanskloof.

Atalaya capensis Wing-nut

capensis = from the Cape; referring to the geographical origin of the species

Reseeding, single-stemmed small tree to 6 m. Bark pale grey, smooth. **Leaves** 3–5 pairs of opposite leaflets, elliptic-oblong, 5 cm long × 1.5 cm wide, margins wavy. **Flowers** Cream, numerous in axillary clusters near the branch tips. Flowering in summer. Fruit a winged nut. **Habitat** Infrequent on rocky koppies in Thicket. **Distribution** Narrow distribution from Baviaanskloof to Gqeberha (Port Elizabeth). **Other names** Cape krantz ash (E), krans-esseboom (A). **Status** Protected in South Africa.

Rooihoek, Kouga Valley 10.03.2012

DODONAEA

Named after Rembert Dodoens, 1517–1585, a Flemish doctor and botanist. He wrote a plant reference book, *Cruydtboeck*, illustrated with 715 images, published in 1554; in the 16th century, it was the most translated book after the Bible.

± 60 species, mostly in Australia; 1 species with 2 subspecies in South Africa.

Dodonaea viscosa **subsp.** *angustifolia*   Ysterhout, sandolien

viscosa = sticky; the fresh growth is covered with a shiny, sticky substance, said to be flavonoids.

Resprouting, often multi-stemmed dioecious tree to 4 m. **Leaves** Linear-oblanceolate, resinous, 8 cm long × 2 cm wide. **Flowers** Greenish yellow, in axillary and terminal clusters. Flowering in winter and spring. Fruit a winged capsule. **Habitat** Common and abundant on sandy loams in Grassy Fynbos and Transitional Shrubland. **Distribution** Very widespread from Namaqualand to tropical Africa. Often associated with biome boundaries. **Notes** Sought after by locals for use as cooking fuel.

Grasnek, Kouga 17.04.2012

HIPPOBROMUS   Basterperdepis

hippos = horse, *bromos* = oats; smelling of horse urine, this is another case of mistaken identity because the plant does not smell bad. *Clausena anisata* is known to smell like horse urine and the leaves of the 2 species look similar, which probably led to the confusion.

1 species in South Africa.

Hippobromus pauciflorus   False horsewood

pauciflorus = few-flowered; this species is a few-flowered relative to horsewood (*Clausena anisata*), of the closely related family Rutaceae.

Resprouting, small upright tree to 5 m. **Leaves** Compound, pinnate, shiny, glabrous, leaflets obovate, toothed near tips, main leaf midrib winged. **Flowers** Numerous, in panicles to 7 cm long, white. Flowering in winter and spring. Fruit round, 0.8 cm wide, black. **Habitat** Mesic Thicket and Forest margins. **Distribution** Widespread from Baviaanskloof to Mpumalanga. **Notes** Too thin for timber but good for yoke pins (*jukskei* stumps) and walking sticks. **Other names** Umfazi onengxolo (X), umhlwathile, umnquma, umnungumabele (X, Z), umfazothethayo (Z).

Kouga Valley, Kareedouw 21.01.2022

PAPPEA

Named after Karl Wilhelm Ludwig Pappe, 1803–1862, a German-born doctor and botanist. He came to the Cape in 1831 and became the University of Cape Town's first professor of botany. Pappe's collections were included in the original herbarium at the South African Museum. In 1956, they were moved to the Compton Herbarium at Kirstenbosch National Botanical Garden.

1 species in Africa.

Pappea capensis Jacket-plum

capensis = from the Cape; referring to the geographical origin of the species.
The common name pertains to the covering that splits open, revealing the fruit.

Single-stemmed, umbrella-forming tree to 6 m, with smooth grey bark. **Leaves** Crowded at branch tips, minutely toothed, hard-textured and sometimes wavy. **Flowers** Numerous in the axils, greenish, attracting bees. Flowering in summer and autumn. Fruit round, coat velvet-green, splitting open to show bright red 1-seeded fruit, flesh edible. **Habitat** Rocky slopes in Thicket. **Distribution** Widespread from Namibia to the Little Karoo to tropical Africa. **Notes** An important tree for the ecosystem. It can tolerate severe droughts by dropping its leaves and 'shutting down' until conditions improve. The yellow oil extracted from roasted seeds has been used medicinally to treat ringworm, baldness, nosebleeds, chest complaints, eye infections and venereal disease, and as a purgative. Also utilised to lubricate rifles. The flesh of the fruit is used to make jam, jelly, vinegar and alcohol. Browsed by game. **Other names** Pruimboom, doppruim (A), ilitye (X).

Zandvlakte, Kouga 08.01.2013

SMELOPHYLLUM

Buig-my-nie, bobbejaanbessie

smelo = undetermined, *phyllon* = leaf; derivation unresolved as this species bears no resemblance to the genus *Smelowskia* or the Russian botanist TA Smielowski

1 species in South Africa.

Smelophyllum capense Buig-my-nie

capense = of the Cape; referring to the geographical origin of the species

Tree to 7 m with spreading branches and smoothish grey bark. **Leaves** Compound, with 3 pairs of leaflets, leaflets scalloped, wavy, leathery, shiny. **Flowers** In small panicles, greenish. Flowering in summer. Fruit coat green, becoming hard and brown, splitting open to show orange-fleshed seed. **Habitat** Forest in the narrow ravines and on steep rocky ridges and cliffs in Fynbos Woodland and Grassy Fynbos. **Distribution** Nearly endemic from Baviaanskloof to Groendal and Zuurberg. **Notes** Favoured and dispersed by baboons. Branches are often snapped in attempts to reach the fruit.

Kouga Wildernis 14.05.2021

SAPOTACEAE

Evergreen trees and shrubs with ± 65 genera and 800 species worldwide. The family name is derived from the Mexican name for the sapote tree (*Manilkara*); the name was later Latinised to sapota. Several species in the family produce fruits of economic importance, such as sapota and star apple. Other economic uses include oil extraction from the seeds (e.g. argan oil) and the production of cleaning products from the latex of certain species. 9 genera and 22 species in southern Africa; 1 species in Baviaanskloof.

SIDEROXYLON   Bully trees

sideros = iron, *xylon* = wood; referring to the hard wood. It is probable that this taxon was brought to the attention of Linnaeus long before other savanna trees with much harder wood.

± 100 species in tropical and subtropical parts of the world; 1 species in South Africa.

Sideroxylon inerme subsp. *inerme*   Milkwood

inerme = unarmed; pertaining to the absence of spines or thorns

Resprouting rounded tree to 10 m tall, with a sturdy trunk and rough dark bark. **Leaves** Leathery, dark green above, paler below, petiole exuding white latex. **Flowers** Small, greenish white, with a strong and unpleasant musky smell. Flowering in summer and autumn. Fruit round, purple-black, 1 cm wide. **Habitat** Common along dry stream beds in the narrow ravines in Forest and Thicket. **Distribution** Very widespread throughout southern Africa. **Other names** Melkhout (A), uqwashu (X). **Status** A protected tree in South Africa.

Zandvlakte, Baviaans 17.09.2010

SCROPHULARIACEAE   Sutera, figwort

Recent phylogenetic work has greatly reduced the size of this family. The Scrophulariaceae used to contain ± 275 genera and over 5,000 species, but now it has 62 genera and 1,680 species worldwide. Genera previously assigned here have been moved into the families Plantaginaceae or Orobanchaceae, amongst others. South African species consist mostly of small, short-lived herbaceous plants, with the bushy genus *Buddleja* being an exception. Flowers tend to be small and asymmetrical, with fused sepals and petals. The fruits have 2-chambered capsules containing many seeds. 47 genera and 825 species in southern Africa; ± 15 genera and 45 species in Baviaanskloof.

APTOSIMUM   Karoo violet

a = not, *ptosimos* = deciduous; pertaining to the capsules, which stay on the plant after the seeds have fallen

Gnarled, dwarf shrublets, with dark blue or purple trumpet-shaped flowers. ± 40 species in southern and tropical Africa; 20 species in South Africa; 1 species in Baviaanskloof.

Aptosimum procumbens Karoo violet

procumbens = procumbent; referring to the prostrate habit, with stems spreading horizontally on the ground

Resprouting from a woody base, prostrate, gnarled, mat-forming shrublet to 1 m wide. **Leaves** Spoon-shaped, petiolate, 1.5 cm long × 0.5 cm wide. **Flowers** 1.5 cm wide, bright blue with whitish throat. Flowering in early spring and summer. **Habitat** Common on the valley floor, also on stony flats and hillsides on non-sandstone soils in open patches of Thicket and Succulent Karoo. **Distribution** Widespread through drier parts of southern Africa from Botswana to the Little Karoo and Baviaanskloof. **Other names** Karoo carpet flower (E), brandbossie (A).

Zandvlakte 17.10.2011

BUDDLEJA   Sagewood

Named after the Reverend Adam Buddle, 1662–1715, an English botanist and plant collector. Linnaeus named the genus in his honour.

Bushy shrubs or small trees that flower prolifically with sweet-smelling blossoms. 100 species, tropical and subtropical; 7 species in South Africa; 3 species in Baviaanskloof.

Buddleja glomerata Karoo sagewood

glomeratus = collected closely together into a head; referring to the inflorescence

Reseeding, bushy shrub to 4 m. **Leaves** Ovate, petiolate, grey-felted, discolorous, margins distinctively incised, wavy and crenate. **Flowers** Numerous in dense panicles, cup-shaped, yellow, smelling of cockroaches (best described as a stale oil smell). Flowering in spring, summer and autumn. **Habitat** Fairly common in the dry stream beds of the side ravines. **Distribution** Widespread from Baviaanskloof to the Little and Great Karoo. **Other name** Kakkerlak (A).

Speekhout, Baviaans 14.09.2010

Buddleja saligna   Witolienhout

salignus = willowy, willow-like; referring to the leaves as well
as the habit, givng the plant a slightly willowy look

Reseeding, willowy tree or shrub to 7 m. **Leaves** Opposite,
lanceolate to narrowly elliptic, 8 cm long × 1 cm wide, dark
green and smoothish above, paler below with dense, stellate
hairs, net-veining prominent below. **Flowers** In many-flowered
paniculate cymes 12 cm long, honey-scented, each flower 0.4 cm
long, cream with orange throat, anthers exserted. Flowering in

Gert Smitskloof, Baviaans 17.01.2022

spring and summer. Fruit a capsule 0.2 cm long. **Habitat** Valley floor on dry stream beds or in or near
Forest. **Distribution** Very widespread from the Richtersveld to tropical Africa. **Notes** Used in traditional
medicine for treating coughs and colds, and as a purgative. Good timber for fencing poles, assegai
shafts and fuel. **Other names** False olive (E), umnceba, igqange (X).

Buddleja salviifolia   Sagewood

Salvia = plant genus, *folia* = leaf; leaves resemble those
of some species of *Salvia*.

Resprouting, often many-stemmed, willowy tree up to 8 m. Young
branches covered in dense woolly hairs. **Leaves** Auriculate at
the base, lanceolate, to 14 cm long × 4 cm wide, discolorous,
margins crenate. **Flowers** In attractive, many-flowered paniculate
cymes 12 cm long, each flower 1 cm long, white to purple with

Vleikloof, Baviaans 20.09.2010

orange throat, anthers included. Flowering in spring. **Habitat** Damp stream beds in the side ravines in
or near Forest. **Distribution** Very widespread from Kamiesberg to tropical Africa. **Notes** The wood is
used in traditional medicine, mainly to treat coughs. **Other names** Saliehout (A), cwangi, igqange (X),
lelothwane (S), iloshana (Z).

CHAENOSTOMA   Skunk bush, stinkbossie

chaino = gaping, *stoma* = mouth; referring to the flower tube that opens widely

Shrublets often with unpleasantly scented toothed leaves that are usually opposite. Flowers are
tubular with 5 similar lobes and 4 anthers, with all or only 2 protruding from the tube.
± 46 species, mostly in southern Africa, also in tropical Africa; 9 species in Baviaanskloof.

Chaenostoma cinereum

cinereus = ash-grey; referring to the grey-felted, cobwebby leaves

Resprouting, slightly woody at the base, silvery, diffuse
shrublet to 40 cm. **Leaves** Woolly, narrow, 2.5 cm long
× 0.3 cm wide, margins rolled under. **Flowers** In short
axillary or terminal racemes, each flower funnel-shaped, 0.7 cm
long, pink or purple with a yellow tube. Flowering in spring and
summer. **Habitat** Infrequently encountered in Grassy Fynbos and
Transitional Shrubland. **Distribution** Endemic to Baviaanskloof.

Elandsvlakte, Baviaans 03.11.2011

Chaenostoma decipiens

decipiens = deceiving; describing the difficulty in making an identification
of this taxon, as differences between species in the genus are slight

Reseeding, twiggy, spreading shrublet to 50 cm. **Leaves** and stems
softly glandular-hairy. **Flowers** Axillary, forming racemes, white with
yellow throat, each with 2 anthers included. Flowering in spring
and summer. **Habitat** Sheltered kloofs in Thicket. **Distribution**
Widespread from Gifberg to Baviaanskloof.

Skuinspadkloof, Baviaans 01.02.2020

Chaenostoma denudatum

denudatus = denuded, stripped; referring to the hairless, glabrescent leaves

Resprouting from woody base, erect, twiggy shrublet to 45 cm.
Leaves Glabrous, narrow, 1.5 cm long × 0.1 cm wide. **Flowers**
In a raceme or diffuse panicle, each flower 1.5 cm long × 1 cm
wide, pink to purple with a yellow tube. Flowering all year round.
Habitat Common in Transitional Shrubland and Thicket and in dry,
rocky stream beds. **Distribution** Nearly endemic to Baviaanskloof,
occurring from Uniondale to Humansdorp

Uitspan, Baviaans 20.10.2018

Chaenostoma integrifolium

integri = entire, *folium* = leaf; inaccurate name since the
leaves are sometimes toothed

Reseeding, relatively large, twiggy, spreading shrublet
to 60 cm. **Leaves** Untidily arranged, elliptic, 1.8 cm long
× 0.4 cm wide, margins rolling under, rarely obscurely
toothed near the tip. **Flowers** Axillary, relatively long
and narrow tubed, white with yellow to orange throat.
Flowering all year round. **Habitat** Occasional on steep south-facing
slopes in open areas in Fynbos Woodland or Grassy Fynbos at lower
altitudes. **Distribution** Widespread from Albertinia to Humansdorp.

Riverside, Kouga 24.05.2014

Chaenostoma marifolium

Teucrium marum = plant species, *folium* = leaf;
the leaves resemble those of cat mint (*T. marum*).

Resprouting, slightly woody at base, small, fragile,
cobwebby-woolly shrublet to 20 cm. **Leaves** Roundish,
toothed, 1 cm long, grey-green. **Flowers** Axillary,
forming a paniculate inflorescence, narrow, funnel-shaped, tube
yellowish, petals pink to mauve. Flowering all year round. **Habitat**
Rock ledges and cliffs in Fynbos. **Distribution** Fairly narrow from
Kouga and Baviaanskloof Mountains to Gqeberha (Port Elizabeth).

Bakenkop, Baviaans 10.12.2011

Chaenostoma polyanthum

poly = many, *anthum* = flower; referring to the floriferous nature of this species

Erect, soft, annual or short-lived perennial, to 30 cm. **Leaves** Usually glabrous, 2.5 cm long × 1 cm wide, toothed. **Flowers** 1 cm long × wide, white, pink to mauve with yellow tube, with calyx slightly glandular-hairy. Flowering from spring to autumn. **Habitat** Sandy areas in Transitional Shrubland and in the valleys. **Distribution** Widespread from Knysna to Peddie.

Bosrug, Baviaans 29.05.2010

Chaenostoma revolutum

revolutus = revolute, rolled back; referring to the revolute margins of the leaf

Reseeding, densely leafy, rounded shrublet to 60 cm. **Leaves** Narrow, 2.5 cm long × 0.5 cm wide, margin rolled under, covered with tiny, almost granular, glandular hairs. **Flowers** Axillary, forming racemes, tube relatively short, and petals to 1.5 cm wide, mauve or pink with yellow throat, 2 anthers protruding, 2 included. Flowering from autumn to early summer. **Habitat** Occurring at the base of the mountains in non-Fynbos habitats, often on shale. **Distribution** Widespread from Botrivier to the Little Karoo and Baviaanskloof.

Rus en Vrede, Baviaans 20.05.2010

DIASCIA   Twinspurs, horinkies

di = two, *askion* = wineskin, bladder, belly; referring to the 2 lateral pouches or spurs on the corolla. The flower has 4 small upper petals, a larger lower petal, a very short tube and bears 2 spurs that produce oil for collection by the specially adapted bee *Rediviva neliana*.

± 70 species in southern Africa; 3 species in Baviaanskloof.

Diascia parviflora   Persbokhorinkie

parviflora = small-flowered; referring to the relatively small flowers

Erect annual to 40 cm. **Leaves** Ovate, 2.5 cm long × 1.5 cm wide, margins serrated. **Flowers** In racemes, 1 cm long, purplish, 2 spurs are 0.3 cm long, with small yellow spots below the 2 upper lobes. Capsules oblong-ovate. Flowering mainly in spring. **Habitat** Clay or loam soils on the valley floor, absent in Fynbos. **Distribution** Widespread from Worcester to Baviaanskloof.

Cockscomb 13.08.2011

Diascia patens

patens = spreading; referring to the opposite leaves that diverge from the axis of the stem at almost 90 degrees

Resprouting from a woody base, much-branched, sprawling shrublet to 1 m. **Leaves** Linear to ovate, hairless, 2.5 cm long × 1 cm wide, entire or sparsely toothed. **Flowers** In racemes, 1.5 cm long, pinkish, with a single yellow window, 2 spurs 0.5 cm long. Capsules ovate. Flowering in winter and spring. **Habitat** Occasional and uncommon on south-facing slopes on shale bands in Grassy Fynbos and Transitional Shrubland. **Distribution** Fairly widespread from Bredasdorp to Baviaanskloof.

Endkrale, Kouenek, Kouga 29.05.2010

FREYLINIA
Bell bush, klokkiesbos

Named after Count L. de Freylino, owner of a famous garden in Italy in the 19th century

Shrubby plants with some species growing into small trees, with clusters of tubular flowers that are usually sweetly scented and attract a large variety of insects. 9 species in southern and central Africa; 3 species in Baviaanskloof.

Freylinia crispa   Kouga honey-bells

crispa = irregularly waved and twisted; referring to the crisped leaves

Resprouting, slender, with few stems from the base, upright bush to 2.5 m. **Leaves** Strongly recurved and twisted, folded, hard, glabrous, to 0.7 cm long × 0.3 cm wide, pale green margins crisped. **Flowers** Several hanging down on short racemes, 2.5 cm long, purplish pink, style exserted. Flowering in winter and spring. Capsule 1 cm long, held upright, woody, tipped with 4 tiny black mucrons, splitting to release many small, 0.2 cm-wide, flat, round, wind-dispersed seeds. **Habitat** Rocky sandstone scree in Transitional Shrubland. **Distribution** Known from 2 populations near Joubertina in the Kouga Valley. **Status** Vulnerable.

Braamrivier, Joubertina 03.01.2004

Freylinia lanceolata   River honey-bells

lanceolata = lanceolate; referring to the shape of the leaves

Resprouting, much branched from the base, small tree to 5m tall. **Leaves** Linear-lanceolate, slightly hairy or glabrous, flat, 12 cm long × 1 cm wide, olive-green, margins slightly rolled under. **Flowers** In a densely branched panicle, flowers honey-scented, 1.5 cm long, cream-yellow fading orange to brown. Flowering in autumn and winter. **Habitat** Boulder beds and on stream banks. **Distribution** Very widespread from Namaqualand to Groot Winterhoek Mountains. **Other name** Heuningklokkies (A).

Rooikloof, Kouga 28.04.2018

Freylinia undulata   Renosterveld honey-bells

undulata = undulate; referring to the undulating margins of the leaves

Resprouting, erect, rigid shrub to 2 m. **Leaves** Ascending, ovate-lanceolate, closely packed, tough, glabrous or slightly hairy, 1.5 cm long × 0.8 cm wide, margins sometimes undulating and crisped. **Flowers** In a narrow raceme, flowers slightly hairy at the throat, 2 cm long, white, pink or purplish. Flowering from winter to summer. Capsule 1 cm long, ovoid, woody, without mucrons at the top. **Habitat** Clay and loam soils in Transitional Shrubland and Grassy Fynbos. **Distribution** Widespread from Grabouw to Gqeberha (Port Elizabeth).

Engelandkop, Kouga 13.07.2011

HEBENSTRETIA   Slugwort, slakblom

Named after Johann Christian Hebenstreit, 1720–1791, a German professor of medicine at Leipzig and of botany at St Petersburg

25 species in South Africa; 1 species in Baviaanskloof.

Hebenstretia integrifolia

integri = entire, *folium* = leaf; inappropriate name as the leaves are slightly toothed

Erect annual herb with ascending branches to 60 cm. **Leaves** Linear, 3 cm long × 0.1 cm wide, sparsely toothed, often fascicled. **Flowers** In dense spikes to 20 cm long, calyx glabrous, flowers 1 cm long, white with orange or red markings. Flowering in spring and summer. Fruit oblong, 0.5 cm long. **Habitat** A variety of habitats, often in grassy Transitional Shrubland after fire. **Distribution** Very widespread from Namibia to Hogsback.

Zandvlakte, Baviaans 10.09.2010

JAMESBRITTENIA

Named after James Britten, 1846–1924, an English botanist who worked at the British Museum

Shrublets usually glandular-haired and potent-smelling. Flowers have a narrow tube, widening suddenly with a kink near the top of the tube. 83 species in Africa; 56 species in South Africa; 5 species in Baviaanskloof.

Jamesbrittenia argentea

argentea = silvery; the leaves tend to shine with a silvery glint in the sunlight.

Reseeding, glandular-leafy shrub to 1 m. **Leaves** Spathulate and toothed, 1.5 cm long × 0.5 cm wide. **Flowers** In the axils, tube hairy outside, inflated above, 2 cm long × 1.5 cm wide, white or lilac with yellow throat. Flowering all year round. **Habitat** Rocky or stony soils at the margins of Thicket and Forest. **Distribution** Widespread from Montagu to Gqeberha (Port Elizabeth).

Grootrivierpoort 03.02.2020

Jamesbrittenia atropurpurea Verfblommetjie, saffraanbossie

atro = dark, black, *purpurea* = purple; flowers are dark purple to brownish in colour.

Reseeding, wiry, glandular-hairy shrublet to 1 m. **Leaves** Small and fascicled, 0.5 cm long × 0.1 cm wide. **Flowers** 2 cm long × 1 cm wide, orange to chocolate-brown and/or black. Flowering in summer and autumn. **Habitat** Frequently found in open, sunny areas in Thicket and Succulent Karoo. **Distribution** Widespread from Botswana to Caledon and to Baviaanskloof. **Notes** The flowers have been used to make a dye, thus the common names.

Geelhoutbos, Kouga 20.04.2012

Jamesbrittenia tenuifolia

tenui = slender, thin, *folia* = leaves; referring to the thin leaves

Reseeding, glandular-hairy shrublet to 60 cm. **Leaves** Thin, fascicled and sparsely toothed, to 0.8 cm long × 0.2 cm wide. **Flowers** 2 cm long × 1 cm wide, white with pink markings, sometimes purple, pink or blue. Flowering all year round, especially after rain. **Habitat** Grassy Transitional Shrubland. **Distribution** Widespread from Mossel Bay to Kouga Mountains and Humansdorp.

Grasnek, Kouga 17.04.2012

Jamesbrittenia tortuosa

tortuosa = bent or twisted in different directions; referring to the fascicled leaves and perhaps also the flowers that tend to face in different directions

Reseeding, glandular-hairy shrublet to 30 cm. **Leaves** Small and fascicled, to 1 cm long × 0.3 cm wide. **Flowers** In racemes, 1 cm long × 1.5 cm wide, white to mauve with pink markings, throat yellowish. Flowering in spring and summer. **Habitat** Thicket and Succulent Karoo. **Distribution** Prince Albert to Baviaanskloof and Jansenville.

Rus en Vrede 4X4, Kouga 21.05.2010

MANULEA
Finger-phlox, vingertjies

manulea = a small hand; referring to the 5 petals of the flower that spread upwards like fingers

Endemic to Africa. ± 68 species in South Africa, mostly in the Western Cape; 2 species in Baviaanskloof.

Manulea cheiranthus

Cheiranthus = plant genus; named by Linnaeus after *Cheiranthus*, the wallflower from the northern hemisphere, popular in gardens for its showy flowers, which he thought bore some resemblance to this species

Annual, glandular-hairy, to 30 cm. **Leaves** Toothed, mostly in a basal rosette, to 4 cm long × 2 cm wide. **Flowers** In racemes, 0.5 cm wide, yellowish to brown, lobes very narrow and curving back. Flowering in winter and spring. **Habitat** The lowlands on gravel flats in Succulent Karoo. **Distribution** Widespread from Piketberg to Baviaanskloof. **Notes** *M. chrysantha* is a species found on gravel flats in Baviaanskloof.

Rus en Vrede, Valley 28.08.2010

NEMESIA
Cape snapdragon, leeubekkies

Nemesion = plant genus; a name used by Dioscorides for a similar-looking plant; the derivation is unresolved.

Soft herbs with opposite leaves and snapdragon-like flowers that have a single spur at the back. ± 60 species in southern Africa; 5 species in Baviaanskloof.

Nemesia deflexa

deflexus = deflected, bent or turned abruptly downwards; referring to the pedicel that bends downwards when in fruit

Glandular-hairy annual or short-lived perennial, to 20 cm. **Leaves** Shortly petiolate, broadly ovate, 3 cm long × 2 cm wide, toothed. **Flowers** In racemes, white, 1 cm long, with about 11 red lines at base of upper petals, lower lip with a raised, yellow palate, spur 0.4 cm long. Flowering from spring, through summer to autumn. Capsules barely longer than wide, on bent pedicels. **Habitat** Localised on cool, shaded cliffs in the narrow ravines. **Distribution** Fairly narrow distribution from Langeberg and Swartberg to Baviaanskloof.

Zandvlaktekloof, Baviaans 18.02.2016

Nemesia denticulata

denticulatus = with very small teeth; referring to the toothed leaves

Annual or perennial, much-branched shrublet, to 30 cm. **Leaves** Ovate, toothed, 2 cm long × 0.8 cm wide. **Flowers** In short racemes or solitary in the leaf axils, 1.5 cm wide, pinkish with purple lines at base of upper petals, lines more distinct on the reverse, lower petal with a raised and relatively elongated yellow palate. Flowering from early summer to autumn. **Habitat** The valley floor in open areas of Thicket. **Distribution** Widespread from Baviaanskloof to KwaZulu-Natal.

Vero's Shop, Baviaanskloof River 14.09.2010

Nemesia fruticans   Stinkleeubekkie

fruticans = becoming shrubby; compared with other nemesias, this species can become a well-formed shrublet.

Shrublet to 50 cm tall, becoming shrubby. **Leaves** Petiole almost absent, linear-lanceolate, 2.5 cm long × 0.5 cm wide, obscurely toothed, margins slightly revolute, smelling extremely unpleasant. **Flowers** In racemes, 1 cm wide, white, pink or violet, with a raised yellow palate, throat hairy, upper lobes oblong, spur 0.4 cm long. Flowering mainly in spring, but also all year after rain. Capsules slightly longer than wide, or squarish. **Habitat** Common in stony, dry riverbeds in the valley floor. **Distribution** Widespread throughout southern Africa.

Nooitgedacht, Kouga 26.09.2011

PHYLLOPODIUM   Capewort, opslag

phyllon = leaf, *podium* = foot; indicating the base of the leaf where it joins the stem; the derivation is unclear.

26 species in Namibia and South Africa; 1 species in Baviaanskloof.

Phyllopodium elegans

elegans = elegant; referring to the neat erect habit and delicate white flowerheads

Annual or biennial, erect, sparsely branched to 45 cm. **Leaves** Linear-oblanceolate, 3 cm long × 0.3 cm wide, in neat bunches. **Flowers** In a dense terminal spike with numerous small white flowers, each 0.3 cm wide, head tends to hang over in flower, but is erect in fruit. Flowering in spring and summer. **Habitat** Encountered mostly in the first few years after fire, in Fynbos. **Distribution** Widespread from Montagu to Humansdorp.

Bosrug, Baviaans 12.10.2016

SELAGO

Selago = ancient name for *Lycopodium*, a genus of moss; Linnaeus thought a *Lycopodium* specimen resembled the genus *Selago*.

Generally long-lived shrubs with fascicled leaves. ± 190 species in southern Africa; 9 species in Baviaanskloof.

Selago corymbosa

corymbosus = corymbose; referring to the inflorescence type

Reseeding, densely leafy, erect shrub, to 60 cm tall, branches pubescent. **Leaves** In neatly arranged tufts, linear, spreading, shortly-hairy to almost glabrous, 1 cm long × 0.1 cm wide, margins rolled under. **Flowers** Numerous, in rounded, flat-topped, corymbose panicles, each to 0.5 cm wide, white, 1 petal lobe longer than the other 4, calyx toothed. Flowering mainly in autumn. **Habitat** Fairly common in more open, disturbed or heavily grazed areas in Transitional Shrubland and Grassy Fynbos. **Distribution** Widespread from Cape Peninsula to Baviaanskloof.

Guerna, Kouga 16.02.2016

Selago geniculata Waterfinder

geniculatus = bent abruptly like a knee, geniculate; referring to the spike-like inflorescence that is often curving or slightly bent

Reseeding, twiggy, much-browsed shrublet, to 60 cm. **Leaves** In scruffy tufts, linear, 1.5 cm long × 0.1 cm wide. **Flowers** In elongate, narrow, bendy spikes, each 0.3 cm wide, magenta, sometimes white, calyx 3–5-lobed. Flowering in summer and autumn. **Habitat** Open areas in Transitional Shrubland and Thicket, often where soil is less poor in nutrients. **Distribution** Widespread from Worcester to the Eastern Cape and Free State. **Other name** Pers-aarbos (A).

Zandvlakte, Kouga 18.01.2022

Selago glomerata

glomeratus = collected closely together into a head; referring to the tightly packed inflorescence

Reseeding, erect, densely leafy shrub to 1 m tall, branches shortly-hairy. **Leaves** Linear-oblong, in dense tufts, glossy and glabrous, 0.8 cm long × 0.2 cm wide. **Flowers** An aggregation of spikes into a rounded corymbose panicle 6 cm wide, each flower 0.5 cm wide, white, bracts glabrous, veins sunken. Flowering from spring to autumn. **Habitat** Sandy-loam soils in Transitional Shrubland and Fynbos. **Distribution** Fairly narrow from Outeniqua to Groot Winterhoek mountains.

Onder Kouga road, Joubertina 08.08.2009

Selago gracilis Honde-kak-en-pisbos

gracilis = thin, slender; referring to the thin leaves;
the stems are also thin, tending not to branch, other
than from the base.

Resprouting, much-branched from the base,
finely leafy shrub to 30 cm. **Leaves** In tufts, linear,
0.5 cm long × 0.1 cm wide, shortly-hairy. **Flowers**
In oblong spikes or small panicles, at the tips of
short branchlets off the main stems,
each 0.5 cm wide, white. Flowering
in winter and spring. **Habitat** Stony
slopes in Arid and Grassy Fynbos.
Distribution Widespread from
Caledon and Worcester to Gqeberha
(Port Elizabeth).

Verberg, Kouga 12.02.2016

Selago luxurians

luxurians = luxuriant, abounding; referring to the many-flowered heads

Reseeding, upright, slender shrub to 1.5 m. **Leaves**
In bunches, evenly spaced on the stems, glabrous,
relatively thick, 0.5 cm long × 0.1 cm wide. **Flowers**
Crowded in dense heads on a well-branched
corymbose panicle, each 0.5 cm wide, white.
Flowering in summer and autumn. **Habitat** Common
in Grassy Fynbos. **Distribution** Widespread from
Swellendam to Gqeberha (Port Elizabeth).

Kouga Wildernis 4X4 16.05.2021

Selago myrtifolia (= *S. nitida; Walafrida myrtifolia*)

myrtifolia = myrtle-like leaves; leaves are said to resemble those of myrtle.

Resprouting, relatively robust
erect shrub to 1 m. **Leaves** Ovate,
not tufted, ascending, overlapping,
leathery, 2 cm long × 1.2 cm wide.
Flowers Arranged in neat, dense,
ovoid spikes at the branch tips,
attractive, corolla tube 1 cm long,
magenta-pink. Flowering mainly
in spring. **Habitat** Uncommon and
localised in Transitional Shrubland
and Grassy Fynbos on the gently
sloping ridges. **Distribution** Fairly
narrow from Baviaanskloof
and Langkloof to Gqeberha
(Port Elizabeth).

Bergplaas, Combrink's Pass 01.09.2010

SUTERA
Named after Johann Rudolf Suter, 1766–1827, Swiss botanist and physician

Most of the genus *Sutera* has been incorporated into *Chaenostoma*. 3 species in South Africa; 1 species in Baviaanskloof.

Sutera foetida
foetidus = stinking, evil-smelling; the leaves give off a very unpleasant smell when crushed.

Reseeding annual or short-lived shrublet, to 60 cm. **Leaves** Soft, smelling bad, 3.5 cm long × 2 cm wide, coarsely toothed, petiole 2 cm long. **Flowers** In few-flowered racemes or panicles in the leaf axils, each 1.5 cm long × 0.8 cm wide, white, pink or violet with orange throat, tube narrowly funnel-shaped. Flowering in early spring and summer. **Habitat** Sheltered rocky areas, often in dry boulder beds in the narrow ravines. **Distribution** Widespread from Kamiesberg to Malmesbury and Baviaanskloof.

Waterkloof, Bokloof 26.04.2018

TEEDIA
Named after Johann Georg Teede, a German botanist who collected in Portugal and Surinam in the late 1700s

2 species in South Africa, including Baviaanskloof.

Teedia lucida
lucida = shining, clear, transparent; referring to the glabrous leaves that can be shiny in the sun

Sprawling shrub, tending to be wider than tall, to 1 m tall, with square branches. **Leaves** Glabrous, shiny, often partly folding to be keeled, 4 cm long, finely serrated. **Flowers** In compact racemes at the branch tips or axillary, each 1 cm long × 0.6 cm wide, white to purplish, tube with dark purple lines. Flowering in spring and summer. Fruit 1 cm wide, a yellow or brown berry becoming purple. **Habitat** Rocky outcrops at middle to upper elevations in Fynbos. **Distribution** Widespread from Namaqualand to Eswatini. **Notes** Dispersed mainly by baboons.

Nuwekloof Pass, Baviaans 16.01.2022

ZALUZIANSKYA
Drumstick flower, verfblommetjie
Named after Adam Zalusiansky von Zaluzian, 1558–1613, a physician and botanist from Prague

Usually small, short-lived herbs with 5 colourful petals that are sometimes said to resemble drumsticks. 55 species in southern Africa, mostly in the Western Cape; 2 species in Baviaanskloof.

Zaluzianskya synaptica

synaptica = like a synapse; named by Olive Mary Hilliard, 1925–2017, botanist and taxonomist from Durban; the shape of the petal tips reminded Hilliard of the axon terminal of the presynaptic synapse – the place where neurons are transmitted in the human brain.

Erect, annual herb to 25 cm tall, with spreading hairs on the stems. **Leaves** Glandular-hairy, 1 cm long × 0.3 cm wide, recurved at the tip, margins sometimes obscurely and bluntly toothed. **Flowers** Diurnal, in spikes, floral tube to 2.5 cm long, covered with glandular hairs, flower white, with a yellow-orange ring at the throat including a circular row of erect hairs, lobes sometimes entire, reverse rusty-orange, 4 anthers. Flowering in winter and spring. **Habitat** Valley floor on gravelly soils in Thicket or Succulent Karoo. **Distribution** Widespread from Worcester to Graaff-Reinet and Baviaanskloof.

Zandvlaktekloof, Baviaans 14.09.2010

SOLANACEAE   Potatoes, nightshades

An important family for agriculture, with 98 genera and 2,700 species. The greatest diversity is in South and Central America. Well-known members include tomatoes, potatoes and eggplant (all members of the genus *Solanum*), chilli peppers (*Capsicum*) and tobacco (*Nicotiana*). Many species are used medicinally and some produce psychotropic effects, while others are poisonous. *Solanum* is the largest genus with ± 1,330 species. 4 genera and ± 60 species in southern Africa, plus a number of naturalised alien invasives.

LYCIUM
Derived from the name for a shrub from Lycia, Turkey

± 100 species worldwide; ± 23 species in southern Africa; 5 species in Baviaanskloof.

Lycium ferocissimum   Cape box thorn

ferocissimum = most ferocious; referring to the thorns that help to protect the plant from browsers

Woody spinescent shrub to 2 m. **Leaves** In bunches at the nodes, oval to elliptic, slightly fleshy, 2 cm long × 0.5 cm wide. **Flowers** 1 cm wide with a short tube covered by the calyx, anthers protruding, white or pale mauve, with purple markings at the throat. Flowering in winter and spring or after rain. Berry orange-red to bright red, egg-shaped. **Habitat** Common in lowland areas especially Succulent Karoo and Thicket. **Distribution** Very widespread from Namaqualand to Eastern Cape. Invasive in Australia. **Other names** Slangbessie (A), umbovu, idywadi (X).

Uitspan, Baviaans 30.01.2020

Lycium oxycarpum   **Honey thorn**

oxy = pointed, *carpum* = fruit; referring to the pointed tips of the fruits

Woody, thorny shrub or small tree to 3 m. **Leaves** Clustered in tufts, oblanceolate, thin, 3 cm long × 0.5 cm wide. **Flowers** mauve, with a yellowish tube 2 cm long, anthers included. Flowering in winter and spring. Berries 0.6 cm long, red. **Habitat** Frequent on dry riverbeds in Thicket. **Distribution** Widespread from Clanwilliam to Makhanda. **Other name** Wolwedoring (A).

Waterkloof, Bokloof 26.04.2018

SOLANUM   Nightshade

solamen = soothing, to comfort; pertaining to the narcotic and/or medicinal properties of some species

± 1,330 species worldwide, 26 species indigenous to South Africa and several exotics; 5 indigenous and 2 exotic species in Baviaanskloof.

Solanum guineense

guineense = of Guinea; named after the country in West Africa, erroneously since the species has not been recorded there

Unarmed, sprawling shrub to 1 m. **Leaves** Ovate to elliptic, glabrous, soft, 7 cm long, petiolate. **Flowers** 2 cm wide, mauve to light blue with purple markings. Flowering mostly in autumn. Berries 1.5 cm wide, yellow, orange or red. **Habitat** Found on stream banks in Thicket. **Distribution** Widespread from Namaqualand to Graaff-Reinet.

Uitspankloof, Baviaans 30.01.2020

Solanum linnaeanum   **Bitter apple**

Named after Carl Linnaeus, 1707–1778, the father of modern taxonomy, who formalised the system of binomial nomenclature

Upright, single-stemmed spiny shrub to 1 m. Spines to 1 cm long, yellowish, on all parts of the plant. **Leaves** 15 cm long, bright green, usually spiny, margins lobed. **Flowers** Single or clustered at branch tips, 3.5 cm wide, purple. Flowering in winter and spring. Berry 3.5 cm wide, green mottled white, becoming yellow. **Habitat** Disturbed areas in Thicket. **Distribution** Widespread from Namaqualand to KwaZulu-Natal. **Notes** A problematic invasive weed in many parts of the world.

Baviaanskloof River, Zandvlakte 13.10.2010

Solanum pseudocapsicum* Jerusalem cherry

pseudo = false, *Capsicum* = plant genus;
referring to chilli peppers in the genus
Capsicum with bright orange fruits, but
they are not round

Upright, soft, unarmed shrub to 1 m.
Leaves Sword-shaped, 12 cm long,
bright green. **Flowers** Few in clusters,
0.5 cm wide, white. Flowering in
summer and autumn. Berry 1 cm
wide, orange-red. **Habitat** Common under Forest
in the narrow ravines. **Distribution** Native to South
America. Now a weed in many parts of the world.
Notes The berries are eaten and dispersed by
baboons. **Other names** Bosgifappel (A), utshesi,
umthuma, umthumawezinja (X).

Damsedrif, Baviaans 01.05.2018

Solanum retroflexum

retroflexum = bent back, reflexed; referring to the
recurved petals on the flowers; the sepals also
bend back when in fruit.

Small, upright, softly-hairy annual to 0.8 m. **Leaves**
Hairy, 10 cm long, with irregularly lobed margins,
petiole winged. **Flowers** Few drooping in clusters,
1 cm wide, white with purple keels. Flowering
all year round. Berry less than 1 cm wide, dark
purple. **Habitat** Common in disturbed areas in
Thicket and Forest. **Distribution** Widespread
throughout southern Africa. **Notes** The berries can
be made into jam. **Other names** Umvumadoda (X),
limomonyane (S), umsobosobo (X, Z).

Waterkloof, Bokloof 26.04.2018

Solanum tomentosum

tomentosum = hairy; referring to the felted hairs
that cover most parts of the plant

Upright, yellowish-green-felted, spiny shrub to
1 m. **Leaves** Ovate, seldom spiny, rough to touch,
8 cm long. **Flowers** Few in clusters, 1.5 cm wide,
mauve or purple. Flowering mainly in spring. Berry
2 cm wide, orange. **Habitat** Occurring at thicket
margins in Grassy Fynbos, Transitional Shrubland
and Thicket. **Distribution** Very widespread in South
Africa from the Richtersveld to the Cape Peninsula,
Makhanda and KwaZulu-Natal.

Zandvlaktekloof, Baviaans 09.03.2010

WITHANIA

Named after Henry Thomas Maire Witham, 1779–1844, a British landowner and philanthropist, acclaimed for his achievements as an amateur palaeobotanist and geologist. He obtained slices of rocks containing fossilised plant remains, and examined the internal structures of these plants under a microscope. He subsequently published a landmark book about fossilised plant remains.

23 species in Africa, parts of Europe and Asia, and the Canary Islands.

Withania somnifera   Poison gooseberry

somnifera = sleep-bearing; referring to the sleep-inducing and healing properties of the plant

Unarmed, softly-hairy shrub to 2 m tall, smelling of green tomatoes. **Leaves** Ovate to oblong, 6 cm long × 3 cm wide, green above, silver-hairy below. **Flowers** Few in clusters, 0.6 cm wide, yellowish. Fruit an orange-red berry, 0.6 cm wide, hidden by a 5–10-ribbed membranous calyx. Flowering in summer and autumn. **Habitat** Occasional in disturbed areas in Succulent Karoo and Thicket. **Distribution** Very widespread throughout southern Africa and other parts of the world. **Notes** Famous for its healing properties for various ailments. The Hindi name *Ashwagandha* is often used for medicines derived from the compounds in this plant. **Other names** Bitterappelliefie, geneesbossie (A), ubuvumba, ucwethikazi (X), bofepha (S), ubuvimba (X, Z), umaqhunsula (Z).

Nuwekloof Pass, Baviaans 16.01.2022

STILBACEAE

A small family of 12 genera and 27 species, found mostly in the Western Cape, South Africa. 3 genera (*Anastrabe*, *Halleria* and *Nuxia*) have species in Africa, Madagascar and the Arabian Peninsula. *Nuxia* is the largest genus with 40 species, followed by *Halleria* with 10 species. 2 genera and 2 species in Baviaanskloof.

HALLERIA   Tree fuschia, notsung

Named after Albrecht von Haller, 1708–1777, a Swiss naturalist, doctor, author and professor, sometimes called 'the father of modern physiology'

± 10 species; 3 species in southern Africa; 1 species in Baviaanskloof.

Halleria lucida   Umbinza

lucida = shining; pertaining to the glossy leaves

Resprouting, straight-stemmed shrub or small tree to 10 m. **Leaves** Opposite, ovate, shiny, margins toothed. **Flowers** In clusters in the axils or on the old wood, tubular, curved, brick-red or orange, attracting sunbirds. Fruit a berry, 1 cm wide, green turning black. Flowering in spring and summer. **Habitat** Common in Forest in the narrow ravines. **Distribution** Widespread from Gifberg to tropical Africa. **Other names** Witolienhout (A), lebetsa (S), indomela (Z).

Van Stadensberg 18.07.2021

NUXIA

Named after Jean Baptiste Francois de Lanux, 1702–1772, a French amateur naturalist who left France as a young man to live on Réunion Island. He described the now-extinct Lesser Mascarene Flying Fox (*Pteropus subniger*), a species of megabat.

± 40 species in Africa, Madagascar and the Mascarene Islands; 5 species in southern Africa; 1 species in Baviaanskloof.

Nuxia floribunda   Forest elder

floribunda = profusely flowering; pertaining to the habit of producing numerous flowers

Large tree with a rounded crown to 10 m. **Leaves** Simple, petiolate, tapering to a point, 10 cm long × 4 crn wide. **Flowers** Numerous in large panicles at the branch tips, white and sweetly scented. Flowering in winter. **Habitat** Frequent in Forest in the narrow ravines. **Distribution** Widespread from Knysna to tropical Africa. **Other names** Vlier (A), ingqota, isikhali (X), ithambo (Z).

Kamassiekloof, Zandvlakte 27.02.2010

THYMELAEACEAE   Strippers

50 genera and 900 species of mainly trees and shrubs, mostly in Africa but also Australia. *Gnidia* is the largest genus, with ± 160 species. The common name refers to the stripping bark of all family members.

GNIDIA   Saffron bush, saffraan

Possibly named for Knidos, an ancient Greek city located in modern-day Turkey

Flowers are always at the branch tips. ± 160 species in Africa and India; 100 species in South Africa; 4 species in Baviaanskloof.

Gnidia juniperifolia

juniperifolia = juniper-like leaves

Resprouting, multi-stemmed, non-hairy shrublet to 30 cm. **Leaves** Wider near the flowers, 0.5 cm long × 0.3 cm wide, grey-green, often purple on margins. **Flowers** Always 2 at branch tips, yellow, with 4 membranous floral scales. Flowering in summer. **Habitat** Occasional in Grassy and Mesic Fynbos. **Distribution** Widespread from Cape Peninsula to Baviaanskloof.

Rus en Vrede 4x4 25.02.2011

Gnidia nodiflora

nodiflora = flowers at the nodes; derivation unclear since the flowers are at the branch tips

Resprouting, silky-hairy, slender shrublet to 30 cm. **Leaves** Linear-oblong and silky-hairy, with older leaves less hairy. **Flowers** Several at the branch tips, surrounded by a whorl of leaves, silky-hairy, with 8 floral scales. Flowering mostly in spring. **Habitat** Uncommon on loamy-clay soils in Grassy Fynbos and Transitional Shrubland. **Distribution** Widespread from Bredasdorp to KwaZulu-Natal.

Bergplaas, Combrink's Pass 26.10.2018

LACHNAEA

Mountain carnation, bergangelier

lachne = woolly hair; pertaining to the woolly calyx and flowers of some species. The flowers never have long tubes.

40 species in South Africa; 2 species in Baviaanskloof.

Lachnaea burchellii

Named after William John Burchell, 1781–1863, an English explorer who collected plants in southern Africa and Brazil. He amassed over 50,000 plant specimens in southern Africa and donated them to Kew Gardens, each record with meticulous, detailed notes on habit and habitat. He described his journey in *Travels in the Interior of Southern Africa*, a 2-volume work published in 1822 and 1824.

Reseeding shrublet to 30 cm. **Leaves** Opposite, spreading, lanceolate, 1 cm long × 0.3 cm wide, hairless. **Flowers** In umbels, cream to pale pink, hairy. Flowering all year round except autumn. **Habitat** Sporadic on south-facing slopes in Mesic Fynbos. **Distribution** Widespread from Langeberg to Gqeberha (Port Elizabeth).

Enkeldoring, Baviaans 21.09.2011

Lachnaea glomerata

glomerata = collected into a head; pertaining to the groups of flowers

Reseeding shrublet to 45 cm. **Leaves** Opposite, ascending, linear-elliptic. **Flowers** In umbels on short lateral branches, pale pink or cream, hairy. Flowering in spring. **Habitat** Upper south-facing slopes in Mesic Fynbos. **Distribution** Distributed on mountains between Rooiberg and Baviaanskloof.

Bosrug, Baviaans 12.07.2011

LASIOSIPHON

Yellow-head, kerrieblom

lasio = woolly, *siphon* = tube; referring to the tubular flowers that are usually covered with silky hairs

± 30 species in Africa and Madagascar; 2 species in Baviaanskloof.

Lasiosiphon anthylloides (= *Gnidia anthylloides*) **Brandbossie**

Anthyllis = genus in the pea family, *oides* = resembling; resembling *Anthyllis* flowers

Resprouting, slender, upright shrub to 1.5 m. **Leaves** Alternate, elliptic, silky below. **Leaves** Smaller under the flowers. **Flowers** In dense heads at the branch tips, hanging down, silky-hairy tubes to 2 cm long, yellow with 5 tiny floral scales. Flowering all year round except late autumn. **Habitat** Occurring sporadically in Grassy Fynbos, grassy Transitional Shrubland and Sour Grassland. **Distribution** Widespread from Riversdale to KwaZulu-Natal. **Other names** Intozwane (X), indolo (Z).

Geelhoutboom,
Patensie 05.02.2020

Lasiosiphon meisnerianus (= *Gnidia cuneata*) **Koorsbossie**

Named after Swiss botanist, Carl Meissner, 1800–1874, professor of botany at the University of Basel; he discovered many plants, especially Proteaceae in Australia.

Resprouting, multi-stemmed, hardy shrub to 1,5 m. **Leaves** Spathulate-oblong, 2 cm long × 0.4 cm wide, with adpressed silky hairs. **Flowers** In clusters at the branch tips, tube to 2.5 cm long, yellowish and rusty or ochre, finely hairy. Flowering in autumn and winter. **Habitat** Common on rocky, lower slopes in Succulent Karoo and Thicket. **Distribution** Widespread from Baviaanskloof to KwaZulu-Natal. **Notes** Used to treat fevers, hence the Afrikaans common name. **Other name** Isidikili (X, Z).

Rus en Vrede 4X4,
Kouga 21.05.2010

PASSERINA

Gonna bush

passer = a sparrow; the black seeds are reminiscent of a sparrow's beak.

Ericoid shrubs with decussate leaves, often with pendulous branches. The flowers have no scent and are wind-pollinated. The bark breaks into strips when branches are broken, more so in this genus than in other members of the family. The long strips are robust and durable, and have been used in thatch and baskets. 21 species in southern Africa; 6 species in Baviaanskloof.

Passerina obtusifolia Gonna, bakkersbos

obtusifolia = having obtuse leaves; referring to the shape of the leaf tips

Reseeding, upright, with ascending branches to 1.5 m. **Leaves** 0.6 cm long, ribbed below, slightly enlarged below the flowers. **Flowers** In dense spikes, tube 0.3 cm long, yellowish pink to dull red. Flowering in spring, especially October. Common and abundant on dry, rocky sandstone slopes in Grassy and Arid Fynbos, Fynbos Woodland and Transitional Shrubland. **Distribution** Widespread from Worcester to Makhanda. **Notes** 1 of the most common shrubs in Baviaanskloof, dominating the shrub cover in favourable habitat.

Nuwekloof Pass, Baviaans 13.10.2016

Passerina pendula

pendula = hanging downwards; referring to the flowers that tend to droop down

Reseeding, slender, upright shrub to 1.5 m. **Leaves** Ovate-lanceolate, adpressed, 0.3 cm long, bracts larger and rhombic. **Flowers** In spikes, dull red. Flowering in spring or after rain in summer and autumn. **Habitat** Occurring sporadically on upper mountain ridges in Mesic and Grassy Fynbos. **Distribution** Narrow distribution from Baviaanskloof to Gqeberha (Port Elizabeth).

Witruggens, Kouga 03.11.2011

Passerina ternata

ternata = ternate, in threes; referring to the leaves arranged in whorls of 3

Resprouting strongly from a woody base, erect shrub to 1.5 m. **Leaves** Shiny, sticky, relatively thick and stiff, 0.6 cm long, ascending and curving in towards the stem. **Flowers** Cream-white, with bracts tinged red and very sticky. Flowering in late spring. **Habitat** Localised on south-facing slopes of higher peaks in Mesic Fynbos. **Distribution** Endemic to Baviaanskloof Mountains. **Notes** Discovered as new to science in 2016 and recently described. **Status** Rare.

West of Bosrug, Baviaans 12.10.2016

STRUTHIOLA

strouthion = a small bird, or *strouthos* = ostrich; pertaining to the pointed seeds that resemble a small bird's beak, or alluding to the long necks (floral tubes) of the flowers

Identification of *Struthiola* species is aided by counting the number of petaloid scales, which appear as reduced — sometimes fleshy — petals at the mouth of the floral tube, surrounded by 4 sepals. There are 4, 8 or 12 petal-like scales. The flowers are scented, some only at night. 40 species in South Africa; 5 species in Baviaanskloof.

Struthiola argentea Aandgonna

argentea = silvery, describing the hairs on the leaf margins; the common name refers to the flowers, which only produce a scent in the evenings or at night, thus attracting the nocturnal moths that pollinate them.

Reseeding shrub to 2m tall, branches hairy. **Leaves** Oval, overlapping, slightly ribbed below, 1.5 cm long × 1 cm wide, silver-hairy on margins. **Flowers** Tube 2 cm long, yellow, tinged pink or reddish orange, obscurely hairy, with 12 floral scales. Flowering all year round but not in autumn. **Habitat** Loamy soils in Grassy Fynbos and Transitional Shrubland. **Distribution** Widespread from Caledon to Makhanda.

Bosrug, Baviaans 27.07.2011

Struthiola macowanii

Named after Peter MacOwan, 1830–1909, a British botanist who started out as a teacher, before becoming a professor of chemistry. He immigrated to the Cape Colony in 1861, settling originally in Grahamstown (Makhanda). He later moved to Cape Town, where he blossomed as a botanist and lecturer, publishing academic papers, attending to the Botanic Garden, consulting on agricultural and horticultural matters, and collecting specimens, which can be found in many herbaria around the world. Twenty plants have been named after him.

Reseeding slender shrub to 1.5 m. **Leaves** Ascending, opposite or arranged in whorls of 3 around the stem, glabrous, 1 cm. **Flowers** 8 floral scales, white or pale yellow to reddish, tube glabrous outside. Flowering all year round. **Habitat** South-facing slopes in Grassy and Mesic Fynbos. **Distribution** Widespread from Still Bay to Gqeberha (Port Elizabeth) and Zuurberg, Eastern Cape.

Krakeel, Langkloof 03.02.2013

Struthiola parviflora

parvi = short, *flora* = flower; referring to the relatively small and short flowers

Reseeding, slender shrublet to 60 cm tall, with branches square in cross section. **Leaves** Opposite, lanceolate, 0.5 cm long, mostly hairless. **Flowers** 8 floral scales, tube 0.8 cm long, cream, greenish yellow or red, glabrous outside. Flowering all year round. **Habitat** Loamy or clay soils in Grassy Fynbos and Transitional Shrubland. **Distribution** Widespread from Swellendam to Makhanda.

De Hoek, Joubertina 07.11.2011

URTICACEAE

Nettles

Familiar to most, the stinging nettle (*Urtica dioica*) is naturalised in most parts of the world. The stinging hairs for which it is infamous are present in most species. This plant family is well known for its medicinal properties. 53 genera and 2,625 species worldwide.

URTICA

Nettle

urere = to burn; pertaining to the stinging hairs on the plant

± 40 species worldwide; 1 species in South Africa.

Urtica lobulata

lobulata = lobed; referring to the coarsely toothed, oddly lobed leaves

Causing stinging if touched, finely prickly, short-lived shrublet (weed) to 0.8 m. **Leaves** Broadly ovate, toothed and petioled, with stinging hairs. **Flowers** Clustered in panicles in the axils, small, white. Flowering in spring and summer. **Habitat** Uncommon in open, damp, rocky areas under Forest. **Distribution** Widespread from Kamiesberg to Free State. **Other name** Uralijane wamanxiwa (X).

Uitspankloof, Baviaans 19.02.2016

VERBENACEAE

Verbena

A mainly tropical family of trees, shrubs and herbs, notable for their attractive, scented flowers. 35 genera and 1,200 species.

CHASCANUM

khasme = gaping, wide open, *kanum* = off-white, ash; pertaining to the off-white flowering tube that flares open at the mouth

± 25 species in Africa; 14 species in South Africa; 1 species in Baviaanskloof.

Chascanum cuneifolium (= *Plexipus cuneifolius*)

cunei = cuneate, wedge-shaped, *folium* = leaf; referring to the wedge-shaped leaves

Resprouting, gnarled, twiggy shrublet to 30 cm. **Leaves** Opposite, wedge-shaped, curvy, shiny, glabrous, 1.5 cm long, 5–7-toothed. **Flowers** In dense spikes, 1 cm wide, white or mauve, tubular and 5-lobed, petal lobes notched. Flowering in winter and spring. **Habitat** Sporadic on loamy and red soils in Grassy Fynbos and Transitional Shrubland. **Distribution** Widespread from Robertson to KwaZulu-Natal.

Grasnek, Kouga 17.04.2012

LANTANA*

Lantana = derived from the Latin name for a similar-looking plant, *Viburnum lantana*

± 150 species in the tropics, but several are naturalised and invasive in other parts of the world. Notable for the multi-coloured flowering heads, with flowers changing colour as they mature. 2 native and 1 invasive species in South Africa; 1 indigenous and 1 exotic species in Baviaanskloof.

Lantana rugosa Bird's brandy

rugosa = wrinkled; referring to the wrinkly leaf surface

Resprouting, unarmed shrublet to 1.5 m tall, with squarish, hairy branchlets. **Leaves** Opposite, wrinkled above, margins toothed, odorous when crushed. **Flowers** Several in dense clusters at the branch tips, pink or purple. Fruit a purple berry, 0.5 cm wide, edible. Flowering in spring and summer. **Habitat** Grassy Fynbos and Transitional Shrubland. **Distribution** Widespread throughout southern Africa to tropical Africa. **Notes** A paste made from the leaves and fruit is used to treat sore eyes and as an antiseptic for cleaning wounds. Popular in gardens. **Other names** Wildesalie (A), utywala bentaka (X), mabele-mabutsoa-pele, molutoane (S), impema (Z).

Damsedrif, Baviaans 17.01.2022

VITACEAE   Vines

Grapes used to make wine are fruit of the common grape vine (*Vitis vinifera*). An economically important family with ± 14 genera and 900 species. 5 genera and ± 80 species in southern Africa; 1 genus and 3 species in Baviaanskloof.

RHOICISSUS   Bosdruif

rhoia = pomegranate, *kissos* = ivy; referring to the asymmetrical fruit (much smaller than the pomegranate) and the ivy-like climbing habit of the tendrils

10 species in southern and tropical Africa; 4 species in Baviaanskloof.

Rhoicissus digitata Wild grape

digitata = finger-like; referring to the leaflets that radiate out from the tip of the petiole

Scrambling vine with tendrils. **Leaves** Trifoliate, entire, the lateral leaflets without petioles. **Flowers** Inconspicuous, yellow-green. Fruit almost round, 1 cm wide, red-brown to dark purple. Flowering in summer and autumn. **Habitat** Wooded vegetation in Thicket and Forest, but also in thicket clumps in Transitional Shrubland and Grassy Fynbos. **Distribution** Widespread from Kogelberg to Mozambique. **Notes** Jam can be made from the fruit. **Other names** Bobbejaandruif (A), intwanza, itsolo lendumbo, uchithibhunga, umgcebha (X).

Zandvlakte bergpad, Kouga 13.07.2016

Rhoicissus kougabergensis

kougabergensis = from the Kouga Mountains; referring to the area in which the plant was first recorded

Scrambling vine. **Leaves** Simple, narrowly obovate, younger leaves hairy below. **Flowers** Small, opposite the leaves, greenish yellow, in rusty-haired cymes. Fruit round, 0.8 cm wide. Flowering in spring and summer. **Habitat** Climbing over thicket on rocky hills in Thicket and Transitional Shrubland. **Distribution** Narrow distribution in the Gamtoos Valley from Kouga Dam to Patensie. **Status** Rare.

Mistkraal, Patensie 05.02.2020

ZYGOPHYLLACEAE

Twinleaf, caltrop

A family of flowering trees, shrubs or herbs, most prevalent in arid habitats, with 22 genera and 285 species worldwide. The leaves are usually opposite. An individual creosote bush (*Larrea tridentata*), growing in the Mojave Desert, California, is affectionately known as King Clone, and has been estimated at 11,700 years old, making it 1 of the oldest living organisms in the world. 7 genera and 54 species in southern Africa; 3 genera and 6 species in Baviaanskloof.

AUGEA

Augeas = a name from Greek mythology, meaning 'bright'; referring to how the plants show up brightly in barren areas where little else grows; Augea is also the name of a village in France.

A monotypic genus with 1 species.

Augea capensis Bobbejaankos

capensis = from the Cape; referring to the geographical origin of the species

Short-lived, upright, succulent bush to 50 cm, with swollen roots. **Leaves** 3 cm long × 1 cm wide, grey-green to yellowish, with salty-watery sap. **Flowers** 1 cm wide, dull white. Flowering in spring. **Habitat** Overgrazed, saline floodplains in Succulent Karoo. **Distribution** Widespread from Namibia to Steytlerville. Absent from Baviaanskloof, but common in Steytlerville Karoo.

Steytlerville Karoo 30.04.2014

ROEPERA (= *Zygophyllum*)

Twinleaf, spekbos

Named after German botanist Johannes August Christian Roeper, 1801–1885. (The old name derives from *zygote* = a fertilised ovum, *phyllon* = leaf; referring to the 2-lobed leaves that resemble a 2-celled zygote.)

Succulents that tolerate arid and saline conditions. Leaves and seeds are poisonous, yet certain species are used medicinally for various ailments. ± 100 species in Africa, Asia and Australia; 39 species in South Africa; 4 species in Baviaanskloof.

Roepera foetida (= *Zygophyllum foetidum*)   Slymbos

foetida = bad-smelling; referring to the foul-smelling leaves

Sprawling, smothering, foetid scrambler to over 2 m. **Leaves** Paired, softly fleshy, leaflets broad, 2.5 cm long × 1.5 cm wide, smelling bad when broken. **Flowers** 2 cm wide, yellow marked with red. Fruit roundish when fresh, 5-ribbed when dry. Flowering in winter and spring. **Habitat** Common and abundant on alluvial plains under sweet thorn (*Vachellia karroo*) in Succulent Karoo and Thicket. **Distribution** Very widespread from Namibia to Makhanda. **Notes** Unbrowsed.

Baviaanskloof River, Beacosnek 28.10.2018

Roepera fulva (= *Zygophyllum fulvum*)

fulva = yellowish brown; referring to the colour of the flowers

Upright, hairless shrub to 1 m. **Leaves** Paired, fleshy, rigid, 1.5 cm long × 0.5 cm wide, margins bony. **Flowers** 1.5 cm wide, cream or yellow with red markings. Flowering in winter and spring. **Habitat** Frequent but uncommon in rocky areas in Transitional Shrubland, and Grassy and Arid Fynbos. **Distribution** Widespread from Nieuwoudtville to Gqeberha (Port Elizabeth). **Notes** Browsed by stock and game.

Stompie se Nek, Kasey 14.09.2010

Illustrated glossary of terms

Leaf parts

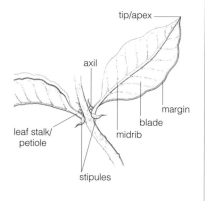

tip/apex

axil

leaf stalk/
petiole

stipules

midrib

blade

margin

Leaf tips/apices

piercing
(pungens)

narrowly acute
(acutus)

broadly acute
(acutus)

awn/beard
(aristatus)

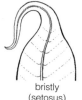
with hairy tip
(piliferus)

broadly acuminate
(obtusus cum acumine)

tapering to a point
(cuspidatus)

short, sharp point
(mucronatus)

short point
(apiculatus)

hooked
(uncinatus)

broadly tapering
(obtusus cum acumine)

curved
(rostratus)

tapering
(acuminatus)

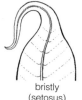
bristly
(setosus)

tendrilled
(cirrhosis)

rounded
(obtusus)

blunt
(retusus)

notched
(emarginatus)

cut off
(truncatus)

bitten off
(praemorsus)

three teeth
(tridentatus)

Leaf shapes

oval

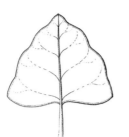
deltate

obovate

orbicular

linear

cuneate

rhombic

spathulate

elliptic

lanceolate

Leaf margins

entire

crisped

crenate

scalloped

serrate

toothed

sinuate

undulate

Inflorescence types

Key to age of flowers

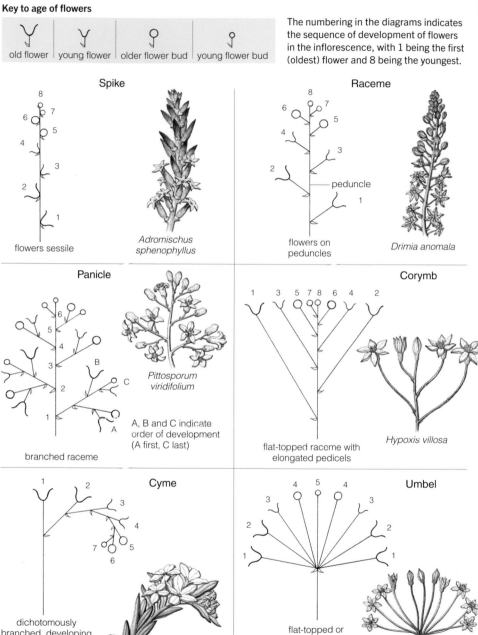

| old flower | young flower | older flower bud | young flower bud |

The numbering in the diagrams indicates the sequence of development of flowers in the inflorescence, with 1 being the first (oldest) flower and 8 being the youngest.

Spike

8
7
6
5
4
3
2
1

flowers sessile

Adromischus sphenophyllus

Raceme

8
6 7
4 5
3
2
peduncle
1

flowers on peduncles

Drimia anomala

Panicle

6
5
4
3 B
2 C
1 A

branched raceme

Pittosporum viridifolium

A, B and C indicate order of development (A first, C last)

Corymb

1 3 5 7 8 6 4 2

flat-topped raceme with elongated pedicels

Hypoxis villosa

Cyme

1 2
3
4
7 5
6

dichotomously branched, developing on one side only

Lithospermum papillosum

Umbel

4 5 4
3 3
2 2
1 1

flat-topped or rounded, with pedicels originating from one point

Strumaria gemmata

Glossary of terms

A

abscission natural separation or detachment, such as leaves dropping off

acaulescent having no stem, or appearing to have no stem

achene dry, single-seeded fruit

acuminate tapering to a point

adpressed closely pressed or lying on the adjacent part

alluvial fan an accumulation or deposition of sediments in a fan shape

angiosperms flowering plants; a large group of plants that produce flowers, and seeds enclosed in a seed coat

apiculate with a pointy tip

areole modified branch on a cactus, from whence the spikes grow

aril an appendage or covering on a seed, usually fleshy and brightly coloured, serving as a reward to animals that disperse the seed

aristate with a pointy or bristly tip

attenuate tapering, gradually becoming thinner

auricle an ear-shaped appendage

auriculate ear-shaped

awn hair-like structure at the tip of a flower part, bract or leaf, often seen in spikelets of certain grass, sedge or restio species

axil upper angle between a leaf stalk or branch, and the stem on which it is growing

axillary growing from an axil, lateral

B

bipinnate when the leaflets of a compound leaf are further subdivided (twice pinnate)

bract leaf-like structure at the base of a flower or inflorescence

bracteole small bract

bulbil a small bulb-like structure produced in place of flowers or on the stem or bulb, falling off and growing into a new plant

C

calycle a group of bracts on the outside of the calyx

calyx (plural **calyces**) outermost part of a flower, comprising sepals; it covers and protects the petals when the flower is closed or in bud

capitulum a dense flowerhead; a racemose flowerhead made up of a dense cluster of florets; seen in some members of the daisy family

caudex stem or rootstock, often thickset or swollen; seen on palms and cycads

cauline arising from the stem

cilia hairs

ciliate hairy, often along the leaf margin

ciliolate covered with tiny hairs

colliculate with small bumps

coppicing describing plant growth that is stimulated when a plant is cut back, burnt or damaged

corolla petals of a flower

corymbose applied to a flower cluster with proportionally longer lower stalks, so that the flowers form a flat or slightly convex head

costule midrib or midvein of a pinnule

crenulate with a scalloped or notched edge

cryptic serving to camouflage in natural environment, easily overlooked

cryptogamic hidden, referring to the gametophytes of ferns and bryophytes

culm stem of grass, restio or sedge

cuneate wedge-shaped

cupulum (adj. **cupulate**) a small bract-like structure positioned on the peduncle at the base of the calyx; found in some members of *Psoralea*

cusp a horn-like pointed end where two curves meet

cyme a cluster of flowers with the flower at the tip of the group being the first to open

cypsela a fruit in the daisy family consisting of a seed (or achene) and pappus bristles

cystoliths enlarged cells containing crystals of calcium carbonate

D

decumbent lying on the ground, sprawling, with extremity curving upwards

decurrent extending down the stem, stipe or rachis, beyond the point of attachment

decussate with each pair of opposite leaves at right angles to the one below

deflexed curving backwards

diadema in the shape of a crown; bristles on *Trichodiadema* leaf tips

dichasium the basic unit of a cymose inflorescence in which each branch with a flower produces two other flowering branches

dichotomous divided into two equal parts

diffuse spread out over a large area

digitate having finger-like projections

dimorphic occurring in two different forms

dioecious having male and female reproductive parts in separate individuals

discolorous having upper and lower surfaces of different colours

distichous alternating arrangement or position

divaricate branching widely

doctrine of signatures a dated pseudoscientific belief that a plant was able to heal the body part that it resembled

drupe fleshy fruit with thin skin and a central stone containing the seed

E

elaiosome fleshy structure attached to a seed; a food source for ants, facilitating seed dispersal

elliptic shaped like a rugby ball

emarginate with a notched margin

enations scaly, leaf-like structures without vascular tissue, mostly created by leaf diseases, sometimes occurring naturally on early terrestrial plants

F

falcate hooked, curved, sickle-shaped

fascicle close cluster of leaves or flowers

filiform thread-like, thin

frond collective name for the stipe and leafy part of a fern

fructan fructose residues present in particular plants, present in 12 per cent of angiosperms globally; used in the food industry as a prebiotic dietary fibre

G

gametophyte the dominant form in bryophytes; in the life cycle of plants with alternating generations, this is the gamete-producing (usually haploid) phase, producing the zygote from which the sporophyte arises

garden escape a non-native plant species that has accidentally spread from a garden and established itself in the wild

geniculate bent at an angle, like a knee

glabrous hairless, smooth

glands modified hairs that secrete a substance, sometimes sticky, usually acting as a form of defence against herbivores (snails, worms, insects, other animals)

glaucous dull greyish-green colour

glumes the two basal or outer bracts of a spikelet in a grass or sedge flower

graminoid plant with a grass-like form; all members of the grass, restio and sedge families are graminoids, but this growth form is also present in some other families

H

hemiparasitic applied to a plant that grows both from photosynthesis and from attaching its roots to other plants

hispid covered with stiff hairs

hyaline translucent

hygrochastic opening with moisture

hypanthium floral cup, structure where basal parts of the calyx, corolla and stamens form a cup-shaped tube

I

imbricate overlapping

indehiscent not splitting open to release the seeds when ripe

indicator species a species whose status provides information on the overall condition of the ecosystem and of other species in that ecosystem, reflecting the quality and changes in environmental conditions as well as aspects of community composition

indusium thin membranous covering over a sorus on a fern frond (pseudo-indusium is formed by modification of the lamina margin)

involucre a group of bracts at the base of a cluster of flowers or flowerhead; seen in some members of the daisy family

involucral at the base of an inflorescence

involute rolling inwards

K

keel the two fused lower petals in flowers of the pea family, a median longitudinal ridge, like the keel of a boat

L

lamina the blade of a fern frond, not including the stipe (stalk)

lanceolate lance-shaped

leaflet the individual division of a compound leaf

lemma inner and lower 'bract' of a grass flower

lenticel raised pore, wart-like protuberance

ligule a small outgrowth at the junction of the leaf sheath and the blade, present in most grasses and some sedges

lobe division of a fern frond, when the pinnules or pinnae are too big to be called teeth

localised with a distribution that is limited to habitats that are infrequent or cover small areas

locule a small, separate cavity or chamber in a seed capsule where seeds are held

M

monoecious having male and female reproductive parts in a single individual

mucron short, sharp point at the tip of a leaf or flower part

mucronate ending abruptly in a short, sharp point

N

narrow endemic a plant with a distribution covering only a very small part of a defined area

naturalised describing plants that have spread into non-native environments and are able to reproduce in their new home, eventually establishing a new population there

non-serotinous describing plants that release their seeds during the inter-fire period; they do not need fire to release their seeds (see **serotinous**)

nurse plants plants that facilitate the growth and development of other plant species beneath their canopy

nut a 1-seeded fruit with a hard shell (especially in grasses)

nyctinasty the opening and closing of flowers or leaves in response to light and/or temperature, often referred to as 'sleep movements'

O

obcordate heart-shaped, but attached to the stalk by the pointy end

oblanceolate lance shaped, but attached to the stalk by the pointy end

obtuse with a blunt or rounded tip, not pointed

overstorey the uppermost canopy level of a forest, formed by the tallest trees

P

paleate covered with scales

palmate having five or more lobes with midribs radiating from a central point

palmatisect applied to leaves that are laterally divided to just before the petiole

panicle a much-branched inflorescence

paniculate growing in panicles

papillate with small pimply bumps

pappus the fine bristles that are the modified calyx in the daisy family, attached to the top of the achene and aiding dispersal

pedicel stalk supporting a flower

peduncle flowering stalk or stem (see **raceme**)

perianth a collective term for the outer parts of the flower, usually consisting of calyx and petals

petiolate applied to a leaf that has a stalk (see **petiole**)

petiole the stalk that joins a leaf to a stem

pilose covered with soft, straight hairs

pinna (plural **pinnae**) primary division of a compound fern frond

pinnate with leaflets attached opposite each other on either side of the common rachis or midrib; in ferns, pertaining to any divided frond or leaf where the division is complete to the rachis or midrib

pinnatifid pertaining to the division of a fern frond, where the division extends less than halfway to, but not quite joining the costa

pinnatisect pertaining to the division of a fern frond, where the division reaches/almost reaches the costa

pinnule the smallest division of a compound fern frond, or the ultimate division, sometimes called a lobe

plumose finely hairy, with a feathery appearance

polyploidy the heritable condition of having more than two sets of chromosomes

procumbent growing along the ground without sending out roots

propagule part of a plant that can detach and give rise to a new plant

pseudo-indusial in a fern, applied to a covering over the sori that is formed by modification of the lamina margin

puberulous covered with fine soft hairs; finely pubescent

pubescent hairy

R

raceme a flower cluster with individual flowers attached by short stalks along a central stem or flowering stalk (see **peduncle**)

racemose in the form of a raceme

rachis main axis or stem of an inflorescence or a compoundly divided fern frond (see **stipe**)

recurved bent or curved backward or downward

reflexed bent sharply back

reniform kidney-shaped

reseeder a plant that, when destroyed (e.g. by fire) regenerates from seed

resprouting able to survive fire by coppicing from underground storage organs

rhizomatous with or like a rhizome

rhizome root-like horizontal stem on or under the ground and capable of producing shoots and roots of a new plant

rugose wrinkled, rough, creased or ridged; usually referring to stem or leaf surfaces

S

sagittate shaped like an arrowhead

scale small piece of thin tissue that resembles a loose piece of skin, usually found on the rhizome, stipe and rachis on a fern

scape long flower stalk coming directly from the root

serotinous describing plants that store seeds, often for long periods of time, until a disaster such as the heat of a fire causes their release

sessile immobile, fixed in position without a stalk or peduncle

sheath protective structure, tubular

sorus (plural sori) fertile body on a fern frond composed of spore-producing sporangia

spadix type of inflorescence, small flowers borne on a fleshy stem, usually surrounded by a leaf-like curved bract (see spathe)

spathe curved bract

sporangium (plural sporangia) a receptacle in which asexual spores are formed (in ferns and other lower plants)

spore asexual reproductive body, usually powder-like and wind-borne (in ferns)

sporophyte the asexual, usually diploid phase and dominant generation in the life cycle of vascular plants; producing spores from which the gametophyte generation arises

spur slim tubular projection from the base of a flower

stipe petiole or stalk of a fern frond; this stalk becomes the rachis in the green part of the frond (see rachis)

stipule outgrowth at the base of the leaf stalk, typically in pairs

stomata minute breathing pores for the exchange of gases on the photosynthetic parts of most plants

striated streaked with parallel lines

strigose covered with short, stiff hairs

subdigitate somewhat but not completely digitate

subspecies a taxonomic category formed by subdividing a species into different groups or races; they usually have separate geographic distributions

subsucculent a plant that has slightly fleshy leaves that are not completely succulent

subterete not completely rounded

synflorescence a compound inflorescence; an inflorescence made up of several flowerheads; seen in some members of the daisy family

T

temperature inversion a layer in the atmosphere in which the air temperature is warmer at higher elevations (usually the air is cooler higher up)

temperature regime the temporal and spatial distribution of temperature; used to describe the climate of an area based on the maximum, minimum and average temperatures

terete cylindrical (see subterete)

trifid split into three parts

trifoliate applied to a leaf divided into three leaflets

truncate short, ending abruptly

type genus the genus from which the name of the family or subfamily is derived, e.g. Protea is the type genus of the Proteaceae

V

variety a taxonomic category formed by subdividing a species into different groups; these plants usually have overlapping distributions

verticillate arranged in whorls

villous densely fluffy, shaggy

viscid sticky

Photographic credits

Bibliography

Bello, A., Stirton, C.H., Chimphango, S.B.M. & Muasya, A.M. 2022. Phylogenetic relationships and biogeography of Psoraleeae (Fabaceae). *Botanical Journal of the Linnean Society* 200(1), pp. 39–74.

Bromilow, C. 2001. *Problem Plants of South Africa*. Pretoria: Briza Publications.

Clarke, H. & Charters, M. 2016. *The Illustrated History of Southern African Plant Names*. Cape Town: Jacana Media.

Coates Palgrave, K., Drummond, R.B., Moll, E.J. & Coates Palgrave, M. 2003. *Trees of Southern Africa*. Cape Town: Struik Nature.

Cohen, J.I., Manning, J.C. & Euston-Brown, D.I.W. 2019. *Lithospermum sylvestre* (Boraginaceae): A new species from the Baviaanskloof, Eastern Cape, South Africa. *Bothalia* 49(1), pp. 1–5.

Crouch, N.R., Klopper, R.R., Burrows, J.E. & Burrows, S.M. 2011. *Ferns of Southern Africa: A Comprehensive Guide*. Cape Town: Struik Nature.

Curtis-Scott, O., Goulding, M., Helme, N., McMaster, R., Privett, S. & Stirton, C. 2020. *Field Guide to Renosterveld of the Overberg*. Cape Town: Struik Nature.

Dold, A.P. & Cocks, M.L. 1999. Preliminary list of Xhosa plant names from the Eastern Cape, South Africa. *Bothalia* 29(2), pp. 267-292.

Euston-Brown, D.I.W. 1995. Environmental and dynamic determinants of vegetation distribution in the Kouga and Baviaanskloof mountains, Eastern Cape. Master's thesis, University of Cape Town (Botany Department).

Frye, H.A., Aiello-Lammens, M.E., Euston-Brown, D.I.W., Jones, C.S., Mollmann, H.K., Merow, C., Slingsby, J.A., van der Merwe, H., Wilson, A.M. & Silander Jr., J.A. 2021. Plant spectral diversity as a surrogate for species, functional and phylogenetic diversity across a hyper-diverse biogeographic region. *Global Ecology and Biogeography* 30(7), pp. 1403–1417.

Gledhill, E. 1981. *Veldblomme van Oos-Kaapland*. Cape Town: The Department of Nature and Environmental Conservation of the Cape Provincial Administration.

Goldblatt, P. & Manning, J. 2000. *Cape Plants: A Conspectus of the Cape Flora of South Africa*. *Strelitzia 9*. Pretoria & USA: National Botanical Institute & Missouri Botanical Garden Press.

Haaksma, E.D. & Linder, H.P. 2000. *Restios of the Fynbos*. Newlands: Botanical Society of South Africa.

Hattingh, L. 2011. *Baviaanskloof – A World Apart*. Port Elizabeth: Baviaans Tourism.

Henderson, L. 2001. Alien weeds and invasive plants. A complete guide to declared weeds and invaders in South Africa. *Plant Protection Research Institute Handbook* 12.

Johnson, S. & Bytebier, B. 2015. *Orchids of South Africa: A Field Guide*. Cape Town: Struik Nature.

Linder, H.P. 2001. 2018. African Restionaceae, Version 7. University of Cape Town: Contributions from the Bolus Herbarium 20.

Manning, J. 2001. *Eastern Cape: South African Wild Flower Guide 11*. Newlands: Botanical Society of South Africa.

Manning, J. 2018. *Field Guide to Fynbos*. Cape Town: Struik Nature.

Manning, J. 2019. *Field Guide to Wild Flowers of South Africa*. Cape Town: Struik Nature.

Manning, J.C., Magee, A.R. & Boatwright, J.S. 2018. *Felicia douglasii* (Asteraceae—Astereae), a distinctive new species from the Cape Floristic Region, South Africa. *South African Journal of Botany* 114, pp. 188–191.

McCarthy, T. & Rubidge, B. 2005. *The Story of Earth & Life*. Cape Town: Struik Nature.

Mucina, L. & Rutherford, M.C. (eds.), 2006. *The Vegetation of South Africa, Lesotho and Swaziland*. *Strelitzia 19*. Pretoria: South African National Biodiversity Institute (SANBI).

Norman, N. 2013. *Geology Off the Beaten Track: Exploring South Africa's Hidden Treasures*. Cape Town: Struik Nature.

Oliver, E.G.H. & Forshaw, N. 2018. Genus *Erica*: An Identification Aid. *Bolus Herbarium* 22(4).

Rebelo, A.G. 1995. *Sasol Proteas: A Field Guide to the Proteas of Southern Africa*. Vlaeberg: Fernwood Press, in assoc. with the National Botanical Institute.

Ruiters, A.K, Tilney, P.M, Van Wyk, B-E. & Magee, A.R. 2016. Taxonomy of the Genus *Phymaspermum* (Asteraceae, Anthemideae). *Systematic Botany* 41(2), pp. 430–456.

Schumann, D., Kirsten, G. & Oliver, E.G. 1992. *Ericas of South Africa*. Vlaeberg: Fernwood Press.

Shearing, D. 1994. *Karoo: Veldblomgids van Suid-Afrika 6*. Kirstenbosch: Botanical Society of South Africa.

Smith, G.F., Chesselet, P., Van Jaarsveld, E.J., Hartmann, H., Hammer, S., Van Wyk, B-E., Burgoyne, P., Klak, C. & Kurzweil, H. 1998. *Mesembs of the World*. Pretoria: Briza Publications.

Smith, G.F., Crouch, N.R. & Figueiredo, E. 2017. *Field Guide to Succulents in Southern Africa*. Cape Town: Struik Nature.

Snijman, D.A. (ed.). 2013. *Plants of the Greater Cape Floristic Region, Vol. 2: The Extra Cape Flora. Strelitzia 30*. Pretoria: South African National Biodiversity Institute (SANBI).

Stearn, W.T. 2004. *Botanical Latin*. Oregon: Timber Press.

Stirton, C.H. & Muasya, A.M. 2016. Seven new species and notes on the genus *Aspalathus* (Crotalarieae, Fabaceae). *South African Journal of Botany* 104, pp. 35–46.

Van der Walt, J.J.A. 1979. *Pelargoniums of Southern Africa*. Cape Town: Purnell & Sons.

Van Ginkel, C.E., Glen, R.P., Gordon-Gray, K.D., Cilliers, C.J., Muasya, M. & Van Deventer, P.P. 2011. *Easy Identification of Some South African Wetland Plants (Grasses, Restios, Sedges, Rushes, Bulrushes, Eriocaulons and Yellow-eyed Grasses)*. Gezina: Water Research Commission.

Van Oudtshoorn, F. & Van Wyk, E. 1999. *Guide to Grasses of Southern Africa*. Pretoria: Briza Publications.

Van Wyk, B-E., Van Oudtshoorn, B. & Gericke, N. 1997. *Medicinal Plants of South Africa*. Pretoria: Briza Publications.

Vanderplank, H.J. 1999. *Wildflowers of the Port Elizabeth Area: Gamtoos to Swartkops Rivers (The Coastal Bush and Fynbos Region)*. Port Elizabeth: Bluecliff Publishing.

Vlok, J.H.J, Euston-Brown, D.I.W. & Cowling, R.M. 2003. Acocks' Valley Bushveld 50 years on: new perspectives on the delimitation, characterisation and origin of subtropical thicket vegetation. *South African Journal of Botany* 69(1), pp. 27–51.

Vlok, J. & Schutte-Vlok, A.L. 2015. *Plants of the Klein Karoo*. Pretoria: Umdaus Press.

Zide, N., Euston-Brown, D.I.W. & Magee, A.R. 2022. *Passerina ternata* (Thymelaeaceae): A new species from the Baviaanskloof, South Africa. *South African Journal of Botany* 150, pp. 44–47.

Useful websites

Biodiversity explorer biodiversityexplorer.info/
Presents research about life on Earth in an understandable and integrated way

Plants of southern Africa posa.sanbi.org/sanbi
Access to South African plant names, herbarium sheets and observations made in the field

PlantZAfrica pza.sanbi.org
Hundreds of articles about southern African plants

Red List of South African Plants redlist.sanbi.org/
Current information on the national conservation status of South Africa's indigenous plants

iNaturalist www.inaturalist.org/
Global community observations of the natural world

Operation Wildflower
www.operationwildflower.org.za/
Dedicated to growing affinity for South African plants by presenting them to the public

Plants of the World Online powo.science.kew.org
Global plant names, descriptions and images

Acknowledgements

I would like to thank the following organisations and people for their support and involvement throughout my research and writing process:

- The University of Cape Town and the Bolus Herbarium: Professor William Bond introduced me to Baviaanskloof in 1992 and supervised my Master of Science degree.
- Eastern Cape Nature Conservation (ECNC) and, more recently, The Eastern Cape Parks & Tourism Agency (ECPTA): many staff members of these organisations contributed to my research. Special thanks go to Denis Laidler, Jan Briers, Japie Buckle, Tertius Schoeman, Cor van den Berg, Dereck Clarke, Hein Gerber, Rodger Smith, and Philip and Doreen Arends who lived at Geelhoutbos, and more recently, Brian Reeves, Wayne Erlank, and Gareth and Lindi Eloff.
- The Gamtoos Irrigation Board (GIB), notably Edwill Moore
- Jan Vlok and Annelise Schutte-Vlok, for assistance with plant identification
- The Wilderness Foundation: Andrew Muir commissioned a vegetation mapping project.
- Andrew Skowno and Anisha Dayaram for digital mapping, a special thank you
- The South African National Biodiversity Institute (SANBI), especially many staff members at the Compton Herbarium
- Professor John Silander Jr, University of Connecticut
- Jasper Slingsby, together with other students (you know who you are!)
- Numerous individuals on the privately owned farms, for access and support
- Living Lands, notably Otto Beukes
- Trávníček Pavel and Zuzi Chumova, Institute of Botany of the Czech Academy of Sciences, Czech Republic
- Plant taxonomists (in alphabetical order by surname): Robert Archer (Celastraceae and others); Anne Bean (Rutaceae); Nicola Bergh (Asteraceae); Josephine Beyers (†) (Asteraceae and *Lachnaea*); Stephen Boatwright (Fabaceae); Peter Bruyns (Euphorbiaceae and Apocynaceae); Mathew Buys (Boraginaceae); Pascale Chesselet (Aizoaceae); Christopher Cupido (Campanulaceae); James Deacon (*Albuca*); Tony Dold (various); Graham Duncan (*Lachenalia*); Tammy Elliot (*Schoenus*); David Gwynne-Evans (*Hermannia*); Cornelia Klak (Aizoaceae); Hubert Kurzweil (various); Peter Linder (Restionaceae); Anthony Magee (Apiaceae and Thymelaeaceae); John Manning (various); Muthama Muasya (Cyperaceae); Kenneth Oberlander (*Oxalis*); Ted and Inge (†) Oliver (*Erica*); Tony Rebelo (Proteaceae); Koos Roux (†) (ferns and others); Brian Schrire (*Indigofera*); Annelise Schutte-Vlok (Fabaceae); Dee Snijman (Amaryllidaceae); Keith Steiner (*Diascia*); Charles Stirton (Fabaceae); Terry Trinder-Smith (Rutaceae); Ross Turner (*Erica*); Ernst van Jaarsveld (various); Jan Vlok (various) and Pieter Winter (*Heliophila* and others).
- All the photographers (photographic credits on page 498)
- Megan Louise Potgieter, for the exceptionally brilliant illustrations
- iNaturalist
- The team at Struik Nature: Pippa Parker, Roelien Theron, Dominic Robson and Natalie Bell
- Emsie du Plessis, for her sterling proofreading at the end of the project
- Geologists: Terence McCarthy and Cameron Penn-Clarke

And finally, thank you to family and friends for love, support and encouragement.

Douglas Euston-Brown

I am grateful to Cedric Oktober and Jan (Tielie) Swarts, who climbed these mountains with me for so many years; also a massive thank you to my husband, Pieter Kruger, for his endless encouragement, patience, support, assistance and dedication, and for being my guide and partner for all these years. To our children, Abri, Johann and Maretha, my heartfelt gratitude for being part of this amazing journey. And to all the landowners and inhabitants of the area, who supported this dream, this book is also for you. Thank you to everyone who loves and treasures this hidden paradise.

Magriet Kruger

Index to scientific names

CAPS = family names; **bold** = genus names; *CAPS ITALICS* = alternative family names; ***bold italics*** = alternative genus names; *italics* = alternative species names; * asterisk = alien/exotic genus or species

A

Abutilon 403
A. sonneratianum 403
*Acacia 358
*A. cyclops 358
*A. mearnsii 358
ACANTHACEAE 180
*Acanthospermum 243
*A. australe 243
ACHARIACEAE 185
Achyranthemum 243
A. paniculatum 243
A. striatum 244
Achyranthes 210
A. aspera 210
Acokanthera 226
A. oppositifolia 226
Acrolophia 133
A. capensis 133
Acrostemon fourcadei 342
Adenocline 349
A. acuta 349
A. pauciflora 350
Adiantum 48
A. capillus-veneris 48
Adromischus 316
A. cristatus var. cristatus 316
A. sphenophyllus 317
A. subdistichus 317
Afroaster 244
A. hispida 244
Afrocarpus 57
A. falcatus 57
AGAPANTHACEAE 62
Agapanthus 62
A. praecox 62
Agathosma 450
A. blaerioides 450
A. capensis 450
A. kougaense 451
A. mucronulata 451
A. mundtii 451
A. ovata 452
A. pilifera 452
A. puberula 452
A. pungens 453
A. recurvifolia 453
A. unicarpellata 453
A. venusta 453
AGAVACEAE 63
*Agave 63
*A. americana 63
AIZOACEAE 188
Aizoon 192
A. africanum 192
A. glinoides 192
A. portulacaceum 192
A. procumbens 193

Albuca 107
A. bracteata 107
A. canadensis 108
A. cremnophila 108
A. longipes 108
A. schoenlandii 109
A. setosa 109
A. virens 109
Alepidea 220
A. capensis 220
A. delicatula 220
ALLIACEAE 64
Allophylus 462
A. decipiens 463
Aloe 77
A. africana 77
A. comptonii 79
A. ferox 78
A. lineata var. lineata 78
A. longistyla 78
A. microstigma subsp.
    microstigma 79
A. perfoliata 79
A. pictifolia 79
A. pluridens 80
A. speciosa 80
A. striata 80
Aloiampelos 81
A. gracilis 81
AMARANTHACEAE 210
AMARYLLIDACEAE 65
Amellus 245
A. tridactylus 245
Amphiglossa 245
A. callunoides 245
Amphithalea 359
A. phylicoides 359
A. williamsonii 359
Anacampseros 214
A. arachnoides 214
A. telephiastrum 214
ANACAMPSEROTACEAE 214
ANACARDIACEAE 215
Anchusa 300
A. capensis 300
Androcymbium 88
Anemia 42
A. caffrorum 42
ANEMIACEAE 42
Anemone 438
A. vesicatoria subsp. humilis 438
Anginon 221
A. difforme 221
Anisodontea 404
A. scabrosa 404
Anthospermum 447
A. aethiopicum 447
A. galioides 447

A. spathulatum 447
APIACEAE 220
APOCYNACEAE 226
Aponogeton 72
A. desertorum 72
A. distachyos 72
APONOGETONACEAE 72
Aptenia cordifolia 203
Aptosimum 466
A. procumbens 467
AQUIFOLIACEAE 236
ARALIACEAE 236
Arctopus 221
A. echinatus 221
Arctotheca 246
A. calendula 246
A. prostrata 246
Arctotis 247
A. acaulis 247
A. arctotoides 247
A. linearis 248
Argyrolobium 359
A. collinum 359
A. incanum 360
A. parviflorum 360
A. pumilum 360
Aristea 118
A. anceps 118
A. cuspidata 119
A. ensifolia 119
A. pusilla 119
Aristida 144
A. congesta 144
A. vestita 144
Artemisia 248
A. afra 248
Arundinella 145
A. nepalensis 145
*Arundo 145
*A. donax 145
Aspalathus 361
A. aciphylla 361
A. chortophila 361
A. collina subsp. collina 362
A. fourcadei 362
A. glabrescens 362
A. hirta subsp. stellaris 363
A. hystrix 363
A. kougaensis 363
A. modesta 364
A. nivea 364
A. rubens 364
A. setacea 364
A. sp. nov. (chlorina ms.) 365
A. sp. nov. (dupreezii ms.) 365
A. sp. nov. (vulpicaudata ms.) 365
A. subtingens 366
A. usnoides 366

ASPARAGACEAE 73
Asparagus 73
A. aethiopicus 73
A. africanus 73
A. asparagoides 74
A. burchellii 74
A. capensis 74
A. crassicladus 75
A. densiflorus 75
A. multiflorus 75
A. setaceus 76
A. striatus 76
A. suaveolens 76
A. subulatus 77
ASPHODELACEAE 77
ASPLENIACEAE 43
Asplenium 43
A. capense 43
A. rutifolium 43
Aster 244
A. bakerianus 244
ASTERACEAE 238
Atalaya 463
A. capensis 463
Athanasia 248
A. linifolia 249
A. pinnata 249
A. tomentosa 249
A. vestita 250
Athrixia 250
A. heterophylla 250
*Atriplex 210
*A. lindleyi subsp. inflata 210
*A. nummularia 211
*A. semibaccata 211
Augea 490
A. capensis 490
Azima 459
A. tetracantha 459

B

Babiana 120
B. sambucina 120
Balotta 396
B. africana 396
Barleria 180
B. obtusa 180
B. pungens 181
Bartholina 133
B. burmanniana 133
Berkheya 251
B. angustifolia 251
B. carduoides 251
B. cruciata 251
Berzelia 304
B. intermedia 304
BIGNONIACEAE 300
BLECHNACEAE 44
Blechnum 44
B. attenuatum 44
B. punctulatum 44
Blepharis 181
B. capensis 181
B. integrifolia 182
Bobartia 120

B. orientalis 120
Boophone 65
B. disticha 65
BORAGINACEAE 300
Boscia 309
B. oleoides 309
Bowiea 110
B. volubilis subsp. volubilis 110
Brachiaria 165
B. serrata 165
Brachycorythis 134
B. macowaniana 134
Brachylaena 252
B. glabra 252
B. ilicifolia 252
BRASSICACEAE 303
Brunia 305
B. noduliflora 305
BRUNIACEAE 304
Brunsvigia 65
B. josephinae 66
B. striata 66
*Bryophyllum 317
*B. delagoense 317
Buddleja 467
B. glomerata 467
B. saligna 468
B. salviifolia 468
Bulbine 81
B. abyssinica 81
B. cremnophila 82
B. frutescens 82
B. latifolia 82
B. narcissifolia 83
Bulbostylis 90
B. humilis 91

C

Cacalia 257
*CACTACEAE 305
Cadaba 309
C. aphylla 309
Caesia 106
C. contorta 107
Calobota 366
C. pungens 366
Calodendrum 454
C. capense 454
Calpurnia 367
C. aurea subsp. aurea 367
CAMPANULACEAE 307
CANNABACEAE 308
Cannomois 165
C. scirpoides 166
C. virgata 166
Canthium 448
C. inerme 448
Capeochloa 146
C. arundinacea 146
CAPPARACEAE 309
Capparis 310
C. sepiaria var. citrifolia 310
Caputia 252
C. pyramidata 253
C. scaposa 253

Carex 91
C. aethiopica 91
C. capensis 91
C. glomerabilis 92
C. uhligii 92
Carissa 226
C. bispinosa 226
C. haematocarpa 226
Carpha 92
C. glomerata 92
Carpobrotus 193
C. deliciosus 193
C. edulis 193
CARYOPHYLLACEAE 310
Cassytha 60
C. ciliolata 60
CELASTRACEAE 313
Celtis 308
C. africana 308
Cenchrus 146
C. caudatus 146
C. ciliaris 147
*C. clandestinum 147
*C. setaceus 147
Centella 222
C. asiatica 222
C. virgata 222
Ceratiosicyos 185
C. laevis 185
Ceterach capense 43
Chaenostoma 468
C. cinereum 468
C. decipiens 469
C. denudatum 469
C. integrifolium 469
C. marifolium 469
C. polyanthum 470
C. revolutum 470
Chaetacanthus 182
Chascanum 488
C. cuneifolium 488
Chasmanthe 121
C. aethiopica 121
Cheilanthes 49
C. contracta 49
C. multifida subsp. multifida 49
C. parviloba 49
C. viridis 50
Chenopodium 212
C. carinatum 212
Chironia 381
C. baccifera 381
C. linoides 381
C. melampyrifolia 381
Chlorophytum 63
C. comosum 63
C. crispum 64
Chondropetalum 167
Christella gueinziana 52
Chrysanthemoides monilifera 280
Chrysitrix 93
C. capensis 93
Chrysocoma 253
C. ciliata 253
C. valida 254

*Cichorium 254
*C. intybus 254
Cineraria 254
C. lobata subsp. platyptera 254
C. saxifraga 255
Clausena 454
C. anisata 454
Clematis 439
C. brachiata 439
Cliffortia 443
C. cervicornu 443
C. drepanoides 444
C. graminea 444
C. ilicifolia 444
C. linearifolia 444
C. neglecta 445
C. paucistaminea var. australis 445
C. polita 445
C. ramosissima 445
C. stricta 446
C. strobilifera 446
Clutia 350
C. affinis 350
C. alaternoides 350
C. daphnoides 351
C. ericoides 351
C. polifolia 351
C. pulchella 351
Coccinia 328
C. quinqueloba 328
COLCHICACEAE 88
Colchicum 88
C. longipes 88
Coleonema 454
C. aspalathoides 455
Coleus 394
C. madagascariensis 394
C. subspicatus 394
Colpoon 460
C. compressum 460
Commelina 89
C. africana 89
C. benghalensis 89
COMMELINACEAE 89
Conophytum 194
C. truncatum 194
CONVOLVULACEAE 316
Convolvulus 316
C. farinosus 316
Cordia 301
C. caffra 301
Corymbium 255
C. africanum 255
C. glabrum 255
Cotyledon 318
C. campanulata 318
C. gloeophylla 318
C. orbiculata 318
C. tomentosa 319
C. velutina 319
C. woodii 319
Crassothonna 256
C. cacalioides 256
C. capensis 256
Crassula 320

C. atropurpurea 320
C. biplanata 320
C. capitella subsp. thyrsiflora 320
C. cotyledonis 321
C. cremnophila 321
C. cultrata 321
C. ericoides 321
C. expansa subsp. expansa 322
C. hemisphaerica 322
C. lactea 322
C. mollis 322
C. muscosa 323
C. nemorosa 323
C. nudicaulis var. platyphylla 323
C. orbicularis 324
C. ovata 324
C. pellucida 324
C. perfoliata 325
C. perforata 325
C. pubescens subsp. radicans 325
C. pyramidalis 326
C. rogersii 326
C. rubricaulis 326
C. rupestris 326
C. saxifraga 327
C. tecta 327
C. tetragona subsp. acutifolia 327
C. umbella 327
CRASSULACEAE 316
Crotalaria 367
C. capensis 367
CRUCIFERAE 303
Ctenomeria capensis 357
CUCURBITACEAE 328
Cullumia 256
C. bisulca 256
C. decurrens 257
Cunonia 330
C. capensis 330
CUNONIACEAE 330
CUPRESSACEAE 54
Curio 257
C. acaulis 257
C. articulatus 257
C. crassulifolius 258
C. ficoides 258
C. radicans 258
Curtisia 330
C. dentata 330
CURTISIACEAE 330
Cussonia 236
C. gamtoosensis 237
C. paniculata 237
C. spicata 237
Cyanella 179
C. lutea subsp. lutea 179
Cyanotis 90
C. speciosa 90
Cyathula 211
C. uncinulata 211
Cyclopia 368
C. intermedia 368
Cyclosorus 52
C. gueinzianum 52
C. interruptus 53

Cymbopogon 148
C. plurinodis 148
C. pospischilii 148
Cynanchum 227
C. ellipticum 227
C. gerrardii 227
C. viminale 227
Cynodon 148
C. dactylon 148
CYPERACEAE 90
Cyperus 93
C. congestus 93
C. difformis 93
C. mundii 94
C. polystachyos 94
C. semitrifidus 94
C. sphaerospermus 94
C. textilis 95
C. uitenhagensis 95
Cyphia 399
C. digitata 400
C. sylvatica 400
Cyrtanthus 66
C. collinus 66
C. flammosus 67
C. labiatus 67
C. montanus 67
C. obliquus 67
C. sanguineus 68
C. smithiae 68

D
Dalechampia 352
D. capensis 352
Delosperma 194
D. echinatum 194
D. esterhuyseniae 195
D. multiflorum 195
D. prasinum 195
D. stenandrum 195
DENNSTAEDTIACEAE 45
Dianthus 311
D. caespitosus 311
D. thunbergii 311
Diascia 470
D. parviflora 470
D. patens 471
Dicerothamnus 259
D. rhinocerotis 259
Dicliptera 182
D. cernua 182
Dicoma 259
D. picta 259
DIDIEREACEAE 331
Dietes 121
D. iridioides 121
Digitaria 149
D. eriantha 149
Dimorphotheca 260
D. cuneata 260
D. ecklonis 260
Dioscorea 106
D. elephantipes 106
DIOSCOREACEAE 106
Diosma 455

D. apetala   455
D. prama   455
D. rourkei   456
D. sp. nov.   456
Diospyros   336
D. austro-africana   337
D. dichrophylla   337
D. lycioides   337
D. whyteana   338
Dipcadi   110
D. viride   110
Dipogon   368
D. lignosus   368
DIPSACACEAE   335
Disa   134
D. bifida   134
D. comosa   135
D. cornuta   135
D. harveyana subsp. harveyana   135
D. lugens var. lugens   136
D. porrecta   136
D. spathulata subsp. tripartita   137
Disparago   260
D. tortilis   261
Disperis   137
D. macowanii   137
Dodonaea   464
D. viscosa subsp. angustifolia   464
Drimia   111
D. anomala   111
D. capensis   111
D. ciliata   112
D. haworthioides   112
D. intricata   112
D. uniflora   113
Drosanthemum   196
D. hispidum   196
D. lique   196
D. wittebergense   197
Drosera   336
D. aliciae   336
D. capensis   336
DROSERACEAE   336
Duvalia   228
D. caespitosa subsp. caespitosa   228
Dyschoriste   182
D. setigera   183
Dysphania   212
D. carinata   212

E

EBENACEAE   336
Echinopsis   305
E. oxygona   305
E. spachiana   306
Egeria   116
E. densa   116
Ehretia   301
E. rigida   301
Ehrharta   149
E. calycina   149
E. dura   150
E. erecta   150
E. ramosa   150
ELAPHOGLOSSACEAE   46

Elaphoglossum   46
E. acrostichoides   46
Elegia   167
E. capensis   167
E. filacea   167
E. juncea   168
E. vaginulata   168
Eleocharis   95
E. limosa   95
Elytropappus   259
E. rhinocerotis   259
Empodium   117
E. plicatum   117
Encephalartos   58
E. horridus   58
E. lehmannii   58
E. longifolius   58
Enneapogon   151
E. scoparius   151
Epischoenus quadrangularis   101
Eragrostis   151
E. capensis   151
E. curvula   152
E. obtusa   152
E. racemosa   152
Erica   341
E. affinis   341
E. andreaei   342
E. angulosa   342
E. articularis   342
E. bolusanthus   343
E. caffra   343
E. chamissonis   343
E. curviflora   344
E. demissa   344
E. diaphana   344
E. discolor subsp. speciosa   345
E. flocciflora   345
E. kougabergensis   345
E. nabea   345
E. newdigateae   346
E. passerinae   346
E. pectinifolia   346
E. rosacea subsp. rosacea   347
E. sp. nov.   347
E. sparrmanii   347
E. thamnoides   348
E. uberiflora   348
E. umbelliflora   348
E. viridiflora subsp. primulina   349
ERICACEAE   340
Eriocephalus   261
E. africanus   261
E. capitellatus   262
E. ericoides   262
Eriospermum   177
E. capense   177
E. ciliatum   177
E. paradoxum   178
E. zeyheri   178
Eucalyptus   415
E. botryoides   415
E. camaldulensis   416
Euchaetis   456
E. cristagalli   456

E. vallis-simiae   457
Euclea   338
E. crispa subsp. crispa   338
E. daphnoides   339
E. natalensis subsp. natalensis   339
E. polyandra   339
E. schimperi   340
E. undulata   340
Eulophia   138
E. tenella   138
E. tuberculata   138
Euphorbia   352
E. atrispina   353
E. caerulescens   352
E. esculenta   353
E. grandidens   353
E. heptagona   353
E. mammillaris   354
E. polygona   354
E. procumbens   354
E. silenifolia   355
E. spartaria   355
E. triangularis   355
EUPHORBIACEAE   349
Euryops   262
E. euryopoides   262
E. lateriflorus   263
E. munitus   263
E. pinnatipartitus   263
E. rehmannii   264
E. spathaceus   264
Eustachys   153
E. paspaloides   153

F

FABACEAE   357
Felicia   264
F. douglasii   264
F. filifolia   265
F. joubertinae   265
F. linifolia   265
F. microcephala   265
F. muricata   266
F. ovata   266
Ficinia   96
F. acuminata   96
F. albicans   96
F. brevifolia   96
F. deusta   97
F. fascicularis   97
F. gracilis   97
F. nigrescens   98
F. ramosissima   98
F. stolonifera   98
F. trispicata   99
Ficus   413
F. burtt-davyi   413
F. sur   413
Fingerhuthia   153
F. africana   153
Flueggea   423
F. verrucosa   423
Freesia   122
F. corymbosa   122
Freylinia   471

F. crispa 471
F. lanceolata 471
F. undulata 472
Fuirena 99
F. hirsuta 99

G

Galenia africana 192
G. portulacacea 192
G. procumbens 193
Galium 448
G. tomentosum 448
Gasteria 83
G. brachyphylla 83
G. camillae 83
G. glomerata 84
G. pulchra 84
G. rawlinsonii 84
Gazania 266
G. krebsiana 266
G. lichtensteinii 267
Geissorhiza 122
G. bracteata 122
G. roseoalba 122
GENTIANACEAE 381
Geochloa 154
G. decora 154
GERANIACEAE 382
Geranium 382
G. canescens 382
Gerbera 267
G. ambigua 267
G. ovata 268
G. piloselloides 268
GESNERIACEAE 390
Gethyllis 68
G. spiralis 68
Gibbaria scabra 281
Gisekia 391
G. pharnaceoides 391
GISEKIACEAE 391
Gladiolus 123
G. floribundus 123
G. geardii 123
G. huttonii 123
G. leptosiphon 124
G. maculatus 124
G. patersoniae 124
G. permeabilis subsp. edulis 125
G. stellatus 125
G. virescens 125
Gleichenia 46
G. polypodioides 46
GLEICHENIACEAE 46
Glottiphyllum 197
G. depressum 197
Gnidia 483
G. anthylloides 485
G. cuneata 485
G. juniperifolia 483
G. nodiflora 484
Gomphocarpus 228
G. cancellatus 228
G. fruticosus 229
Gonioma 229

G. kamassi 229
Graderia 418
G. scabra 418
Grewia 404
G. occidentalis 404
G. robusta 404
Gunnera 392
G. perpensa 392
GUNNERACEAE 392
Gymnosporia 313
G. buxifolia 313
G. linearis subsp. linearis 313

H

Habenaria 139
H. arenaria 139
Haemanthus 69
H. albiflos 69
H. coccineus 69
H. sanguineus 70
Halleria 482
H. lucida 482
Haplocarpha 268
H. lyrata 269
H. scaposa 269
Harveya 418
H. purpurea 418
Haworthia 85
H. cooperi 85
H. decipiens 85
H. monticola 86
Haworthiopsis 86
H. fasciata 86
H. longiana 86
H. scabra 87
H. viscosa 87
Hebenstretia 472
H. integrifolia 472
Helichrysum 269
H. anomalum 269
H. cymosum 270
H. excisum 270
H. felinum 270
H. herbaceum 271
H. lancifolium 271
H. nudifolium var. nudifolium 271
H. nudifolium var. oxyphyllum 272
H. odoratissimum 272
H. rosum 272
H. rugulosum 273
H. teretifolium 273
H. zeyheri 273
Heliophila 303
H. cornuta var. squamata 303
H. elongata 303
H. suavissima 304
H. subulata 304
HEMEROCALLIDACEAE 106
Hermannia 405
H. althaeifolia 405
H. althaeoides 405
H. coccocarpa 405
H. filifolia 406
H. flammea 406

H. gracilis 406
H. holosericea 406
H. mucronulata 407
H. saccifera 407
H. salviifolia 407
H. stipulacea 407
H. velutina 408
Hermas 223
H. ciliata 223
Hertia 274
H. kraussii 274
Heteromorpha 223
H. arborescens 223
Heteropogon 154
H. contortus 154
Hibiscus 408
H. aethiopicus 408
H. aridus 408
H. pusillus 409
Hilliardiella 274
H. pinifolia 274
Hippobromus 464
H. pauciflorus 464
Holothrix 139
H. burchellii 139
H. mundii 140
H. pilosa 140
H. schlechteriana 140
Huernia 230
H. echidnopsioides 230
H. longii subsp. echidnopsioides 230
H. thuretii 230
HYACINTHACEAE 107
Hydnora 59
H. africana 59
HYDNORACEAE 59
HYDROCHARITACEAE 116
Hyobanche 419
H. sanguinea 419
HYPERICACEAE 392
Hypericum 392
H. lalandii 393
Hypertelis 393
Hypodiscus 169
H. aristatus 169
H. striatus 169
H. synchroolepis 170
Hypoestes 183
H. aristata 183
H. forskaolii 183
Hypolepis 45
H. villoso-viscida 45
HYPOXIDACEAE 117
Hypoxis 117
H. villosa 117

I

Ilex 236
I. mitis 236
Indigofera 368
I. capillaris 369
I. declinata 369
I. denudata 369
I. disticha 369

I. flabellata 370
I. sp. nov. (vlokii ms.) 370
I. sulcata 370
IRIDACEAE 118
Isoglossa 184
I. ciliata 184
Isolepis 99
I. hystrix 99
I. prolifera 100
I. sepulcralis 100
Ixia 126
I. orientalis 126

J

Jamesbrittenia 472
J. argentea 472
J. atropurpurea 473
J. tenuifolia 473
J. tortuosa 473
Jatropha 356
J. capensis 356
JUNCACEAE 131
Juncus 131
J. acutus 131
J. capensis 131
Justicia 184
J. cuneata 184
J. leptantha 184
J. tubulosa subsp. late-ovata 184

K

Kalanchoe 328
K. rotundifolia 328
Kedrostis 329
K. capensis 329
K. nana 329
Kewa 393
K. salsoloides 393
KEWACEAE 393
Kiggelaria 188
K. africana 188
Kleinia 257
K. articulatus 257
Kniphofia 87
K. uvaria 87

L

Lachenalia 113
L. ensifolia 113
L. latimerae 113
Lachnaea 484
L. burchellii 484
L. glomerata 484
Lacomucinaea 460
L. lineata 460
LAMIACEAE 394
Lampranthus 198
L. affinis 198
L. coralliflorus 198
L. haworthii 199
L. spectabilis 199
L. stayneri 199
Lanaria 132
L. lanata 132
LANARIACEAE 132

*Lantana 489
L. rugosa 489
Lasiosiphon 485
L. anthylloides 485
L. meisnerianus 485
LAURACEAE 60
Ledebouria 114
L. revoluta 114
Leonotis 394
L. leonurus 395
L. ocymifolia 395
Lessertia 371
L. frutescens 371
Leucadendron 429
L. album 429
L. eucalyptifolium 430
L. loeriense 430
L. nobile 430
L. pubibracteolatum 431
L. rourkei 431
L. rubrum 431
L. salignum 432
L. sorocephalodes 432
L. uliginosum subsp.
   glabratum 432
Leucospermum 433
L. cuneiforme 433
L. wittebergense 433
Lichtensteinia 224
L. latifolia 224
Lightfootia divaricata 308
L. rigida 307
LIMEACEAE 398
Limeum 398
L. aethiopicum 398
LINACEAE 398
Lindernia 399
L. parviflora 399
LINDERNIACEAE 399
Linum 398
L. africanum 398
L. esterhuyseniae 399
Lithops 200
L. localis 200
Lithospermum 302
L. papillosum 302
L. sylvestre 302
Lobelia 400
L. anceps 400
L. erinus 401
L. linearis 401
L. neglecta 401
L. tomentosa 401
LOBELIACEAE 399
Lobostemon 302
L. marlothii 302
LORANTHACEAE 402
Lotononis 371
L. azurea 371
L. pungens 371
Loxostylis 215
L. alata 215
Lycium 479
L. ferocissimum 479
L. oxycarpum 480

M

Machairophyllum 200
M. bijliae 200
Maerua 310
M. cafra 310
*Maireana 212
*M. brevifolia 212
Malephora 201
M. uitenhagensis 201
MALVACEAE 403
Manulea 474
M. cheiranthus 474
Marchantia 40
M. berteroana 40
MARCHANTIACEAE 40
Mastersiella 170
M. purpurea 170
M. spathulata 171
Maytenus 314
M. acuminata 314
M. heterophylla 313
M. oleoides 314
M. undata 314
Melhania 409
M. didyma 409
MELIACEAE 410
MELIANTHACEAE 411
Melianthus 411
M. comosus 411
Melica 155
M. racemosa 155
Melinis 155
*M. repens 155
Melolobium 372
M. candicans 372
Mentha 395
M. longifolia 395
Merxmuellera 146, 154, 162
M. stricta 162
MESEMBRYANTHEMACEAE 188
Mesembryanthemum 202
M. aitonis 202
M. articulatum 202
M. bicorne 203
M. cordifolium 203
M. granulicaule 203
M. guerichianum 204
M. splendens 204
Mestoklema 204
M. tuberosum 204
Metalasia 275
M. acuta 275
M. aurea 275
M. massonii 275
M. pallida 276
M. pungens 276
M. trivialis 276
Microchloa 156
M. kunthii 156
Microglossa 277
M. mespilifolia 277
Microloma 230
M. tenuifolium 231
Mikania 277
M. capensis 277

M. chenopodiifolia   277
Miscanthus   156
M. capensis   156
M. ecklonii   156
Mohria   42
M. caffrorum   42
MOLLUGINACEAE   411
Monechma   185
M. spartioides   185
Monopsis   402
M. unidentata   402
Monsonia   382
M. emarginata   382
Montinia   412
M. caryophyllacea   412
MONTINIACEAE   412
Moquiniella   402
M. rubra   403
MORACEAE   413
Moraea   126
M. algoensis   126
M. bipartita   126
M. exiliflora   127
M. falcifolia   127
M. fugacissima   127
M. lewisiae   127
M. polyanthos   128
M. ramosissima   128
M. spathulata   128
M. tricuspidata   129
M. unguiculata   129
Morella   414
M. serrata   414
Muraltia   426
M. ericifolia   426
M. juniperifolia   426
M. spinosa   426
M. squarrosa   427
Myrica   414
MYRICACEAE   414
MYRSINACEAE   414
Myrsine   414
M. africana   414
MYRTACEAE   415

N

Nanobubon   224
N. capillaceum   224
Nemesia   474
N. deflexa   474
N. denticulata   475
N. fruticans   475
Nerine   70
N. humilis   70
N. peersii   70
*****Nerium**   231
*N. oleander   231
Nidorella   277
N. ivifolia   278
Noltea   439
N. africana   439
Notobubon   225
N. ferulaceum   225
N. gummiferum   225
Nuxia   483

N. floribunda   483
Nylandtia spinosa   426
Nymania   410
N. capensis   410
Nymphaea   61
N. nouchali var.
   caerulea   61
NYMPHAEACEAE   61

O

Ochna   416
O. serrulata   416
OCHNACEAE   416
Oedera   278
O. calycina   278
O. decussata   278
O. genistifolia   279
O. imbricata   279
O. speciosa   279
O. squarrosa   279
Olea   417
O. capensis subsp. capensis   417
O. europaea subsp. africana   417
OLEACEAE   417
Oligocarpus calendulaceus   280
Olinia   422
O. ventosa   422
*****Opuntia**   306
*O. ficus-indica   306
ORCHIDACEAE   133
Ornithogalum   114
O. dubium   114
O. juncifolium   115
O. longibracteatum   107
Ornithoglossum   88
O. undulatum   88
OROBANCHACEAE   418
OSMUNDACEAE   47
Osteospermum   280
O. calendulaceum   280
O. junceum   280
O. moniliferum   280
O. polygaloides   281
O. scabrum   281
Osyris   460
O. compressa   460
Otholobium   373
O. acuminatum   373
O. candicans   374
O. pictum   375
O. polyphyllum   375
O. prodiens   375
O. sericeum   376
O. stachyerum   376
Othonna   281
O. carnosa   256
O. gymnodiscus   281
O. parviflora   282
O. retrofracta   282
O. triplinervia   282
OXALIDACEAE   419
Oxalis   419
O. caprina   419
O. ciliaris   420
O. fergusoniae   420

O. imbricata var. violacea   420
O. obtusa   421
O. psilopoda   421
O. punctata   421
O. purpurea   421
O. smithiana   422
O. stellata   422

P

Pachycarpus   231
P. dealbatus   231
Pachypodium   232
P. bispinosum   232
P. succulentum   232
Panicum   157
P. maximum   157
Pappea   465
P. capensis   465
Paranomus   434
P. esterhuyseniae   434
Paspalum   157
*P. dilatatum   157
*P. distichum   158
*P. urvillei   158
Passerina   485
P. obtusifolia   485
P. pendula   486
P. ternata   486
Pauridia   118
P. trifurcillata   118
Pegolettia   283
P. baccaridifolia   283
P. retrofracta   283
Pelargonium   383
P. alchemilloides   383
P. carneum   383
P. carnosum   384
P. cordifolium   384
P. exhibens   384
P. inquinans   385
P. laevigatum   385
P. laxum   385
P. myrrhifolium   386
P. odoratissimum   386
P. ovale subsp. ovale   386
P. ovale subsp.
   veronicifolium   387
P. panduriforme   387
P. peltatum   387
P. pulverulentum   387
P. quercifolium   388
P. radens   388
P. reniforme   388
P. ribifolium   389
P. schizopetalum   389
P. sidoides   389
P. tetragonum   390
P. zonale   390
Pellaea   50
P. calomelanos   50
P. leucomelas   50
PENAEACEAE   422
Pennisetum   146
P. clandestinum   147
P. macrourum   146

P. setaceum   147
Pentameris   158
P. distichophylla   158
P. dregeana   158
P. eriostoma   159
P. macrocalycina   159
P. pallida   159
P. rigidissima   160
Pentaschistis   158
Pentzia   284
P. dentata   284
P. incana   284
Pharnaceum   411
P. dichotomum   411
Phragmites   160
P. australis   160
Phylica   440
P. abietina   440
P. alba   440
P. axillaris var. axillaris   440
P. axillaris var. hirsuta   441
P. debilis   441
P. lachneaeoides   441
P. lanata   441
P. paniculata   442
PHYLLANTHACEAE   423
Phyllanthus   423
P. incurvus   423
P. verrucosus   423
Phyllopodium   475
P. elegans   475
Phymaspermum   285
P. oppositifolium   285
Phytolacca   424
*P. dioica   424
PHYTOLACCACEAE   424
PITTOSPORACEAE   424
Pittosporum   424
P. viridiflorum   425
Platycaulos   171
P. callistachyus   171
Plectranthus   394
P. madagascariensis   394
P. spicatus   394
Plectranthus   396
P. verticillatus   396
Plexipus cuneifolius   488
PLUMBAGINACEAE   425
Plumbago   425
P. auriculata   425
POACEAE   142
Podalyria   372
P. burchellii   372
P. myrtillifolia   373
PODOCARPACEAE   57
Podocarpus   57
P. falcatus   57
P. latifolius   57
Pollichia   311
P. campestris   311
Polygala   427
P. asbestina   427
P. ericaefolia   427
P. fruticosa   428
P. myrtifolia var. myrtifolia   428

P. myrtifolia var. pinifolia   428
P. virgata   428
POLYGALACEAE   426
POLYPODIACEAE   48
Polypodium   48
P. ensiforme   48
Portulacaria   331
P. afra   331
Printzia   285
P. polifolia   285
Prionium   132
P. serratum   132
Protea   434
P. cynaroides   434
P. eximia   435
P. intonsa   435
P. lorifolia   435
P. mundii   436
P. neriifolia   436
P. nitida   436
P. punctata   437
P. repens   437
P. rupicola   437
P. tenax   438
PROTEACEAE   429
Psammotropha   412
P. quadrangularis   412
Pseudodictamnus   396
P. africanus   396
Pseudoschoenus   100
P. inanis   100
Psilocaulon articulatum   202
Psoralea   373
P. acuminata   373
P. axillaris   373
P. candicans   374
P. imminens   374
P. kougaensis   374
P. oligophylla   375
P. picta   375
P. polyphylla   375
P. prodiens   375
P. sericea   376
P. stachyera   376
P. trullata   376
Ptaeroxylon   457
P. obliquum   457
PTERIDACEAE   48
Pteridium   45
P. aquilinum   45
Pteris   51
P. dentata   51
*P. tremula   51
Pterocelastrus   315
P. tricuspidatus   315
Pteronia   285
P. adenocarpa   286
P. camphorata   286
P. glauca   286
P. glomerata   286
P. incana   287
P. membranacea   287
P. paniculata   287
P. staehelinoides   288
P. teretifolia   288

Pulicaria   288
P. scabra   288
Pupalia   213
P. lappacea   213
Putterlickia   315
P. pyracantha   315

R
Rafnia   377
R. elliptica   377
RANUNCULACEAE   438
Rapanea   415
R. melanophloeos   415
Relhania calycina   278
R. decussata   278
R. speciosa   279
Restio   172
R. andreaeanus   172
R. capensis   172
R. gaudichaudianus   173
R. hystrix   173
R. paniculatus   173
R. sejunctus   174
R. triticeus   174
R. vallis-simius   174
RESTIONACEAE   165
RHAMNACEAE   439
Rhamnus   442
R. prinoides   442
Rhigozum   300
R. obovatum   300
Rhodocoma   175
R. capensis   175
R. fruticosa   175
Rhoicissus   489
R. digitata   489
R. kougabergensis   490
Rhombophyllum   205
R. dolabriforme   205
Rhus   216
Rhynchosia   377
R. caribaea   377
*Ricinus   356
*R. communis   356
Roepera   491
R. foetida   491
R. fulva   491
ROSACEAE   443
Rothmannia   449
R. capensis   449
Rubia   449
R. petiolaris   449
RUBIACEAE   447
Rubus   446
R. rigidus   446
RUSCACEAE   177
Ruschia   205
R. cradockensis   205
R. inclusa   206
R. knysnana   206
R. multiflora   206
R. perfoliata   207
R. pungens   207
R. tenella   207
RUTACEAE   450

S

SALICACEAE 458
Salix 458
S. mucronata 458
SALVADORACEAE 459
Salvia 396
S. namaensis 396
S. stenophylla 397
Sansevieria 179
S. aethiopica 179
SANTALACEAE 460
SAPINDACEAE 462
SAPOTACEAE 466
Sarcostemma viminale 227
Satyrium 141
S. bracteatum 141
S. jacottetiae 141
S. longicolle 141
Scabiosa 335
S. albanensis 335
S. columbaria 335
Scadoxus 71
S. puniceus 71
*Schinus 215
*S. molle 215
Schizaea 52
S. pectinata 52
Schoenoxiphium
  ecklonii 91
S. lehmannii 92
SCHIZAEACEAE 52
Schoenus 101
S. graciliculmis 101
S. megacarpus 101
S. quadrangularis 101
S. schonlandii 102
S. selinae 102
S. submarginalis 102
Schotia 378
S. afra 378
S. latifolia 378
Scirpus inanis 100
Scolopia 459
S. zeyheri 459
SCROPHULARIACEAE 466
Scutia 443
S. myrtina 443
Searsia 216
S. dentata 216
S. incisa var. effusa 216
S. lancea 216
S. longispina 217
S. lucida 217
S. pallens 218
S. pterota 218
S. pyroides 218
S. rehmanniana 219
S. rosmarinifolia 219
S. tomentosa 219
Secamone 233
S. alpini 233
S. filiformis 233
Selago 476
S. corymbosa 476
S. geniculata 476

S. glomerata 476
S. gracilis 477
S. luxurians 477
S. myrtifolia 477
S. nitida 477
Senecio 289
S. angulatus 289
S. chrysocoma 289
S. coronatus 289
S. crassiusculus 290
S. crenatus 290
S. deltoideus 290
S. erubescens 291
S. glastifolius 291
S. hollandii 291
S. junceus 292
S. lineatus 292
S. linifolius 292
S. othonniflorus 293
S. pauciflosculosus 293
S. rhomboideus 293
S. robertiifolius 294
S. serratuloides 294
S. speciosus 294
Senecio 252, 257
S. articulatus 257
S. lanifer 291
S. oligantha 293
Seriphium 295
S. burchellii 295
S. plumosum 295
Setaria 161
S. sphacelata var. torta 161
Sideroxylon 466
S. inerme subsp. inerme 466
Silene 312
S. burchellii 312
S. undulata 312
Simocheilus spp. 348
Siphonoglossa leptantha 184
Smelophyllum 465
S. capense 465
SOLANACEAE 479
Solanum 480
S. guineense 480
S. linnaeanum 480
*S. pseudocapsicum 481
S. retroflexum 481
S. tomentosum 481
Sparrmannia 409
S. africana 409
Stachys 397
S. aethiopica 397
Stapelia 234
S. grandiflora 234
S. obducta 234
S. paniculata subsp.
  kougabergensis 234
Sterculia 410
S. alexandri 410
STILBACEAE 482
Stipa 161
S. dregeana 161
Stoebe 296
S. aethiopica 296

S. microphylla 296
S. plumosa 295
Streptocarpus 390
S. meyeri 391
Strumaria 71
S. gemmata 71
Struthiola 487
S. argentea 487
S. macowanii 487
S. parviflora 487
Suaeda 213
S. plumosa 213
Sutera 478
S. foetida 478
Syncarpha 296
S. canescens 297
S. eximia 297
S. ferruginea 297
S. milleflora 298
S. paniculata 243
S. striata 244

T

Tarchonanthus 298
T. littoralis 298
TECOPHILAEACEAE 179
Teedia 478
T. lucida 478
Tenaxia 162
T. stricta 162
Tephrosia 379
T. capensis 379
T. grandiflora 379
Tetragonia 208
T. echinata 208
T. fruticosa 208
T. spicata 208
Tetraria 103
T. bromoides 103
T. burmannii 103
T. capillacea 104
T. compar 101
T. cuspidata 101
T. fimbriolata 104
T. maculata 104
T. triangularis 105
T. ustulata 105
Teucrium 397
T. africanum 397
Thamnochortus 176
T. rigidus 176
Thamnus multiflorus 348
Theilera guthriei 307
THELYPTERIDACEAE 52
Thelypteris 53
T. confluens 53
T. gueinziana 52
Themeda 162
T. triandra 162
Thesium 461
T. flexuosum 461
T. scandens 461
T. squarrosum 461
T. strictum 461
Thesium 460

T. lineatum   460
Thoracosperma nanum   343
T. rosaceum   347
THYMELAEACEAE   483
Todea   47
T. barbara   47
Tragia   356
T. capensis   357
Tragus   163
*T. berteronianus   163
Tribolium   163
T. hispidum   163
*Trichocereus   306
*T. spachianus   306
Trichodiadema   209
T. barbatum   209
T. densum   209
T. mirabile   209
Triraphis   164
T. andropogonoides   164
Tristachya   164
T. leucothrix   164
Tritonia   129
T. linearifolia   129
Tritoniopsis   130
T. antholyza   130
Tromotriche   235
T. baylissii   235
Tulbaghia   64
T. violacea   64

U

UMBELLIFERAE   220

Urochloa   165
U. serrata   165
Ursinia   298
U. anethoides   299
U. anthemoides   299
U. heterodonta   299
U. scariosa   299
Urtica   488
U. lobulata   488
URTICACEAE   488

V

Vachellia   380
V. karroo   380
Veltheimia   115
V. bracteata   115
V. capensis   116
Venidium arctotoides   247
Vepris   458
V. lanceolata   458
V. undulata   458
VERBENACEAE   488
Vernonia   274
Virgilia   380
V. divaricata   380
Viscum   462
V. capense   462
V. rotundifolium   462
VITACEAE   489

W

Wahlenbergia   307
W. cinerea   307

W. guthriei   307
W. neorigida   307
W. rubens   308
W. thunbergii   308
W. uitenhagensis   308
Walafrida myrtifolia   477
Watsonia   130
W. knysnana   130
W. schlechteri   131
Widdringtonia   56
W. nodiflora   56
W. schwarzii   56
Willdenowia   176
W. teres   176
Wimmerella   402
W. hederacea   402
Withania   482
W. somnifera   482

X

Xysmalobium   235
X. gomphocarpoides   235

Z

Zaluzianskya   478
Z. synaptica   479
ZAMIACEAE   58
Zehneria   329
Z. scabra   329
ZYGOPHYLLACEAE   490
Zygophyllum   491
Z. foetidum   491
Z. fulvum   491

Index to common names

CAPS = family names; * asterisk = alien/exotic genus or species

A

Aalwyn   77
Aambeibessie   381, 397
Aandgonna   487
Aarbossie   311, 398
Aasblom   234
ACANTHUS FAMILY   180
African bluebell   307
African boxwood   414
African daisy   267
African flax   398
African holly   236
African mallow   404
African stachys   397
Afrikaner   123
Agapantha   62
AGAPANTHUS FAMILY   62
AGAVE FAMILY   63
Agretjie   129
Agtdaegeneesbos   302
Aloe   77
ALOE FAMILY   77
AMARANTH FAMILY   210

AMARYLLIS FAMILY   65
Angelier   311
Angled ice plant   202
Ankerkaroo   284
April fool   69
Arctotis   247
Asbossie   287
Asbosvygie   202
ASPARAGUS FAMILY   73
Asparagus fern   75
Assegai tree   330
Astertjie   264, 245
Augustusbos   428
*Australian bluebush   212
Autumn star   117

B

Bakbesembossie   278
Bakkersbos   485
Balloon pea   371
*Bangalay   415
Bankrotbos   285
Baroe   399

Bastard shepherd's tree   309
Basterboegoe   452
Bastergousblom   268
Basterpendoring   315
Basterperdepis   464
Basterslangbos   260
Bastertaaibos   463
Baviaanskers   339
BEAN FAMILY   357
BEDSTRAW FAMILY   447
Beesbos   253
Beesgras   145
Beeskloutjie   200
Bekruipmelkbos   355
*Belhambra   424
Bell bush   471
Belletjie   371
BELLFLOWER FAMILY   307
Bergangelier   484
Bergankerkaroo   272
Bergbamboe   166
Bergflap   128
Bergklapper   412

Bergmagriet 298
Bergpalmiet 103
Bergplakkie 321
Bergroos 404
Bergtee 368
Besemkraaibessie 217
Besemriet 166
Bietou 280
BIGNONIA FAMILY 300
Bindweed 316
BINDWEED FAMILY 316
Bird's brandy 489
Bitter aloe 78
Bitter apple 480
Bitter turpentine grass 148
Bitterblaar 252
Bitterbos 253
Bitterbossie 381
Bittergousblom 247
BITTERSWEET
   FAMILY 313
Bitterwortel 381
Blaas-ertjie 371
*Black wattle 358
Blackberries 446
Bladdernut 338
Bleekkoeniebos 218
Bliksembos 350
Blinkblaar 442
Blinktaaibos 217
Blinkvygie 195, 199
Blombiesie 120
Blombos 275
Blood lily 71
Bloublommetjie 266
Bloubos 184, 337
Bloughwarrie 338
Bloukeur 373
Blouklokkie 307
Bloulelie 62
Blounaaldeteebossie 274
Bloupoeierkwassie 90
Blouselblommetjie 89
Blousuurkanol 118
Blouteebos 274
Bloutulp 128
Blouvingertjies 258
Blue bonnet disa 136
Blue chalk-sticks 258
Blue cycad 58
*Blue dandelion 254
Blue fingers 258
Blue grass-lily 106
Blue lily 62
Blue mountain sage 397
Blue pea 373
Blue tulp 126
Blue water lily 61
Bobbejaanarm 309
Bobbejaanbessie 465
Bobbejaankomkommer 328
Bobbejaankool 281, 282
Bobbejaankos 59, 490
Bobbejaanstert 132
Bobbejaansuring 421

Bobbejaantjie 120
Bobbejaantoontjies 258
Boegoe 450
Boegoebossie 280
Boegoekaroo 286
Boerboon 378
Bog fern 53
Bohobe-ba-balisana 323
Bokbaardgras 162
Bokhoring 227
Bokhorings 230
Boneseed 280
BORAGE FAMILY 300
Bosbaroe 400
Bosboerboon 378
Bosdruif 489
Bosklimop 368
Bosluisbessie 280
Bosmagriet 260
Bosui 109
Bosvygie 204
Botterblom 246, 266
Bottlebrush grass 151
Boxwood 229
BRACKEN FAMILY 45
Bracken fern 45
Bracket satyre 141
Brakbos 213
BRAKE FAMILY 48
Brakopslag 192
Braksuring 393
Brakvygie 203
Brambles 446
Brandblaar 438
Brandbossie 485
Brandgras 150
*Brazilian waterweed 116
Bristle grass 161
Broad-leaved bulbine 82
Broodboom 58
Broom needlegrass 164
BROOMRAPE FAMILY 418
Bruinafrikaner 124
Buchu 450
BUCHU FAMILY 450
Buckthorn 442
BUCKTHORN FAMILY 439
Buig-my-nie 465
Bulbine 81
Bully trees 466
Bunchy sedge 94
*Burr grass 163
Burweed 211, 213
Bush asparagus 73
Bush cherry 310
Bush violet 180
Bushy bulbine 81
BUTTERCUP FAMILY 438
Butterspoon tree 330
Bylvygie 205

C
CABBAGE FAMILY 303
Cabbage tree 236, 237
*CACTUS FAMILY 305

CALTROP FAMILY 490
Campfire 320
Camphor tree 298
Campion 312
Canary creeper 290
Candelabra lily 65, 66
Cape beech 415
Cape box thorn 479
Cape cheesewood 425
Cape chestnut 454
Cape fig 413
Cape forget-me-not 300
Cape gardenia 449
Cape gorse 361
Cape ivy 289
Cape may 454
Cape myrtle 414
CAPE PONDWEED FAMILY 72
Cape smilax 74
Cape snapdragon 474
Cape snow bush 261
Cape snowflake 71
Cape star chestnut 410
Cape stock rose 409
Cape sumach 460
Cape sweet pea 368
Cape weed 246
Caper 310
CAPER FAMILY 309
Capewort 475
CARNATION FAMILY 310
CARPET WEED FAMILY 411
Carrion flower 234
CARROT FAMILY 220
Carrot fern 43
CASHEW FAMILY 215
*Castor oil plant 356
Cat-thorn 443
Catchfly 312
Catmint 396
Cedar 56
CEDAR FAMILY 54
CELERY FAMILY 220
Centaury 381
CENTURY PLANT FAMILY 63
Chaff flower 210, 243
CHAIN FERN FAMILY 44
*Chandelier plant 317
Cheesewood 424
Cherrywood 315
*Chicory 254
Chincherinchee 114
Chinese lantern bush 410
Christmas aloe 79
Christmas berry 381
CITRUS FAMILY 450
Climber's friend 443
Climbing aloe 81
Climbing onion 110
Clockflower 127
Coastal camphor bush 298
Cobra lily 121
COFFEE FAMILY 447
COMB FERN FAMILY 52
Common bush cherry 310

Common dwarf wild hibiscus   408
Common gazania   266
Common lip fern   50
Common needle-leaf   221
Common reed   160
Common spikethorn   313
Common sugarbush   437
Common sunshine
   conebush   432
Common turkey-berry   448
Concertina plant   325, 326
Cone bush   429
Cone plant   194
Confetti bush   454
Copper mesemb   204
Coral aloe   80
Coral fern   46
CORAL FERN FAMILY   46
Cornflower sceptres   434
COTTON FAMILY   403
Cottonseed   177
Cotyledon   318
Couch grass   148
Cranesbill   382
CRASSULA FAMILY   316
Creeping fern   46
*Creeping saltbush   211
Cross-berry   404
CROWFOOT FAMILY   438
CRUCIFEROUS PLANT
   FAMILY   303
CUCURBIT FAMILY   328
Cup-and-saucer   88
CYANELLA FAMILY   170
CYCAD FAMILY   58

D

Dabagrass   156
DAFFODIL FAMILY   65
DAISY FAMILY   238
*Dallis grass   157
Dark-eyed dwarf hibiscus   409
Dassievygie   198
DAY LILY FAMILY   106
Dayflower   89
DAYFLOWER FAMILY   89
DEER FERN FAMILY   44
DEER TONGUE FERN
   FAMILY   46
Devil's horsewhip   210
Devil's tresses   60
Dew grass   152
Diadem vygie   209
Diktamarak   109
Dikvoet   232
Disseldoring   251
Dodder   60
DOGBANE FAMILY   226
Doily tinder-leaf   223
Doll's bush   434
Doll's powderpuff   90
Doll's rose   405
Donkievygie   204
Doringblaarvygie   209
Doringboom   380

Doringkroonvygie   209
Doringtaaibos   217
Doringtee   444
Doringvygie   205
Dotted sorrel   421
Dottypea   373
Doublaarvygie   196
Douvygie   196
Draaibos   283
Draaibossie   265
Dronkgras   155
Drumstick flower   478
Duikernoors   354
Duikervygie   199
Dwarf nerine   70
Dyebossie   423
Dysentery herb   382

E

Eared rattle pod   367
Eastern Cape flame   123
EBONY FAMILY   336
Edging lobelia   401
Eendjies   371
Elephant's foot   106
EUCALYPTUS FAMILY   415
EUPHORBIA FAMILY   349
Everlasting   296

F

Fairy snowdrop   113
False buchu   455
False gerbera   269
False horsewood   464
False spike-thorn   315
FAN FERN FAMILY   46
Feathery asparagus   76
Fig   413
FIG FAMILY   413
FIGWORT FAMILY   466
Finger grass   149
Finger-phlox   474
Fire lily   66
Fire-sticks   337
Five fingers   179
Flax   398
FLAX FAMILY   398
Fleabane   288
'FLOWERING FERN' FAMILY   42
Fonteinbos   305, 373
Forest elder   483
Forest iris   121
Forest sand-lily   115
Forest sugarbush   436
FORGET-ME-NOT FAMILY   300
FORKING FERN FAMILY   46
*Fountain grass   147
Foxtail buffalo grass   147
Foxtail sedge   92
French aloe   80

G

Gamtoos cabbage tree   237
Gansiebos   228
Gansies   371

*Garingboom   63
Gazania   266
Geelbergdraaibos   283
Geelblombos   275
Geelbos   192
Geelgazania   267
Geelkeurtjie   367
Geeloogsuring   421
Geeltjienk   114
Gemsboksuring   214
GENTIAN FAMILY   381
Geranium   382
Gewone kapokbossie   262
Gewonetaaibos   218
Ghoukum   193
*Giant paspalum   158
*Giant reed   145
Gifbol   65
Gifboom   226
Giftou   227
GINSENG FAMILY   236
Glossy guarri   340
Goat's foot   419
Goeieboegoe   453
Golden Timothy   161
Golden winter star   117
Gombos   285
Gombossie   286, 287
Gonna   485
*Goosefoot   212
Gortjie   228
GOURD FAMILY   328
Gousblom   247, 266
GRASS FAMILY   142
Grass lily   63
Grass-leaved
   chincherinchee   115
Greater groundsel   294
Green cliff brake   50
Green-bearded disa   136
Groenblomvygie   203
Gromwell   302
Grootbloublommetjie   266
Grootskaapkaroo   284
GROUND FERN FAMILY   45
Groundsel   289
Guarri   340
GUAVA FAMILY   415
Guinea grass   157
*Gum   415
Gum-leaf conebush   430
Gunpowder plant   312

H

Haakgras   155
Haasklossie   178
Hairy guarri   339
Hard fern   50
HARD FERN FAMILY   44
Hard pear   422
Hardepeer   422
Harlequin orchid   138
Harpuisbos   262, 264
Hartblaarmalva   384
Hartebeeskaroo   296

Hartjiegras 151
Heart-leaved pelargonium 384
HEATH FAMILY 340
*Hedgehog cactus 305
Hedgehog sedge 93
Helmet orchid 134
HEMP FAMILY 308
Hen-and-chickens 63
Heuningbos 255
HIBISCUS FAMILY 403
Hoenderspoor 153
HOLLY FAMILY 236
Honde-kak-en-pisbos 477
Honey thorn 480
Honeybush 368
Hongerblom 289
Horehound 396
Horinkies 470
Horse mint 395
Horseshoe geranium 390
Horsetail restio 167
Horsewood 454
HYACINTH FAMILY 107

I

Ice plant 204
ICE PLANT FAMILY 188
Idambiso 294
Idwara 293
Igabushe 204
Igcakriya 205
Ikhakhakhaka 251
Impepho 270
Ink flower 418
Inkblom 418
Intelezi 194
Intlolokotshane 216
Intlungunyembe 226
Intlwathi 224
Inyamayamakhwenkwe 324
Inzula 352
IRIS FAMILY 118
Isichwe 267
Isihlungu 201
Isikholokotho 394
Isilawu 335
ISIQAJI 226
Itshilizi 400
Ivy-leaved pelargonium 387
Iyeza logezo 250
Iyeza lomoya 280

J

Jacket-plum 465
Jakkalsbos 260
Jakkalspisbos 451
Jankoensedoring 444
*Jerusalem cherry 481
Jongmansknoop 335
Justafina 67

K

Kaapse boekenhout 415
Kalkoentjie 125
Kalmoes 220, 224

Kalossie 126
Kalwerbossie 389
Kamassie 229
Kammetjie 122
Kammiebos 446
Kandelaar 65
Kanferboegoe 453
Kanferbos 298
Kankerbos 371
Kannetjies 231
Kapok-lily 132
Kapokbossie 261
Karee 216
Kareemoer 209
Kariega aloe 77
Karkaarblom 130
Karkai 327
Karoo aloe 78
Karoo bluebush 337
Karoo cabbage tree 237
Karoo conebush 430
Karoo cycad 58
Karoo gold 300
Karoo sagewood 467
Karoo violet 466, 467
Karoo-aalwyn 79
Karooboerboon 378
Karookruisbessie 404
*Kasterolieboom 356
Kasuur 425
Katbossie 397
Katdoring 74
Katnaels 419
Katoenbos 228
Katstertkanet 175
Katstertsteekgras 144
Kattekruie 396
*Keeled goosefoot 212
Kei lily 68
Kerrieblom 485
Kersbos 339
Kershout 315
Keurboom 380
Kiepersol 236
*Kikuyu 147
King fern 47
KING FERN FAMILY 47
King protea 434
Kinkelbos 208
Klaaslouwbos 248
Klapperbos 410
Klapperbrak 208
Klappiesbrak 208
Klawersuring 422
Kleefgras 448
Klein ysterhout 417
Klein-kopervygie 195
Kleinblomvygie 195
Kleinperdekaroo 279
Klimopkinkelbossie 208
Klipblom 324
Klipdagga 395
Klipkershout 314
Klipnoors 353
Klipsalie 396

Klipsterretjie 118
Klipvygie 195
Klokkiesbos 471
Knoppiesdoringbossie 259
Knoppiessewejaartjie 298
Knoppiesslangbos 296
Knysna lily 67
Knysnavygie 206
Koedoekos 87
Koggelmandervoet 398
Kokoboom 314
Kol-kol 304
Kooigoed 272
Koorsbossie 485
Kopervygie 204
Kopieva 81
Kopseerblom 65
Kopseerbossie 245
Koraalvygie 198
Korentebos 216, 219
Kouga cannomois 166
Kouga flame lily 67
Kouga honey-bells 471
Kougabergtee 368
Kougoed 267
Kousklits 163
Kraakbos 287
Kraalbos 192
Krantz sugarbush 437
Kritikom 337
Kruidjie-roer-my-nie 411
*Kruipsterklits 243
Kruisbessie 404
Kruisblaartjie 412
Kukumakranka 68

L

Lady's hand 179
Large deer fern 44
Large woolly three-awn 144
Large-flowered campion 312
LAUREL FAMILY 60
LEADWORT FAMILY 425
Leeubekkies 474
LEGUME FAMILY 357
Lekoto-la-litsoene 293
Lidjiesvygie 202
Lightning bush 351
Lilac disa 135
LIMEUM FAMILY 398
*Lindley's saltbush 210
Linear-leaf conebush 429
Lip ferns 49
Little pickles 256
LIVERWORT FAMILY 40
Living stones 200
Lizard's foot 398
Lizard's tail 323
LOBELIA FAMILY 399
Loogbossie 203
Love grass 151

M

Madder 449
MADDER FAMILY 447

Maerman 111
Magriet 260
MAHOGANY FAMILY 410
MAIDENHAIR FERN FAMILY 48
Maidenhair ferns 48
Makghaap 234
Makhanda heath 343
Maklikbreekbos 185
Malbar 252
MALLOW FAMILY 403
Malva 383
Marguerite 260
MARSH FERN FAMILY 52
Mat sedge 95
Matjiesgoed 93
Melkbol 354, 355
Melkbos 352
Melktou 227
Men-in-a-boat 88
MESEMB FAMILY 188
Mickey mouse bush 416
MILKWEED FAMILY 226
Milkwood 466
MILKWORT FAMILY 426
Minaret flower 395
Mint 395
MINT FAMILY 394
Mistletoe 462
Moederkappie 405
Monkey rope 227, 233
Monkey-tail everlasting 271
MORNING GLORY
    FAMILY 316
*Mother of millions 317
Mountain carnation 484
Mountain cedar 56
Mountain cypress 56
Muishondbossie 388, 389
Muisvygie 196
MULBERRY FAMILY 413
MUSTARD FAMILY 303
MYRSINE FAMILY 414
MYRTLE FAMILY 415

N

Nanabessie 216
Narrow heart love grass 152
Narrow-leaved crassula 324
Narrow-leaved sugarbush 436
Natal laburnum 367
*Natal red top 155
Needle bush 459
Needlegrass 164
Nentabos 328
*Nerium 231
Nettle 488
NETTLE FAMILY 488
Nieshout 457
Nightshade 480
NIGHTSHADE FAMILY 479
Noors 352
Noseburns 356
Notsung 482
Nou-tou 413
Num-num 226

O

Old man's beard 439
*Old man's saltbush 211
*Oleander 231
Olienhout 417
Olifantsgras 146
Olifantsriet 166
OLIVE FAMILY 417
*Ombu 424
ONION FAMILY 64
Oondbos 278
Opslag 475
Opslagkinkelbossie 208
*Orach 210
ORCHID FAMILY 133
Oumeisieknie 350
Oupa-met-sy-pyp 137
Outeniqua yellowwood 57
Oxbane 65

P

Paddaklou 397
Paintbrush lily 69
Palmiet 132
Pampoenblaar 392
Panic grass 157
*Paraguayan starburr 243
Parasol 382
PARSLEY FAMILY 220
Parsley tree 223
Patrysblom 88
Patrysuintjie 127
PEA FAMILY 357
Pelargonium 383
PELARGONIUM FAMILY 382
PENAEA FAMILY 422
Pendoring 313
Pendoringtaaibos 218
Pennywort 222
*Pepper tree 215
Perdekaroo 278
Perdepis 454
Perdevy 193
Persbokhorinkie 470
Persimmon 336
Pickle plant 194
Pienksewejaartjie 297
Pig's ears 318
Pincushion 433
Pincushion flowers 335
Pincushion grass 156
Pineapple reed 169
Pink 311
Pink dollsrose 405
PINK FAMILY 310
Pink ground-bells 418
Pink trailing pelargonium 383
PISTOL BUSH FAMILY 180
Plampers 255
Platdoring 221
Plectranthus 396
Ploegtydblommetjie 117
PLUMBAGO FAMILY 425
Poeierkwassie 427
Poison gooseberry 482

Poison squill 111
Poison star-apple 337
Pokeweed 424
POKEWEED FAMILY 424
Polgras 149
POLYGRAM FERN
    FAMILY 48
Poppiesbos 434
Poprosie 405, 406
POTATO FAMILY 479
Pregnant onion 107
*Prickly pear 306
*Prostrate starburr 243
Protea 434
PROTEA FAMILY 429
Purple-gorse 426
Puzzle bush 301
Pylgras 154
Pypie 123
Pypsteelbos 446
Pypsteelgras 150

Q

Quick grass 148

R

Rabassam 388
Ragwort 289
Ramenas 78
Rankbol 110
Rankmalva 383, 387
Raspberry 446
Rattle pod 367
Real yellowwood 57
Red broomrape 419
Red pagoda 320
Red Rhodes grass 153
Red-hot poker 87
Red-spined aloe 78
Red-stem crassula 326
Red-topped signal grass 165
*Redeye 358
Renosterbos 259
Renostergousblom 247
Renosterveld honey-bells 472
RESTIO FAMILY 165
RHUBARB FAMILY 392
Ribbokvygie 207
Ribbon bush 183
Rietpypie 130
Rietuintjie 129
River honey-bells 471
River pumpkin 392
*River redgum 416
Riverbed grass 146
Riviergras 145
Riviernaboom 355
Rondeblaar-brosplakkie 317
Rooi-ertjie 379
Rooiels 330
Rooigazania 266
Rooigras 149, 162
Rooikershout 315
*Rooikrans 358
Rooirabas 388

Rooisaadgras 164
Rooistingelplakkie 326
Rooisuring 422
Roosmaryntaaibos 219
ROSE FAMILY 443
ROYAL FERN FAMILY 47
Rub-rub berry 216
Ruigtegras 156
Ruigtevygie 199
Ruikpeperbossie 304
RUSH FAMILY 131
Rush iris 120
Rygbossie 326

S

Sabelvygie 200
Saffraan 483
Saffraanbossie 473
Saffron bush 483
SAGE FAMILY 394
Sagewood 467, 468
*Saltbush 210
Sambreeltjie 382
SANDALWOOD FAMILY 460
Sandbur 146
Sandgombos 286
Sandlelie 116
Sandolien 464
Sandroos 404
Satinflower 122
Satyr orchid 141
SCABIOUS FAMILY 335
Scaly lady fern 53
Scarlet geranium 385
Scented fern 42
SCENTED FERN FAMILY 42
Scented geranium 386
Scrambling aloe 81
Sea spinach 202
*Sea urchin cactus 305
Sea-blites 213
SEDGE FAMILY 90
Seepblinkblaar 439
Seepweeds 213
Seeroogblom 71
Sekopo 323
Septee-tree 301
September bush 428
Sewejaartjie 243, 271, 296
Shade ehrharta 150
SHOWY MISTLETOE FAMILY 402
Sieketroos 221
*Sigorei 254
Silky bark 314
*Sisal 63
Sjambokbos 292
Skaamblommetjie 110
Skaapbos 447
Skaapkaroo 284
Skaapvygie 194
Skilpadbos 426
Skilpadkos 197, 280
Skoenlapperbos 428
Skunk bush 468
Slakblom 472

Slangbos 295
Slangui 110
Slime-lily 107
Slugwort 472
Slymbos 491
Slymlelie 107
Slymstok 108
Small bitterleaf 252
*Small carrot-seed grass 163
Small forest iris 121
Small ironwood 417
Small-flowered nutsedge 93
Snake lily 88
Sneeuvygie 411
Sneezewood 457
SOAPBERRY FAMILY 462
Soetdoring 380
Soetharpuisbos 263
Sokkiesblom 107
Sorrels 419
Sosaties 325, 326
Sour fig 193
Spaansaalwyn 80
*Spanish reed 145
Spekboom 331
Spekbos 491
Spekvygie 192
Speldedoring 459
Speldekussing 433
Spider orchid 133
Spider plant 63
SPIDERWORT FAMILY 89
Spinnekoporgidee 133
Spinning top 431
Spleenwort 43
SPLEENWORT FAMILY 43
Sporrie 303
Springbokbos 274
Springbokvygie 201
Spur-flower 396
Spurge 352, 356
SPURGE FAMILY 349
Square-stemmed
   pelargonium 390
St John's wort 392
ST JOHN'S WORT FAMILY 392
STAFF VINE FAMILY 313
STAR GRASS FAMILY 117
Starburr 243
Stargrass 117
Steekgras 144
Steenbokboegoe 450
Stekelvygie 207
Sterkastaiing 410
Sticky-leaved cotyledon 318
Stinkbossie 468
Stinkdoring 313
Stinkleeubekkie 475
Stompie 305
Stonecrop 320
STONECROP FAMILY 316
Storksbill 383
Strap-leaf sugarbush 435
Strap-leaved bulbine 83
Strawberry everlasting 297

Strawflower 269
STRIPPER FAMILY 483
Strooiblom 269
Stroopbos 372
Sugarbush 434
SUMAC FAMILY 215
Sundew 336
SUNDEW FAMILY 336
Surings 419
SUTERA FAMILY 466
Suurgras 151
Suurtaaibos 219
Suurvy 193
Swartbasboom 339
Swartberg pincushion 433
Swartstormbos 309
Swarttee 268
*Sweet prickly pear 306
Sweet thorn 380
Sybas 314
Sybossie 289
Syselbos 425
Sysie 122

T

Taai-astertjie 266
Taaibos 216
Tamarak 107
TAPE GRASS FAMILY 116
Taylor's parches 322
TEASEL FAMILY 335
Teerhout 215
Tenacious sugarbush 438
Teringbos 461
Thimble grass 153
Thorn pear 459
Thread orchid 139
Three-awn wiregrass 144
Thunberg's cycad 58
Tjienk 114
Tokkelossiebos 425
Tolbol 71
Tolbos 336, 429
Tongue ferns 46
Tongued thick-fruit 231
Tontelblaar 223
Toontjies 194
Toothed brake 51
*Torch cactus 306
Tortoise berry 426
Traanbos 356
Trailing daisy 277
Traveller's joy 439
Tree euphorbia 353, 355
Tree fuschia 482
Trewwa 141
Trident grass 164
Tsitsikamma conebush 432
Tufted sugarbush 435
Tulp 126
Turpentine grass 148
Twinleaf 491
TWINLEAF FAMILY 490
Twinspurs 470
Two-day cure 294

U

Ubuhlungu 114
Ubukhubele 382
UBULAWU 238
UBUNGXANI 226
Udlatshana 244
Uintjie 126
Umbinza 482
Umbrella liverwort 40
Umbrella sedge 95
Umncwane 419
Umnunge 123
Umsola 263
Umthithimbili wentaba 289
Ungcana 111, 311
Ungcilikinde 304
Uniondale conebush 431
Unomatyumtyuma 198
Utangazana 329
Utyuthu 208
Uvelemonti 291
Uzandokwa 272

V

Vaalbergkaroo 273
Vaalbliksembos 351
Vaalstorm 460
Vanstaden's daisy 260
Varkoor 318
Varkoortjies 222
Veld fig 413
Velskoenblaar 70
VERBENA FAMILY 488
Verfblommetjie 473, 478
Vergeetmynietjie 300
Vetplantmalva 384
Vetstingelvygie 207
Vierkantperdekaroo 279
VINE FAMILY 489
Vingerhoedgras 153
Vingerpol 353
Vingersuring 420
Vingertjies 474
Vlaktebaroe 400
Vlaktevygie 206
Vleibiesie 92

Vleibos 446
Vleikruid 277
Vleiriet 92
Vleirooigras 444
Vlei-uintjie 128
Vlerkiesriet 168
Voëlent 462
Volstruisuintjie 127
Vuurhoutjies 403
Vuurpyl 87
VYGIE FAMILY 188

W

Wagon tree 436
Wandering Jew 89
Warty-stemmed
    pincushion 433
*Water couch 158
Water heath 343
WATER LILY FAMILY 61
Water sugarbush 437
Waterblommetjie 72
Waterbos 344
Waterdissel 291
Waterfinder 476
Waterheide 344
Waterlelie 72
Waternael 222
Waterolier 414
WAX-MYRTLE
    FAMILY 414
Waxberry 311
Weeping love grass 152
White bramble 446
White ironwood 458
White stinkwood 308
White-stem guarri 339
Wide-leaf sugarbush 435
Wild asparagus 73
Wild clove bush 412
Wild garlic 64
Wild gloxinia 390
Wild grape 489
Wild iris 121
Wild lobelia 401
Wild olive 417

Wild peach 188
Wild penstemon 418
Wild rosemary 261
Wild spearmint 395
Wild thistle 251
Wild violet 402
Wilde-als 248
Wilde-amarilla 68
Wilde-ertjie 444
Wildedagga 395
Wildedissel 251
Wildegranaat 300
Wildekarmedik 259
Wildekastaiing 454
Wildemalva 386, 403
Wildeseldery 221
Wildewortel 179
Willow 458
WILLOW FAMILY 458
Willowmore cedar 56
Wing-nut 463
Witolienhout 468
Witstam 339
Wittamarak 108
Wittebergvygie 197
Witvygie 206
Witysterhout 458
Wolweklits 211
Wolwekos 419
WOOD SORREL FAMILY 419
Wood's cotyledon 319
Woolly conebush 432
Woolly finger grass 149
Wormwood 248
Worsies 257
Wortelgras 222

Y

Yam 106
YAM FAMILY 106
Yellow-head 485
Yellowwood 57
YELLOWWOOD FAMILY 57
Yendlebe 403
Ysterhout 464
Ystervarkpatat 329

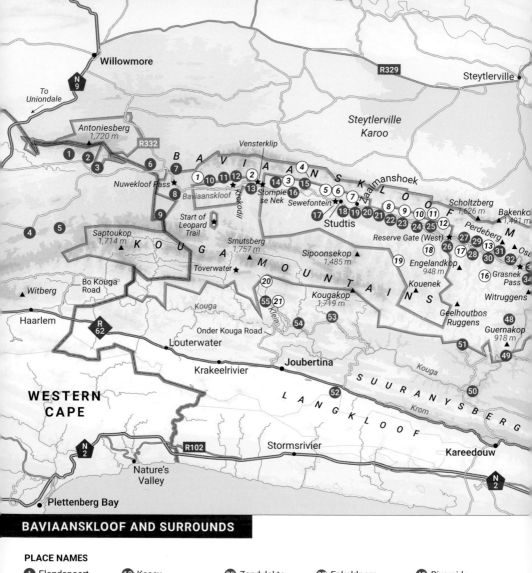

BAVIAANSKLOOF AND SURROUNDS

PLACE NAMES

1. Elandspoort
2. Misgunst
3. Verberg
4. Hartbeesrivier
5. Blue Hill Escape
6. Vaalwater
7. Adamskraal
8. Nuwekloof
9. Quaggasvlakte
10. Uitspan
11. Duiwekloof
12. Speekhout
13. Kasey
14. Verlorenrivier
15. Bokloof
16. Vleikloof
17. Beacosnek
18. Tchnuganoo
19. Kleinpoort
20. Damsedrif
21. Klipfontein
22. Doringkloof
23. Rus en Vrede
24. Joachemskraal
25. Zandvlakte
26. Rooikloof
27. Coleskeplaas
28. Geelhoutbos
29. Miskraal
30. Akkerdal
31. Apieskloof
32. Keerom
33. Doringkraal
34. Smitskraal
35. Rooihoek
36. Doodsklip
37. Enkeldoorn
38. Bergplaas
39. Cambria
40. Bruintjieskraal
41. Geelhoutboom
42. Groendal
43. Gonjah
44. Mistkraal
45. Andrieskraal
46. Komdomo
47. Wittekliprivier
48. Guerna
49. Riverside
50. Kouga Kliphuis
51. Nooitgedacht
52. Heights
53. Braamrivier
54. Brandhoek
55. Kouga Wildernis